NEW RESEARCH ON ACOUSTICS

NEW RESEARCH ON ACOUSTICS

BENJAMIN N. WEISS
EDITOR

Nova Science Publishers, Inc.
New York

For permission to use material from this book please contact us:
Telephone 631-231-7269; Fax 631-231-8175
Web Site: http://www.novapublishers.com

LIBRARY OF CONGRESS CATALOGING-IN-PUBLICATION DATA

New research on acoustics / Benjamin N. Weiss, editor.
p. cm.
ISBN 978-1-60456-403-7 (hardcover)
1. Sound--Equipment and supplies. 2. Acoustics. I. Weiss, Benjamin N.
TK7881.4.N48 2008
620.2--dc22 2008000024

Published by Nova Science Publishers, Inc. ✢ New York

CONTENTS

PREFACE

Acoustics is the science concerned with the production, control, transmission, reception, and effects of sound. Its origins began with the study of mechanical vibrations and the radiation of these vibrations through mechanical waves, and still continue today. Research was done to look into the many aspects of the fundamental physical processes involved in waves and sound and into possible applications of these processes in modern life. The study of sound waves also leads to physical principles that can be applied to the study of all waves.

The broad scope of acoustics as an area of interest and endeavor can be ascribed to a variety of reasons. First, there is the ubiquitous nature of mechanical radiation, generated by natural causes and by human activity. Then, there is the existence of the sensation of hearing, of the human vocal ability, of communication via sound, along with the variety of psychological influences sound has on those who hear it. Such areas as speech, music, sound recording and reproduction, telephony, sound reinforcement, audiology, architectural acoustics, and noise control have strong association with the sensation of hearing. That sound is a means of transmitting information, irrespective of our natural ability to hear, is also a significant factor, especially in underwater acoustics. A variety of applications, in basic research and in technology, exploit the fact that the transmission of sound is affected by, and consequently gives information concerning, the medium through which it passes and intervening bodies and inhomogeneities. The physical effects of sound on substances and bodies with which it interacts present other areas of concern and of technical application.

This new book presents recent and important research in the field.

Measurements of thermal acoustic radiation (TAR) from a model object were carried out in the first Short Communication, for reconstruction of temperature inside the object. This experiment presented the simulation of the temperature control in medical applications (e.g. in hyperthermia and thermoablation). The model object was a plasticine cylinder placed in water and water solution of glycerine (in this medium an acoustic absorption is similar to that of soft tissues of a human body). The plasticine was heated for 4-6 minutes. Its temperature was controlled with the help of an IR-thermograph. The TAR measurements in 1.4-2.2 MHz region were carried out for 8-10 minutes with five acoustothermometers placed on different sides of the model object. The changes of the TAR intensity during both the heating and the cooling of the object were registered. By using these experimental data the reconstruction of the 2-D temperature distribution in the model object was carried out. The authors averaged data over several seconds and reconstructed the object temperature, i.e. they calculated parameters of the temperature distribution: position (x and y coordinates), size and temperature of thermal source. So they obtained temporal changes of the temperature distribution. The authors used different methods for the reconstruction and investigated effect of the average time on the reconstruction quality. To calculate the object temperature by using the TAR measurements it is necessary to have information about the ultrasound absorption in the object. For estimation of the absorption coefficient they used the data obtained in the same experiment.

The generalization of the Hamilton's and Osager's variational principles for dissipative hydrodynamical systems is represented in terms of the mechanical and thermal displacement fields, as explained in the second Short Communication. A system of equations for these fields is derived from the extreme condition for action with a Lagrangian in the form of the difference between the kinetic and the free energies minus the time integral of the dissipation function. The generalized hydrodynamic equation system is then evaluated on the basis of the generalized variational principle. At low frequencies this system corresponds to the traditional Navier – Stokes equation system and in the high frequency limit it describes propagation of acoustical and thermal modes with the finite propagation velocities.

Further, the system of generalized Biot's equations, describing waves propagation in a multi-phase or multi-component medium in the presence of heat exchange between phases, is derived on the basis of generalized variational principle. It is shown that in the presence of N phases 2N propagating eigen-modes can exist in this medium. At high frequencies N modes are of the mechanical (acoustical) type and N modes are of the diffusive (thermal) type of propagation. At low frequencies there is the single acoustical (wave) mode and the rest 2N-1 modes possess the diffusion (thermal) type of behavior. For a two-component medium without temperature exchange the developed approach is reduced to the well known Biot's model. The account of the temperature field yields the generalized Biot's model for two components medium.

Chapter 1 considers application of different types of wavelet transform for analyzing acoustic fields and signals in elastic media. It is shown that different types of the discrete wavelet analysis, namely, fast wavelet transform, stationary wavelet transform and wavelet packets, are powerful and effective tools for denoising the signals and images. Using the continuous wavelet transform allows to realize the spatio-temporal spectral analysis of nonstationary processes. It makes possible to estimate the instantaneous frequency of signal, determine the frequency modulation law, reveal the relations between oscillations of different frequencies and so on. One of the main advantages of wavelet analysis is its ability to adapt itself to parameters of analyzed signals and images.

The purpose of Chapter 2 is the experimental research of properties for acoustic antenna arising during braking of an intensive beam of accelerated protons in a water environment. Research was conducted in a near-field zone that had allowed us to allocate signals from separate elements of the antenna and to carry out the analysis of such parameters of signals, as amplitude, width and time of their propagation.

As a source of protons the external beam of the Institute of Theoretical and Experimental Physics (ITEP, Moscow) accelerator with energy of 200 MeV and the time duration of 70 ns has been used. The beam intensity was supported at the level of $4{\cdot}10^{10}$ protons per pulse and supervised by the current transformer. The experiment was carried out in the parallelepiped plexiglas basin of a square section 95 cm in length and with the volume of 250 liters filled 85% with water. Input of the proton beam inside the volume was realized through a pipe with the diameter of 59 mm, 46 cm length and wall thickness of 1.5 mm inserted into a lateral side of the basin and closed by a plug made from organic glass with a thickness of 2 mm. The average ionizing range of protons in water was 25.2 cm. So, the sizes of the basin and the applied equipment have allowed us to study the un deformed structure of a hydroacoustic field induced by the proton beam.

Measurements of an acoustic field were made by means of a relocatable hydrophone in two mutual-perpendicular directions. Along the beam axis hydrophone movement was carried out with the step of 8.9 mm at the distance of 3.5 cm from the beam axis. In the cross-cut direction the trace passed in the horizontal plane passing through the beam axis at the distance of 35.6 cm from the point of the entrance of the proton beam into the water. In this case the scanning step was equal to 4.45 mm.

According to a thermoacoustic model in the area of beam action for time, comparable with the action time, an acoustic antenna arises. In the present work the problem of reconstruction of the form of the antenna using the experimental results is being solved. The technique of calculation of the hydrophone response to the radiation of separate elements of the acoustic antenna has been developed. The dependences of amplitude of the signals and their time parameters on the relative position of the antenna and the hydrophone have been obtained. The angular distribution of the field created by the terminal area of the radiation zone has been obtained. This characteristic, generally speaking, is similar to the directional diagram of an audio antenna.

To test the experimental results, the full-scale simulation of set-up geometry and the physical processes accompanying the propagation of protons in water had been carried out using GEANT-3.21 package. The simulation of the generation process of an acoustic signal was performed as a first approximation in the assumption of proportionality of the signal intensity to the energy which is generated at the ionization of water atoms by a proton without taking into account heat conductivity and the elastic properties of environment, leading to relaxation. The model calculations confirm the qualitative conclusions and the results obtained at the processing of experimental data.

Along with the advancement of virtual auditory displays (VAD), which render three-dimensional auditory perceptual space by controlling sound paths drawn from a sound source to a listener's ears, auditory games with VAD have been attracting increasing interest. The VAD games offer advantages over visual action-video-games in that both sighted persons and visually impaired people can enjoy them. Although some previous studies have attempted to apply auditory virtual reality games to the auditory education of visually impaired people, few studies have investigated the transfer effects of playing virtual auditory games. Chapter 3, presents VAD games as an effective training tool for skills related to auditory information processing in daily life situations. The studies investigated transfer effects on various auditory skills from playing VAD games. The authors particularly confirmed transfer effects in the following human aspects: sound localization performance, communication behaviors in face-to-face situations, and avoidance behavior from approaching objects. Finally, they propose new perspectives and future applications of auditory games which use the VAD system.

Friction-induced vibration and noise emanating from car disc brakes is a source of considerable discomfort and leads to customer dissatisfaction. The high frequency noise above 1 kHz, known as squeal, is very annoying and very difficult to eliminate. There are typically two methods available to study car disc brake squeal, namely complex eigenvalue analysis and dynamic transient analysis. Although complex eigenvalue analysis is the standard methodology used in the brake research community, transient analysis is gradually gaining popularity. In contrast with complex eigenvalues analysis for assessing only the stability of a system, transient analysis is capable of determining the vibration level and in theory may cover the influence of the temperature distribution due to heat transfer between brake components and into the environment, and other time-variant physical processes, and nonlinearities. Wear is another distinct aspect of a brake system that influences squeal generation and itself is affected by the surface roughness of the components in sliding contact.

Chapter 4 reports recent research into car disc brake squeal conducted at the University of Liverpool. The detailed and refined finite element model of a real disc brake considers the surface roughness of brake pads and allows the investigation into the contact pressure distribution affected by the surface roughness and wear. It also includes transient analysis of heat transfer and its influence on the contact pressure distribution. Finally transient analysis of the vibration of the brake with the thermal effect is presented. These studies represent recent advances in the numerical studies of car brake squeal.

Recent findings indicate one of major causes of damages, which is attributed to the resonant behaviours, in a railway track and its components. Basically, when a railway track is excited to generalised dynamic loading, the railway track deforms and then vibrates for certain duration. Dynamic responses of the railway track and its components are the key to evaluate the structural capacity of railway track and its components. If a dynamic loading resonates the railway track's dynamic responses, its components tend to have the significant damage from excessive dynamic stresses. For example, a rail vibration could lead to defects in rails or wheels. The track vibrations can cause the crack damage in railway sleepers or fasteners, or even the breakage of ballast support. Therefore, the identification of dynamic properties of railway track and its components is imperative, in order to avoid any train operation that might trigger such resonances. Chapter 5 deals with the vibration measurement techniques and the dynamic behaviours of ballasted railway tracks, and in particular their major components. It describes the concept of vibration measurements and the understanding into the dynamic behaviour of ballasted railtrack sleepers. It discusses briefly on the track structures and track components in order to provide the foundation of understanding ballasted railway tracks. The highlight in this paper is the state-of-the-art review of dynamic properties of railway track and its components. It summarises the non-destructive acoustic methods, the identification processes, and the properties of each rail track element.

The acoustics of open-air performance spaces, in particular ancient theatres of Classic, Hellenistic and Roman periods, has gained much attention in the past years. However, earlier theatres, situated in the courtyards of the Minoan palaces, in unique shapes, dated around 1500 B.C., have only recently been examined from the acoustic viewpoint. In this chapter, the examination of six identified types of ancient theatre has revealed that theatres evolved architecturally and acoustically through the centuries. Moreover, the excavation of ancient theatre sites in the last century allowed the revival of ancient drama and instituted drama festivals in southern Europe. Scenery is an important component of those drama performances, both from the aesthetic and, as new studies show, the acoustic viewpoint. In Chapter 6, the effects of temporary scenery design, classified into four generic categories, on the acoustic environment of open-air theatres are investigated, aiming at providing guidelines for architects and scenery designers. In addition to the above two key issues, a general literature review on the acoustics of ancient Greek/Roman theatres is given and relevant research methodology on the subject is discussed.

Porous and fibrous materials provide sound absorption within a frequency band which lower limit depends mainly on its thickness. In general, low frequency absorbers require such a large thickness than they are not installed in practice, except perhaps for large anechoic chambers. Active systems, on the other hand, allow to control the input impedance of multilayer absorbers, this affording absorption in the low frequency range. Combining appropriately the properties of the passive material with those of the active system, it is possible to design efficient absorbers for broadband noise, including low frequencies, with a reduced thickness. These systems are named hybrid passive-active absorbers. Chapter 7 describes different implementations of such hybrid absorbers using porous and microperforated panels (MPP) as the passive material. Both *pressure-release* and *impedance-matching* are analyzed as the active control condition. Experimental results measured both in 1D (impedance tube) and 2D (anechoic chamber), which validate the predictions of the theoretical model, are presented.

In animals, recognition between individuals is essential to the settlement of sexual and social relationships. Due to their physical properties and their potentiality to encode any kind of information, sounds are an effective mean to reliably transmit the identity of the emitter. Nevertheless, in colonial animals, the vocal signal produced by an adult seeking its young or its partner among thousands of individuals is transmitted in a particularly noisy context generated by the colony. Such a background noise drastically reduces the signal-to-noise ratio and masks the signal by a noise with similar spectral and temporal characteristics. In Chapter 8, the authors report how this extreme acoustic environment constrains the transfer of information by sounds. To illustrate the problem of precise acoustic identification in the noise, they have chosen two representative biological models: the penguins and the otariids. The authors examine solutions found at the level of the emitter to improve the efficiency of communication and they report how the receiver can optimize the collected information. On the basis of the results obtained in numerous field studies, they show that penguins and otariids use a particularly efficient "anti-confusion" and "anti-noise" acoustic coding system, allowing a quick and accurate identification and localization of individuals on the move in a noisy crowd.

The study of these biological models allows us to highlight the basic rules that govern the identification of a precise acoustic message in the noise. The differences in the coding-decoding strategies are also discussed with respect to the social structure and the environment of the different studied species.

As explained in Chapter 9, the spectral finite element method is an advanced implementation of the finite element method in which the solution over each element is expressed in terms of a priori unknown values at carefully selected spectral nodes. These methods are naturally chosen to solve problems in regular rectangular, cylindrical or spherical regions. However in a general irregular region it would be unwise to turn away from the finite element method since models defined in such regions are extremely difficult to implement and solve with a spectral method. Hence for a complex waveguide the method uses the efficiency and accuracy of the spectral method and is combined with the flexibility of finite elements to produce a high–performance engineering tool. Contemporary examples from engineering including fluid-filled pipes, tyre acoustics, silencers and waveguides. Some of these will be reviewed, presented and analysed. From simple examples to complex mixed materials configurations the study will highlight the strengths of the method with respect to standard methods.

Structure-borne intensity fields indicate the magnitude and direction of structure-borne sound in vibraioning structures. The structural intensity fields can be used to identify the energy sources, sinks and indicate the distribution of the energy in structures. It can guide the application of vibration and noise control treatments. In this research, the structrual intensiy concept is utilized to investigate the cracked plate's vibratino characteristics. Generally speaking, a existing crack may change the dynamic characteristics of a structure, therefore the existance of the crack in a structure will change vibrational wave in the structure and substantially affect the power flow or structural intensity characteristics. As a result, the investigation of the structural intensity in cracked structures will have the potential to crack detection. In Chapter 10, the structure-borne intensity fields of a simply-supported thin aluminium plate with a surface crack are investigated by using solid finite elements. The structural intensity conectp is introduced at first and the formulas of basic structural elements (beam, shell and solid) are given in detail. The structural intensity streamline is introduced to visualize the structural intensity fields. To describe the internal element stress fields more accurate, the isoparametric solid elements are used to model the plate. The intact plate's structural intensity patterns obatined by shell element and solid element are respectively computed, which validates the accuracy of the solid element calculation for structural intensity. The crack is modelled by quarter point crack tip element. Based on solid finite element, the cracked plate's displacement vector field, structural intensity vector field and structural intensity streamline field are obtained under a point excitation harmonic force applied at the centre of the plate. The calcuations show that the structural intensity field is dependent on the vibraion mode, and the vibrating source can be successfully indicated. The cracked plate's intensity vector and streamline patterns show that the existance of crack changes the structural intensity in the plate. At the location of the crack, the intensity vector and streamline have abruptly changes in magnitude and direction. Cases of different crack location are considered to investigate the relationship bewteen crack's location with the structural intensity patterns, which indicates that the structural intensity pattern can successfully identify the location of the crack in the plate.

As a form of energy, diagnostic ultrasound (DUS) has the potential to have effects on living tissues, e.g. bioeffects. The two most likely mechanisms for bioeffects are heating and cavitation. The thermal index (TI) expresses the potential for rise in temperature at the ultrasound's focal point. Since an output of TI over 1.5 is considered hazard, the question is what the settings in which such hazardous exposure occurs are. The mechanical index (MI) indicates the potential for the ultrasound to induce inertial cavitation in tissues. Nevertheless, cavitation has not been documented in mammalian fetuses, since there is not an air-water interface, which is needed for the cavitation mechanism.

Chapter 11 presents data regarding ultrasound end-users familiarity with safety issues, acoustic output and safety of obstetrics ultrasound.

There are scarce data on instruments acoustic output (nor patient acoustic exposure) for routine clinical ultrasound examinations. Ultrasound end-users are poorly informed regarding safety issues during pregnancy. While first trimester ultrasound is associated with negligible rise in the thermal index, increased TI levels are reached while performing obstetrical Doppler studies. In particular, TI levels may reach 1.5 and above. Acoustic exposure levels during 3D/4D ultrasound examination, as expressed by TI are comparable to the two-dimensional B-mode ultrasound. However, it is very difficult to evaluate the additional scanning time needed to choose an adequate scanning plane and to acquire a diagnostic 3D volume.

SHORT COMMUNICATIONS

In: New Research on Acoustics
Editor: Benjamin N. Weiss, pp. 3-20
ISBN: 978-1-60456-403-7

EXPERIMENTAL ACOUSTICAL THERMOTOMOGRAPHY OF MODEL OBJECT

Andrej A. Anosov [a,b,*], ***Alexander S. Kazanskij*** [a,c] ***and Anton S. Sharakshane*** [d]

[a]Institute of Radioengineering and Electronics of RAS, Mochovaja street 11, Moscow 125009, Russia

[b]Sechenov Moscow Medical Academy, B.Pirogovskaja street 2/6, Moscow 119992, Russia

[c]Moscow State Institute of Radio Engineering, Electronics and Automatics, Vernadskogo prospect 78, Moscow 119454, Russia

[d]Institute of biochemical physics of RAS, Kosygina street 4, Moscow 117997, Russia

Abstract

Measurements of thermal acoustic radiation (TAR) from a model object were carried out for reconstruction of temperature inside the object. This experiment presented the simulation of the temperature control in medical applications (e.g. in hyperthermia and thermoablation). The model object was a plasticine cylinder placed in water and water solution of glycerine (in this medium an acoustic absorption is similar to that of soft tissues of a human body). The plasticine was heated for 4-6 minutes. Its temperature was controlled with the help of an IR-thermograph. The TAR measurements in 1.4-2.2 MHz region were carried out for 8-10 minutes with five acoustothermometers placed on different sides of the model object. The changes of the TAR intensity during both the heating and the cooling of the object were registered. By using these experimental data the reconstruction of the 2-D temperature distribution in the model object was carried out. We averaged data over several seconds and reconstructed the object temperature, i.e. we calculated parameters of the temperature distribution: position (*x* and *y* coordinates), size and temperature of thermal source. So we obtained temporal changes of the temperature distribution. We used different methods for the reconstruction and investigated effect of the average time on the reconstruction quality. To calculate the object temperature by using the TAR measurements it is necessary to have

[*] Address for correspondence: A. A. Anosov, Goncharova street 16-57, Moscow 127254, Russia. Tel.: +7 495 639 02 06; E-mail: aanosov@atom.ru

information about the ultrasound absorption in the object. For estimation of the absorption coefficient we used the data obtained in the same experiment.

Keywords: thermal acoustic radiation, acoustical thermotomography, temperature reconstruction

Introduction

The temperature control of human body tissues is necessary when different thermal therapies (for example, hyperthermia or thermoablation in oncology) are carried out. It is desirable for this control to use non-invasive methods. For this purpose one can use acoustical thermotomography (or acoustotermography) namely measurements of thermal acoustic radiation (TAR) from an object and then reconstruction of the temperature distribution inside this object. The acoustotermography method was proposed independently by Babii, 1974 and Bowen, 1981. At 1987 Bowen suggested the use of multifrequency measurements of TAR for the reconstruction of temperature at an depth of an object. In Passechnick's scientific group these measurements were carried out (see Anosov et al., 1998a). Let us note that our calculations show that the multifrequency measurements allow to reconstruct with acceptable accuracy only monotonous temperature distributions. Anosov et al., 1998b reconstructed the 1-D temperature distribution in a human hand from palm to backside. In this experiment only one acoustothermometer was used. Mansfeld's scientific group used a 12-channel acoustic radiometer (Ksenofontov et al., 1997, Krotov et al., 1999) and focused acoustothermometer (Vilkov et al., 2005) for location of the heated region in model medium. In those experiments the authors didn't calculate the temperature but only detected the heated region position. Passechnick et al., 1999b reconstructed the 2-D temperature distribution in model object (water solution of glycerin). In this experiment only one fixed acoustothermometer was used and the object under investigation was shifted. In this work principles of X-ray tomography were used. Hessemer et al., 1983 suggested to use the correlation reception of TAR for temperature detection. In Passechnick's group the first experiment with correlation reception of TAR was carried out (see Anosov et al., 2000). Mirgorodskij with colleagues (see Gerasimov et al., 1999, Mirgorodskij et al., 2006) suggested to use 4-order correlation function of TAR pressure. But this technique needs measurements taken during a long period of time that is unlikely in medical applications. Burov et al, 2002 suggested "active-passive" correlation tomography namely both to measure TAR and to use an external "active" ultrasound source for detection of medium absorption. Vilkov et al., 2005 used the focused acoustothermometer for the correlation reception of TAR. Theoretical estimations (see Passechnick et al., 1999a) show that the acoustotermography method permits the detection of the internal temperature at a depth of up to 5-10 cm with an accuracy of approximately up to 0.5 – 1 K in the volume of about 1 cm^3. Let us note that in all these cases only stationary distributions were considered. It connects with essential average time (about 1 min) of TAR noise signal. But if the object temperature changes at a slower rate (several degrees per minute) one can use averaging over about minute and reconstruct the temperature. In the present work we are going to detect the temporal change of the 2-D temperature distribution in other words we expect to solve the problem of dynamic mapping of the internal temperature. Along with the measurements of TAR we are going to use measurements of the infrared thermal electromagnetic radiation of the object (see Anosov et al., 2008a). Combined usage of two non-invasive methods for the temperature control raises reliability of the measurements.

Scheme of Acoustical Measurements

The scheme of the experiment is shown in Figure 1. The measurements were carried out in a thermostat tank ($43 \times 43 \times 15\text{cm}^3$) filled with water. Mercury and electronic thermometers controlled the temperature of the tank with precision of a 0.2 K. The TAR measurements were carried out with five acoustothermometers (AT1 – 5) constructed by Mansfeld's group from the Institute of Applied Physics of RAS, Nizhny Novgorod (see Anosov et al., 2008b). The frequency region of the receiver was 1.8±0.4 MHz and the diameter was 10 mm. The acoustothermometers registered the pressure of acoustic waves from the tank, transformed it to voltage and amplified it. Then electrical signal passed through a square detector and was averaged over 30 ms. The output signal was recorded in 14-bit analog-to-digital converter L-780M developed by LCard Inc. (www.lcard.ru). The frequency of the digitised signal was 1 kHz. The signal was averaged by the computer over 0.5 s. A cavity C (base square 13 x 18 cm^2) with water or water solution of glycerin was placed into the tank. The side walls of the cavity were made from acoustically transparent film (from thin polyethylene). As a model object we used a plasticine cylinder T placed vertically in the cavity. Three different cylinders (8 cm high, 6, 10 and 22 mm base diameter) were used in our experiments. A metallic rod (4 mm diameter) was placed inside the cylinder at its axis. As a rod a soldering iron (25 W power) was used. The rod was heated up to high temperature (about 70 – 100 ^{0}C). In the plasticine cylinder there was significant transversal temperature gradient. The cylinder surface temperature was close to the temperature cavity. We supposed that the temperature along the cylinder was constant. In the different experiments the temperature, size and location of the heated source varied. The source was in water or a glycerin solution. Acoustothermometers were placed in a horizontal plane on different sides of the plasticine cylinder. We could change the distance between the acoustothermometers and the angle between their acoustical axes. In all cases we had the 2-D measurement scheme.

Acoustobrightness Temperature

The TAR pressure is a noise signal with a mean value of zero. To obtain the temperature it is necessary to detect the mean value of the pressure square. This value is proportional to the temperature of the investigated object. If we calibrate this value in temperature degrees we obtain the acoustobrightness temperature T_A of the investigated object (Passechnick 1994). For the 2-D case the acoustobrightness temperature is as follows:

$$T_A = \int_0^{\infty} dy \int_{-\infty}^{+\infty} dx \ \alpha(x,y) A(x,y) T(x,y) \exp[-\int_0^{y} \alpha(x,y) dy], \tag{1}$$

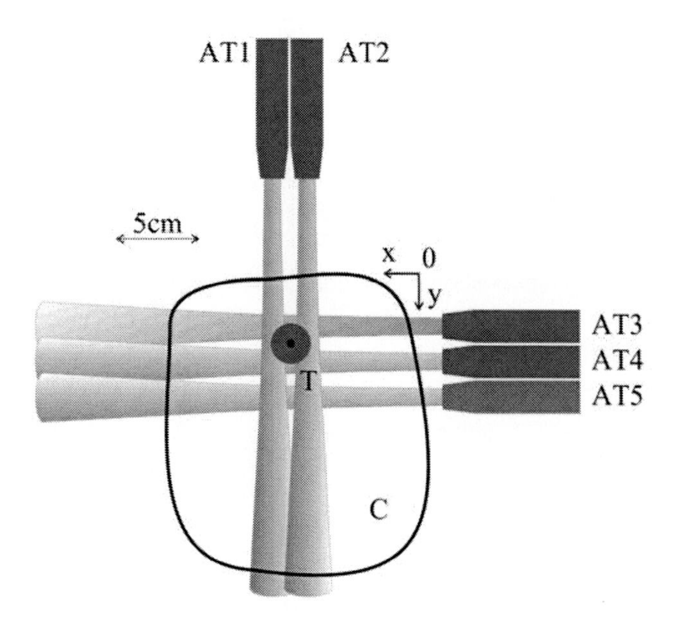

Figure 1. Scheme of experiment: AT1 – 5 are acoustothermometers, C is cavity, T is heated object. Directional patterns of acoustothermometers are shown in grey.

where $\alpha(x, y)$ is the absorption coefficient distribution, $T(x, y)$ is the internal temperature distribution ($T = 0$ is the environment temperature), $A(x, y)$ is the directive pattern of the acoustothermometer, axis x (y) is directed across (along) the acoustothermometer acoustical axis. Let us note the properties of the expression (1). The normalization of the directive pattern is $\int_{-\infty}^{+\infty} A(x,y)dx = 1$ for any y. The acoustobrightness temperature of the medium with the constant temperature T_0 is equal to T_0 too. It is correct for any functions $\alpha(x, y)$ and $A(x, y)$. To calibrate (in degrees) the mean value of the pressure square (the acoustobrightness temperature) we used the heated plasticine plate (20 mm thickness). Plasticine has significant absorption coefficient: α=5 cm^{-1} at 2 MHz (Passechnick 1994) and the plasticine plate can be considered as a black body. The acoustobrightness temperature of a black body is equal to its thermodynamic temperature. Details of the calibration one can see in work of Anosov et al., 2007.

Measurements of TAR

The TAR measurement results of the heated object are presented in Figure 2. The experiments were carried out during 8 – 10 min. Heating was switched on in about 1 min and off in about 5 – 7 min after experiment started. The temporal dependences of measured acoustobrightness temperatures presented in Figure 2 a-c (d) were obtained for the heated source placed in water (in water solution of glycerin). Data presented in Figure 2 a, b (c, d) were obtained for the heated source with diameter 10 mm (22 mm). The all curves were obtained for different positions of the heated source. The temporal changes of the acoustobrightness temperatures were connected with the experiment script. Before switching on the signal was equal to zero, after switching on the signal rised (after some time delay), after switching off the signal were reduced (again after some time delay). These delays were connected with the heat transfer inside the plasticine cylinder.

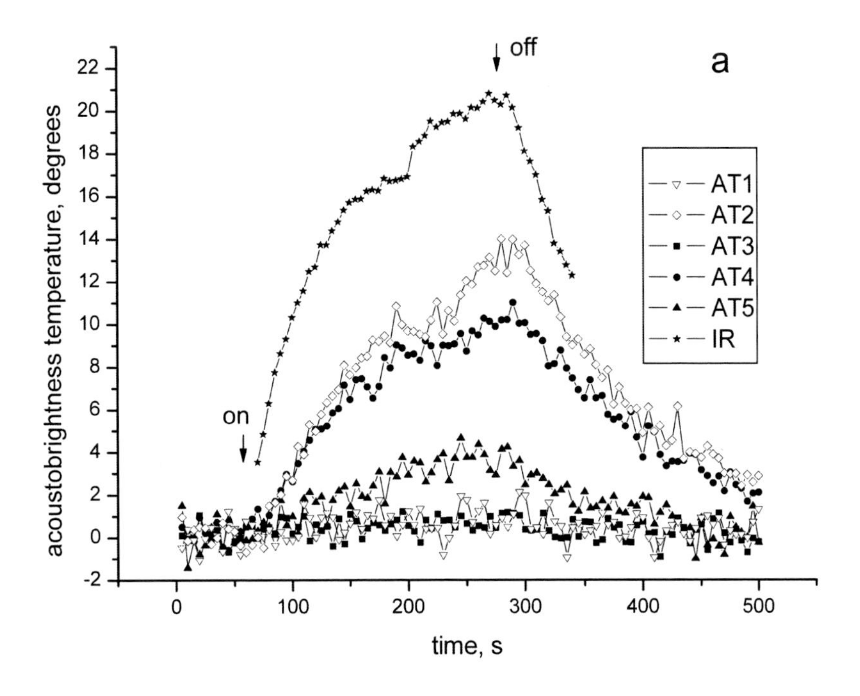

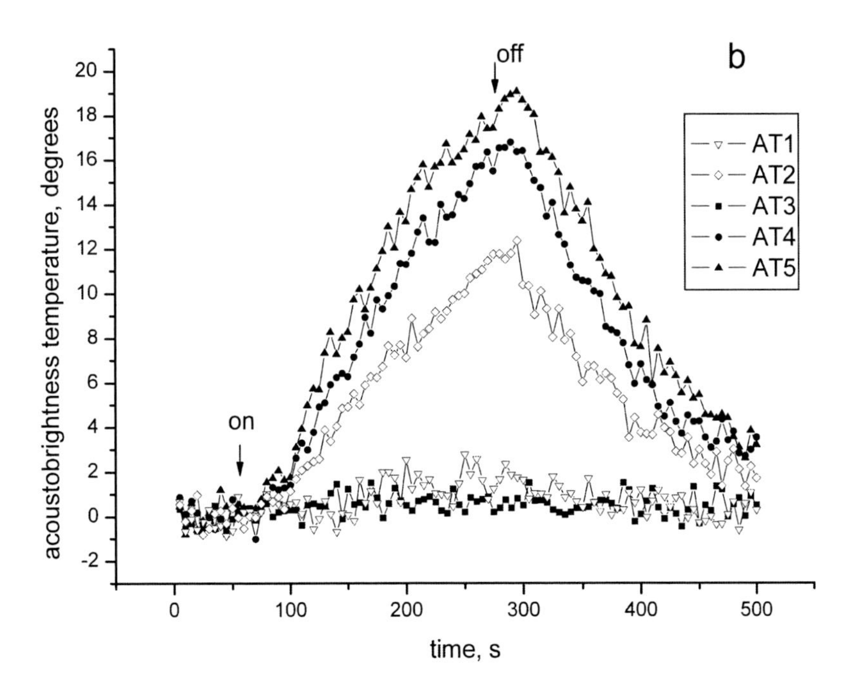

Figure 2. Continued on next page.

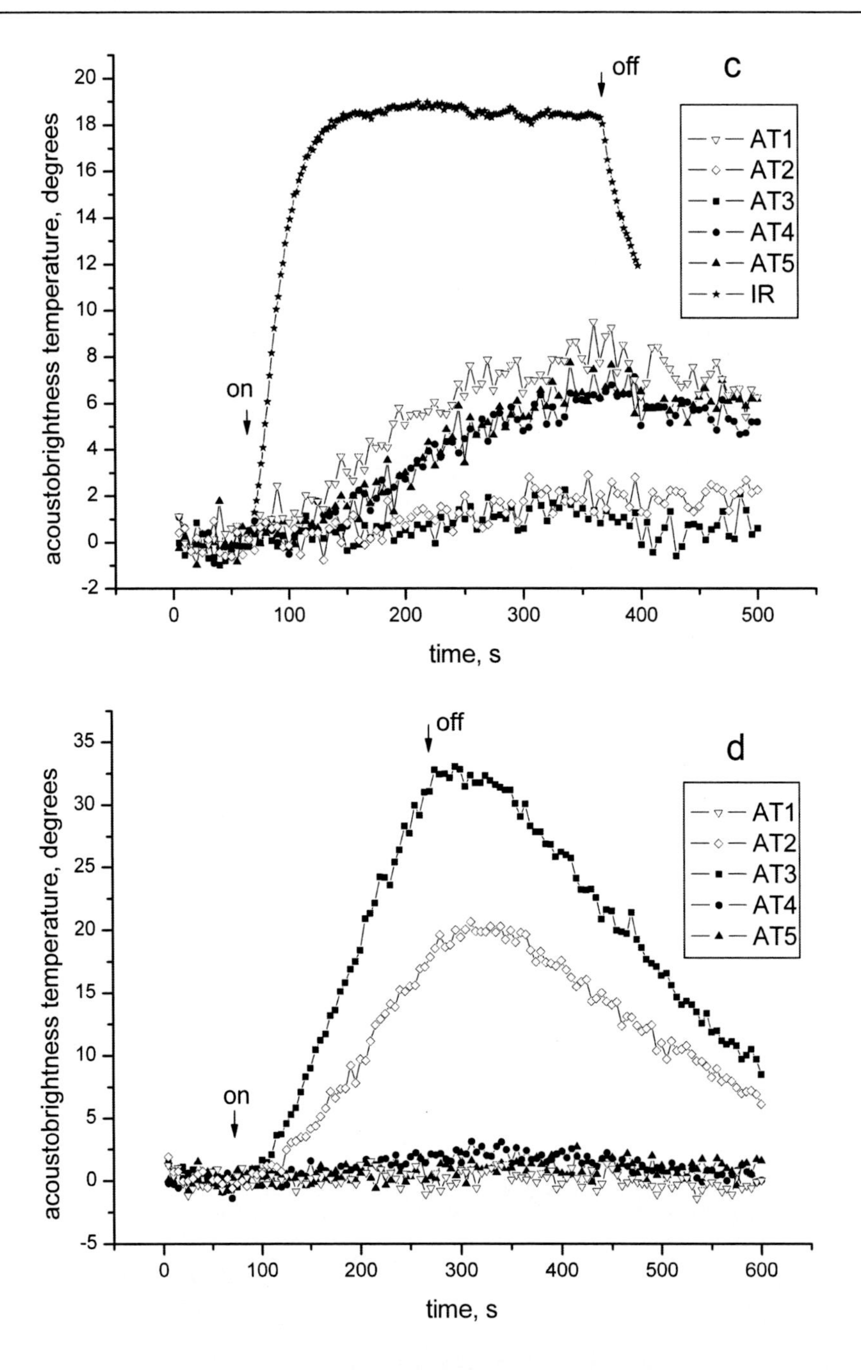

Figure 2. Temporal dependences of acoustobrightness temperatures measured in four experiments (a – d) with 5 acoustothermometers (AT1 – 5). Zero corresponds to the tank temperature (25 ^{0}C). The signals were averaged over 5 s. The moments of the heat switching on and off are shown with arrows. For the experiment 1 and 3 the radiobrightness temperatures (IR) measured with IR-thermograph are presented.

Measurements of Absorption Coefficient

As well known the intensity of TAR depends on the ultrasound absorption inside an object. In our conditions the absorption is negligible in water and very significant in plasticine. We measured the absorption in water solution of glycerin. The heated plasticin plate with thickness 20 mm was placed in the tank and the TAR measurements were caried out. The cavity (13 cm thickness) with water solution of glycerin was placed between the acoustothermometer and the plasticin plate. The thermodynamic and hence acoustobrightness temperatures of the heated plate were decreasing during the experiment. By neglecting ultrasoud dispersion we can detect the absorption coefficient α using the known expression:

$$\alpha=\ln(T_A/T_{AG})/d, \tag{2}$$

where T_{AG} (T_A) is the plasticine plate acoustobrightness temperature if water solution of glycerin (only water) is in the cavity, d=13 cm is the cavity thickness. The experiment results are presented in Figure 3. The acoustobrightness temperatures of the plasticin plate were aproximated with exponents. The coefficient α calculated with the help of expression (2) was equal to 0.11±0.01 cm^{-1}.

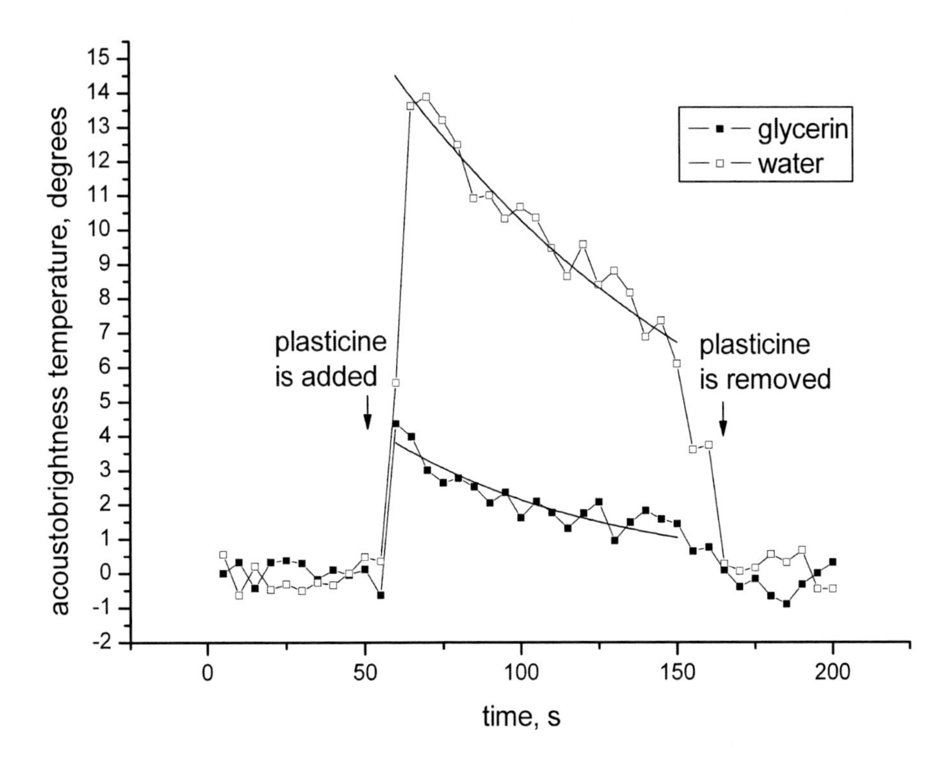

Figure 3. Acoustobrightness temperatures of heated plasticin plate placed in water and water solution of glycerin. The moments when the plasticine was placed into tank and removed from it are shown with arrows. The approximations by exponent are presented with solid curves.

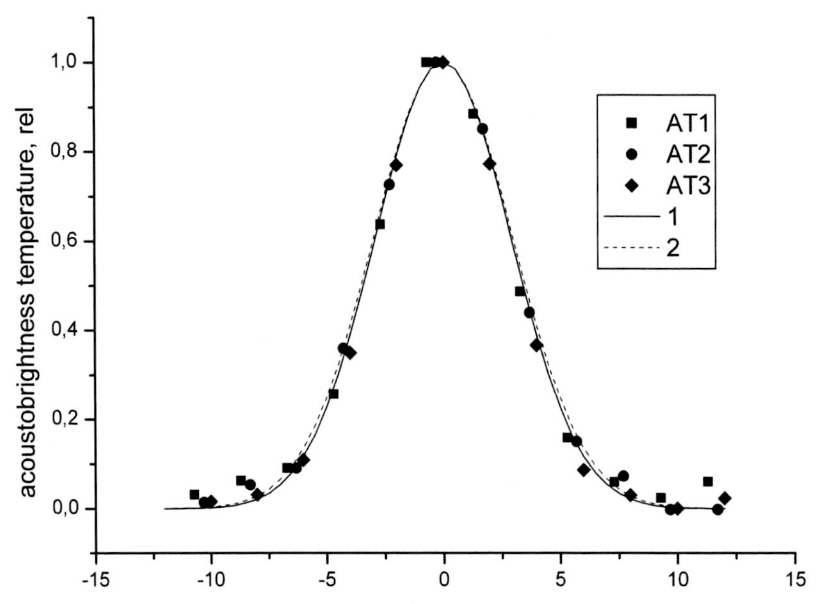

Figure 4. Measurement and calculation (1) results (normed by the unit) of the directive pattern for 3 acoustothermometers (AT1 – 3). Distance between the heated source and acoustothermometers is 7 cm. Curve (2) is calculated for a case when the directive pattern is ray.

Measurements of Acoustothermometer Directive Pattern

For detection of the acoustobrightness temperature it is necessary to measure the acoustothermometer directive pattern. The source of TAR was the plasticine cylinder with 6 mm diameter. Inside the cylinder the heated metallic rod with 2 mm diameter was placed. This source was at a distance 7 cm from the acoustothermometers. The source was shifted accros the acoustical axis of the acoustothermometers. The measured signals (normed by the unit) for 3 acoustothermometers are shown in Figure 4. The position of the source center equal to zero corresponds the case when the source center is located at the acoustical axis of the acoustothermometer. For TAR measurements we used wideband receiver (the relation of the frequency region to the mean frequency was equal to 44 %). Therefore we could approximate the acoustothermometer directive pattern by Gaussian function

$$A(x) = \frac{1}{d_{AT}\sqrt{2\pi}} \exp\left(\frac{-x^2}{2d_{AT}^{\ 2}} \right), \tag{3}$$

where x is the source center position relatively the receiver acoustical axis, d_{AT} is the transversal size of the directive pattern. The value d_{AT} was to detect. The radial temperature distribution $T(r)$ inside the plasticine cylinder was calculated with the help of expression: $T(r)=T_R \ln(r_2/r) / \ln(r_2/r_1)$, where r_1 = 1 mm is the rod radius, r_2 = 3 mm is the plasticine cylinder radius, T_R is the rod temperature. This radial temperature distribution is solution of stationary heat equation. Substitutions the radial temperature distribution and the acoustothermometer directive pattern (3) in the expression (1) allows to calculate the acoustobrightness temperature as function of the distance between the source center and the acoustical axis of the receiver. The acoustobrightness temperature (normed by the unit) calculated with d_{AT}=2.6 mm is shown in Figure 4 (curve 1). One can see that the calculated data are close to the experimental data. We used other approximation too. We detected the temperature distribution as Gaussian function

$$T = T_0 \exp\left(\frac{-r^2}{2d^2}\right), \tag{4}$$

where T_0 and d are the effective temperature and size of the heated source. We suppose that the acoustothermometer directive pattern is the ray. In this case the acoustobrightness temperature calculated with d=3 mm is shown in Figure 4 (curve 2). One can see that the calculated data are close to the experimental data and we can use this approximation in next calculations.

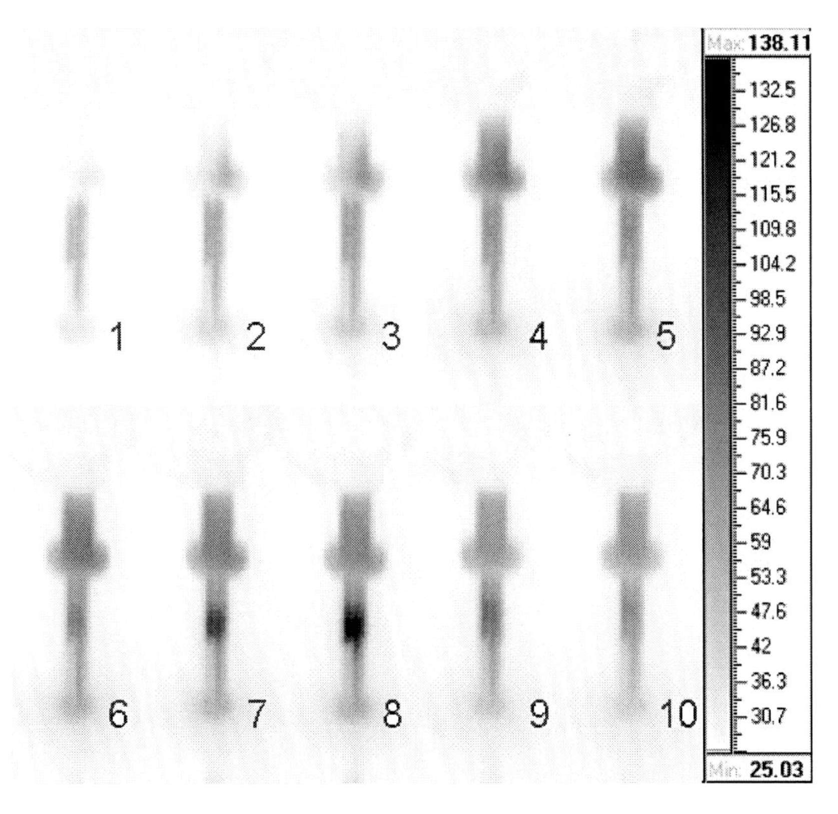

Figure 5. Infrared imaging of the rod (soldering iron) upper the water surface after switching on (1 – 8) and off (9, 10) of the heated source. Data of this experiment are presented in Figure 2a. The temperature scale is shown.

Measurements of Infrared Thermal Electromagnetic Radiation

Along with the TAR registration we carried out the measuments of infrared (IR) thermal electromagnetic radiation. We used portable computer thermograph IRTIS-2000 developed by «IRTIS» Ltd. (www.irtis.ru). The threshold sensitivity of this thermograph is equal to 0.05 K. The IR measuments give information about the surface temperature of an object. In our experiments the plasticine cylinder (the sourse of TAR) was inside water and we could not measure its temperature. With the help of this method we measured the surface temperature of the metallic rod upper the water surface. The experimental results are presented in Figure 2 a, c and Figure 5. The measured radiobrightness temperature can be considered as the one closed to maximum temperature of the heated source.

Reconstruction Algorithm of Temperature Distribution

Let us consider that the temperature distribution is given with four parameters. Coordinates x_0 and y_0 detect center position of the heated region, T_0 is its maximum temperature (zero is the tank temperarure) and d is its size. We consider the directive patterns of the acoustothermometers as the rays. Therefore the acoustobrightness temperatures can be calculated as follows:

$$T_{Ai} = T_0 \exp\left[-\alpha\, y_0 - \frac{(x_{ATi} - x_0)^2}{2d^2}\right], \text{ if } i = 1, 2;$$

$$T_{Ai} = T_0 \exp\left[-\alpha\, x_0 - \frac{(y_{ATi} - y_0)^2}{2d^2}\right], \text{ if } i = 3, 4, 5 \qquad (5)$$

where i is the acoustothermometer number, x_{ATi} and y_{ATi} are the coordinates of the acoustothermometers centers, α is the absorption coefficient in the cavity (if water is in the cavity then $\alpha = 0$).

The algorithm of the temperature reconstruction consists of three steps:

- The first step is detection of coordinates x_0 and y_0 of the heated region center. We propose that these coordinates are inside region limited the acoustothermometer directive patterns: $\begin{cases} x_{AT1} \le x_0 \le x_{AT2} \\ y_{AT3} \le y_0 \le y_{AT5} \end{cases}$. If the signal of the middle acoustothermometer is not equal to zero and the signals of two neighbouring acoustothermometers are equal to zero then we can consider that the source center coordinate is equal to the midlle acoustothermometer center coordinate. If the signals of two neighbouring acoustothermometers are equal then the source center coordinate is in the middle between these acoustothermometers. These rules allow us to calculate the source coordinates:

$$x_0 = \frac{x_{i+1} + x_i}{2} + \frac{T_{Ai+1} - T_{Ai}}{T_{Ai+1}} \cdot \frac{x_{i+1} - x_i}{2},$$

$$y_0 = \frac{y_{i+1} + y_i}{2} + \frac{T_{Ai+1} - T_{Ai}}{T_{Ai+1}} \cdot \frac{y_{i+1} - y_i}{2}, \qquad (6)$$

where x_{i+1} and x_i (y_{i+1} and y_i) – x (y) coordinates of the neighbouring acoustothermometer centers ($T_{Ai+1} \geq T_{Ai}$).

- The second step is the detection of the absorption coefficient inside the cavity with a glycerin solution. We used two methods for calculation of the absorption coefficient. We measured this value separately(see expression (2) and Figure 3). We estimated the result with the help of the method suggested by Anosov and Gavrilov, 2005 as well. This method allows us not to carry out additional measurements but to use only the information about the measured acoustobrightness temperatures. As one can see in Figure 2 d the acoustobrightness temperature measured with the third acoustotermometer is greater than the one measured with the second acoustotermometer. This difference of signals from the same source is connected with the path difference of acoustic waves from the source to the receivers. The acoustic radiation passes in the absorptive medium (inside the cavity with water solution of glycerin) a smaller distance from the source to the third acoustotermometer than to the second acoustotermometer (see Figure 1). These acoustobrightness temperatures allow us to estimate the absorption coefficient in the cavity with expression:

$$\alpha = \frac{\ln\left(T_{A1} / T_{A3} \right)}{x_0 - y_0} \qquad (7)$$

- The third step is minimization of function $F(T_0, d)$ to detect the temperature T_0 and size d of the source:

$$F(T_0, d) = \sum_{i=1}^{5} \left[T_{AiEXP} - T_{Ai}(T_0, d)\right]^2 \rightarrow \min, \qquad (8)$$

where T_{AiEXP} are the experimental acoustobrightness temperatures, T_{Ai} are the acoustobrightness temperatures determined with expressions (5).

Reconstruction Results

Reconstruction results of the heated region center position, source size and temperature are shown in Figure 6. To obtain these results we averaged the experimental data over 50 s. The heating was switched on in about 60 s after the measurements started. The acoustobrightness temperature was equal to zero (see Figure 2) before switching on and during approximaly 40 s after switching on when the plasticine was being heated inside the cylinder. So we began to use our algorithm in 100 s after the experiment started. In Figure 6 a we can see that the reconstructed positions of the heat source changed insignificantly (inside the region about 1 – 2 mm^2). Therefore the algorithm allows us to reconstruct the heat source position quite well. The reconstructed sizes of the heat source are shown in Figure 6 b. We can see that they also show insignificant changes (in limits of about 1 mm). It argues for our algorithm. But there are systematical errors for the source with 0.5 cm radius (in the first case the size is equal to about 0.2 cm, in the second case about 0.7 cm). We think it is connected with small size of source in comparison with the distance between the receivers (see Figure 1). For the cylinder with 1.1 cm radius the reconstructed effective sizes (0.8 – 1.0 cm) are close to the real value. The reconstructed effective temperatures are shown in Figure 6c. Let us note that the temporal dependencies of these temperatures correlate with the experimental acoustobrightness temperatures very well. We cannot detect the temperature distribution inside the plasticine cylinder. We calculate only two parameters of this distribution: the effective size and temperature. We didn't measure the temperature inside the cylinder but we registered the metallic rod temperature upper water suface with the help of IR-thermograph. We can only maintain that the reconstructed temperatures are less than the rod temperature.

The values of the reconstructed absorption coefficient are shown in Figure 6 d. These values are close to the measured absorption coefficient. The results of the temperature reconstruction obtained with the help both the reconstructed and measured absorption are

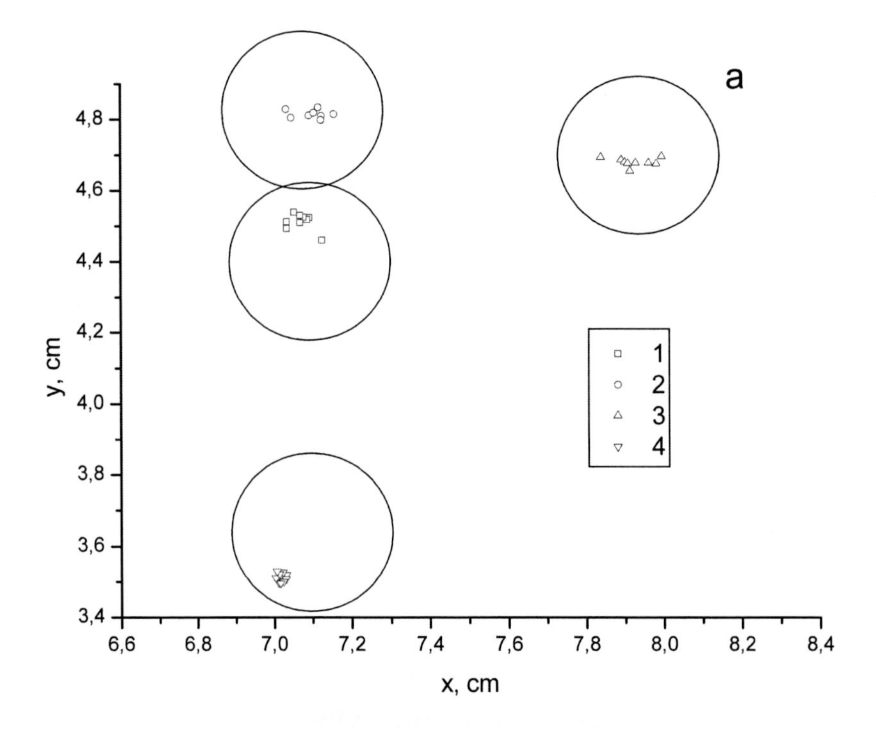

Figure 6. Continued on next page.

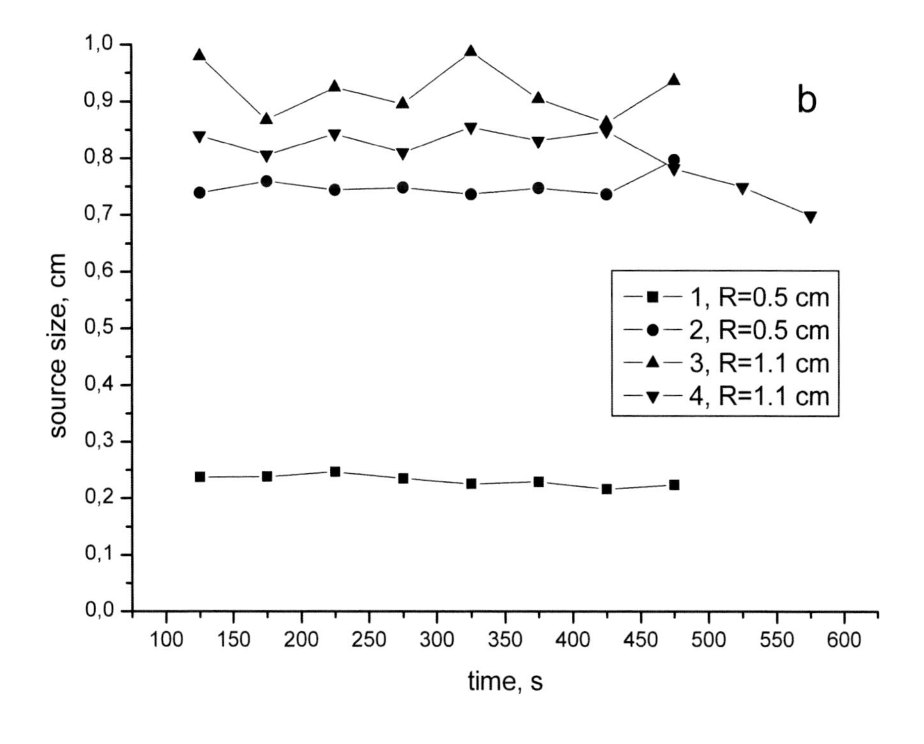

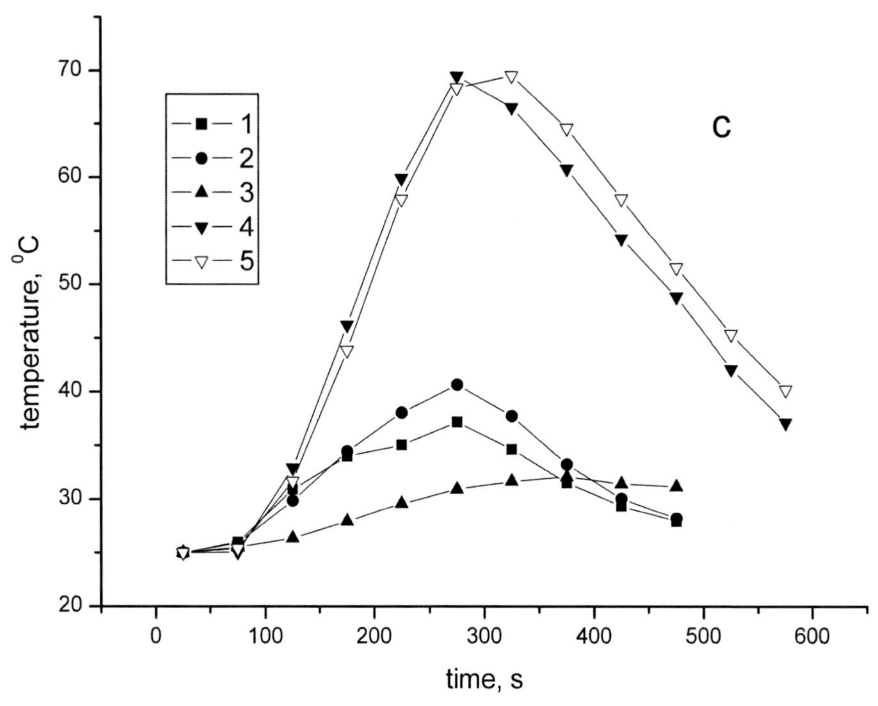

Figure 6. Continued on next page.

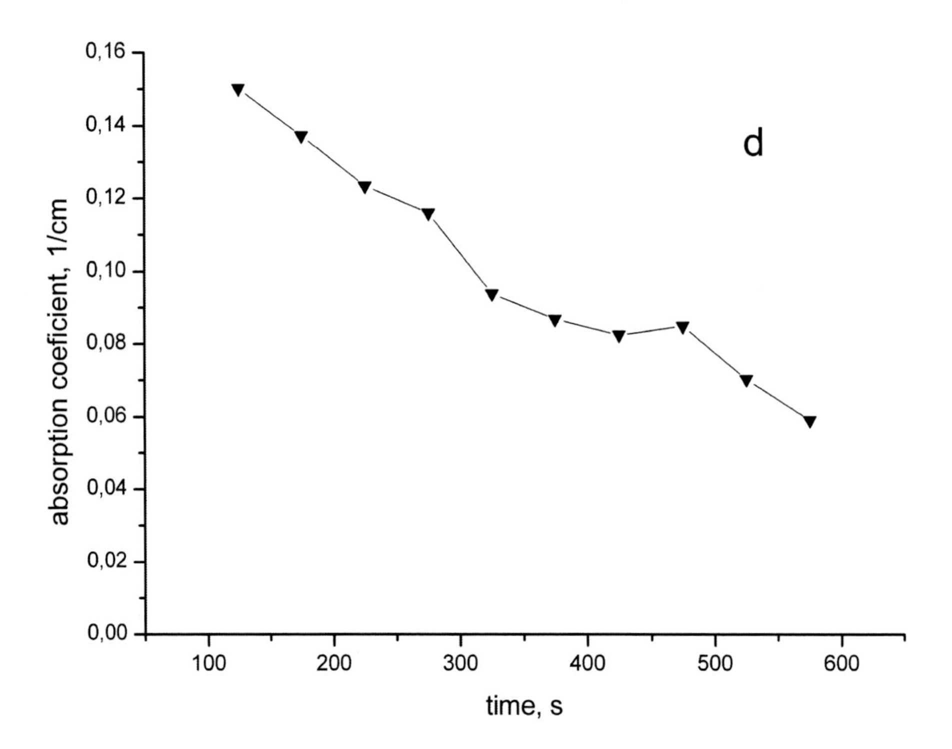

Figure 6. Reconstruction results of the heated region center position (a), source size (b), temperature (c) and absorption coefficient (d). 1 – 4 are numbers of experiments. In experiments 1 and 2 (3 and 4) the real source radius is R=0.5 cm (1.1 cm). In experiments 1 – 3 (4) the absorpting medium is absent (present). Circles show the real rod positions. The temperature curve 5 is obtained for α=0.11 cm^{-1}.

shown in Figure 6c (curves 4 and 5) and these results are close. It speaks well of the method of the absorption detection (see the second step of the reconstruction algorithm). We can see that the reconstructed values of the absorption coefficient were decreasing during the experiment. It can be compared to the decrease of the absorption in the water solution of glycerin when the solution temperature rises. Similar results were obtained by Passechnick *et al*., 1999b.

The reconstruction results presented in Figure 6 were obtained with averaging over 50 s but the experimental results in Figure 2 were obtained with averaging only over 5 s. We tried to reconstruct the temperature distribution parameters using different average times. The *x* and *y* coordinates, size and temperature of the thermal source obtained with averaging over 5, 10, 25 and 50 s are shown in Figure 7. We can see that the averaging over 10 s and more gives quite good results.

Thus we caried out the experimental reconstruction of the temperature distribution inside the model object. We got information about the location, size and temperature of the model object using data of five acoustotermometers. In addition the temporal changes of the temperature distribution parameters were detected. For the temperature control we used the measurements of both TAR and infrared radiation. These results allow us to advance to acoustical thermotomography of biological objects.

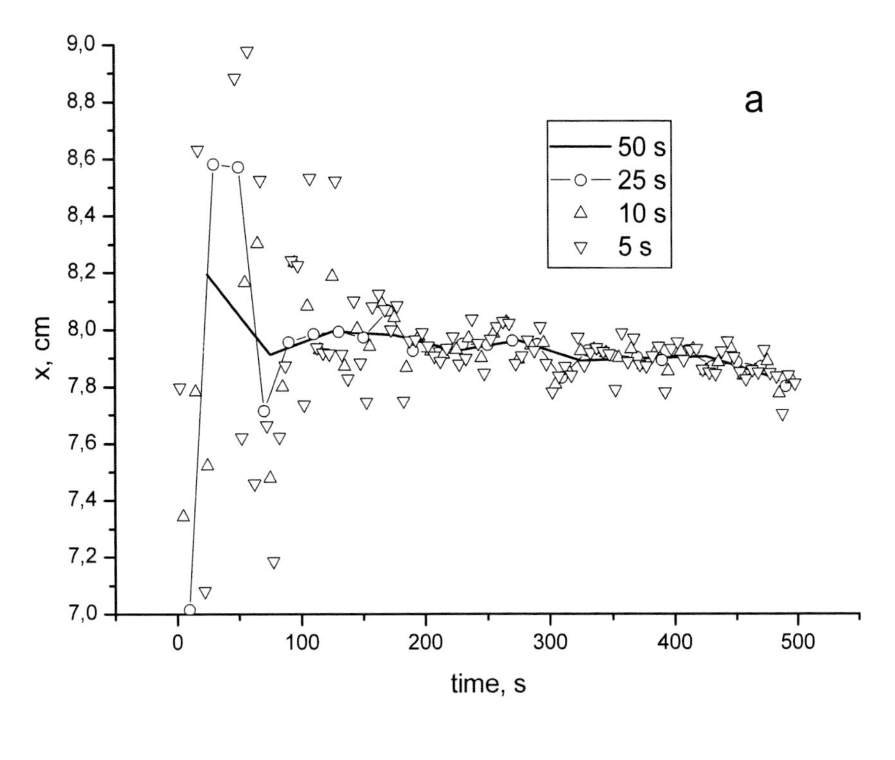

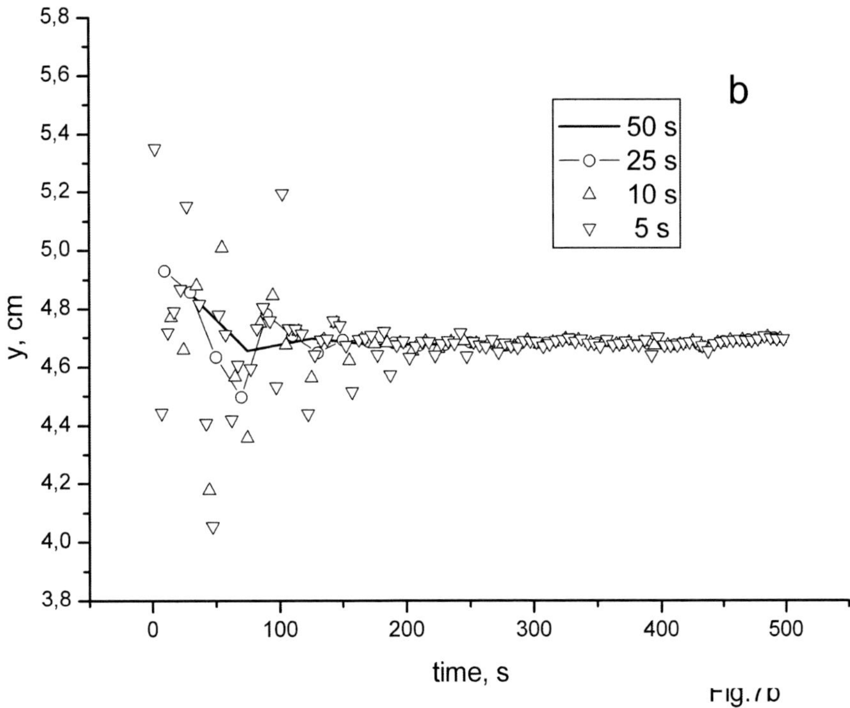

Figure 7. Continued on next page.

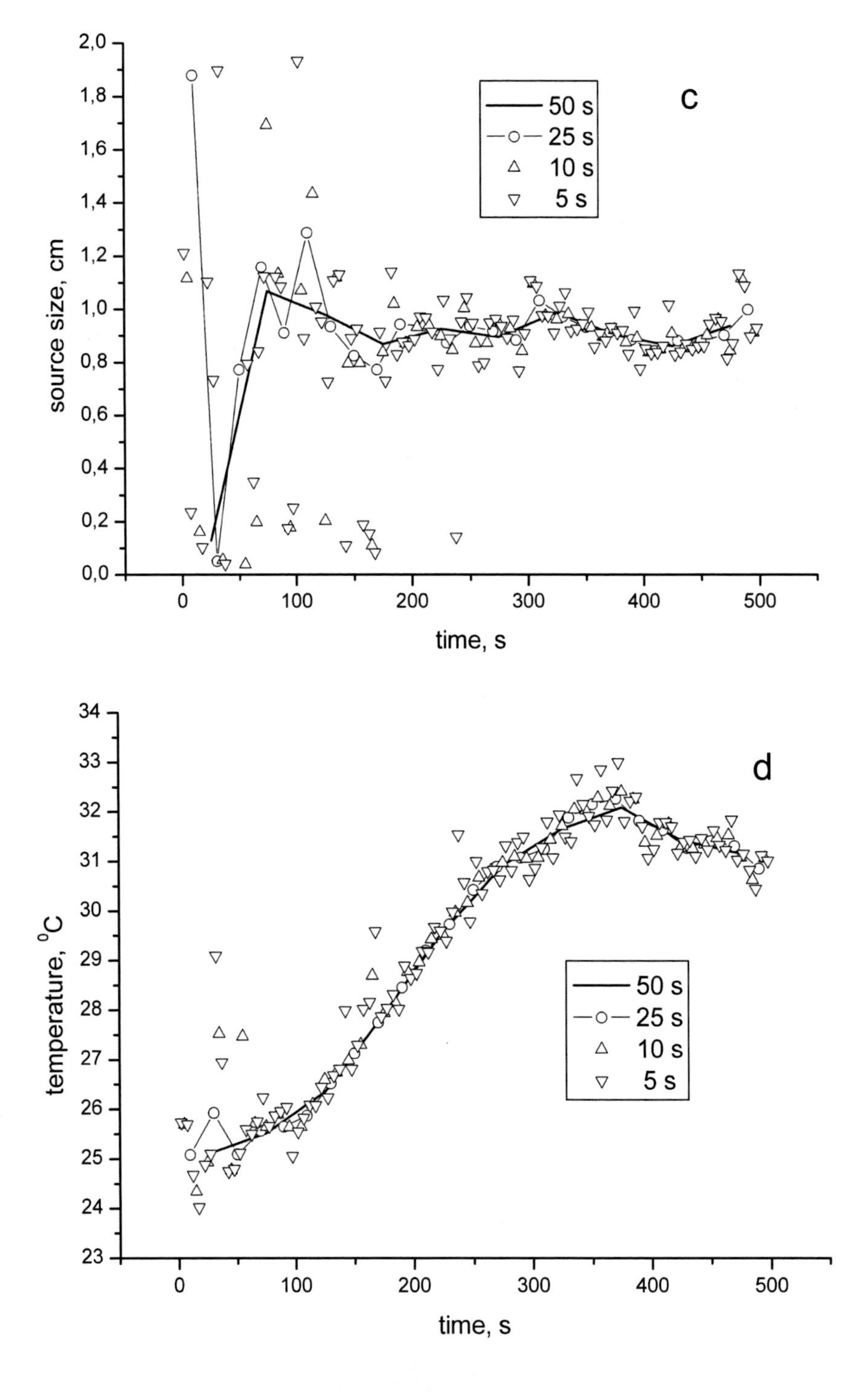

Figure 7. Reconstruction results of x and y coordinates, source size and temperature with averaging over 5, 10, 25 and 50 s for experimental data presented in Figure 2 c.

Acknowledgement

This work was supported by Russian Foundation for Basic Research.

References

Anosov, AA; Bograchev, KM; Pasechnik, VI. (1998a) Measurement of the Thermal Acoustic Radiation from a Human Hand. *Acoustical Physics* **44**, 248-254.

Anosov, AA; Bograchev, KM; Pasechnik, VI. (1998b) Passive Thermoacoustic Tomography of a Human Hand. *Acoustical Physics* **44**, 629-634.

Anosov, AA; Antonov, MA; Pasechnik, VI. (2000) Measurement of the Correlation Properties of Thermal Acoustic Radiation. *Acoustical Physics* **46**, 21-26.

Anosov, AA; Gavrilov, LR. (2005) Reconstruction of the In-Depth Temperature Distribution for Biological Objects by Linear Phased Arrays. *Acoustical Physics* **51**, 376-384.

Anosov, AA; Kazanskij, AS; Less, YuA; Sharakshane, AS. (2007) Thermal Acoustic Radiation in Model Membranes at Phase Transition of Lipids. *Acoustical Physics* **53**, 843-848.

Anosov, AA; Barabanenkov, YuN; Bograchev, KM; Garskov, RV; Kazanskij, AS; Sharakshane, AS. (2008a) Simultaneous usage of acoustothermography and IR-thermovision for temperature control at model object heating. *Acoustical Physics* **54**, in press.

Anosov, AA; Bel'aev, RV; Vilkov, VA; Kazanskij, AS; Mansfel'd, AD; Sharakshane, AS. (2008b) Detection of temperature change dynamics in model object with acoustothermography method. *Acoustical Physics* **54**, in press.

Babii, VI. (1974) The transfer of acoustic energy in absorptive and radiating medium. *Sea hydrophysical studies, Sevastopol, USSR,* **65**, 189-193 (in russian)..

Bowen, T. (1981) Passive remote temperature sensor system. United States patent 4,246,784, Jan.27..

Bowen, T., (1987) Acoustic Radiation Temperature for Noninvasive Thermometry. *Automedica (UK)* **8** (4), 247-267..

Burov, VA; Darialashvili, PI; Rumyantseva, OD. (2002) Active–Passive Thermoacoustic Tomography. *Acoustical Physics* **48**, 412-421.

Hessemer R, Perper T, Bowen T. (1983) Correlation thermography. United States patent 4,416,552, Nov.22..

Gerasimov, VV; Gulyaev, YuV; Mirgorodskii, AV; Mirgorodskii, VI; Peshin, SV. (1999) Spatial Resolution of the Passive Location Based on Fourth-order Correlation Processing. *Acoustical Physics* **45**, 433-438.

Krotov, EV; Ksenofontov, SYu; Mansfel'd, AD; Rejman, AM; Sanin, AG; Prudnikov, MB. (1999) Experimental investigations of posibilities of multichannel acoustical thermotomography. *Izv. VUZov Radiofizika* **42**, 479-484 (in russian)..

Mirgorodskij, VI; Gerasimov, VV; Peshin, SV. (2006) Experimental Studies of Passive Correlation Tomography of Incoherent Acoustic Sources in the Megahertz Frequency Band. *Acoustical Physics* **52**, 606-612. .

Passechnick, VI. (1994) Verification of the Physical Basis of Acoustothermography. *Ultrasonics* **32**, 293-299.

Passechnick, VI; Anosov, AA; Isrefilov, MG. (1999a) Potentialities of Passive Thermoacoustic Tomography of Hyperthermia. *Int. J. Hyperthermia* **15**, 123-144.

Passechnick, VI; Anosov, AA; Isrefilov, MG; Erofeev, AV. (1999b) Experimental Reconstruction of Temperature Distribution at a Depth through Thermal Acoustic Radiation. *Ultrasonics* **37**, 63-66.

Vilkov, VA; Krotov, EV; Mansfel'd, AD; Rejman, AM. (2005) Application of Focusing Arrays to the Problems of Acoustic Brightness Thermometry. *Acoustical Physics* **51**, 63-70.

In: New Research on Acoustics
Editor: Benjamin N. Weiss, pp. 21-61
ISBN: 978-1-60456-403-7

GENERALIZED VARIATIONAL PRINCIPLE FOR DISSIPATIVE HYDRODYNAMICS AND ITS APPLICATION TO THE BIOT'S EQUATIONS FOR MULTICOMPONENT, MULTIPHASE MEDIA WITH TEMPERATURE GRADIENT

G.A. Maximov
Moscow Engineering Physics Institute, Moscow, Russia

Abstract

The generalization of the Hamilton's and Osager's variational principles for dissipative hydrodynamical systems is represented in terms of the mechanical and thermal displacement fields. A system of equations for these fields is derived from the extreme condition for action with a Lagrangian in the form of the difference between the kinetic and the free energies minus the time integral of the dissipation function. The generalized hydrodynamic equation system is then evaluated on the basis of the generalized variational principle. At low frequencies this system corresponds to the traditional Navier – Stokes equation system and in the high frequency limit it describes propagation of acoustical and thermal modes with the finite propagation velocities.

Further, the system of generalized Biot's equations, describing waves propagation in a multi-phase or multi-component medium in the presence of heat exchange between phases, is derived on the basis of generalized variational principle. It is shown that in the presence of N phases 2N propagating eigen-modes can exist in this medium. At high frequencies N modes are of the mechanical (acoustical) type and N modes are of the diffusive (thermal) type of propagation. At low frequencies there is the single acoustical (wave) mode and the rest 2N-1 modes possess the diffusion (thermal) type of behavior. For a two-component medium without temperature exchange the developed approach is reduced to the well known Biot's model. The account of the temperature field yields the generalized Biot's model for two components medium.

Part I. Genaralized Variational Principle for Dissipative Hydrodynamics

Introduction

A system of hydrodynamic equations for a viscous, heat conducting fluid is usually derived on the basis of the mass, momentum and energy conservation laws [1]. Certain assumptions about forms of viscous stresses tensor and energy density flow vector are made to complete it. This system is considered presently as the one describing quite adequately a large set of hydrodynamical phenomena. However, there are some aspects which suggest that this system is an approximation only.

For example, if we consider propagation of small perturbations described by this system, then it is possible to separate formally the longitudinal, shear and heat or entropy waves. The coupling of the longitudinal and heat waves results in their splitting into independent acoustic-thermal and thermal-acoustic modes. For these modes the limits of phase velocities tends to the infinity at high frequencies so that the system is in formal contradiction with the requirements for a finite propagation velocity of any perturbation. By this reason it is possible to suggest that the traditional hydrodynamical equation system is a mere low frequency approximation.

In particular, an account of the viscosity relaxation phenomenon [2] allows us to provide the limit for the propagation velocity of the shear mode and the introduction of the heat relaxation term [3-5] provides the finite propagation velocities of the acoustic-thermal and thermal-acoustic modes.

However, the introduction of such relaxation processes requires serious efforts especially for more complicated cases, for example, for a medium possesses additional internal degrees of freedom or in the case of a multi-component or multi-phase media.

Classical mechanics provides us with the Lagrange's variational principle which allows to derive easily the equations of motion for a mechanical system knowing the forms of kinetic and potential energies. The difference between these energies is determined the Lagrange's function. This approach translates easily into continuum mechanics by introduction of the Lagrangian density for non-dissipative media. In this approach the dissipation forces can be accounted for by the introduction of the dissipation function derivatives into the corresponding equations of motion in accordance with Onsager's principle of symmetry of kinetic coefficients [6]. There is an established opinion that for a dissipative system it is impossible to formulate the variational principle analogously to Hamilton's principle of the least action [6]. At the same time there are successful approaches [7-10] in which the variational principles for heat conduction theory and for irreversible thermodynamics are applied to account explicitly for the dissipation processes. The thermodynamic approach of Mandelshtam and Leontovich [11] formulated the thermodynamic forces also can be considered along these lines.

Therefore, there are good reasons to attempt to formulate the generalized Hamilton's variational principle for dissipative systems which argue against its established opposition [1]. It is shown in [12-14] and in the first part of this section that such variational principle can be formulated in terms of the displacements of the mechanical and thermal fields.

Hydrodynamical Equation System

As it was mentioned the convenient hydrodynamical system for viscous, heat conductive fluid is derived on the basis of conservation laws of mass, momentum and energy in application to continue medium. It is represented by the system of the continuity equation, the motion equation (Navier-Stokes equation), and energy balance equation (in ordinary or entropy forms), as well as the state equation [1]

$$\frac{\partial \rho}{\partial t} + div(\rho \vec{v}) = 0$$

$$\rho\left(\frac{\partial \vec{v}}{\partial t} + (\vec{v}\nabla)\vec{v}\right) = -\nabla P + \nabla \sigma' \tag{1.1}$$

$$\rho T\left(\frac{\partial S}{\partial t} + \vec{v}\nabla S\right) = div(\kappa \nabla T) + \sigma'_{ik}\frac{\partial v_i}{\partial x_k}$$

$$P = P(\rho, S) \text{ or } \rho = \rho(P, T)$$

The certain assumptions have to be made about form of momentum flow density tensor due to viscous stresses

$$\Pi_{ik} = \rho v_i v_k + \delta_{ik} P - \sigma'_{ik}$$

where

$$\sigma'_{ik} = \eta\left(\frac{\partial v_i}{\partial x_k} + \frac{\partial v_k}{\partial x_i} - \frac{2}{3}\delta_{ik}\frac{\partial v_l}{\partial x_l}\right) + \zeta \delta_{ik}\frac{\partial v_l}{\partial x_l}$$

as well about vector of heat energy flow density in accordance to Fourier's law

$$\vec{q} = -\kappa \nabla T$$

In the represented form (1.1) the system of hydrodynamical equations describes quite adequately a lot of hydrodynamical phenomena. However there are the cases when this description occurs to be not quite correct ones. In particular, that question appears at analysis of short acoustical pulses propagation [15-17], when the essential interest is attracted to the pulse behavior on its forward front, which is formed by high frequencies. Experimental works on ultra short acoustic wave propagation [18,19] show the finite velocity of sound propagation at high frequencies. At the same time the propagation velocity of small perturbation, described by the system (1.1), occurs unbounded in the high frequency limit, that is in formal contradiction with general relativity, limiting propagation velocity of signals of any nature. In application to electromagnetic pulses analogous question does not even arise due to postulation of light speed propagation in vacuum as the maximal one.

Propagation of Small Perturbations

After linearization the system of hydrodynamical equations (1.1) can be reduced to the form

$$\frac{\partial \rho'}{\partial t} + \rho_0 div(\vec{v}) = 0$$

$$\rho_0 \frac{\partial \vec{v}}{\partial t} = -\nabla P' + \eta \Delta \vec{v} + \left(\zeta + \frac{\eta}{3}\right) graddiv(\vec{v})$$

$$\rho_0 T_0 \frac{\partial S'}{\partial t} = div(\kappa \nabla T')$$

and linearization of the state equation gives the following relations between thermodynamical values

$$\rho' = \left(\frac{\partial \rho}{\partial P}\right)_T P' + \left(\frac{\partial \rho}{\partial T}\right)_P T', \text{ where } \left(\frac{\partial \rho}{\partial P}\right)_T = c_0^{-2}, \left(\frac{\partial \rho}{\partial T}\right)_P = -\rho_0 \alpha$$

and

$$S' = \left(\frac{\partial S}{\partial P}\right)_T P' + \left(\frac{\partial S}{\partial T}\right)_P T', \text{ where } \left(\frac{\partial S}{\partial T}\right)_P = \frac{C_P}{T}, \left(\frac{\partial S}{\partial P}\right)_T = -\frac{C_P - C_V}{\alpha T \rho_0 c_0^2} = -\alpha$$

These relations allow us to obtain the following system, describing propagation of small hydrodynamical perturbations

$$c_0^{-2} \frac{\partial P'}{\partial t} + \rho_0 div(\vec{v}) = \alpha \rho_0 \frac{\partial T'}{\partial t}$$

$$\rho_0 \frac{\partial \vec{v}}{\partial t} = -\nabla P' + \eta \Delta \vec{v} + \left(\zeta + \frac{\eta}{3}\right) graddiv(\vec{v}) \tag{1.2}$$

$$\rho_0 C_P \frac{\partial T'}{\partial t} - div(\kappa \nabla T') = \frac{C_P - C_V}{\alpha c_0^2} \frac{\partial P'}{\partial t}$$

If to neglect by the thermo-elastic connection (the right parts in the first and the third equations), then in this approximation the well known wave equation can be obtained for a medium with dissipation in the Voight's model and the heat conductivity equation. All three equations in this approximation have parabolic type of solutions behavior with the infinite propagation speed of perturbations.

Velocity field as a vector field can be expanded onto longitudinal and transversal parts, which correspond to conditions

$$\vec{v} = \vec{v}_l + \vec{v}_s,\ rot(\vec{v}_l) = 0,\ div(\vec{v}_s) = 0$$

This expansion allows us to split the system on the equation for the shear mode

$$\rho_0 \frac{\partial \vec{v}_s}{\partial t} = \eta rotrot(\vec{v}_s) \tag{1.3}$$

and the equations for the thermo-elastic subsystem

$$\left(1 + \frac{\zeta + \frac{4}{3}\eta}{\rho_0 c_0^2} \frac{\partial}{\partial t}\right) \Delta P' = \alpha \rho_0 \left(\frac{\partial}{\partial t} - \frac{1}{\rho_0}\left(\zeta + \frac{4}{3}\eta\right)\Delta\right) \frac{\partial T'}{\partial t}$$

$$\rho_0 C_P \frac{\partial T'}{\partial t} - \kappa \Delta T' = \frac{C_P - C_V}{\alpha c_0^2} \frac{\partial P'}{\partial t} \tag{1.4}$$

Further, it is useful to fulfill the diagonalization of the thermo-elastic subsystem in the frequency representation. Taking a linear combination of pressure and temperature variables of the form

$$\psi = \left(1 + i\omega \frac{\zeta + \frac{4}{3}\eta}{\rho_0 c_0^2}\right) P' - \left(\beta\kappa + i\omega\alpha\left(\zeta + \frac{4}{3}\eta\right)\right) T', \tag{1.5}$$

where parameter β satisfies to the quadratic equation

$$\beta^2 \left(\kappa \frac{C_P - C_V}{\alpha}\right) - \beta \rho_0 c_0^2 C_P \left(1 - i\omega\left(\frac{\kappa}{\rho_0 c_0^2 C_P} - \frac{C_V}{C_P} \frac{\zeta + \frac{4}{3}\eta}{\rho_0 c_0^2}\right)\right) - i\omega\alpha\rho_0 c_0^2 = 0, \tag{1.6}$$

it is possible to see that variable ψ will satisfy to the Helmholtz's equation

$$\Delta\psi + K^2(\omega)\psi = 0 \tag{1.7}$$

with wave number

$$K^2(\omega)=\frac{\dfrac{\omega^2}{c_0^2}-i\omega\beta\dfrac{C_P-C_V}{\alpha c_0^2}}{1+i\omega\dfrac{\zeta+\frac{4}{3}\eta}{\rho_0 c_0^2}} \tag{1.8}$$

Thus, two different modes: the acoustical (or thermo-acoustical) mode and the heat (or acousto-thermal) mode will correspond to two roots of the quadratic equation (1.6). Propagation of the acoustical mode is accompanied mainly by pressure variation, while for the heat mode – by temperature variation.

In the low frequency limit wave number asymptotics for acoustical and heat modes have the forms

$$K_1(\omega)=\frac{\omega}{c_\infty}\left(1-\frac{i\omega}{2}\left(\frac{\zeta+\frac{4}{3}\eta}{\rho_0 c_0^2}+\chi\left(\frac{1}{c_0^2}-\frac{1}{c_\infty^2}\right)\right)\right),\ K_2^2(\omega)=-\frac{i\omega}{\chi} \tag{1.9}$$

which wholly correspond to the known results for diffusion character of the heat mode behavior and to the wave sound propagation (with adiabatic velocity) and its absorption due to viscosity and heat conductivity.

In the high frequency limit the wave number asymptotics for the acoustical and the heat modes have the forms

$$K_1^2(\omega)=-\frac{i\omega\rho_0}{\zeta+\frac{4}{3}\eta},\ K_2^2(\omega)=-\frac{i\omega}{\chi}\frac{C_V}{C_P} \tag{1.10}$$

Thus, in the high frequency limit the both modes have a parabolic or diffusion character, when phase velocity and attenuation coefficient tend to infinity. It means that at high frequencies a signal propagates with infinite speed, that contradicts to relativism, limiting a signal propagation speed. Hence the reason question is appeared: In what point the system of the Navier - Stokes hydrodynamical equations becomes to be incorrect? And how this system should be improved to obtain the finite propagation speed of high frequencies?

The answer on this question many times discussed since middle of the last century [3-5,20-23], and, nevertheless, it continues to be actual up to current time [24,25]. Physical mechanisms providing finite propagation velocity of hydrodynamical perturbations are also well known [3-5,20-23]. In particular, account of viscosity relaxation in the Maxwell's model [2]

$$\eta\to\frac{\eta}{1+i\omega\tau_\eta} \tag{1.11}$$

allows us to provide finite propagation velocity of the shear mode and the longitudinal acoustic mode at high frequencies:

$$c_{S\infty} = \sqrt{\frac{\eta}{\rho_0 \tau_\eta}}, \; c_{l\infty} = c_0 \sqrt{1 + \frac{\zeta + \frac{4}{3}\eta}{\rho_0 c_0^2 \tau_\eta}}. \tag{1.12}$$

Account of heat flow relaxation, generalizing the Fourier's law [3-5]

$$\tau_T \dot{\vec{q}} + \vec{q} = -\kappa \nabla T \tag{1.13}$$

allows to provide the finite propagation velocity of the heat mode at high frequencies

$$c_{T\infty} \approx \frac{c_0}{\sqrt{\Delta + \frac{\tau_\chi}{\chi} c_0^2}} \tag{1.14}$$

In spite of the mentioned opportunity to provide the finite propagation velocity of all hydrodynamical perturbations by an account of the viscosity relaxation and the heat flow relaxation, there are questions as about generality of such approach, as about correspondence of the obtained generalized hydrodynamical equation system to, for example, the second principle of thermodynamics [6].

Beside, the introduction of such relaxation processes requires, at least, a serious motivation, especially, if it is needed to write corresponding equation system for more complex cases, for example, at presence of internal degrees of freedom, or for multi-components, multi-phases medium.

For this case there is the Lagrange's formalism in classical mechanics, which allows to obtain easily the motion equations of mechanical system by knowledge of kinetic and potential energies which difference represents Lagrange's function. The same approach translates easily to continuum mechanics by introduction of the Lagrangian density for a dissipationless media.

Hamilton's Variational Principle

The non-dissipative case of the Hamilton's variational principle can be formulated for a continuous medium in the form of extreme of the action functional $\delta S = 0$:

$$S = \int_{t_1}^{t_2} dt \int_V d\vec{r} L, \tag{1.15}$$

where by analogy with the mechanics the Lagrangian density is represented as the difference between the kinetic and potential energies:

$$L(\dot{\vec{u}}, \nabla\vec{u}) = C(\dot{\vec{u}}) - U(\nabla\vec{u}) . \qquad (1.16)$$

Expression (1.2) implies that the Lagrangian can be considered as a function of the velocities of displacements and deformations.

The motion equations derived from variational principle (1.15), (1.16) have the following form

$$\frac{d}{dt}\frac{\partial L}{\partial \dot{\vec{u}}} + \nabla \frac{\partial L}{\partial \nabla\vec{u}} = 0 . \qquad (1.17)$$

In the simplest case, when the kinetic and potential energies are determined by the quadratic forms

$$2K(\dot{\vec{u}}^2) = \rho_0 \dot{\vec{u}}^2 ,\ 2U = \lambda\varepsilon_{ll}^2 + 2\mu\varepsilon_{ik}^2 ,\ \varepsilon_{ik} = \frac{1}{2}\left(\frac{\partial u_i}{\partial x_k} + \frac{\partial u_k}{\partial x_i}\right), \qquad (1.18)$$

the well-known motion equation for an elastic medium can be derived:

$$\rho_0 \frac{d}{dt}\dot{\vec{u}} - \mu\Delta\vec{u} - (\lambda + \mu) graddiv(\vec{u}) = 0 . \qquad (1.19)$$

Onsager's Variational Principle

Further, if we consider the quasi-equilibrium systems, then the Onsager's variational principle for the least energy dissipation can be formulated [7]. This principle is based on the symmetry of the kinetic coefficients and can be formulated as the extremum of the functional constructed as the difference between the entropy increase rate $\dot{s}$, and the dissipation function D, which are considered as functions of the rate of the thermodynamical relaxation processes, $\dot{\alpha}$, i.e.

$$\delta_{\dot{\alpha}}\left[\dot{s}(\alpha) - D(\dot{\alpha})\right] = 0 \qquad (1.20)$$

The kinetic equation can then be derived from the variational principle (1.6) to describe the relaxation of a thermodynamic system to its equilibrium state, i.e.:

$$\frac{d}{dt}s(\alpha) = 2D(\dot{\alpha}) . \qquad (1.21)$$

The above equation satisfies the symmetry principle for the kinetic coefficients.

Variational Principle for Mechanical Systems with Dissipation

As it was mentioned before, the generalization of motion equation (1.3) in the presence of dissipation is satisfied by introducing the dissipation function derivative with respect to the velocities into the right hand side of equation (1.3). Therefore, in accordance with the Onsager's symmetry principle for the kinetic coefficients [6] we have

$$\frac{d}{dt}\frac{\partial L}{\partial \dot{\vec{u}}} + \nabla \frac{\partial L}{\partial \nabla \vec{u}} = -\frac{\partial D}{\partial \dot{\vec{u}}}. \tag{1.22}$$

Such approach for dissipation introduction into the motion equation satisfies only to the Onsager's symmetry principle for kinetic coefficients and is nothing related with the initial Hamilton's variational principle, which is the basis for derivation of initial motion equation for dissipationless case. Moreover, in [6] on the page 448 there is a sentence with affirmation about impossibility to formulate such principle:

"Because internal motion of body atoms depends not only on motion in the given time moment, but on previous history of this motion, not only body macroscopic coordinates and their first and second time derivatives will enter, generally speaking, into motion equations, but all higher derivatives (more accurately some integral operator by coordinates). The Lagrange's function for macroscopic motion of a system, of course, does not exist, and motion equations for different cases will have perfectly different character."

It is possible to show, however, that the equation of motion can be derived in the form (1.22) if the Hamilton's variational principle is adapted with the following form of the Lagrangian:

$$L(\dot{\vec{u}}, \nabla \vec{u}) = C(\dot{\vec{u}}) - U(\nabla \vec{u}) - \int_0^t D(\dot{\vec{u}})dt', \tag{1.23}$$

where the time integral of the dissipation function is added to Lagrangian in difference from (1.16).

It needs, however, to pay attention that at variation of dissipative term in such approach an additional item appears, which has to be neglected by hands. Indeed, variation of the last term in (1.23) leads us to result

$$\delta \int_0^t D(\dot{\vec{u}})dt' = \int_0^t \frac{\partial D(\dot{\vec{u}})}{\partial \dot{\vec{u}}} \delta \dot{\vec{u}}\, dt' = \int_0^t \frac{d}{dt'}\left(\frac{\partial D(\dot{\vec{u}})}{\partial \dot{\vec{u}}}\delta \vec{u}\right)dt' - \int_0^t \frac{d}{dt'}\left(\frac{\partial D(\dot{\vec{u}})}{\partial \dot{\vec{u}}}\right)\delta \vec{u}\, dt'$$

If to neglect by the last item in this expression

$$\delta \int_0^t D(\dot{\vec{u}}(t'))dt' = \frac{\partial D(\dot{\vec{u}})}{\partial \dot{\vec{u}}}\delta \vec{u}(t) - \int_0^t \frac{d}{dt'}\left(\frac{\partial D(\dot{\vec{u}})}{\partial \dot{\vec{u}}}\right)\delta \vec{u}\, dt' \approx \frac{\partial D(\dot{\vec{u}})}{\partial \dot{\vec{u}}}\delta \vec{u}(t)$$

then the result gives us the same term $\frac{\partial D(\dot{\vec{u}})}{\partial \dot{\vec{u}}}$, which we need artificially to introduce in the motion equation (1.22) for account of dissipation. From the one hand this approach can be considered as some rule at variation of integral term, because it leads us to the required form of motion equation, but not as simple adding as it is usually suggested to do, but on the basis of formulated variational principle. From the other hand the following supporting basement can be proposed. Variation of action containing all terms in Lagrangian (1.23) with account of initial and boundary conditions cab be written in the form

$$\int_{t_1}^{t_2} dt \int dV \left\{ \left(-\frac{d}{dt}\frac{\partial C(\dot{\vec{u}})}{\partial \dot{\vec{u}}} + \nabla \frac{\partial U(\nabla \vec{u})}{\partial \nabla \vec{u}} - \frac{\partial D(\dot{\vec{u}})}{\partial \dot{\vec{u}}} \right) \delta \vec{u} + \int_0^t \frac{d}{dt'} \left(\frac{\partial D(\dot{\vec{u}})}{\partial \dot{\vec{u}}} \right) \delta \vec{u}\, dt' \right\} =$$

it is seen from here that the required form of the motion equation with dissipation arises due to zero value of coefficient at arbitrary variation of the displacement field $\delta \vec{u}$. The last additional item, containing variation $\delta \vec{u}$ under time integral, prevents to the strict conclusion in the given case. Nevertheless, if to rewrite the first term in this expression in the same integral form as the additional term

$$= \int_{t_1}^{t_2} dt \int dV \int_0^t dt' \left\{ \delta(t-t') \left(-\frac{d}{dt'}\frac{\partial C(\dot{\vec{u}})}{\partial \dot{\vec{u}}} + \nabla \frac{\partial U(\nabla \vec{u})}{\partial \nabla \vec{u}} - \frac{\partial D(\dot{\vec{u}})}{\partial \dot{\vec{u}}} \right) + \frac{d}{dt'} \left(\frac{\partial D(\dot{\vec{u}})}{\partial \dot{\vec{u}}} \right) \right\} \delta \vec{u}$$

then, due to the same reason of arbitrary variation $\delta \vec{u}$ the multiplier in brackets at this variation has to be equal to zero. It is possible to see now, that, if the function $\frac{d}{dt'}\left(\frac{\partial D(\dot{\vec{u}})}{\partial \dot{\vec{u}}} \right)$ is not singular one in the point $t' = t$, then its contribution can be neglected in this point in comparison with singular contribution from the delta-function. The presented arguments can be considered as a basis for variation rule of integral term in Lagrangian.

On the basis of variational principle with Lagrange's function (1.23), in particular, it is possible to obtain linearized Navier-Stokes equation. The dissipation function should be considered in this case as a quadratic form of deformation velocities

$$D(\nabla \dot{\vec{u}}) = \eta \left(\frac{\partial \dot{u}_i}{\partial x_k} + \frac{\partial \dot{u}_k}{\partial x_i} \right)^2 + \varsigma \left(\frac{\partial \dot{u}_l}{\partial x_l} \right)^2, \tag{1.24}$$

then the derived motion equation with the account (1.18) corresponds wholly to the linearized Navier-Stokes equation.

$$\rho \frac{d}{dt}\dot{\vec{u}} - (\lambda + \mu)\Delta \vec{u} - \lambda\, grad div(\vec{u}) = (\eta + \varsigma)\Delta \dot{\vec{u}} + \varsigma\, grad div(\dot{\vec{u}}) \tag{1.25}$$

Generalized Variational Principle for Dissipative Hydrodynamics

The above example with derivation of the motion equation for dissipative systems on the basis of Hamilton's variational principle with Lagrange's function (1.24) shows that there is possibility for formulation of the generalized variational principle for dissipative hydrodynamical systems, which can be obtained by simple combination (summation) of Hamilton's variational principle (1.15), (1.16) and Onsager's variational principle (1.20), if to integrate the last principle by time and to multiply the result by temperature. Indeed, if to integrate the expression (1.20) by time and to multiply the result by temperature, we obtain

$$T\left[s - \int_0^t D dt'\right] \tag{1.26}$$

In this case the Lagrangian density can be written as difference between kinetic K and internal E energies plus expression (1.26)

$$L = K - E + T\left[s - \int_0^t D dt'\right] \tag{1.27}$$

Further, with account of expression for the free energy F: $F = E - Ts$, the Lagrange's function can be written in the final form

$$L = K - F - T\int_0^t D dt' \tag{1.28}$$

which is equivalent by structure to the expression (1.23) with the only difference that the free energy stays instead potential energy and the integral with dissipation function is multiplied by temperature.

The generalized variational principle itself for dissipative hydrodynamical systems [12-14] is formulated in the same form as Hamilton's variational principle. Namely, let the considered mass of continue, dissipative medium takes in the time moments t_1 and t_2 volumes, restricted by surfaces σ_1 and σ_2, then the real trajectory of medium motion among all possible motion trajectories of continue medium will correspond to that, which gives extremum to the action functional

$$\delta S = 0\,,\; S = \int_{t_1}^{t_2} dt \int_V d\vec{r} L\,, \tag{1.29}$$

which is built on Lagrangian

$$L = K - F - T\int_0^t D dt'. \quad (1.30)$$

Here by K and F the kinetic and the free energies of a given mass of continue medium are denoted, and D - is its dissipative function, T is temperature.

It is possible to show [12-14], that for dissipationless systems the generalized vaiation principle is reduced to conventional Hamilton's variational principle and for static thermodynamical systems it is reduced to the Onsager's principle of the least dissipation. For dissipative hydrodynamical systems the equations of the desired forms are derived also, in particular, the linearized Navier – Stokes equation.

Independent Variables

When the generalized variational principle is formulated in the form (1.28) we need to determine variables in which terms the Lagrange's function has to be expressed. To answer on this question let's return to hydrodinamical equations and look at variables for their description.

In absence of dissipation, as it easy to see, these variables are velocity, density, pressure and entropy $\vec{v}, \rho, P, s$. Because for the dissipationless case entropy holds to be constant for given material point, then a pressure can be considered, for example, as a function of solely density $P(\rho, s), s = const$. The density of the given mass of medium is expressed in terms of its volume. Hence a variation of density can be expressed in terms of volume variation or through divergence of the displacement field $\rho = \rho(div\vec{u})$. In particular, the linearization of the continuity equation leads to relation

$$\rho = \rho_0(1 - div\vec{u}) \quad (1.31)$$

Velocity by definition is a derivative by time from a displacement

$$\vec{v} = \dot{\vec{u}}$$

Thus, the displacement field $\vec{u}$ can be considered as the principal hydrodinamical variable for dissipationless case.

At presence of dissipation besides the mentioned variables a temperature additionally appears in hydrodynamical equations: $\vec{v}, \rho, P, s, T$. If to consider now a pressure and an entropy as functions of density and temperature $P(\rho, T)$, $s = s(\rho, T)$ in accordance to the state equation, then the displacements and temperatures fields: $\vec{u}, T$ can be considered as the principle hydrodynamical variables.

Further, let's use the idea of Biot [9], and instead a temperature as independent variable we will use some vector field $\vec{u}_T$ (some vector potential), which we will call in accordance with Biot as the heat displacement field $\vec{u}_T$, which divergence determines temperature, namely in analogy with (1.31)

$$T = T_0(1 - \theta\, div\vec{u}_T) \tag{1.32}$$

where θ is some constant. Thus, a divergence of the heat displacement field $\vec{u}_T$ determines temperature deviation from its equilibrium level.

$$\theta\, div\vec{u}_T = (T - T_0)/T_0 .$$

Variational Principle. Thermodynamical Dissipative Systems.

Accordingly to the general approach, the Lagrangian (1.30) should be considered as a function of the both mentioned displacement fields $\vec{u}$ and $\vec{u}_T$:

$$L(\dot{\vec{u}}, \nabla\vec{u}, \nabla\vec{u}_T) = C(\dot{\vec{u}}) - F(\nabla\vec{u}, \nabla\vec{u}_T) - T_0 \int_0^t D(\dot{\vec{u}}, \dot{\vec{u}}_T)dt' . \tag{1.33}$$

The motion equations derived by variation of action with Lagrangian (1.33), can be expressed in the forms

$$\frac{d}{dt}\frac{\partial K}{\partial \dot{\vec{u}}} - \nabla \frac{\partial F}{\partial \nabla \vec{u}} = -T_0 \frac{\partial D}{\partial \dot{\vec{u}}}, \tag{1.34}$$

$$T_0 \frac{\partial D}{\partial \dot{\vec{u}}_T} - \nabla \frac{\partial F}{\partial \nabla \vec{u}_T} = 0 . \tag{1.35}$$

The kinetic energy of averaged motion is a quadratic function of the mean velocity

$$2K(\dot{\vec{u}}) = \rho_0 \dot{\vec{u}}^2 , \tag{1.36}$$

the free energy near the thermodynamical equilibrium state has minimum and hence it is represented by conventional quadratic form used in thermoelasticity [2]:

$$2F(\nabla\vec{u}, T) = 2\mu\varepsilon_{ik}^2 + \lambda\varepsilon_{ll}^2 + 2\tilde{\alpha}\varepsilon_{ll}\left(\frac{T - T_0}{\theta T_0}\right) + \tilde{\kappa}\left(\frac{T - T_0}{\theta T_0}\right)^2 , \tag{1.37}$$

and dissipation function has to be quadratic form of velocities which disappears for the state of thermodynamical equilibrium. Taking into account condition $\dot{\vec{u}}_T = \dot{\vec{u}}$, which has to be fulfilled at condition of termodynamical equilibrium, the dissipation function can be written in the form of a square of a difference between mean mass and heat displacement fields:

$$2D(\dot{\vec{u}},\dot{\vec{u}}_T) = \beta(\dot{\vec{u}} - \dot{\vec{u}}_T)^2 . \tag{1.38}$$

We note that dimentionless parameter θ was specially introduced in definition of the heat displacement $\theta\, div\vec{u}_T = (T - T_0)/T_0$, to have a symmetrical form in expression for the dissipation function.

Variation of action with Lagrangian (1.33) by two displacement fields $\vec{u}$ and $\vec{u}_T$ with account of (1.36) - (1.38) leads to the following motion equations for these fields:

$$\rho_0 \frac{d}{dt}\dot{\vec{u}} - \mu\Delta\vec{u} - (\lambda + \mu) graddiv(\vec{u}) - \tilde{\alpha}\, graddiv(\vec{u}_T) = \beta(\dot{\vec{u}}_T - \dot{\vec{u}}), \tag{1.39a}$$

$$\beta(\dot{\vec{u}}_T - \dot{\vec{u}}) - \tilde{\kappa}\, grad\, div\vec{u}_T = \tilde{\alpha}\, grad\, div\vec{u} . \tag{1.40a}$$

If to express the right part of the equation (1.39a) through equation (1.40a), and to take a divergence from equation (1.40a) itself and to express the divergence of the heat displacement through a temperature (1.32), then we obtain the conventional form of motion equations of the thermoelastic medium in terms of mean displacements and temperatures:

$$\rho_0 \frac{d}{dt}\dot{\vec{u}} - \mu\Delta\vec{u} - (\lambda + \mu + \tilde{\alpha}) graddiv(\vec{u}) = (\tilde{\alpha} + \tilde{\kappa})/(\theta T_0)\, gradT , \tag{1.39b}$$

$$\beta(\dot{T} - T_0\theta\, div\dot{\vec{u}}) - \tilde{\kappa}\,\Delta T = \tilde{\alpha}\, T_0\theta\,\Delta\, div\vec{u} . \tag{1.40b}$$

Comparison with the System of Hydrodynamical Equations

Coefficients of quadratic forms in (1.37), (1.38) can be determined by comparison of the equation system (1.39b), (1.40b) with analogous system obtained by linearization of the hydrodynalical equation system in variables $\vec{u}, T$. In these variables the linearized system of hydrodynamical equations can be written in the form

$$\rho = \rho_0 (1 - div\, \vec{u}) ,$$

$$\rho_0 \frac{d^2\vec{u}}{dt^2} - \rho_0 c_0^2 \Delta\vec{u} = -\rho_0 \tilde{\alpha} \nabla T + \eta\Delta\dot{\vec{u}} + \left(\zeta + \frac{\eta}{3}\right) graddiv\left(\dot{\vec{u}}\right), \tag{1.41a}$$

$$\rho_0 C_V \frac{dT}{dt} + \rho_0 T_0 \tilde{\alpha} \nabla \dot{\vec{u}} - \kappa \Delta T' = 0 . \qquad (1.41b)$$

In absence of viscosity $\eta = 0$, $\varsigma = 0$, which was not taken into account in the dissipation function (1.38) at derivation of equations (1.40), the structure of the equations (1.40) practically coincides with the second and the third equations of the system (1.41). The only difference is the additional term in the right part (1.40b) in comparison with (1.41b). We note that the reason for the introduction of this term is discussed in detail in ref. [26].

The direct comparison of coefficients in equations (1.41) and (1.40) at condition $rot\vec{u} = 0$ allows us to determine relations between them. It needs only to take into account the different dimension of the equation (1.40b) and (1.41b), and, hence, the presence of common dimension multiplayer at comparison of coefficients for these equations.

In the explicit form the parameters of quadratic forms are expressed through the known parameters by the following expressions

$$\beta = \frac{\rho_0 c_0^2}{\chi}\left(\gamma^2 - 1\right),\ \theta = -\frac{\gamma - 1}{\alpha T_0},\ \tilde{\alpha} = \rho_0 c_0^2 (\gamma - 1),$$

$$\lambda + 2\mu = \rho_0 c_0^2 \gamma,\ \tilde{\kappa} = \rho_0 c_0^2 \left(\gamma^2 - 1\right). \qquad (1.42)$$

where γ is the specific heat capacity ratio $\gamma = C_P / C_V$, and $\chi = \kappa / \rho_0 C_V$ - is the heat conductivity coefficient. It is remarkably that the coefficient at dissipation function occurs to be inversely proportional to the heat conductivity coefficient.

Let's discuss now the additional term in the right part (1.40b). This term is proportional to Laplacian of density or pressure and it could appear in equation (1.41b), if the heat energy flow vector in the Fourier's law is not proportional to the temperature gradient only, but contains additional term being proportional to gradient of pressure or density. This question is discussed in [1] on pages 274-275. The main argument against that term concludes in that "derivative of entropy by time will be not essentially positive in this case, that is impossible". However they make additional remark that entropy of moving fluid "will be not already true thermodinamical entropy: volume integral from its density will not be, strongly speaking, that value, which has to increase in time. Nevertheless, ... at small gradients of velocity and temperature in the accepted ... approximation (that entropy) coincides with true entropy.

Indeed, at presence of gradients additional terms can, generally speaking, appear in entropy. ... (for example, term is proportional to scalar $div\,\vec{v}$). These terms could have as positive or negative values. Nevertheless they have to be essentially negative because equilibrium state of entropy is maximal of available ones. Hence entropy expansion by small gradients powers can contain (besides the zero term) the terms, beginning from the second order."

It is needed to note, however, that a set of articles [26,27] appeared during last time, where the necessity of introduction of additional term in standard hydrodynamical equations (into the energy balance equation in the entropy form) is discussed just of the same form as

(1.39b). Moreover the similar term can appear in the vector of energy density flow derived in kinetic theory of gases based on Boltzman's kinetic equation [28,29].

Thus the suggested generalized variational principle includes naturally corrections, improving traditional Fourier's law for heat flow.

In connection with the above citation from [1] it is needed to note , that equation (1.40b) is the motion equation for heat displacement field, while equation (1.41b) is linearization of energy balance equation in entropy form, which itself has to be integral of motion for the motion equation system (1.39), (1.40) at existence of the variational principle.

Variational Principle with Account of Internal Parameters in the Framework of Mandelshtam – Leontovich Approach. Relaxation of Viscosity

Let us consider now how it is possible to derive equations with account of viscosity on the basis of generalized variational principle. In the complete analogy with Mandelshtam – Leontovich approach [4] let us consider together with basic thermodynamical parameters – a specific volume and a temperature (in our description the fields of mean $\vec{u}$ and heat $\vec{u}_T$ displacements) some additional (internal) parameters $\{\xi_i\}$, by which the system state is characterized in vicinity of the thermodynamic equilibrium. If there are the only scalar internal parameter, which deviation from its thermodynamically equilibrium value is denoted by ξ, then the quadratic form of the free energy expansion can be written as following

$$2F(\nabla\vec{u},T,\xi)=\mu\varepsilon_{ik}^2+\lambda\varepsilon_{ll}^2+\kappa\left(\frac{T-T_0}{T_0}\right)^2+2\alpha\varepsilon_{ll}\left(\frac{T-T_0}{T_0}\right)+$$
$$+a\xi^2+2b\xi\varepsilon_{ll}+2c\xi\left(\frac{T-T_0}{T_0}\right) \qquad (1.43a)$$

Analogously the dissipation function will have a form:

$$2D(\dot{\vec{u}},\dot{\vec{u}}_T,\dot{\xi})=\beta(\dot{\vec{u}}-\dot{\vec{u}}_T)^2+\gamma\dot{\xi}^2 \qquad (1.44a)$$

Then, taking variation by fields $\vec{u}$ and $\vec{u}_T$ together with variation by the field of the internal parameter ξ, we obtain the following system of motion equations

$$\rho_0\frac{d}{dt}\dot{\vec{u}}-\mu\Delta\vec{u}-(\lambda+\mu)graddiv(\vec{u})-\alpha\, grad\, div(\vec{u}_T)-b\, grad\,\xi=\beta(\dot{\vec{u}}_T-\dot{\vec{u}})$$

$$\beta(\dot{\vec{u}}_T-\dot{\vec{u}})-\kappa\, grad\, div\vec{u}_T=\alpha\, grad\, div\vec{u}+c\, grad\xi \qquad (1.45a)$$

$$\gamma \frac{d\xi}{dt} + a\xi + b\,div\vec{u} + c\,div\vec{u}_T = 0$$

where the last equation represents by itself the linear kinetic equation for internal parameter ξ, which is simply postulated in the Mandelshtam – Leontovich approach [11]. Since in relation to parameter ξ this equation is ordinary differential equation of the first order, it can be integrated, and its solution can be represented in the form:

$$\xi = -\frac{1}{\gamma}\int_0^t e^{-\frac{a}{\gamma}(t-t')}\left(b\,div\vec{u} + c\frac{T-T_0}{T_0}\right)dt' \tag{1.46}$$

It is seen that this solution describes relaxation of the internal parameter to the current values of the basic thermodynamical fields. Transforming as before at derivation (1.39b), (1.40b) the two rested equations of the system (1.45a) to the ordinary form, it is possible to obtain the motion equations, accounting relaxation of the internal parameter.

$$\rho_0 \frac{d}{dt}\dot{\vec{u}} - \mu\Delta\vec{u} - (\lambda + \mu + \alpha)grad\,div(\vec{u}) - (\alpha + \kappa)grad\,T =$$

$$= -\frac{b+c}{\gamma}\int_0^t e^{-\frac{a}{\gamma}(t-t')} grad\left(b\,div\vec{u} + cT\right)dt' \tag{1.47a}$$

$$\beta(\dot{T} - div\dot{\vec{u}}) - \kappa\,\Delta T = \alpha\,\Delta\,div\vec{u} - \frac{c}{\gamma}\int_0^t e^{-\frac{a}{\gamma}(t-t')}\Delta\left(b\,div\vec{u} + cT\right)dt' \tag{1.48a}$$

To show that the obtained equation really describes relaxation of viscosity and at the long times corresponds to the linearized Navier – Stokes equation (1.41a), (1.41b), let us integrate by parts integral in (1.47a), (1.48a). At that the contributions, arising from the top integration limit, can be combined with other terms in this equation.

By introduction of the following notations

$$\tilde{\alpha} = \alpha - b\frac{b+c}{a}, \quad \tilde{\kappa} = \kappa - c\frac{b+c}{a} \tag{1.49}$$

the equation system (1.47a), (1.48a) can be written in the equivalent form

$$\rho_0 \frac{d}{dt}\dot{\vec{u}} - \mu\Delta\vec{u} - (\lambda + \mu + \tilde{\alpha})grad\,div(\vec{u}) - (\alpha + \tilde{\kappa})grad\,T =$$

$$= \frac{b+c}{a}\int_0^t e^{-\frac{a}{\gamma}(t-t')} grad\left(b\,div\dot{\vec{u}} + c\dot{T}\right)dt' \tag{1.47b}$$

$$\beta(\dot{T} - div\dot{\vec{u}}) - \left(\kappa - \frac{c^2}{a}\right)\Delta T = \left(\alpha - \frac{bc}{a}\right)\Delta\, div\vec{u} + \frac{c}{a}\int_0^t e^{-\frac{a}{\gamma}(t-t')}\Delta\left(b\,div\dot{\vec{u}} + c\dot{T}\right)dt' \quad (1.48b)$$

Now, it is easy to see, that at the long times in comparison with relaxation time $\tau = \gamma / a$ (or in the limit $\tau \to 0$) the integrand can be taken out in the vicinity of the top integration limit, and the resulting equation in this limiting case can be written in the form

$$\rho_0 \frac{d}{dt}\dot{\vec{u}} - \mu\Delta\vec{u} - (\lambda + \mu + \tilde{\alpha})grad\,div(\vec{u}) - (\alpha + \tilde{\kappa})grad\,T = \gamma(b + c)grad\left(b\,div\dot{\vec{u}} + c\dot{T}\right)$$

(1.47c)

$$\beta(\dot{T} - div\dot{\vec{u}}) - \left(\kappa - \frac{c^2}{a}\right)\Delta T = \left(\alpha - \frac{bc}{a}\right)\Delta\, div\vec{u} + \gamma c\Delta\left(b\,div\dot{\vec{u}} + c\dot{T}\right) \quad (1.48c)$$

It is easy to see, that at $c = 0$ and $\mu = 0$ the equation (1.47c) coincides by structure with the equation (1.41a), so that together with relations (1.42) we will have also the relation

$$\gamma b^2 = \varsigma + \frac{4}{3}\eta. \quad (1.50)$$

Thus, the account of internal parameters allows us to introduce naturally the viscous terms in the motion equation, that is usually the principal difficulty in other variational approaches. Moreover, the presence of viscosity relaxation occurs to be natural. Introduction of this relaxation for viscosity is required, for example, for finite speed of perturbation propagation in dissipative hydrodynamics.

Let's note also that in general case $c \neq 0$ the term describing viscosity relaxation appears also in the motion equation for the heat field (1.48) and besides in the both equations (1.47), (1.48) the contribution appears, describing relaxation of internal parameter to varying temperature of medium. Opportunity of appearance of such terms even does not mentioned in other approaches.

It should be mentioned also, that additional terms in quadratic forms (1.43a), (1.44a), dealt with internal parameter ξ, assume its scalar nature. However such parameters can possess vector or tensor properties. In the last case relaxation of viscosity will be attributed to the shear viscosity together with the bulk viscosity.

For this case the additional items in the free energy and dissipation function will have the following forms:

$$2\Delta F(\nabla\vec{u}, \nabla\vec{u}_T, \xi_{ij}) = a_1\xi_{ll}^2 + a_2\xi_{ik}^2 + 2b_1\xi_{kk}\varepsilon_{ll} + 2b_2\xi_{ik}\varepsilon_{ki} + 2c\xi_{ll}\left(\frac{T - T_0}{T_0}\right) \quad (1.43b)$$

$$2D(\dot{\vec{u}},\dot{\vec{u}}_T,\dot{\xi}_{ij}) = \beta(\dot{\vec{u}} - \dot{\vec{u}}_T)^2 + \gamma_1\dot{\xi}_{ll}^2 + \gamma_2\dot{\xi}_{ik}^2 \tag{1.44b}$$

Here we implicitly continue to assume that the heat displacement field is a pure longitudinal one. The system (1.45b) in this case will be rewritten in the form

$$\rho_0 \frac{d}{dt}\dot{\vec{u}} - \mu\Delta\vec{u} - (\lambda + \mu) graddiv(\vec{u}) - \alpha\, grad\, div(\vec{u}_T) =$$

$$= \beta(\vec{u}_T - \vec{u}) + b_{\,1} grad\, \xi_{ll} + b_{\,2}\frac{\partial \xi_{ik}}{\partial x_k}$$

$$\beta(\dot{\vec{u}}_T - \dot{\vec{u}}) - \kappa\, grad\, div\vec{u}_T = \alpha\, grad\, div\vec{u} + c\, grad\xi_{ll} \tag{1.45b}$$

$$\gamma_1\delta_{ik}\frac{d\xi_{ll}}{dt} + \gamma_2\frac{d\xi_{ik}}{dt} + a_1\delta_{ik}\xi_{ll} + a_2\xi_{ik} + b_{\,1}\delta_{ik} div\vec{u} + b_{\,2}\varepsilon_{ik} + c\,\delta_{ik} div\vec{u}_T = 0$$

Here in the first equation the tensor notation for a vector, obtained as divergence of internal parameter tensor, is rested for compactness.

By convolution of kinetic equation by indexes it is possible to obtain the separate kinetic equation for a spherical part of internal parameter tensor ξ_{ll}.

$$\gamma_1\delta_\gamma\frac{d\xi_{ll}}{dt} + a_1\delta_a\xi_{ll} + b_1\delta_b div\vec{u} + c\, div\vec{u}_T = 0 \tag{1.51}$$

where dimensionless correction multipliers are denoted by symbols δ_α

$$\delta_\gamma = 1 + \frac{1}{3}\frac{\gamma_2}{\gamma_1},\ \delta_a = 1 + \frac{1}{3}\frac{a_2}{a_1},\ \delta_b = 1 + \frac{1}{3}\frac{b_2}{b_1}$$

Solution of the kinetic equation (1.51) is completely analogous to the solution (1.4b). For other components of the internal parameter tensor ξ_{ik} we also can obtain kinetic equation of the form (1.51) with the only difference that the additional term, being the solution of kinetic equation (1.51) is added to inhomogeneous terms.

$$\gamma_2\frac{d\xi_{ik}}{dt} + a_2\xi_{ik} + \tilde{a}_1\delta_{ik}\xi_{ll} + \tilde{b}_1\delta_{ik} div\vec{u} + b_2\varepsilon_{ik} + \tilde{c}\,\delta_{ik} div\vec{u}_T = 0 \tag{1.52}$$

where the following notations are introduced

$$\widetilde{a}_1 = a_1\left(1-\frac{\delta_a}{\delta_\gamma}\right),\ \widetilde{b}_1 = b_1\left(1-\frac{\delta_b}{\delta_\gamma}\right),\ \widetilde{c} = c\left(1-\frac{1}{\delta_\gamma}\right)$$

Again the solution of the equation (1.52) has a form analogous to (1.46) with account of additional contributions with multipliers $\widetilde{a}_1$ and b_2. This solution has the view:

$$\xi_{ik} = -\frac{1}{\gamma_2}\int_0^t e^{-\frac{a_2}{\gamma_2}(t-t')}\left(\widetilde{b}_1\delta_{ik} div\vec{u} + b_2\varepsilon_{ik} + \widetilde{c}\,\delta_{ik}\frac{T-T_0}{T_0} + \widetilde{a}_1\delta_{ik}\xi_{ll}\right)dt' \qquad (1.53)$$

Taking divergence from tensor (1.53), we obtain the vector of the form

$$\frac{\partial\xi_{ik}}{\partial x_k} = -\frac{1}{\gamma_2}\int_0^t e^{-\frac{a_2}{\gamma_2}(t-t')}\left(\left(\widetilde{b}_1 + b_2\right) graddiv\vec{u} + b_2\Delta\vec{u} + \widetilde{c}\,grad\frac{T}{T_0} + \widetilde{a}_1 grad\xi_{ll}\right)dt' \qquad (1.54)$$

If to consider the particular case $\widetilde{a}_1 = 0$, $c = 0$, and to substitute (1.54) into the first equation (1.45b), then after transformations analogous to (1.47b), it can be written in the form:

$$\rho_0\frac{d}{dt}\dot{\vec{u}} - \mu\Delta\vec{u} - (\lambda+\mu+\alpha)graddiv(\vec{u}) + (\alpha+\kappa)grad\frac{T}{T_0} - b_1 grad\,\xi_{ll} =$$

$$= -\frac{b_2}{\gamma_2}\int_0^t e^{-\frac{a_2}{\gamma_2}(t-t')}\left(\left(\widetilde{b}_1 + b_2\right) graddiv\vec{u} + b_2\Delta\vec{u}\right)dt'$$

In the low frequency limit after the transformations analogous to (1.47c) it is possible to obtain the equation analogous to Navier-Stokes equation with shear and bulk viscosities:

$$\rho_0\frac{d}{dt}\dot{\vec{u}} - \widetilde{\mu}\Delta\vec{u} - (\lambda+\mu+\widetilde{\alpha})graddiv(\vec{u}) - (\alpha+\kappa)gradT = \widetilde{\zeta}graddiv\dot{\vec{u}} + \widetilde{\eta}\Delta\dot{\vec{u}} \qquad (1.47d)$$

where coefficients are expressed as

$$\widetilde{\zeta} = \left(\gamma_1\delta_\gamma b_1^2\delta_b^2 + \gamma_2 b_2\left(\widetilde{b}_1 + b_2\right)\right),\ \widetilde{\eta} = \gamma_2 b_2^2,$$

$$\widetilde{\mu} = \mu - b_2^2 / a_2,\ \widetilde{\alpha} = \alpha - \frac{b_1^2\delta_b^2}{a_1\delta_a} - \frac{b_2(\widetilde{b}_1 + b_2)}{a_1}$$

To have the exact correspondence with the low frequency Navier-Stokes equation for viscous fluid it is needed to put $\tilde{\mu} = 0$, that leads to relation

$$\mu = b_2^2 / a_2$$

It is possible to say now several words about physical sense of the introduced internal parameters. Because in the low frequency limit the majority of gases and fluids, including the simplest from them, is described by the Navier-Stokes equation, then the only available value, which could relax in this situation, and, hence, to be considered as the internal parameter, is the mean distance between molecules in gas or liquid. In the condensed and especially in the solid media the mutual space placement of atoms becomes to be essential, then the space variation of their mutual positions, holding rotational invariance of a body as whole, has to be described by symmetrical tensor value of the second order. Hence the corresponding internal parameter occurs to be the same tensor. Thus, the discrete structure of medium predetermines existence, at least, of the mentioned internal parameters, responsible for relaxation.

Variational Principle with Account of Internal Parameters and Inertia of Heat Displacement Field

Introduction into variational principle of internal parameters, responsible for viscosity relaxation, nevertheless, does not remove parabolic behavior of heat mode at high frequencies. As it was mentioned in the beginning of the article this effect is removed by introduction of finite relaxation time for heat flow. Let's consider how such effect can be introduced into variational principle. Up to now we assumed that heat displacement field does not possess inertia, because of kinetic energy have considered as quadratic function of mean velocities only. However the heat field propagation also possesses certain inertia dealt with necessity to vary kinetic energy of random motion of medium atoms. This fact can be taken into account, considering kinetic energy as quadratic form of the mean and the heat displacements. The free energy and dissipation function will be considered, as dependent on internal parameter, as in the previous section:

$$2K(\dot{\vec{u}}^2, \dot{\vec{u}}_T^2) = \rho_0 \dot{\vec{u}}^2 + \rho_T \dot{\vec{u}}_T^2 \tag{1.55}$$

$$2F(\nabla\vec{u}, T, \xi) = \mu\varepsilon_{ik}^2 + \lambda\varepsilon_{ll}^2 + \kappa\left(\frac{T-T_0}{T_0}\right)^2 + 2\alpha\varepsilon_{ll}\left(\frac{T-T_0}{T_0}\right) + a\xi^2 + 2b\xi\varepsilon_{ll} + 2c\xi\left(\frac{T-T_0}{T_0}\right)$$

$$2D(\dot{\vec{u}}, \dot{\vec{u}}_T, \dot{\xi}) = \beta(\dot{\vec{u}} - \dot{\vec{u}}_T)^2 + \gamma\dot{\xi}^2$$

The motion equation system, obtained with account (1.55) has the form:

$$\rho_0 \frac{d}{dt}\dot{\vec{u}} - \mu\Delta\vec{u} - (\lambda+\mu) graddiv(\vec{u}) - \alpha\, grad\, div(\vec{u}_T) - b\, grad\, \xi = -\beta(\dot{\vec{u}} - \dot{\vec{u}}_T)$$

$$\rho_T \frac{d\dot{\vec{u}}_T}{dt} + \beta(\dot{\vec{u}}_T - \dot{\vec{u}}) - \kappa\, grad\, div\vec{u}_T = \alpha\, grad\, div\vec{u} + c\, grad\xi \qquad (1.56a)$$

$$\gamma\frac{d\xi}{dt} + a\xi + b div\vec{u} + c div\vec{u}_T = 0$$

Taking divergence from the first and the second equations of the system (1.64), and expressing divergence of the displacement field through variations of density and temperature, we obtain

$$\frac{d^2\rho'}{dt^2} + \beta\left(\frac{1}{\rho_0}\frac{\partial\rho'}{\partial t} - \frac{1}{T_0}\frac{\partial T'}{\partial t}\right) - \frac{\lambda+2\mu}{\rho_0}\Delta\rho' = \frac{\alpha}{T_0}\Delta T' - b\,\Delta\xi$$

$$\frac{\rho_T}{T_0}\frac{d^2T'}{dt^2} + \beta\left(\frac{1}{T_0}\frac{\partial T'}{\partial t} - \frac{1}{\rho_0}\frac{\partial\rho'}{\partial t}\right) - \frac{\kappa}{T_0}\Delta T' = \frac{\alpha}{\rho_0}\Delta\rho' - c\,\Delta\xi \qquad (1.56b)$$

$$\gamma\frac{d\xi}{dt} + a\xi = b\frac{\rho'}{\rho_0} + c\frac{T'}{T_0}$$

As it seen, the given system is symmetrical relatively to fields of density and temperature, and hence, it is hyperbolical system in the given case. In particular, at high frequencies, when it is possible to neglect by relaxation of internal parameter and dissipation items, the system (1.56b) is reduced to

$$\frac{d^2\rho'}{dt^2} - \frac{\lambda+2\mu}{\rho_0}\Delta\rho' - \frac{\alpha}{T_0}\Delta T' = 0$$

$$\frac{d^2T'}{dt^2} - \frac{\kappa}{\rho_T}\Delta T' - \frac{\alpha T_0}{\rho_T\rho_0}\Delta\rho' = 0 \qquad (1.57)$$

This system by diagonalization procedure can be reduced to two independent wave equations, describing propagation of acousctical (thermo-acoustical) mode and heat (acousto-thermal) mode, which propagation velocities can be written in the form:

$$c_{1,2}^2 = \frac{1}{2}\left(c_\infty^2 + \frac{\kappa}{\rho_T}\right) \pm \frac{1}{2}\sqrt{\left(c_\infty^2 - \frac{\kappa}{\rho_T}\right)^2 + 4\frac{\alpha^2}{\rho_0 \rho_T}}$$

Thus, it is shown that in the framework of the suggested generalized variational principle it is succeeded firstly to obtain the hydrodynamical system of Navier-Stokes equations, and secondly by natural generalization of quadratic forms of kinetic energy, free energy and dissipation function, accounting additional internal parameter as well as inertia of heat field, it is succeeded to generalize this system to the form which describe propagation of small hydrodynamical perturbations with finite velocity at high frequency limit

Relaxation in the Case of a Number Internal Parameters

Let us consider the more general case of a number set of internal parameters. For simplicity let us consider that all internal parameters are the scalar values and temperature field is neglected. Then instead (1.43a), (1.44a) for positively determined quadratic forms of free energy and dissipation function it is possible to write the expressions

$$2F(\nabla\vec{u}, \{\xi_i\}) = 2\mu\varepsilon_{ik}^2 + \lambda\varepsilon_{ll}^2 + \sum_{i,j} a_{ij}\xi_i\xi_j + 2\varepsilon_{ll}\sum_i b_i\xi_i\,, \tag{1.43c}$$

$$2D(\{\dot{\xi}_i\}) = \sum_{i,j} \gamma_{ij}\dot{\xi}_i\dot{\xi}_j\,. \tag{1.44c}$$

where matrices of coefficients a_{ij} and γ_{ij} are symmetrical and positively determined ones, if all internal parameters possess the same symmetry in respect to time inversion. Then instead the system (1.45a) we will have

$$\rho_0 \frac{d}{dt}\dot{\vec{u}} - \mu\Delta\vec{u} - (\lambda + \mu)graddiv(\vec{u}) - \sum_i b_i\, grad\, \xi_i = 0\,, \tag{1.47e}$$

$$\sum_j \gamma_{ij}\frac{d\xi_j}{dt} + \sum_j a_{ij}\xi_j + b_i\, div\vec{u} = 0\,, \tag{1.58}$$

The system of kinetic equations can be diagonalized by transmission to new variables $\tilde{\xi}_i = \sum_j \lambda_{ij}\xi_j$, being a linear combinations of initial internal parameters $\{\xi_i\}$. In the new variables the system of kinetic equations (1.58) is split on the system of independent kinetic equations, each of them can be written in the form (1.51)

$$\tilde{\gamma}_{ii} \frac{d\tilde{\xi}_i}{dt} + \tilde{a}_{ii}\tilde{\xi}_i + b_i \, div\vec{u} = 0,$$

where

$$\tilde{\gamma}_{ii} = \sum_j \gamma_{ij}\lambda_{ji}, \; \tilde{a}_{ii} = \sum_j a_{ij}\lambda_{ji}$$

The solution of this equation is analogous to (1.46)

$$\tilde{\xi}_i = -\frac{b_i}{\tilde{\gamma}_{ii}} \int_0^t e^{-\frac{\tilde{a}_{ii}}{\tilde{\gamma}_{ii}}(t-t')} div\vec{u} \; dt',$$

so that, the initial variables can be written in the form

$$\xi_i = -\sum_j \frac{\lambda_{ij}^{-1} b_j}{\tilde{\gamma}_{jj}} \int_0^t e^{-\frac{\tilde{a}_{jj}}{\tilde{\gamma}_{jj}}(t-t')} div\vec{u} \; dt' \tag{1.59}$$

With account of represented expression the motion equation (1.47e) can be written as following

$$\rho_0 \frac{d}{dt}\dot{\vec{u}} - \mu\Delta\vec{u} - (\lambda+\mu) graddiv(\vec{u}) + \sum_{ij} \frac{\lambda_{ij}^{-1} b_i b_j}{\tilde{\gamma}_{jj}} \int_{-\infty}^t e^{-\frac{\tilde{a}_{jj}}{\tilde{\gamma}_{jj}}(t-t')} graddiv\vec{u} \; dt' = 0,$$

With notations $\tau_i = \frac{\tilde{\gamma}_{ii}}{\tilde{a}_{ii}}$ and $g_i = \sum_j \frac{\lambda_{ij}^{-1} b_i b_j}{\tilde{\gamma}_{jj}}$ for relaxation times and weights of relaxation processes, motion equation will have the view

$$\rho_0 \frac{d}{dt}\dot{\vec{u}} - \mu\Delta\vec{u} - (\lambda+\mu) graddiv(\vec{u}) + \int_{-\infty}^t graddiv\vec{u} \sum_i g_i e^{-(t-t')/\tau_i} dt' = 0,$$

If to consider the longitudinal waves only, then this equation can be represented in the form.

$$\ddot{\vec{u}} - c_\infty^2 graddiv \int_{-\infty}^t \vec{u}(t') \left(\delta(t-t') - \sum_i \frac{g_i}{\lambda+2\mu} e^{-(t-t')/\tau_i} \right) dt' = 0, \tag{1.60}$$

where the notation $c_\infty^2 = \dfrac{\lambda + 2\mu}{\rho_0}$ is introduced for propagation velocity of high frequencies, when relaxation processes are frozen.

If to consider the plane monochromatic waves, then the wave number, corresponding to equation (1.60) will have the view

$$k^2(\omega) = \frac{\omega^2}{c_\infty^2} \frac{1}{1 - \sum_j \dfrac{g_j}{\lambda + 2\mu} \dfrac{1}{i\omega + 1/\tau_j}} \tag{1.61}$$

It follows from here, that propagation speed of low frequencies $\omega \to 0$ is determined by expression

$$c_0^2 = c_\infty^2 \left(1 - \sum_j \frac{g_j \tau_j}{\lambda + 2\mu} \right)$$

Introducing the notation $\Delta = \dfrac{c_\infty^2 - c_0^2}{c_\infty^2}$ for dispersion jump of velocity, and dimensionless normalized weights of relaxation processes s_i

$$s_i = \frac{g_j \tau_j}{(\lambda + 2\mu)\Delta}, \quad \sum_i s_i = 1 \tag{1.62}$$

the expression for square of wave number (1.61) can be written in convenient for analysis form

$$k^2(\omega) = \frac{\omega^2}{c_\infty^2} \frac{1}{1 - \Delta \sum_j \dfrac{s_j}{1 + i\omega\tau_j}} \tag{1.63}$$

In the case of small values of dispersion jump $\Delta << 1$, which usually take place, the wave number (1.63) can be expanded by this parameter, and it will have a view

$$k(\omega) = \frac{\omega}{c_\infty} \left(1 + \frac{\Delta}{2} \sum_j \frac{s_j}{1 + i\omega\tau_j} \right) \tag{1.64a}$$

Transmission to a continue spectrum of relaxation times produces by change $s_j \to s(\tau)d\tau$, so (1.64a) is rewritten in the form

$$k(\omega) = \frac{\omega}{c_\infty}\left(1 + \frac{\Delta}{2}\int_{\tau_{\min}}^{\tau_{\max}} \frac{s(\tau)d\tau}{1 + i\omega\tau}\right) \tag{1.64b}$$

The real part of wave number determines the phase velocity.

$$c(\omega) = c_\infty\left(1 - \frac{\Delta}{2}\sum_j \frac{s_j}{1 + (\omega\tau_j)^2}\right) \tag{1.65}$$

the imagery part – attenuation coefficient

$$\alpha(\omega) = \frac{\Delta\omega^2}{2c_\infty}\sum_j \frac{s_j\tau_j}{1 + (\omega\tau_j)^2} \tag{1.66}$$

Thus, the frequency behavior of attenuation coefficient and phase velocity in fluids of complex rheological structure, characterized by large number of internal parameters, can be described in terms of normalized spectrum of relaxation times s_i (or $s(\tau)$ in the case of continue spectra) and dispersion jump of velocity Δ.

Part II. Application of Generalized Variational Principle for Description of Multicomponents, Multiphase Media

In the Part I represented above the generalized variational principle have been formulated as a simple sum of the Hamilton's and Onsager's variational principles (see also [12-14]). It allowed us to derive the well known system of equations for dissipative hydrodynamics.

The developed variational principle is used in this section for derivation of the motion equations for multi-component, multi-phase media. In particular, we are concerned with a two-phase, double temperature medium, that is generalization of the Biot's equations which do not take into account the dynamics of the temperature field.

The difference between miticomponent and multiphase media can be explained as following. For the case of multicomponent medium at definition of ensemble of particles contented material point the interaction of separate particle with all other particles, belonging to other components, is taken into account. The convenient image for such medium is a mixture of reared gases, which flaxes penetrate through each other. Each component in this case can possesses own mean velocity, temperature and concentration, as well as specific interaction between its particles.

In difference from multicomponent medium for the multiphase medium separate particles inside material point interact between each other only inside a phase, which they are belonged, and interaction between phases is reduced to interaction through interphase boundary. Each phase can be characterized by its own density, mean velocity, temperature and volume content as well as specific interphase interaction.

Multicomponent Media

In order to describe the behavior of a multi-component, multi-phase medium the mean mass $\vec{u}_i$ and heat $\vec{u}_{Ti}$ displacement fields for each component (phase) can be introduced. These displacement vectors are related to the partial concentrations C_i and temperatures T_i of the each individual component via the following equations:

$$C_i = C_i^0(1 - div\vec{u}_i) \tag{2.1a}$$

$$T_i = T_i^0(1 - \theta_i div\vec{u}_{Ti}). \tag{2.2a}$$

Here C_i^0 and T_i^0 are the initial concentrations and temperatures of the individual components. The dimensionless parameters θ_i are related to the thermodynamical parameters of separate components $\theta_i = -(\gamma_i - 1)/(\alpha_i T_i^0)$, where α_i is the heat expansion coefficient and $\gamma_i = C_{Pi}/C_{Vi}$ is the specific heat ratio. In such variables the kinetic and free energies as well as dissipative function can be written using the following quadratic forms:

$$2K(\dot{\vec{u}}_i^2) = \rho_0 \sum_{i=1}^{N} C_i^0 \dot{\vec{u}}_i^2 \tag{2.3}$$

$$2F(\nabla\vec{u}_i, \nabla\vec{u}_{Ti}) = \sum_{i,j=1}^{N} \lambda_{ij} C_i^0 C_j^0 \left(\nabla\vec{u}_i \nabla\vec{u}_j\right) + \sum_{i,j=1}^{N} \tilde{\kappa}_{ij} C_i^0 C_j^0 \left(\nabla\vec{u}_{Ti} \nabla\vec{u}_{Tj}\right) + \sum_{i,j=1}^{N} \tilde{\alpha}_{ij} C_i^0 C_j^0 \left(\nabla\vec{u}_i \nabla\vec{u}_{Tj}\right) \tag{2.4}$$

$$2D(\dot{\vec{u}}_i, \dot{\vec{u}}_{Ti}) = \sum_{i=1}^{N} \beta_i C_i^0 (\dot{\vec{u}}_i - \dot{\vec{u}}_{Ti})^2 + \sum_{i,j=1}^{N} \tilde{\gamma}_{ij} C_i^0 C_j^0 (\dot{\vec{u}}_i - \dot{\vec{u}}_j)^2 + \sum_{i,j=1}^{N} \tilde{\gamma}_{Tij} C_i^0 C_j^0 (\dot{\vec{u}}_{Ti} - \dot{\vec{u}}_{Tj})^2 \tag{2.5}$$

The kinetic energy (2.3) is merely a sum of kinetic energies of all components. The free energy (2.4) is positively determined quadratic form contain interaction between elastic and heat fields. The dissipation function (2.5) takes into account dissipation canals as inside separate components (the first term) as due to intercomponent interaction of elastic (the second term) and heat (the third term) fields.

The system of 2N motion equations for the concentration and heat fields derived on the basis of the generalized variational principle, can be written in the form:

$$\frac{d}{dt}\frac{\partial K}{\partial \dot{\vec{u}}_i} - \nabla \frac{\partial F}{\partial \nabla \vec{u}_i} = -\frac{\partial D}{\partial \dot{\vec{u}}_i} \tag{2.6}$$

$$\frac{\partial D}{\partial \dot{\vec{u}}_{Ti}} - \nabla \frac{\partial F}{\partial \nabla \vec{u}_{Ti}} = 0 . \tag{2.7}$$

For multi-component medium this system can be written in explicit form

$$\rho_0 C_i^0 \frac{d}{dt}\dot{\vec{u}}_i - \sum_{j=1}^{N} C_j^0 \left(\lambda_{ij} \nabla\nabla \vec{u}_j + \alpha_{ij} \nabla\nabla \vec{u}_{Tj}\right) = -\beta_i \left(\dot{\vec{u}}_i - \dot{\vec{u}}_{Ti}\right) - \sum_{j=1}^{N} \gamma_{ij} C_j^0 \left(\dot{\vec{u}}_i - \dot{\vec{u}}_j\right) \tag{2.8}$$

$$\beta_i \left(\dot{\vec{u}}_{Ti} - \dot{\vec{u}}_i\right) + \sum_{j=1}^{N} \gamma_{Tij} C_j^0 \left(\dot{\vec{u}}_{Ti} - \dot{\vec{u}}_{Tj}\right) = \sum_{j=1}^{N} C_j^0 \left(\kappa_{ij} \nabla\nabla \vec{u}_{Tj} + \alpha_{ij} \nabla\nabla \vec{u}_j\right) \tag{2.9}$$

Have taken divergence from these equations they can be rewritten in terms of concentration and temperature fields

$$\rho_0 \frac{d^2}{dt^2} C_i - \sum_{j=1}^{N} C_i^0 C_j^0 \left(\lambda_{ij} \Delta \left(\frac{C_j}{C_j^0} \right) + \alpha_{ij} \Delta \left(\frac{T_j}{\theta_j T_j^0} \right) \right) =$$

$$= -\beta_i \frac{d}{dt} \left(C_i - \left(\frac{C_i^0 T_i}{\theta_i T_i^0} \right) \right) - \frac{d}{dt} \sum_{j=1}^{N} \gamma_{ij} C_i^0 C_j^0 \left(\left(\frac{C_i}{C_i^0} \right) - \left(\frac{C_j}{C_j^0} \right) \right) \tag{2.10a}$$

$$\beta_i \frac{d}{dt} \left(\left(\frac{C_i^0 T_i}{\theta_i T_i^0} \right) - C_i \right) + \frac{d}{dt} \sum_{j=1}^{N} \gamma_{Tij} C_i^0 C_j^0 \left(\left(\frac{T_i}{\theta_i T_i^0} \right) - \left(\frac{T_j}{\theta_j T_j^0} \right) \right) =$$

$$= \sum_{j=1}^{N} C_i^0 C_j^0 \left(\kappa_{ij} \Delta \left(\frac{T_j}{\theta_j T_j^0} \right) + \alpha_{ij} \Delta \left(\frac{C_j}{C_j^0} \right) \right) \tag{2.11a}$$

Summation by all components gives us

$$\frac{d^2}{dt^2} \rho - \sum_{i,j=1}^{N} C_i^0 C_j^0 \left(\lambda_{ij} \Delta \left(\frac{C_j}{C_j^0} \right) + \alpha_{ij} \Delta \left(\frac{T_j}{\theta_j T_j^0} \right) \right) = -\sum_{i,j=1}^{N} \beta_i \frac{d}{dt} \left(C_i - \left(\frac{C_i^0 T_i}{\theta_i T_i^0} \right) \right)$$

$$\sum_{i,j=1}^{N} \beta_i \frac{d}{dt} \left(\left(\frac{C_i^0 T_i}{\theta_i T_i^0} \right) - C_i \right) = \sum_{i,j=1}^{N} C_i^0 C_j^0 \left(\kappa_{ij} \Delta \left(\frac{T_j}{\theta_j T_j^0} \right) + \alpha_{ij} \Delta \left(\frac{C_i}{C_i^0} \right) \right) \tag{2.12a}$$

which leads to equality

$$\frac{d^2}{dt^2}\rho = \sum_{i,j=1}^{N} C_i^0\left(\left(\lambda_{ij}-\alpha_{ij}\right)\Delta C_j - \left(\kappa_{ij}-\alpha_{ij}\right)\Delta\left(\frac{C_j^0 T_j}{\theta_j T_j^0}\right)\right)$$

Let's consider now different limiting cases described by the system (2.10), (2.11).

Double Component Medium

For the two-component case the system (2.8) – (2.9) consist of four coupled equations

$$\rho_0\frac{d}{dt}\dot{\vec{u}}_1 - \left(C_1^0\lambda_{11}\Delta\vec{u}_1 + C_2^0\lambda_{12}\Delta\vec{u}_2\right) - \left(C_1^0\tilde{\alpha}_{11}\Delta\vec{u}_{T1} + C_2^0\tilde{\alpha}_{12}\Delta\vec{u}_{T2}\right) = -\beta_1\left(\dot{\vec{u}}_1 - \dot{\vec{u}}_{T1}\right) - \tilde{\gamma}_{12}C_2^0\left(\dot{\vec{u}}_1 - \dot{\vec{u}}_2\right)$$

$$\rho_0\frac{d}{dt}\dot{\vec{u}}_2 - \left(C_2^0\lambda_{22}\Delta\vec{u}_2 + C_1^0\lambda_{12}\Delta\vec{u}_1\right) - \left(C_2^0\tilde{\alpha}_{22}\Delta\vec{u}_{T2} + C_1^0\tilde{\alpha}_{12}\Delta\vec{u}_{T1}\right) = -\beta_2\left(\dot{\vec{u}}_2 - \dot{\vec{u}}_{T2}\right) - \tilde{\gamma}_{12}C_1^0\left(\dot{\vec{u}}_2 - \dot{\vec{u}}_1\right)$$

(2.13)

$$\beta_1\left(\dot{\vec{u}}_{T1} - \dot{\vec{u}}_1\right) + \gamma_{T12}C_2^0\left(\dot{\vec{u}}_{T1} - \dot{\vec{u}}_{T2}\right) = \left(C_1^0\tilde{\kappa}_{11}\Delta\vec{u}_{T1} + C_2^0\tilde{\kappa}_{12}\Delta\vec{u}_{T2}\right) + \left(C_1^0\alpha_{11}\Delta\vec{u}_1 + C_2^0\alpha_{12}\Delta\vec{u}_2\right)$$

$$\beta_2\left(\dot{\vec{u}}_{T2} - \dot{\vec{u}}_2\right) + \gamma_{T12}C_1^0\left(\dot{\vec{u}}_{T2} - \dot{\vec{u}}_{T1}\right) = \left(C_2^0\tilde{\kappa}_{22}\Delta\vec{u}_{T2} + C_1^0\tilde{\kappa}_{12}\Delta\vec{u}_{T1}\right) + \left(C_2^0\alpha_{22}\Delta\vec{u}_2 + C_1^0\alpha_{12}\Delta\vec{u}_1\right)$$

Applying divergence operator to these equations and using expressions (2.1), (2.2), they can be written in terms of concentration and heat fields C_1, C_2, T_1, T_2.

$$\rho_0\frac{d}{dt}\dot{C}_1 - C_1^0\left(\lambda_{11}\Delta C_1 + \lambda_{12}\Delta C_2\right) - C_1^0\left(\frac{C_1^0\tilde{\alpha}_{11}}{\theta_1 T_1^0}\Delta T_1 + \frac{C_2^0\tilde{\alpha}_{12}}{\theta_2 T_2^0}\Delta T_2\right) = -\beta_1\left(\dot{C}_1 - \frac{C_1^0}{\theta_1 T_1^0}\dot{T}_1\right) - \tilde{\gamma}_{12}\left(C_2^0\dot{C}_1 - C_1^0\dot{C}_2\right)$$

$$\rho_0\frac{d}{dt}\dot{C}_2 - C_2^0\left(\lambda_{22}\Delta C_2 + \lambda_{12}\Delta C_1\right) - C_2^0\left(\frac{C_2^0\tilde{\alpha}_{22}}{\theta_2 T_2^0}\Delta T_2 + \frac{C_1^0\tilde{\alpha}_{12}}{\theta_1 T_1^0}\Delta T_1\right) = -\beta_2\left(\dot{C}_2 - \frac{C_2^0}{\theta_2 T_2^0}\dot{T}_2\right) - \tilde{\gamma}_{12}\left(C_1^0\dot{C}_2 - C_2^0\dot{C}_1\right)$$

(2.14)

$$-\beta_1\left(\dot{C}_1 - \frac{C_1^0}{\theta_1 T_1^0}\dot{T}_1\right) + \tilde{\gamma}_{T12}C_2^0 C_1^0\left(\frac{\dot{T}_1}{\theta_1 T_1^0} - \frac{\dot{T}_2}{\theta_2 T_2^0}\right) = C_1^0\left(\frac{C_1^0\tilde{\kappa}_{11}}{\theta_1 T_1^0}\Delta T_1 + \frac{C_2^0\tilde{\kappa}_{12}}{\theta_2 T_2^0}\Delta T_2\right) + C_1^0\left(\tilde{\alpha}_{11}\Delta C_1 + \tilde{\alpha}_{12}\Delta C_2\right)$$

$$-\beta_2\left(\dot{C}_{21} - \frac{C_2^0}{\theta_2 T_2^0}\dot{T}_2\right) + \tilde{\gamma}_{T12}C_1^0 C_2^0\left(\frac{\dot{T}_2}{\theta_2 T_2^0} - \frac{\dot{T}_1}{\theta_1 T_1^0}\right) = C_2^0\left(\frac{C_2^0\tilde{\kappa}_{22}}{\theta_2 T_2^0}\Delta T_2 + \frac{C_1^0\tilde{\kappa}_{12}}{\theta_1 T_1^0}\Delta T_1\right) + C_2^0\left(\tilde{\alpha}_{22}\Delta C_2 + \tilde{\alpha}_{12}\Delta C_1\right)$$

The two first equations in this system are the motion equations of the concentration fields and the rest ones are the motion equations of the heat fields.

Double Component Medium at Constant Temperature

If we neglect the heat content of the system ($T_1 = T_2 \equiv T_1^0$, $\beta_1 = \beta_2 = 0$), then equation system (2.14) reduced to the two equations which are analogous to the Biot's equations for two fluids [30]

$$\rho_0 \frac{d^2}{dt^2} C_1 - \lambda_{11} C_1^0 \Delta C_1 - \lambda_{12} C_1^0 \Delta C_2 = -\tilde{\gamma}_{12} \frac{d}{dt}\left(C_2^0 C_1 - C_1^0 C_2\right)$$

$$\rho_0 \frac{d^2}{dt^2} C_2 - \lambda_{22} C_2^0 \Delta C_2 - \lambda_{12} C_2^0 \Delta C_1 = -\tilde{\gamma}_{12} \frac{d}{dt}\left(C_1^0 C_2 - C_2^0 C_1\right) \tag{2.15}$$

Let's note that with account normalization condition $C_1 + C_2 = \rho / \rho_0$, equations (2.15) can be written in terms of one of concentration field and density field

$$\rho_0 \frac{d^2}{dt^2} C_1 - (\beta_1 - \gamma_{12}) \frac{d}{dt} C_1 - C_1^0 (\lambda_{11} - \lambda_{12}) \Delta C_1 = \gamma_{12} \frac{d}{dt} \frac{\rho}{\rho_0} + C_1^0 \lambda_{12} \Delta \frac{\rho}{\rho_0}$$

$$\frac{d^2}{dt^2} \rho + (\beta_2 + \gamma_{12}) \frac{d}{dt} \frac{\rho}{\rho_0} - C_2^0 (\lambda_{22} - \lambda_{12}) \Delta \left(\frac{\rho}{\rho_0}\right) = \tag{2.16}$$

$$= \rho_0 \frac{d^2}{dt^2} C_1 + (\beta_2 + \gamma_{12}) \frac{d}{dt} C_1 - C_2^0 (\lambda_{22} - \lambda_{12}) \Delta C_1$$

In comparison with (2.15), the system (2.16) possesses higher symmetry that is more useful for analysis.

In frequency domain the system (2.15) can be split into two independent equations describing propagation of two acoustic eigen modes. Each of these modes $\Psi_{1,2}$ satisfies to Helmholtz's equation

$$\Delta \Psi_{1,2} + K_{1,2}^2(\omega) = 0 \tag{2.17}$$

with wave number

$$K_{1,2}^2(\omega) = \frac{\rho_0 \omega^2 - i\omega \gamma_{12} C_2^0 (1 - \alpha_{1,2})}{\lambda_{11} C_1^0 + \lambda_{12} C_2^0 \alpha_{1,2}} \tag{2.18}$$

where $\alpha_{1,2}$ are roots of the quadratic equation $A\alpha^2 + B\alpha + C = 0$ with coefficients

$$A = C_2^0 \left(\rho_0 \omega^2 \lambda_{12} - i\omega \gamma_{12} \left(\lambda_{22} C_2^0 + \lambda_{12} C_1^0\right)\right)$$

$$B = \rho_0 \omega^2 \left(\lambda_{11} C_1^0 + \lambda_{12} C_2^0\right) + i\omega\gamma_{12}\left(C_2^0\left(\lambda_{22}C_2^0 + \lambda_{12}C_1^0\right) - C_1^0\left(\lambda_{11}C_1^0 + \lambda_{12}C_2^0\right)\right)$$

$$C = -\rho_0 \omega^2 \lambda_{12} C_1^0 + i\omega\gamma_{12} C_1^0 \left(\lambda_{11} C_1^0 + \lambda_{12} C_2^0\right)$$

For the low frequency limit $\omega \to 0$ the roots of this equation have the following asymptotics

$$\alpha_1 = 1 + \frac{\rho_0 \omega^2}{i\omega\gamma_{12}} \frac{\lambda_{11}C_1^0 - \lambda_{22}C_2^0 + \lambda_{12}\left(C_2^0 - C_1^0\right)}{C_2^0\left(\lambda_{22}C_2^0 + \lambda_{12}C_1^0\right) + C_1^0\left(\lambda_{11}C_1^0 + \lambda_{12}C_2^0\right)},$$

$$\alpha_2 = -\frac{C_1^0\left(\lambda_{11}C_1^0 + \lambda_{12}C_2^0\right)}{C_2^0\left(\lambda_{22}C_2^0 + \lambda_{12}C_1^0\right)} \tag{2.19}$$

the corresponding wave numbers of these modes can be represented in the forms

For the low frequency limit $\omega \to 0$ the wave numbers of these modes can be represented in the forms

$$K_1^2(\omega \to 0) = \frac{\rho_0 \omega^2}{\lambda_{22}\left(C_2^0\right)^2 + \lambda_{11}\left(C_1^0\right)^2 + 2\lambda_{12}C_1^0 C_2^0} \tag{2.20a}$$

$$K_2^2(\omega \to 0) = -i\omega\gamma_{12} \frac{\lambda_{22}\left(C_2^0\right)^2 + \lambda_{11}\left(C_1^0\right)^2 + 2\lambda_{12}C_1^0 C_2^0}{C_1^0 C_2^0\left(\lambda_{11}\lambda_{22} - \lambda_{12}^2\right)} \tag{2.20b}$$

and the independent modes themselves are proportional to the following combinations

$$\Psi_1 = C_1 C_1^0 \lambda_{11} + C_2 C_2^0 \lambda_{22} + \lambda_{12}\left(C_1 C_2^0 + C_2 C_1^0\right) \tag{2.21a}$$

$$\Psi_2 = \left(C_1 C_2^0 - C_2 C_1^0\right) \tag{2.21b}$$

In the high frequency limit, $\omega \to \infty$, the roots of the quadratic equation have asymptotics

$$\alpha_{1,2} = \frac{1}{2\lambda_{12}C_2^0}\left(\lambda_{22}C_2^0 - \lambda_{11}C_1^0 \pm \sqrt{\left(\lambda_{11}C_1^0 + \lambda_{22}C_2^0\right)^2 + \lambda_{12}^2 C_1^0 C_2^0}\right) \tag{2.22}$$

and the corresponding wave numbers have the forms

In the high frequency limit, $\omega \to \infty$, the corresponding wave numbers have the form

$$K_{1,2}^2(\omega \to \infty) = \frac{\rho_0 \omega^2}{\lambda_{11}C_1^0 + \lambda_{22}C_2^0 \pm \sqrt{\left(\lambda_{11}C_1^0 - \lambda_{22}C_2^0\right)^2 + \lambda_{12}^2 C_1^0 C_2^0}} \tag{2.23}$$

It is easy to see from (2.20)-(2.23), that the behavior of the two linearly independent concentration modes are analogous to the behavior of Biot's waves of the first and the second kind [30]. One of these modes represents the acoustic mode with finite propagation velocity at low and high frequencies, while the second mode at low frequencies possesses the characteristics of a diffusion mode and at high frequencies it behaves like a wave mode. Besides, the relations (2.21) show that at low frequencies the concentration fields in the acoustic mode vary in phase while in diffusive mode they vary in anti-phase as it take place for the fast and slow Biot's modes.

Double Temperature Heat Conductivity of Immobile Medium

Let's consider now the case when the both mass displacement components are immobile $\vec{u}_1 = \vec{u}_2 = 0$, so that the total interaction between them is reduced to the heat exchange. In this case only the two last equations remain in system (2.14), which can be written in the following form:

$$\beta_1 \frac{\partial T_1}{\partial t} + \tilde{\gamma}_{T12} C_2^0 \left(\frac{\partial T_1}{\partial t} - \frac{\theta_1 T_1^0}{\theta_2 T_2^0} \frac{\partial T_2}{\partial t} \right) = C_1^0 \tilde{\kappa}_{11} \Delta T_1 + C_2^0 \tilde{\kappa}_{12} \frac{\theta_1 T_1^0}{\theta_2 T_2^0} \Delta T_2$$

$$\beta_2 \frac{\partial T_2}{\partial t} + \tilde{\gamma}_{T12} C_1^0 \left(\frac{\partial T_2}{\partial t} - \frac{\theta_2 T_2^0}{\theta_1 T_1^0} \frac{\partial T_1}{\partial t} \right) = C_2^0 \tilde{\kappa}_{22} \Delta T_2 + C_1^0 \tilde{\kappa}_{12} \frac{\theta_2 T_2^0}{\theta_1 T_1^0} \Delta T_1 \qquad (2.24)$$

The system (2.24) describes the two-temperature heat conductivity in the immobile medium. This system can be also split into two independent equations describing propagation of two heat modes. In the both modes the temperature propagation possesses a diffusion or thermal character and differs only by the value of coefficient of effective heat conductivity.

Double Component Medium with Single Temperature

If the all components are in heat equilibrium state between each other in each space point $T_i = T$, we obtain the multi-component with single temperature conductivity. For this case we will have

$$\rho_0 \frac{d^2}{dt^2} C_i - \sum_{j=1}^{2} C_i^0 C_j^0 \left(\lambda_{ij} \Delta \left(\frac{C_j}{C_j^0} \right) + \alpha_{ij} \Delta \left(\frac{T}{\theta_j T^0} \right) \right) =$$

$$= -\beta_i \frac{d}{dt} \left(C_i - \left(\frac{C_i^0 T}{\theta_i T^0} \right) \right) - \frac{d}{dt} \sum_{j=1}^{2} \gamma_{ij} C_i^0 C_j^0 \left(\left(\frac{C_i}{C_i^0} \right) - \left(\frac{C_j}{C_j^0} \right) \right) \qquad (2.10b)$$

$$\frac{d}{dt}\left(\frac{T}{T_0}\right)\sum_{i=1}^{2}\left(\frac{\beta_i C_i^0}{\theta_i}\right)+\frac{d}{dt}\sum_{i=1}^{2}\beta_i C_i=\Delta\left(\frac{T}{T_0}\right)\sum_{i,j=1}^{2}C_i^0 C_j^0\frac{\kappa_{ij}}{\theta_j}+\sum_{i,j=1}^{2}C_i^0 C_j^0\alpha_{ij}\Delta\left(\frac{C_j}{C_j^0}\right) \quad (2.12b)$$

Again for the simplest case of two-component medium we obtain the following equation system

$$\rho_0\frac{d^2}{dt^2}C_1+\frac{d}{dt}\left(\beta_1+\gamma_{12}\left(C_2^0 C_1-C_1^0 C_2\right)\right)-\lambda_{11}C_1^0\Delta C_1-\lambda_{12}C_1^0\Delta C_2=$$

$$=\frac{\beta_1 C_1^0}{\theta_1 T_0}\frac{d}{dt}T+C_1^0\left(\frac{\alpha_{11}C_1^0}{\theta_1 T_0}+\frac{\alpha_{12}C_2^0}{\theta_2 T_0}\right)\Delta T \quad (2.25a)$$

$$\rho_0\frac{d^2}{dt^2}C_2+\frac{d}{dt}\left(\beta_2+\gamma_{12}\left(C_1^0 C_2-C_2^0 C_1\right)\right)-\lambda_{22}C_2^0\Delta C_2-\lambda_{12}C_2^0\Delta C_1=$$

$$=\frac{\beta_2 C_2^0}{\theta_2 T_0}\frac{d}{dt}T+C_2^0\left(\frac{\alpha_{22}C_2^0}{\theta_2 T_0}+\frac{\alpha_{12}C_1^0}{\theta_1 T_0}\right)\Delta T \quad (2.25b)$$

$$\left(\frac{\beta_1 C_1^0}{\theta_1}+\frac{\beta_2 C_2^0}{\theta_2}\right)\frac{d}{dt}\left(\frac{T}{T_0}\right)-\left(\left(C_1^0\right)^2\frac{\kappa_{11}}{\theta_1}+C_1^0 C_2^0\kappa_{12}\left(\frac{1}{\theta_1}+\frac{1}{\theta_2}\right)+\left(C_2^0\right)^2\frac{\kappa_{22}}{\theta_2}\right)\Delta\left(\frac{T}{T_0}\right)=$$

$$=-\frac{d}{dt}\left(\beta_1 C_1+\beta_2 C_2\right)+\left(\alpha_{11}C_1^0+\alpha_{12}C_2^0\right)\Delta C_1+\left(\alpha_{22}C_2^0+\alpha_{12}C_1^0\right)\Delta C_2 \quad (2.25c)$$

Equations (2.25) represent thermo-diffusion subsystem in single temperature approximation. In frequency representation this equation system can be written in matrix form, if to introduce the vector of variables $\vec{C}=\left(C_1,C_2,T\right)$

$$A(\omega)\Delta\vec{C}=B(\omega)\vec{C} \quad (2.26)$$

where $A(\omega)$ and $B(\omega)$ are matrixes of dimension 3x3. That equation system can be diagonalized by appropriate choice of linear combination of components of variable vector $\vec{C}$, which components $\vec{\alpha}=\left(\alpha_1,\alpha_2,\alpha_3\right)$ are eigen vectors of the matrixes $A(\omega)$ and $B(\omega)$. Three linearly independent eigen vectors $\vec{\alpha}_i$ in this case corresponds to three independent eigen modes $\Psi_i=\left(\vec{\alpha}_i\vec{C}\right)$, which satisfy to the Helmholtz's equations

$$\Delta\Psi_i+K_i^2(\omega)=0 \quad (2.27)$$

Analysis shows that in the low frequency limit one mode has wave character with finite propagation speed, while two other modes have diffusion behavior of propagation. In the high frequency limit two modes propagate with finite frequencies while the third rests a diffusion one.

Analysis of complete equations system (2.14) shows that at the low frequency limit one of the modes is always an acoustical mode possessing the wave characteristics and finite propagation velocity, while the other modes possess the diffusion (or thermal) type of behavior. In the high frequency limit two modes propagate with finite velocities and two modes have diffusion type of \behavior.

This conclusion can be generalized onto the case of a N -component, N -temperature medium. At low frequencies the only acoustical mode exists and $2N-1$ modes possess the diffusion type of behavior. At high frequencies N mainly concentration modes possess the acoustic wave type of behavior and N heat modes possess the diffusion type of behavior.

Multiphase Media

Let's consider now the case of multiphase media. In this case the density of material point is the averaged by volume phase density.

$$\rho = \sum_{i=1}^{N} \rho_i \, m_i^0 \tag{2.28}$$

where by m_i^0 are denoted the volume parts of phases in initial state, ρ_i - are current phase densities, which are related with corresponding mean displacements $\vec{u}_i$, analogically to (2.1). Temperatures of separate phases are related with corresponding heat displacements $\vec{u}_{Ti}$, analogously to (2.2)

$$\rho_i = \rho_i^0 \left(1 - div\vec{u}_i\right) \tag{2.1b}$$

$$T_i = T_i^0 (1 - \theta_i div\vec{u}_{Ti}) \tag{2.2b}$$

It is possible to determine and current volume phase part in accordance to expression

$$m_i = m_i^0 \frac{\rho_i^0}{\rho_i}$$

Kinetic energy is a sum of kinetic energies of all separate phases

$$2K(\dot{\vec{u}}_i^2) = \sum_{i=1}^{N} \rho_i^0 m_i^0 \dot{\vec{u}}_i^2 \tag{2.29}$$

The free energy of multiphase also is a sum of the free energies of separate phases, as well as the free energies of surface interaction between phases, which is proportional to square of interphase boundaries S_{ij}

$$2F(\nabla\vec{u}_i, \nabla\vec{u}_{Ti}) = F(C_i^0, T_i^0) + \sum_{i}^{N} m_i^0\left(\lambda_i(\nabla\vec{u}_i)^2 + 2\alpha_i(\nabla\vec{u}_i\nabla\vec{u}_{Ti}) + \kappa_i(\nabla\vec{u}_{Ti})^2\right) + \sum_{i,j=1}^{N} S_{ij}F_{ij}^S \tag{2.30}$$

The dissipation function includes contributions from heat nonequilibriums inside of separate phases, as well as from nonequilibriums of mean and heat displacements between separate phases. At this the interphase interaction has to be proportional to squares of interphase boundaries S_{ij}

$$2D(\dot{\vec{u}}_i, \dot{\vec{u}}_{Ti}) = \sum_{i=1}^{N} \beta_i m_i^0 (\dot{\vec{u}}_i - \dot{\vec{u}}_{Ti})^2 + \sum_{i,j=1}^{N} \gamma_{ij} S_{ij} (\dot{\vec{u}}_i - \dot{\vec{u}}_j)^2 + \sum_{i,j=1}^{N} \gamma_{Tij} S_{ij} (\dot{\vec{u}}_{Ti} - \dot{\vec{u}}_{Tj})^2 \tag{2.31}$$

Motion equations for multiphase media can be written through introduced potential (2.28)-(2.31) in analogy to the case of multicomponents media (2.6), (2.7). In terms of displacement fields they have the forms (at assumption that interphase boundaries and related surface free energies are unchangeable)

$$m_i^0 \rho_i^0 \frac{d}{dt}\dot{\vec{u}}_i - m_i^0(\lambda_i \nabla\nabla\vec{u}_i + \alpha_i \nabla\nabla\vec{u}_{Ti}) = -m_i^0 \beta_i(\dot{\vec{u}}_i - \dot{\vec{u}}_{Ti}) - \sum_{j=1}^{N} \gamma_{ij} S_{ij}(\dot{\vec{u}}_i - \dot{\vec{u}}_j) - \sum_{j=1}^{N} \gamma_{Tij} S_{ij}(\dot{\vec{u}}_{Ti} - \dot{\vec{u}}_{Tj}) \tag{2.32a}$$

$$m_i^0 \beta_i(\dot{\vec{u}}_{Ti} - \dot{\vec{u}}_i) + \sum_{j=1}^{N} \gamma_{Tij} S_{ij}(\dot{\vec{u}}_{Ti} - \dot{\vec{u}}_{Tj}) = m_i^0(\kappa_i \nabla\nabla\vec{u}_{Ti} + \alpha_i \nabla\nabla\vec{u}_i) \tag{2.33a}$$

Substituting (2.33a) in (2.32a), the last equation can be written in the form

$$m_i^0 \rho_i^0 \frac{d}{dt}\dot{\vec{u}}_i - m_i^0((\lambda_i + \alpha_i)\nabla\nabla\vec{u}_i + (\kappa_i + \alpha_i)\nabla\nabla\vec{u}_{Ti}) = -\sum_{j=1}^{N} \gamma_{ij} S_{ij}(\dot{\vec{u}}_i - \dot{\vec{u}}_j) \tag{2.32b}$$

$$m_i^0 \beta_i(\dot{\vec{u}}_{Ti} - \dot{\vec{u}}_i) - m_i^0(\kappa_i \nabla\nabla\vec{u}_{Ti} + \alpha_i \nabla\nabla\vec{u}_i) = -\sum_{j=1}^{N} \gamma_{Tij} S_{ij}(\dot{\vec{u}}_{Ti} - \dot{\vec{u}}_{Tj}) \tag{2.33b}$$

As it seems from the comparison of the equations (2.32), (2.33) and (2.8), (2.9) without account of changes of interphase boundaries and related changes of free energies, the motion equations for multiphase medium have a few simpler forms then for multicomponent medium. Nevertheless, wholly, they are equivalent in the considered approximation

Taking divergence from these equations, they can be rewritten in terms of phase densities and temperatures:

$$m_i^0\left(\frac{d^2}{dt^2}\rho_i-\frac{\lambda_i+\alpha_i}{\rho_i^0}\Delta\rho_i-\frac{\kappa_i+\alpha_i}{\theta_i T_i^0}\Delta T_i\right)=-\sum_{j=1}^{N}\gamma_{ij}S_{ij}\frac{d}{dt}\left(\frac{\rho_i}{\rho_i^0}-\frac{\rho_j}{\rho_j^0}\right) \quad (2.32c)$$

$$m_i^0\left(\frac{\beta_i}{\theta_i T_i^0}\frac{d}{dt}T_i-\frac{\beta_i}{\rho_i^0}\frac{d}{dt}\rho_i-\frac{\kappa_i}{\theta_i T_i^0}\Delta T_i+\frac{\alpha_i}{\rho_i^0}\Delta\rho_i\right)=-\sum_{j=1}^{N}\gamma_{Tij}S_{ij}\frac{d}{dt}\left(\frac{T_i}{\theta_i T_i^0}-\frac{T_j}{\theta_j T_j^0}\right) \quad (2.33c)$$

Summation of the equation (2.32c) by whole phases with account of cancellation of the right part, it is possible to obtain equation for the mean density

$$\frac{d^2}{dt^2}\rho-\sum_{i=1}^{N}m_i^0\frac{\lambda_i+\alpha_i}{\rho_i^0}\Delta\rho_i-\sum_{i=1}^{n}m_i^0\frac{\kappa_i+\alpha_i}{\theta_i T_i^0}\Delta T_i=0 \quad (2.34)$$

Consider several simplest limiting cases described by the equation system (32) – (33).

Double Phase Medium at Fixed Temperature

Let's begin from the case, when the temperature is considered as constant $T=T_0$, and medium is a double phases. Motion equations have the forms

$$m_1^0\left(\frac{d^2}{dt^2}\rho_1-c_1^2\Delta\rho_1\right)=-\gamma_{12}S_{12}\frac{d}{dt}\left(\frac{\rho_1}{\rho_1^0}-\frac{\rho_2}{\rho_2^0}\right) \quad (2.35a)$$

$$\left(1-m_1^0\right)\left(\frac{d^2}{dt^2}\rho_2-c_2^2\Delta\rho_2\right)=-\gamma_{12}S_{12}\frac{d}{dt}\left(\frac{\rho_2}{\rho_2^0}-\frac{\rho_1}{\rho_1^0}\right) \quad (2.35b)$$

Here we used the notation for isothermal sound velocity

$$c_i^2=\frac{\lambda_i+\alpha_i}{\rho_i^0}$$

Summing equations (2.35) and accounting definition $\rho = m_1\rho_1 + (1-m_1)\rho_2$, we obtain equation for the mean density

$$\frac{d^2}{dt^2}\rho - \left(m_1 c_1^2 \Delta\rho_1 + (1-m_1)c_2^2 \Delta\rho_2\right) = 0$$

Transforming equations (2.35) from time representation to frequency representation ($t \to \omega$), and making diagonalization of the resulting equation system, it is possible to obtain two linearly independent combinations of densities, propagating as independent modes. Each of these modes $\Psi_{1,2}$ satisfies to the Helmholtz's equation

$$\Delta\Psi_{1,2} + K_{1,2}^2(\omega) = 0 \tag{2.36}$$

with the wave numbers

$$K_{1,2}^2(\omega) = \frac{\omega^2}{c_1^2} - \frac{i\omega\gamma_{12}S_{12}}{m_1^0 c_1^2 \rho_1^0}\left(1 - \alpha_{1,2}\frac{m_1^0 c_1^2}{m_2^0 c_2^2}\right) \tag{2.37}$$

where $\alpha_{1,2}$ are the roots of quadratic equation $A\alpha^2 + B\alpha + C = 0$ with coefficients

$$A = 1,\; B = \frac{\rho_1^0}{\rho_2^0} - \frac{m_2^0 c_2^2}{m_1^0 c_1^2} + \frac{\omega^2\left(c_1^{-2} - c_2^{-2}\right)}{i\omega\gamma_{12}S_{12}} m_2^0 c_2^2 \rho_1^0,\; C = -\frac{m_2^0 c_2^2 \rho_1^0}{m_1^0 c_1^2 \rho_2^0}$$

In the low frequency limit $\omega \to 0$ the roots of this equation has the following asymptotics

$$\alpha_1 = \frac{m_2^0 c_2^2}{m_1^0 c_1^2}\left(1 + \frac{i\omega\dfrac{m_2^0 c_2^2 \rho_1^0}{\gamma_{12}S_{12}}\left(c_1^{-2} - c_2^{-2}\right)}{\dfrac{\rho_1^0}{\rho_2^0} + \dfrac{m_2^0 c_2^2}{m_1^0 c_1^2}}\right),\; \alpha_2 = -\frac{\rho_1^0}{\rho_2^0} \tag{2.38}$$

Corresponding wave numbers have the following asymptotics

$$K_1^2(\omega \to 0) = \omega^2 \frac{m_1^0 \rho_1^0 + m_2^0 \rho_2^0}{m_1^0 c_1^2 \rho_1^0 + m_2^0 c_2^2 \rho_2^0} \tag{2.39a}$$

$$K_2^2(\omega \to 0) = -i\omega\gamma_{12}S_{12}\frac{m_1^0 c_1^2 \rho_1^0 + m_2^0 c_2^2 \rho_2^0}{m_1^0 c_1^2 \rho_1^0 m_2^0 c_2^2 \rho_2^0} \tag{2.39b}$$

and the linearly independent modes are represented by the following combinations

$$\Psi_1 = m_1^0 c_1^2 \rho_1 + m_2^0 c_2^2 \rho_2 \tag{2.40a}$$

$$\Psi_2 = \left(\rho_1 \rho_2^0 - \rho_2 \rho_1^0\right) \tag{2.40b}$$

In the high frequency limit $\omega \to \infty$, the roots of the quadratic equation have the asymptotics

$$\alpha_1 = i\omega \frac{\rho_1^0 m_2^0 c_2^2}{\gamma_{12} S_{12}} \left(\frac{1}{c_1^2} - \frac{1}{c_2^2} \right) - \left(\frac{\rho_1^0}{\rho_2^0} - \frac{m_2^0 c_2^2}{m_1^0 c_1^2} \right) \quad \alpha_2 = 0 \tag{2.41}$$

and corresponding wave numbers have the forms

$$K_{1,2}^2(\omega \to \infty) = \frac{\omega^2}{c_{1,2}^2} - i\omega \frac{\gamma_{12} S_{12}}{m_{1,2}^0 \rho_{1,2}^0 c_{1,2}^2} \tag{2.42}$$

Thus, in the high frequency limit the propagation of each of two modes occurs independently by the own phase. At this the mode attenuation coefficient is inversely proportional to the volume phase content, so that for the low concentration of one of the phases the corresponding mode occurs to be highly attenuating one. Independent propagation of modes by phases in high frequency limit originated from the fact that interphase interaction is determined only by dissipation processes, which are small in high frequency limit in comparison with inertion effects.

In the low frequency limit the one of the modes is an acoustical one with effective elastic module is equaled to a sum of phase elastic modules weighted with their volumes. The second mode in low frequency limit becomes the diffusive one with attenuation proportional to a sum of inverse elastic modules.

Conclusion

The main results presented in the article can be formulated as following:

It is shown that conventional system of hydro-dynamical equations in the form of Navier-Stokes, representing by itself the low frequency approximation, is not provided the finite propagation velocity of small perturbations at high frequencies. The required justifications can be introduced with account of viscosity relaxations and relaxation of heat flow. However the generality of such justifications are not clear.

The generalization of the hydrodynamical equations system can be obtained on the basis of variational principle. The generalized variational principle can be formulated in the form of combination of Hamilton's and Onsager's variational principles for dissipative systems in terms of two independent fields: the usual field of mass displacements and the heat displacements field.

The additional account of inertion of the heat displacements field and relaxation of internal parameters in the framework of Mandelshtam-Leontovich approach allows to obtain by natural way the generalization of Navier-Stokes's hydrodynamical equations system, possessed by finite propagation velocity at high frequencies.

The generalized variational principle was applied to description of multicomponents and multiphase media. The complete system of motion equation for concentration and temperature fields is derived on the basis of the generalized variational principle. Analogous results obtained and for multiphase medium.

In the simplest case of double component medium at constant temperature the system of two concentration fields is analogous to the Biot's system for porous permeable medium. Motion of two concentration fields in this case is described by two independent modes, one of them is the acoustical mode at low frequencies (the Biot's wave of the first kind), while the second one has diffusive behavior (the Biot's wave of the second kind).

The same equation system for double component medium can describe double temperature heat conductivity of immobile medium or even double temperature fluid filtration in more general case.

In the case of double components medium with common temperature the single acoustical mode and two diffusive modes exist at low frequencies. This result can be generalized to the common case of N-components, N-temperatures medium: the only acoustical mode and 2N-1 diffusive modes exist at low frequencies. Analogous results are valid for multiphase medium.

Acknowledgements

The work was supported by RFBR (grant №05-02-17670-a)

Referencies

[1] Landau L.D., Lifshitz E.M. Theoretical physics. V.6. Hydrodynamics. Moscow, Nauka, 1986. (in Russian)

[2] Landau L.D., Lifshitz E.M. Theoretical physics. V.7. Theory of elasticity. Moscow, Nauka, 1972. (in Russian)

[3] Deresiewicz H. Plane wave in a termoelastic solids. // *J.Acoust.Soc. Am.* 1957, V.29, p.204-209.

[4] Nettleton R.E. Relaxation theory of thermal conduction in liquids. // *Phys.Fluids* 1960, V.3, p.216-223.

[5] Lykov A.V. Theory of heat conduction. Moscow Vysshaya Shkola, 1967. (in Russian)

[6] Landau L.D., Lifshitz E.M. Theoretical physics. V.5. Statistical physics. Moscow, Nauka, 1964. (in Russian)

[7] Onsager L. Reciprocal relations in irreversible process I, II. // *Phys. Rev. 1931*, V.37, p.405-426. // *Phys. Rev.* 1931, V.38, p.2265-2279.

[8] Glensdorf, P. and Prigogine, I., Thermodynamic Theory. of Structure, Stability, and Fluctuations, New York,. 1971.

[9] Biot M. Variational principles in theory of heat transfer. Moscow. Energy. 1975. (in Russian)

[10]Gyarmati I. Non-equilibrium thermodynamics. Field theory and variational principles. Berlin, Springer-Verlag, 1970

[11]Mandelshtam L.I., Leontovich M.A. To the sound absorption theory in liquids // *Journal of Experimental and Theoretical Physics (ZhETPh)* 1937. V.7, №3. p.438-444. (in Russian)

[12]Maximov G.A. On variational principle for dissipative hydrodynamics. // Preprint 006-2006. Moscow: MEPhI, 2006. – 36p.

[13]Maximov G.A. On variational principle for dissipative hydrodynamics and acoustics. // *Proc. of XVIII session of Russsian Acoustic Society.* –Moscow.:GEOS, -2006. V.1. P. 5-8.

[14]Maximov G.A. On variational principle for dissipative hydrodynamics and acoustics. // *Proc. of the International conference Days on Diffraction* 2006. May 30 – June 2, 2006, St. Petersburg, Russia, pp.173-177.

[15]Maksimov G.A. Application of pulsed acoustic diagnostics to homogeneous relaxation media. // *Acoustical Physics* 1996. V.42. N 4, p.541-550.

[16]Larichev V.A., Maksimov G.A. Acoustic pulse propagation in a medium with two relaxation processes. Analysis of the exact solution. // *Acoustical Physics* 1999. T.45. №6. P.844-856.

[17]Maksimov G.A., Larichev V.A. Propagation of a short pulse in a medium with a resonance relaxation: The exact solution // *Acoustical Physics* 2003. T.49. №5. C. 555-564.

[18]Zozulya P.V., Zozulya O.M. Measurement of the moments of the relaxation time spectrum of a liquid by pulsed acoustic spectroscopy // *Acoustical Physics* 2004. T.50. №4. P. 401-405.

[19]Karabutov A.A., Larichev V.A., Maksimov G.A., Pelivanov I.M., Podymova N.B. Relaxation dynamics of a broadband nanosecond acoustic pulse in a bubbly medium. // *Acoustical Physics* 2006. T.52. №5. N 5, p.582-588.

[20]Truesdell C. Precise theory of the absorption and dispersion of forced plane infinitesimal waves according to the Navier-Stokes equations. // *J. Ration. Mech. Anal.* 1953. V.2. P.659.

[21]Chester M. Second sound in solids // *Phys. Rev.* 1963. V.131. №5. P.2013-2015.

[22]Prohofsky E.W., Krumhansl J.A. Second sound propagation in dielectric solids // *Phys. Rev.* 1964. V.133. №5A. P.A1403-A1410.

[23]Lord H.W., Shulman Y. A generalized dynamical theory of thermoelasticity // *J. Mech. Phys. Solids* 1967. V.15. P.299-309.

[24]Green A.E., Lindsay K.A. Thermoelasicity // J. Elasticity 1972. №2. P.1-7.

[25]Rudgers A.J. Analysis of thermoacoustic wave propagation in elastic media // *J. Acoust. Soc. Am.* 1990. V.88. №2. P.1078-1094.

[26]Martynov G.A. Hydrodanamic theory of sound wave propagation. // *Theor. and Math. Phys.* 2001, V.129, p.1428-1438.

[27]Martynov G.A. General theory of acoustic wave propagation in liquids and gases// *Theor. and Math. Phys.* 2006. V.146. №2. p. 285-294.

[28]Chapman S., Cowling T.G. The mathematical theory of non-uniform gases: An account of the kinetic theory of viscosity, thermal conduction and diffusion in gases. Cambridge Univ. Press, Cambridge. 1952.

[29]Klimontovich Yu.L. Statistical theory of open systems. V.1., Moscow: Yanus, (in Russian)

[30]Biot M.A. Theory of propagation of elastic waves in fluid-saturated porous solid. I. Low-frequency range. // *J. Acoust. Soc. Am.,* 1956, V.28, N2, p.168-178.

RESEARCH AND REVIEW STUDIES

In: New Research on Acoustics
Editor: Benjamin N. Weiss, pp. 65-109
ISBN: 978-1-60456-403-7

Chapter 1

USING THE DIFFERENT TYPES OF WAVELET TRANSFORM FOR ANALYZING ACOUSTIC FIELDS AND SIGNALS IN ELASTIC MEDIA

Dmitry V. Perov** and Anatoly B. Rinkevich
Institute of Metal Physics, Ural Division of the Russian Academy of Sciences
18 Sofia Kovalevskaya St., GSP-170, Ekaterinburg 620041, Russia

Abstract

In this chapter we consider application of different types of wavelet transform for analyzing acoustic fields and signals in elastic media. It is shown that different types of the discrete wavelet analysis, namely, fast wavelet transform, stationary wavelet transform and wavelet packets, are powerful and effective tools for denoising the signals and images. Using the continuous wavelet transform allows to realize the spatio-temporal spectral analysis of nonstationary processes. It makes possible to estimate the instantaneous frequency of signal, determine the frequency modulation law, reveal the relations between oscillations of different frequencies and so on. One of the main advantages of wavelet analysis is its ability to adapt itself to parameters of analyzed signals and images.

Introduction

This chapter is dedicated to research the peculiarities of acoustic fields and signals on the basis of application of different types of wavelet transform for analyzing the processes in elastic media. The results have been obtained during the investigations of the processes of acoustic wave propagation and diffraction in elastic media, which are relevant to ultrasonic nondestructive testing and geoacoustics.

One of the most major areas of wavelet transform application is creating the highly efficient filters for denoising acoustic signals and images of acoustic field distributions. It is common knowledge that the most frequently applied type of filtering is signal processing

* E-mail address: peroff@imp.uran.ru ; rin@imp.uran.ru

based on the use of band pass filters that allows yield a large dividend in the signal-to-noise ratio if a narrow-band signal is to be separated from broadband noise. But the difficulty emerges, however, when we deal with a wave process whose carrier frequency or bandwidth of its spectrum varies essentially from one received signal to other. In this case, it is next to impossible to create an appropriate band pass filter for effective signal denoising because the exact parameters of analyzed signals are unknown a priori. Indeed, if the passband is too narrow for the chosen filter, then the spectrum of analyzed signal will be cut off; alternatively, if the passband is too wide, then the signal-to-noise ratio at the output of filter will be too small. One way to resolve this contradiction is using the wavelet filtration that is based on the discrete wavelet transform. In this chapter, the authors show what the values of wavelet filter parameters should be chosen to optimize the signal-to-noise ratio at the filter output depending upon the type of discrete wavelet transform and kind of analyzed signal.

Another important field in which the wavelet transform can be successfully applied is the spectral analysis of nonstationary signals. For this purpose, the continuous wavelet transform has to be applied. It makes possible to estimate the instantaneous frequency of signal, which dominates at different times, search for modulations and relations between oscillations with different periods. Various aspects of applying the algorithms based on the continuous wavelet transform to different problems of acoustics are shown in this article.

Any wavelet-based algorithm has an advantage that results from its ability to adapt itself in accordance with the parameters of analyzed signals.

Discrete Wavelet Analysis

Fast Wavelet Transform

According to the theory of the orthogonal wavelet transform [1, 2], a function of time t, $f(t) \in L^2(\mathbf{R})$, defined on the entire real axis $R(-\infty; +\infty)$ and possessing a finite norm $\|f\|$ can be represented in terms of the following spectral expansion:

$$f(t) = f_{i_0}^A(t) + \sum_{i=i_0}^{+\infty} f_i^D(t) = \sum_{j=-\infty}^{+\infty} v_j^{(i_0)} \varphi_{i_0 j}(t) + \sum_{i=i_0}^{+\infty} \sum_{j=-\infty}^{+\infty} w_j^{(i)} \psi_{ij}(t) \quad , \tag{1}$$

where $f_i^A(t)$ and $f_i^D(t)$ are, respectively, the large-scale component (approximations) and the small-scale component (details) of the function $f(t)$ on the ith level of decomposition. Here, i_0 refers to a certain initial level corresponding to the lowest time resolution.

In formula (1), $v_j^{(i)}$ and $w_j^{(i)}$ are the coefficients of expansion of the function $f(t)$ in systems of basis functions $\varphi_{ij}(t)$ and $\psi_{ij}(t)$, respectively. The index i is the decomposition level number which determines the size of the domain of the basis functions on the time axis, i.e., their resolution; the index j defines the position of the basis functions on the time axis.

The functions $\varphi_{ij}(t)$ and $\psi_{ij}(t)$ are called scaling functions and wavelets, respectively. They determine the corresponding orthonormal bases in the orthonormal space $L^2(\mathbf{R})$ and are defined

$$\varphi_{ij}(t) = 2^{i/2}\varphi\left(2^i t - j\right) \quad , \tag{2}$$

$$\psi_{ij}(t) = 2^{i/2}\psi\left(2^i t - j\right) \quad , \tag{3}$$

where i and j are integers.

Thus, using the two functions, $\varphi(t)$ and $\psi(t)$, which are called the father scaling function and the mother wavelet, respectively, the scaling procedure with the scale factor equals to an integer power of two, and integer time shifts, one can obtain the orthonormal bases $\{\varphi_{ij}\}$ and $\{\psi_{ij}\}$. The following orthogonality relations are valid for the scaling functions and wavelets [1, 2]:

$$(\varphi_{ij}, \varphi_{il}) = \delta_{jl}\,;\ (\psi_{ij}, \psi_{kl}) = \delta_{ik}\delta_{jl}\,;\ (\varphi_{ij}, \psi_{il}) = 0\,, \tag{4}$$

where i, j, k and l are integers and δ is the Kronecker delta.

The expansion coefficients of the function $f(t)$ in the bases (2) and (3) are given by the following formulas [1, 2]:

$$v_j^{(i)} = \int_{-\infty}^{+\infty} f(t)\,\varphi_{ij}(t)\,dt \quad , \tag{5}$$

$$w_j^{(i)} = \int_{-\infty}^{+\infty} f(t)\,\psi_{ij}(t)\,dt \quad , \tag{6}$$

which are similar to the expressions for spectral coefficients of the generalized Fourier transform. The numerical sequences given by Eqs. (5) and (6) are called the scaling and wavelet coefficients.

Relationships (1) and (4) – (6) express the concept of the multiple-scale analysis, i.e., the multilevel expansion of the function being analyzed in orthonormal systems of basis functions that differ only in the scaling factor of the argument. This expansion represents the space $L^2(\mathbf{R})$ as a sequence of embedded subspaces $V^{(i)}$ such that $\ldots \subset V^{(i-1)} \subset V^{(i)} \subset V^{(i+1)} \subset \ldots$ and $\bigcup_i V^{(i)} = L^2(\mathbf{R})$, $\bigcap_i V^{(i)} = \{0\}$. Then, for example, the approximation $f_{i_0}^{A}(t)$ corresponds to an orthogonal projection of the function $f(t)$ onto the subspace $V^{(i_0)}$ and contains the components of $f(t)$ whose time scale is no smaller than 2^{-i_0}.

The refinement $f_{i_0}^{D}(t)$ is a projection of the function $f(t)$ onto the subspace $W^{(i_0)}$, which is the orthogonal complement of the space $V^{(i_0)}$ in $V^{(i_0+1)}$: $V^{(i_0+1)} = V^{(i_0)} \oplus W^{(i_0)}$, which extracts from $f(t)$ the components whose time scale is on the order of $2^{-(i_0+1)}$. The possibility of using the multiple-scale analysis to construct the orthogonal wavelet transform was first shown by Mallat [3, 4].

It is known [1, 2] that the scaling function and the wavelet must satisfy the following refinement relations:

$$\varphi(t) = \sqrt{2}\sum_{k} h_k \varphi(2t-k) \quad , \tag{7}$$

$$\psi(t) = \sqrt{2}\sum_{k} g_k \varphi(2t-k) \quad , \tag{8}$$

where k is an integer. The numerical sequences h_k and g_k are referred to as the coefficients of the scaling and wavelet filters, respectively. They are related by the expression

$$g_k = (-1)^k h_{1-k} \,. \tag{9}$$

In the applications which use the discrete wavelet transforms certain requirements to the bases employed to realize these transforms are formulated. We will outline some of these requirements [1, 2]:

1. Smoothness of the scaling functions and wavelets which is necessary for solving the problems of approximation, signal shrinkage, filtering, etc.;
2. Presence of symmetry of the scaling functions and wavelets near their definition domain center;
3. Number of zero moments of the wavelet.

In accordance with the definition [1, 2], the wavelet, for which

$$\int_{-\infty}^{+\infty} t^m \psi(t)\, dt = 0 \tag{10}$$

for any $m \in [0, M]$, where m and M are integers, is called M-order wavelet. For $m = 0$ relation (10) must be satisfied for any wavelet (zero mean condition). The wavelets with a large number of zero moments make it possible better to analyze small-scale fluctuations of the analyzed signal and clarify non-differentiable higher-order singularities [1]. The presence of zero moments of the function M means its orthogonality to the piecewise polynomial function of the degree not higher than $M-1$. Consequently, this function, being expanded in terms of wavelets with M zero moments, will have a finite number of decomposition levels at which the function details will be nonzero.

As was shown in the earlier studies [1, 2], one obtains a system of algebraic equations for the coefficients g_k and h_k using the orthogonality relations (4) combined with Eqs. (7) – (9), the normalization condition for the scaling function, $\int_{-\infty}^{+\infty} \varphi(t)\,dt = 1$ and the condition of vanishing of first M moments of the wavelet – (10). The solution of this system yields the sequences of the coefficients for the scaling and wavelet filters. By substituting these results in functional equation (7), one can calculate the scaling function $\varphi(t)$ and then, using Eq. (8), determine the wavelet $\psi(t)$.

This procedure applied to different numbers of the wavelet function moments that are equal to zero: $M = 1, 2, 3, \ldots$; generates a family of Daubechies wavelets denoted as db. Different wavelets of this family corresponding to different M are denoted as db$M$, where M is the wavelet order. Wavelet db1 is also called the Haar wavelet. It is known [1] that the db$M$ wavelet and the corresponding scaling function are localized mostly within the time interval $2M$–1. The sequences of the coefficients for the scaling filter and wavelet filter corresponding to wavelet db$M$ are composed of $2M$ numbers.

The Daubechies wavelets are strongly asymmetrical. In some applications of the wavelet transformation, it is desirable that the basis function be symmetrical, at least approximately. This condition is satisfied to a greater extent by wavelets of the sym family or symlets. The sym$M$ wavelets, where $M = 2, 3, 4, \ldots$; and the corresponding coefficients g_k and h_k are localized just as db$M$ wavelets are. The symlets, however, are more symmetrical since the coefficients of their scaling filter and wavelet filter are corrected with a view to minimizing the deviation of the scaling filter coefficients from linear law for the phase of Fourier components [1, 2].

Figure 1 shows wavelets sym5 and sym8 and the corresponding scaling functions, along with the coefficients for the scaling filter and wavelet filter. A more detailed description of various wavelet families can be found, e.g., in the monographs [1, 2].

A straightforward implementation of the discrete wavelet transform based on Eqs. (1), (5) and (6) presents a heavy burden for a computer. However, the performance of the method can be considerably improved and the method can best be adapted to the analysis of discrete signals by the technique described below. Let us perform the following transformations: substitute Eqs. (7), (8) into definitions (2), (3) and substitute the result into Eqs. (5), (6). Ultimately, we obtain the following recurrence relations [1, 2], which give the fast wavelet transform (FWT) algorithm:

$$v_j^{(i)} = \sum_{k=0}^{k=K-1} h_k^* v_{2j-k}^{(i+1)} \quad , \tag{11}$$

$$w_j^{(i)} = \sum_{k=0}^{k=K-1} g_k^* v_{2j-k}^{(i+1)} \quad , \tag{12}$$

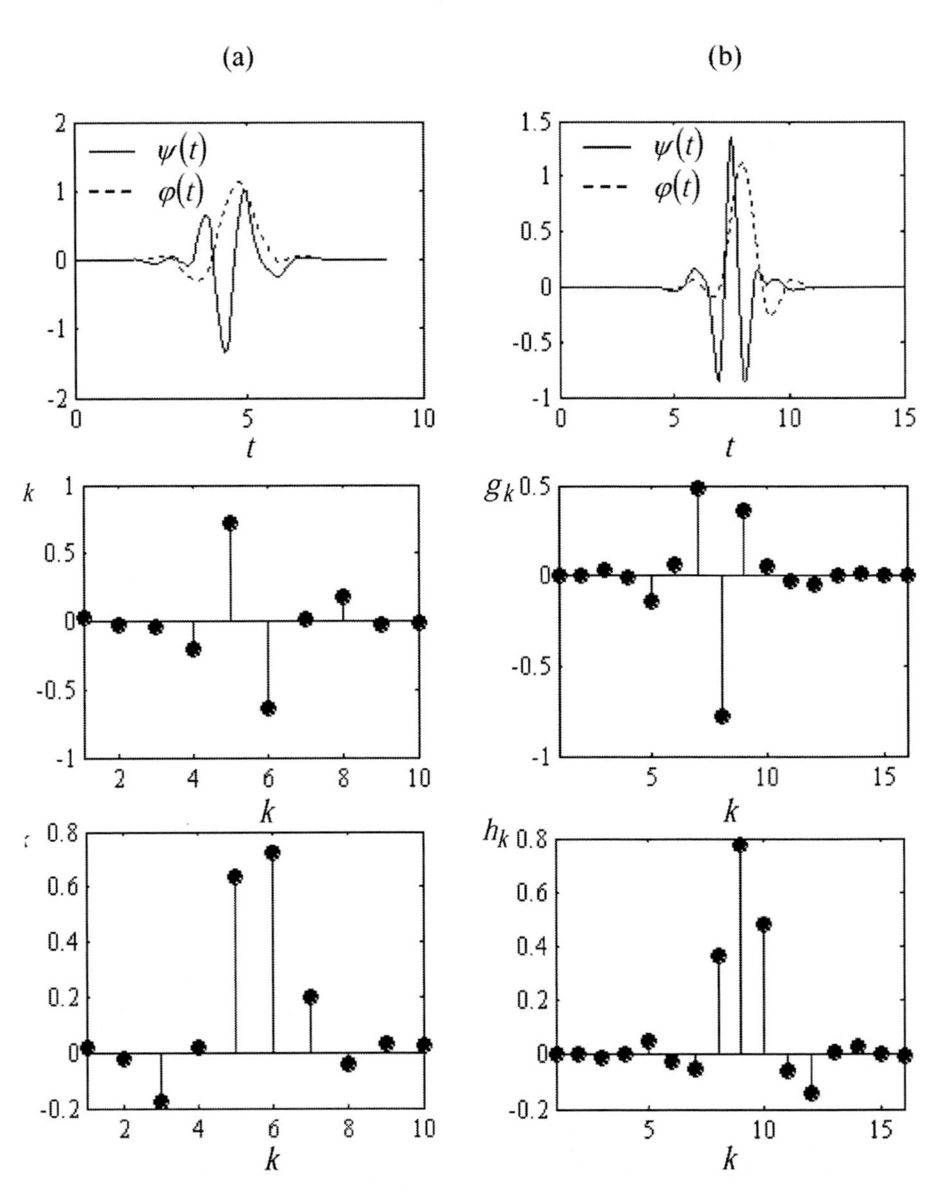

Figure 1. Scaling functions and wavelets as well as corresponding coefficients of scaling and wavelet filters: (a) sym5; (b) sym8.

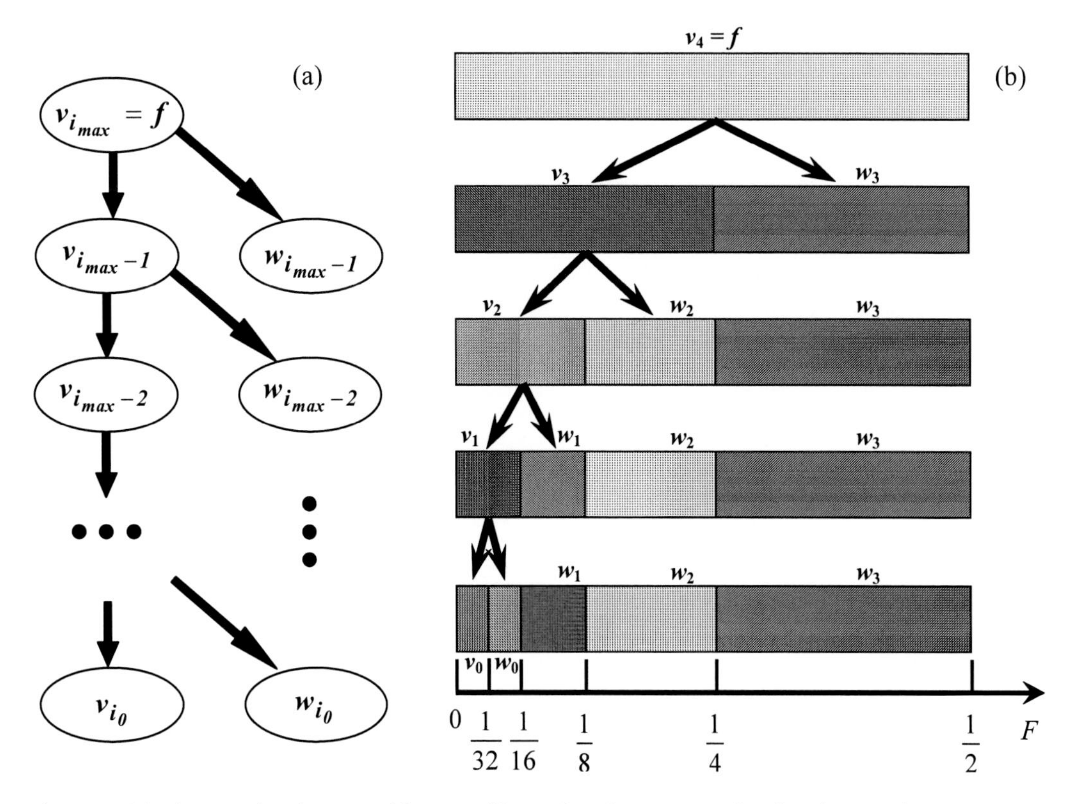

Figure 2. (a) The wavelet decomposition tree illustrating the process of a signal expansion into scaling and wavelet coefficients using the fast wavelet transform algorithm; (b) Recurrent division of the frequency band between the scaling and wavelet coefficients at different decomposition levels (i_0 =0; $i_{\max}$ = 4).

where $h_k^* = h_{K-k-1}$; $g_k^* = g_{K-k-1}$; and K is the length of sequences h^*, g^*, h and g. Is clear that, in practice, calculations by formulas (11) and (12) must be performed a finite number of times and completed at a certain lowest level of decomposition with index i_0. Consequently, the index i in Eqs. (11) and (12) must take the following values: $i = i_{\max} - 1,\ i_{\max} - 2, \ldots, i_0 + 1,\ i_0$. The index j in Eqs. (11) and (12) ranges over the following values: $j = 0, 1, 2, \ldots, J^{(i)}$; $J^{(i)} = \mathrm{Int}\left[\left(N^{(i+1)} + K - 1\right) / 2\right]$, where $N^{(i+1)}$ is the length of the scaling and wavelet coefficient vectors at the decomposition level number i + 1. Here, Int[·] is the integer part of the argument.

It should be noted that, at each step from an ith decomposition level to the next level i + 1, the length of the vectors of the scaling and wavelet coefficients decreases by a factor of two.

As the initial values, formulas (11) and (12) use the scaling coefficients $v_j^{(i_{\max})}$ of the signal being analyzed that correspond to the highest decomposition level of number $i_{\max}$. As a rule, $v_j^{(i_{\max})} = f_j$, where $f_j = f(j\Delta t)$ is the discrete signal being analyzed, which consists of $N = N^{(i_{\max})}$ samples taken with sampling period Δt.

Realization of the FWT algorithm in accordance with formulas (11) and (12) can be schematically presented as the wavelet decomposition tree; see Figure 2a.

It can be shown [1, 2] that the signal can be reconstructed from its scaling and wavelet coefficients by an algorithm called the inverse fast wavelet transform (IFWT), which can also be represented as the recurrence relation

$$v_j^{(i+1)} = \sum_{k=0}^{k=K-1} h_{j-2k} v_k^{(i)} + g_{j-2k} w_k^{(i)} \quad . \tag{13}$$

The IFWT procedure can be illustrated by a diagram similar to that in Figure 2a, the only difference being that the motion along the tree is not from the top to bottom, as in FWT, but from the bottom to top.

The expressions on the right of (11) and (12) can be treated as discrete convolutions [5]. Thus, the FWT algorithm can be interpreted as processing of a signal using high- and low-frequency finite impulse response (FIR) filters with impulse response sequences h^* and g^*, respectively, which form a pair of quadrature mirror filters (QMF) [1, 2]. Similarly, another pair of QMFs with impulse response sequences h and g corresponds to the IFWT algorithm in accordance with the relation (13). It is known [5] that the spectrum of a discrete signal corresponding to positive frequencies is concentrated in the frequency band from 0 to $f_{\max} = 1/\Delta t$. Let us define the normalized dimensionless frequency F: $F = f/f_{\max}$. Obviously, the spectrum of the initial signal occupies the band of $F = 0$ to $F = 0.5$. Each of the pair of QMFs covers one half of this band, the dividing point being at $F = 0.25$. Hereafter these frequencies will be termed dyadic.

An important feature of the FWT algorithm is that the lengths of the vectors of the scaling and wavelet coefficients are halved in each transition to the decomposition level with a number reduced by one unit. Since only scaling coefficients of the $(i + 1)$th level with even numbers are taken into account on the ith level, the frequency band occupied by the spectra of the scaling and wavelet coefficients is contracted by a factor of two in each transition. For example, after the transition from the level number $i_{\max}$ to the level number $i_{\max}$-1, the spectra characterizing $v_j^{(i_{\max}-1)}$ and $w_j^{(i_{\max}-1)}$ occupy the band from $F = 0$ to $F = F^{(i_{\max}-1)}$ = 0.25; and the dyadic frequency is $F = 0.125$. Thus, the FWT algorithm retains only one half of the previous frequency band of the processed signal on each step.

The division of the frequency band between the scaling and wavelet coefficients on different decomposition levels in the FWT algorithm starting with $i_{\max} = 4$ is illustrated by Figure 2b. In this case, the sequence of the dyadic frequencies at i = 4, 3, 2, 1 and 0 is 0.25, 0.125, 0.0625, 0.03125 and 0.015625.

Wavelet Filtering Using the Fast Wavelet Transform

The wavelet coefficients obtained as a result of the wavelet analysis can be modified in a special procedure before the wavelet synthesis with the use of the IFWT. In particular, the scaling and wavelet coefficients can be modified to filter the analyzed signal, i.e., to separate it from the noisy background, in order to make easier its perception by the operator, reduce errors in the signal amplitudes estimation, etc.

The wavelet filtering algorithm which is called the wavelet shrinkage denoising (WSD) was first suggested by Donoho and Johnstone [6,7]. It relies on the FWT algorithm and is performed in the following way. For all vectors of wavelet coefficients obtained at all decomposition levels of the FWT algorithm, the thresholds, whose values are chosen by a certain strategy, are adjusted. The elements that are beyond the thresholds are left unchanged, and all the remaining elements are replaced by zeroes. This procedure is called hard thresholding strategy or simply hard thresholding. If the values of the nonzero vector elements that are left after applying the threshold limitation are additionally shifted towards the zero level by the corresponding threshold values, the procedure is called soft thresholding. Subsequently, IFWT procedure (13) is applied to the modified vectors of wavelet coefficients to obtain the denoised signal.

Thresholds for a vector of wavelet coefficients are set symmetrically with respect to the zero level, and their amplitudes are selected by a special procedure. More detailed information can be found in literature [2, 6, 7]. From the standpoint of noise reduction, one of the most efficient strategies to wavelet filtering is the method of fixed-form thresholding [6, 7]. It is used to separate signals from a noisy background, which is characterized by the normal or almost normal distribution.

Donoho and Johnstone suggested one of the possible rules for choosing the thresholds on the ith decomposition level as follows [6, 7]:

$$T_i = \pm \operatorname{med}\left(\left| w_j^{(i)} \right| \right) \frac{\sqrt{2 \ln N}}{0.6745} , \tag{14}$$

where med(·) is the median of the absolute value of the vector $w_j^{(i)}$; $i = i_{\max} - 1, \colon \colon \colon, 1, 0$. The median of a numerical sequence is determined as follows. First all the components of the array are arranged in the increasing order. Then, if the number of components in the array is odd, the median is equated to the component placed in the middle, but if the number of components is even, the median equals the arithmetic mean of the two components nearest to the middle. The median is an approximate estimate of the mean deviation of the absolute values of the sequence components from zero. It is weakly susceptible to the presence of isolated components whose magnitudes are much larger than the mean level.

The procedure of wavelet filtering is different from traditional methods of noise reduction [5], when a frequency band is selected where the spectrum of a desired signal is localized, and components of frequencies beyond this band are suppressed. This is done with the use of band pass filters.

But the underlying principle of signal denoising in the wavelet filtering is quite different. In the FWT algorithm, specific frequency bands are also defined, but frequency characteristics of QMFs are not determined by the spectral characteristics of analyzed signals. Furthermore, in the process of discrimination, vectors of wavelet coefficients retain mostly components that are beyond the intervals of resolution determined by statistical noise characteristics. If the analyzed process is driven exclusively by noise, all wavelet coefficients are assumed to belong to these intervals with a high probability. But if the detector output contains, in addition to noise, a signal with different statistical characteristics, the corresponding components of vectors of wavelet coefficients should be above the threshold level. The thresholds determined by conditions like (14) are optimal [2, 6, 7] in the sense that the root mean square (RMS) errors in estimates of deterministic signals against the background of white noise with a normal distribution are minimal.

Note that this principle of filtering can be implemented, generally speaking, when a signal is expanded in any basis of orthonormal functions, but not only in the wavelet basis. In particular, the Fourier basis can also be used. Even so, the error in the signal estimate is determined in this case by the capability of this or that basis to contract the analyzed signal, i.e., by the number of the expansion coefficients for the basis functions required to represent the signal in this basis with a predetermined error. In this sense, these are the wavelet bases which are optimal [6].

We will further consider using the wavelet filtering technique, as applied to some typical pulse acoustic signals, to reveal specificity of denoising this type of signals which are widespread in acoustics. For example, we select the parameters corresponding to signals with the frequency range of megahertz and duration of some microseconds. It is well known that various distortions of a signal shape appear during a filtering process, namely, changing the amplitude of signal and time shifting of the signal peak. In this chapter we will try to answer the question: How should we choose the parameters of wavelet filtering algorithm to minimize the distortions?

Let us introduce the following notation: s_0 is a probing signal, which is a mathematical model of a real acoustic echo-signal; s_{in} is the input of a wavelet filter, which contains desired signals, noise and spurious pulses; s_{out} is the output of the wavelet filter. The amplitudes or maximal absolute values of these signals are denoted as A_0, A_{in} and A_{out}, respectively. All the signals are assumed to be numerical sequences because detectors equipped with microprocessors deal with digitized signals.

Let us express the probing signal by the following formula [8]:

$$s_{0_j} = S_0 j^2 \exp\left(-\alpha j^2\right) \cos\left(2\pi F_0 j\right) \quad , \tag{15}$$

which fairly accurately describes real reflected signals received by ultrasonic detectors. Here S_0 is the normalizing factor, α is the parameter that determines the signal duration, F_0 is the normalized carrier frequency and $j = 0; 1; 2; : : : ; N$ is the integer index.

Further we will consider three types of probing signals described by Eq. (15): a long signal with $\alpha = 0{:}0001$ which image contains more than 10 carrier frequency oscillations; a middle signal with $\alpha = 0{:}0005$ – 5 - 7 oscillations; and a short signal with $\alpha = 0{:}005$ – 2 - 3 oscillations. They are shown in Figure 3. The relative width of the signal spectrum, $\Delta F/F_0$ is about 0.15 for the long signal and 0.4 for the short one.

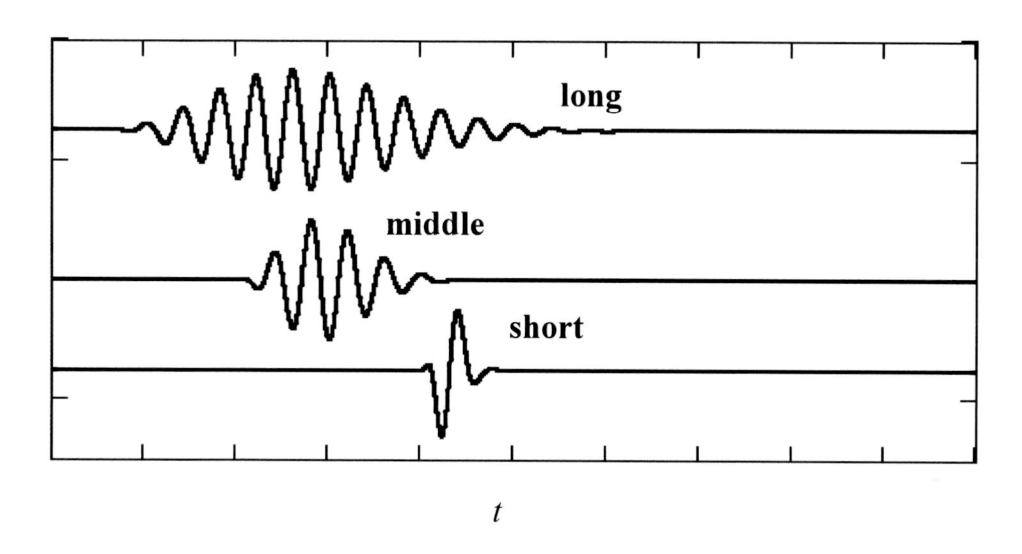

Figure 3. The probing signals.

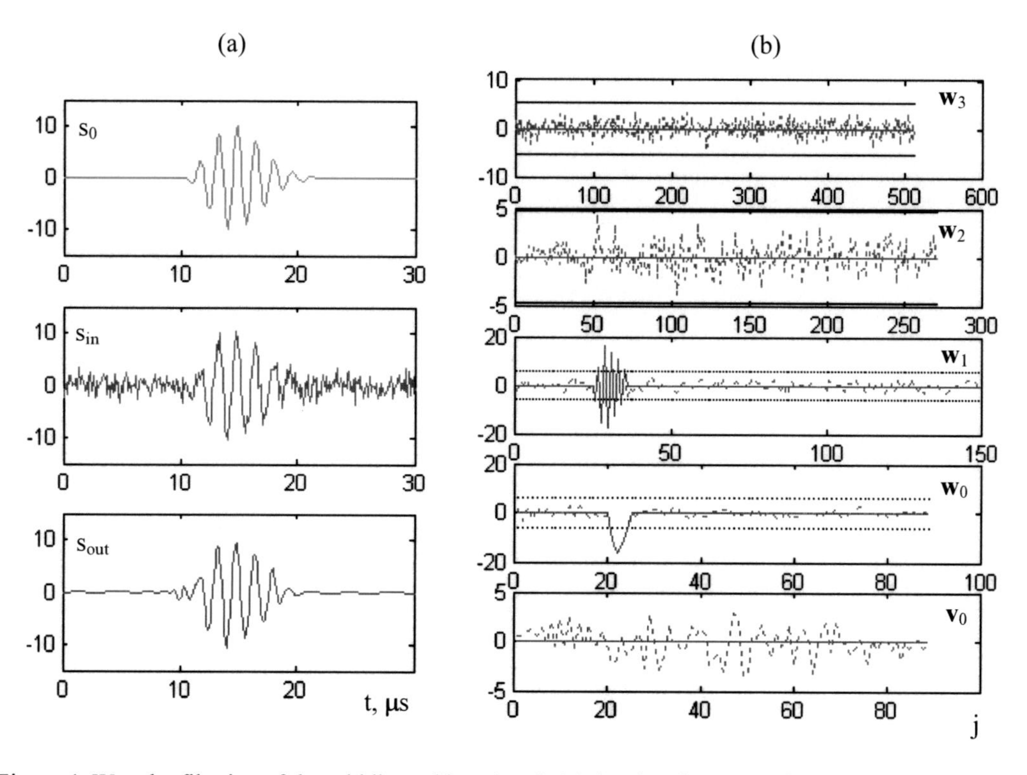

Figure 4. Wavelet filtering of the middle probing signal: (a) the signals s_0, s_{in} and s_{out}; (b) the scaling and wavelet coefficients at different decomposition levels before (red) and after (blue) thresholding; the horizontal black lines show the thresholding levels. The parameters: L = 4; FWT basis – sym8; hard thresholding; signal-to-noise ratio – 16.5 dB.

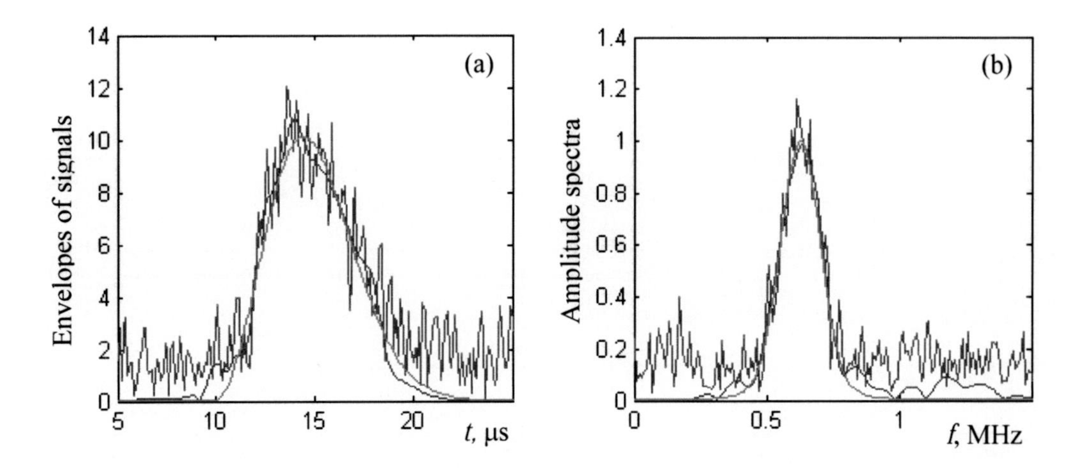

Figure 5. Wavelet filtering of the middle probing signal: (a) the envelopes of the signals; (b) the amplitude Fourier spectra of the signals. The colors correspond to the signals: s_0 (green), s_{in}, (red) and s_{out} (blue).

Let us apply the wavelet filtering procedure to the sum of the middle probing signal at the carrier frequency of 650 kHz and white Gaussian noise. We choose the parameters of the procedure as follows: the number of decomposition levels L = 4; FWT basis: sym8; hard thresholding; the amplitude of the reference signal A_0 = 1; the RMS variation of the noise amplitude $\sigma = 1.5$; signal-to-noise ratio: 16.5 dB. In order to eliminate completely low-frequency oscillations in the signal s_{out} at the output of the wavelet filter, we set $\mathbf{v}_0 = 0$. From here on we will use these parameters elsewhere apart from some specially mentioned cases. The process of the wavelet filtering of the middle probing signal is shown in Figure 4.

In Figure 5 we compare the envelopes of the probing signal and signals before and after wavelet filtering as well as their amplitude Fourier spectra. These pictures allow everyone to estimate visually the wavelet filtering quality.

An example of applying the wavelet filtering to an experimentally obtained echo-signal, reflected from side-drilled hole in a steel block [9], is shown in Figure 6. The diameter of the hole equals 6 mm. The carrier frequency of longitudinal wave is 5 MHz. The experimental studies were performed with the use of an ultrasonic-microprocessor-based flaw detector built on the basis of the PCUS-10 system [8] developed at the Fraunhofer Institute for Nondestructive Testing (Germany). This instrument is built in the form of an ISA-standard board, which is installed in the appropriate slot of a PC. Transducers are connected to its external connectors. Such a system features the capabilities of a conventional ultrasonic flaw detector and can additionally employ the computer for storing and processing the data obtained. The PCUSware™ program is used to control the operation of the device.

Other important characteristics of an acoustic measuring procedure, in addition to the echo-signal amplitude, are the temporal characteristics that affect the instrument's resolution and the accuracy of reflector localization, namely, $\Delta\tau_0$, the time shift of the signal peak after filtering, and $\Delta\tau_e$, the change in the signal edge width due to filtering. The analysis of the mathematical model of the wavelet filtering indicates that the maximal value of $\Delta\tau_0$ is within

0.15 μs and the mean value is lower than 0.1 μs. On the other hand, $\Delta\tau_e$ is no higher than 0.1 μs. Thus, both these parameters remain after filtering within the ranges admissible for ultrasonic testing [8].

Figure 7 illustrates [8] how the time resolution of the wavelet filtering changes as a result of overlapping of two middle testing signals with an amplitude ratio 1:0.7. The picture shows that the signal s_{out} separated from intense noise (the signal-to-noise ratio is –1:1 dB) has the shape that is, by and large, similar to that of the reference signal. The two peaks in the signal are clearly resolved. The change $\Delta\tau_e$ is negligible in comparison with the edge width of the reference signal.

The process of wavelet denoising of the pair of the probing signals [10] is displayed in Figure 8. The ratio of their amplitudes equals to 1:0.4; $F = 0.125$; $\alpha = 0.0005$. The signal-to-noise ratio that is the total power of two narrow-band signals divided by the power of the additive noise is here equal to -1.9 dB. The signals at the output of the Hamming FIR filter are also demonstrated in Figure 8 for comparison. The order of the FIR filter is chosen to be equal to 100 and its pass band is about 20 percent wider than the spectral width of the pair of the probing signals. The parameters of the wavelet filter: L = 4; FWT basis – sym8; hard thresholding.

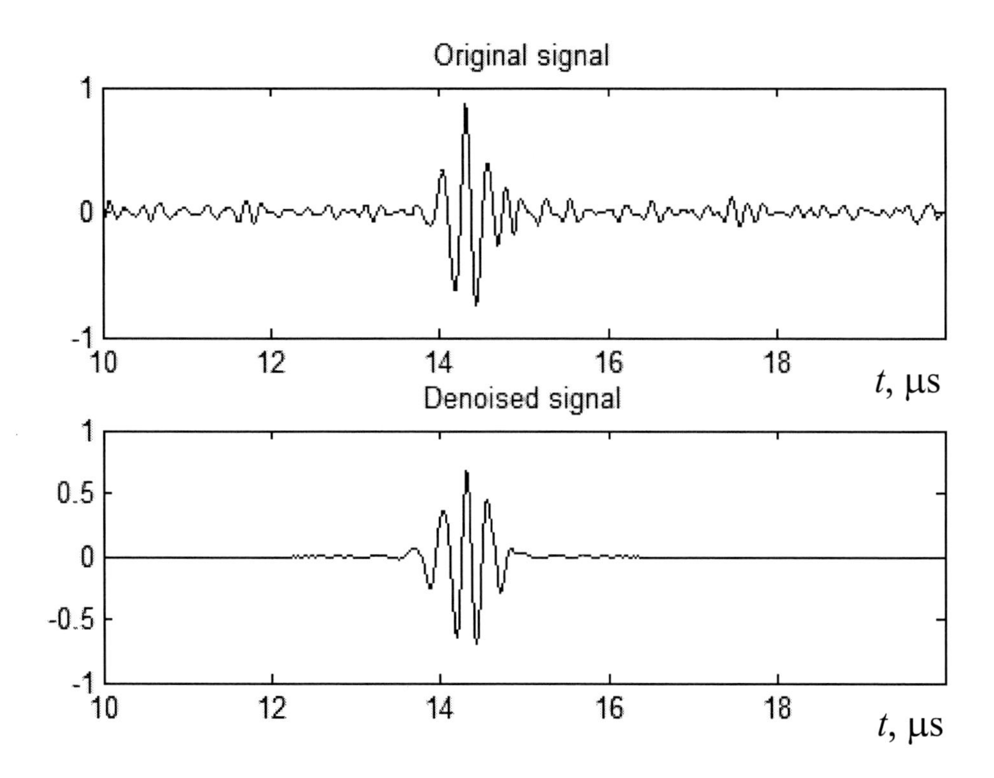

Figure 6. Wavelet filtering of the experimentally obtained echo-signal reflected from the side-drilled hole. The parameters: L = 4; FWT basis – sym10; hard thresholding.

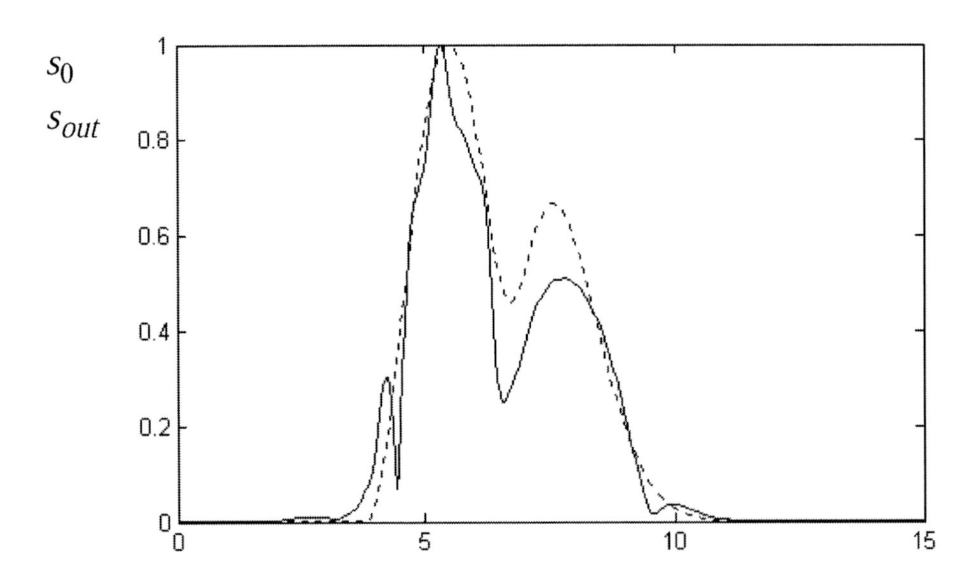

Figure 7. Curves illustrating the resolution of the wavelet filtering for two overlapping signals with the ratio between their amplitudes 1:0.7: the dashed line corresponds to the envelope of the probing signal; the solid line to the envelope of the signal after filtering using the sym15 wavelet, the number of decomposition levels is four; $\alpha = 0{:}0005$, $F_0 = 0{:}0625$, the signal-to-noise ratio is –1:1 dB.

Consider the effect of the wavelet filtering procedure on the parameters of denoised signal. In many cases, the most important of them is the accuracy of estimation of the echo-signal amplitude. It is characterized by the relative error δA of the signal amplitude after filtering with respect to the reference signal amplitude [8]:

$$\delta A = \left|A_{out} - A_0\right| / A_0 \quad . \tag{16}$$

First, we estimate δA using the fast wavelet transform and applying the hard thresholding to fulfill the wavelet filtering in accordance with Eq. (14). The inputs s_{in} of the wavelet filter are special testing signals which are sums of probing signals described by Eq. (15) and the white normally distributed noise produced by the algorithm generating sequences of random numbers.

Figure 9 plots relative errors in the probing signal amplitudes after wavelet filtering. Because δA is a random number, it may be characterized by its mean value $< \delta A >$ as well as standard deviation $\Delta(\delta A)$ which are obtained by averaging over a hundred independent samplings of random numbers for each normalized frequency. In order to obtain more accurate estimates of the average value δA, the delays for the reference signals are set differently for each sampling since the relative errors may differ considerably for various positions of the same signal on the time axis [1].

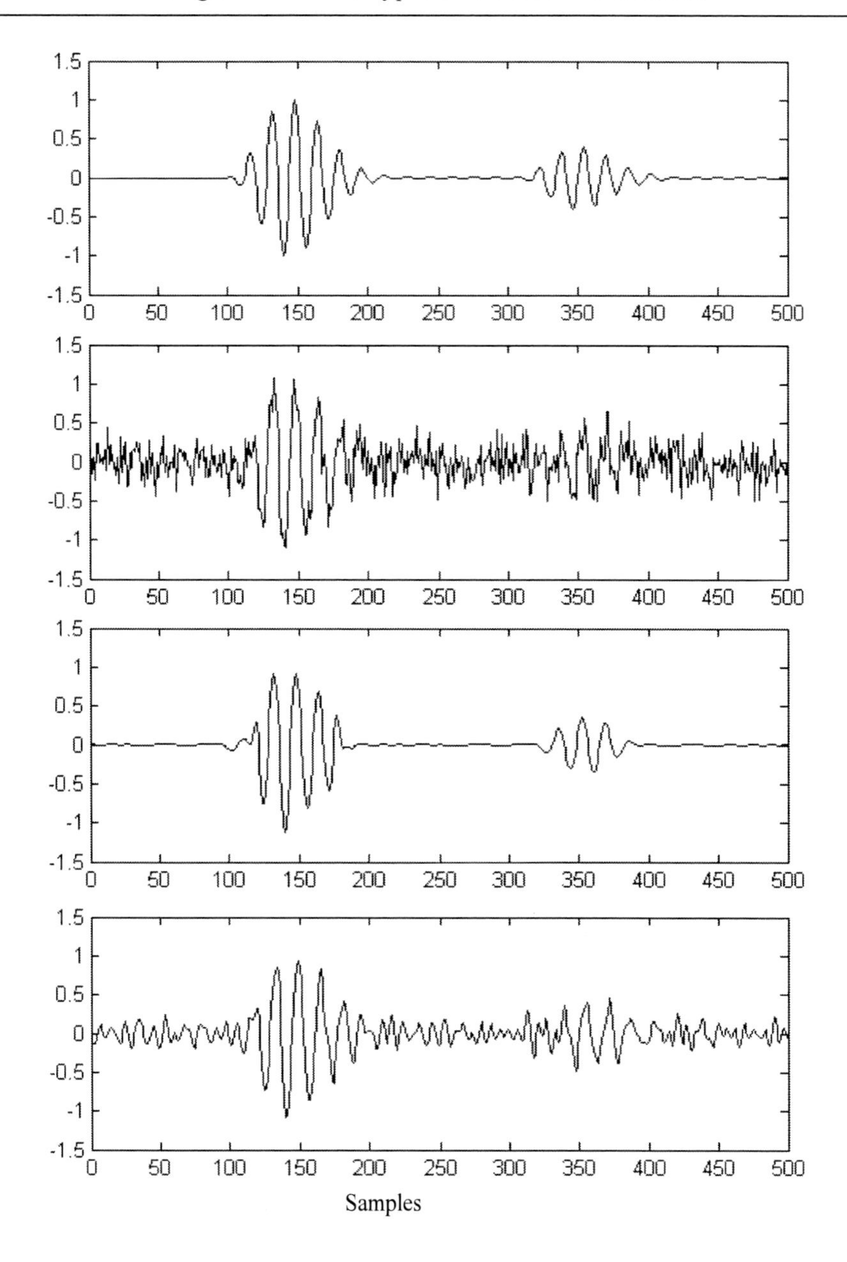

Figure 8. Denoising of the probing signals using FWT: (a) the pristine probing signals; (b) the noised probing signals; (c) the signals after wavelet denoising; (d) the signals at the output of the Hamming FIR filter.

The curves given in Figure 9 indicate that δA has local minima near the dyadic points, where the normalized frequency F equals 0.0625, 0.125 and 0.25. Note that the errors at $F \approx$ 0:25 are remarkably larger than at the other two points. It is noteworthy that the regularity observed in these distributions, namely, the smaller relative error in the signal amplitude after filtering around the dyadic frequencies and the larger error on the intervals between these frequencies, is universal. This regularity manifests itself more or less strongly when different wavelets and signal durations are used.

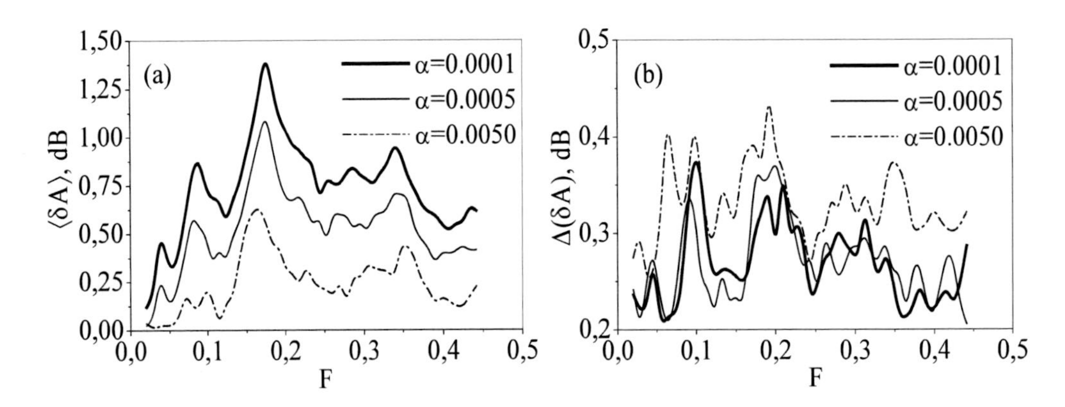

Figure 9. The relative error in the probing signal amplitudes after wavelet filtering using the FWT algorithm: (a) mean value; (b) standard deviation. The parameters: L = 4; FWT basis – sym8; hard thresholding; signal-to-noise ratio – 16.5 dB.

Since the normalized frequency F is related to frequency f by the formula $f = F f_{\max}$, and $f_{\max} = 1/\Delta t$, then $f = F/\Delta t$. On the other hand, the duration T of a signal to be analyzed is given by the formula $T = N\Delta t$. The duration T and the number N of elements in a sampling can be set by an operator. First he has to select the required signal duration T and then, given the rated resonant frequency f_0 of the transducer and using the obvious relation $N = fT/F$, to determine N corresponding to F = 0:125. For example, for f = 5 MHz and T = 20 μs the optimal dimension of the sampling is N = 800.

An important point in the wavelet filtering procedure is the correct selection of the order of the wavelet used in filtering and the number of decomposition levels in the FWT scheme [8]. As was noted above, the dimension of the region where the scaling functions and wavelets are localized is directly proportional to the wavelet order M. For the wavelet families db and sym it is $2M$–1. Consequently, as M grows, the wavelets become longer, which has a detrimental effect on the resolution of the algorithm based on these wavelets for signals like those shown in Figure 7. On the other hand, the widths of the impulse response characteristics of QMFs are also proportional to M: for the wavelet families db and sym it is $2M$. This means that, when the wavelet order is too low, the filters have relatively flat frequency characteristics around the dyadic frequencies, which degrades the quality of the wavelet filtering. Another detrimental effect of the low orders is in the presence of discontinuity points in the first derivatives of wavelets of families db and sym [1], as a result, the signals contains quite large oscillations after the wavelet filtering.

The relative errors in the amplitudes of the testing signals of different lengths after the wavelet filtering against the order of wavelets of sym family are plotted in Figure 10. The normalized carrier frequency F_0 is 0.125. The curves in Figure 10 lead us to the conclusion that the errors are approximately constant for M >5. Basing on the limitations on the wavelet order discussed previously, we recommend for filtering radio-frequency pulses the wavelets of the sym family with M ranging from 8 to 15. This recommendation is also valid for the db family.

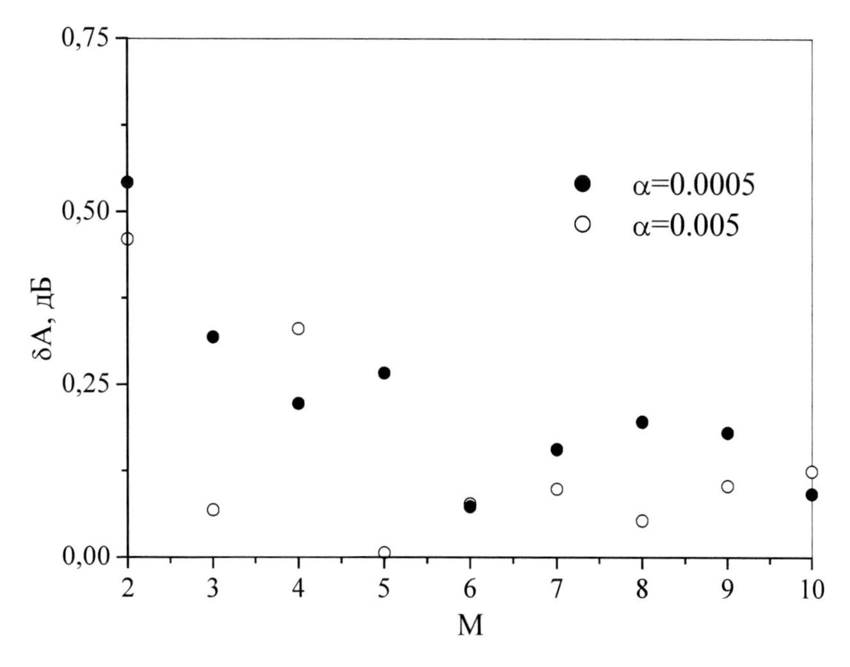

Figure 10. The relative error in the probing signal amplitudes after wavelet filtering using the FWT algorithm versus the order of wavelets of sym family. The parameters: $L = 4$; hard thresholding; signal-to-noise ratio – 16.5 dB; $F_0 = 0.125$.

Now let us discuss the selection of the number L of the decomposition levels. It seems obvious that the filtering quality should improve with L since at larger L the wavelet coefficients of the noise are distributed over larger numbers of levels than are the coefficients of desired signals. A look at Figure 11, however, is enough to make it clear that, starting with certain L, the number of decomposition levels does not affect the quality of the wavelet filtering. At the same time, the number of operations of addition and multiplication of real numbers prescribed by the FWT and IFWT is proportional to $L \times M \times N$. Therefore, the correct decision is to select a reasonable value of L, alongside that of M, at which the quality of the wavelet filtering sufficient for practical needs is achieved.

The relative errors in the amplitudes of testing signals of various lengths after the wavelet filtering as functions of the number of decomposition levels with sym8 wavelets are shown in Figure 11. The normalized carrier frequency F_0 is set to 0.125. The results of Figure 11 lead to the conclusion that the error varies little for $L \geq 3$. Consequently, when sym8 wavelets are used, the value of L equal to four is sufficient for filtering radio-frequency signals. This conclusion is also valid for other wavelets of families sym and db.

Note also that the methods of statistical averaging used in obtaining the data plotted in Figs. 10 and 11 are the same as those used in calculating the points for Figure 9.

Equation (14) was obtained for white noise. But a question arises about the efficiency of the filtering procedure for noise with a different spectrum. As an example of a colored noise, we can consider noise with the spectral power density of $1/f$. Such a noise can be acoustic in nature; i.e., it can arise in the medium through which the wave travels. Apart from the acoustic noise, there may be electric noise and interferences whose spectral density maximum is displaced from the maximum of the acoustic signal spectrum. The action of such interference within the bandwidth of receiver can be modeled by colored noise. As it was shown in [11], applying the thresholding procedure (14) to the sum of an acoustic signal and $1/f$ noise give quite satisfactory results.

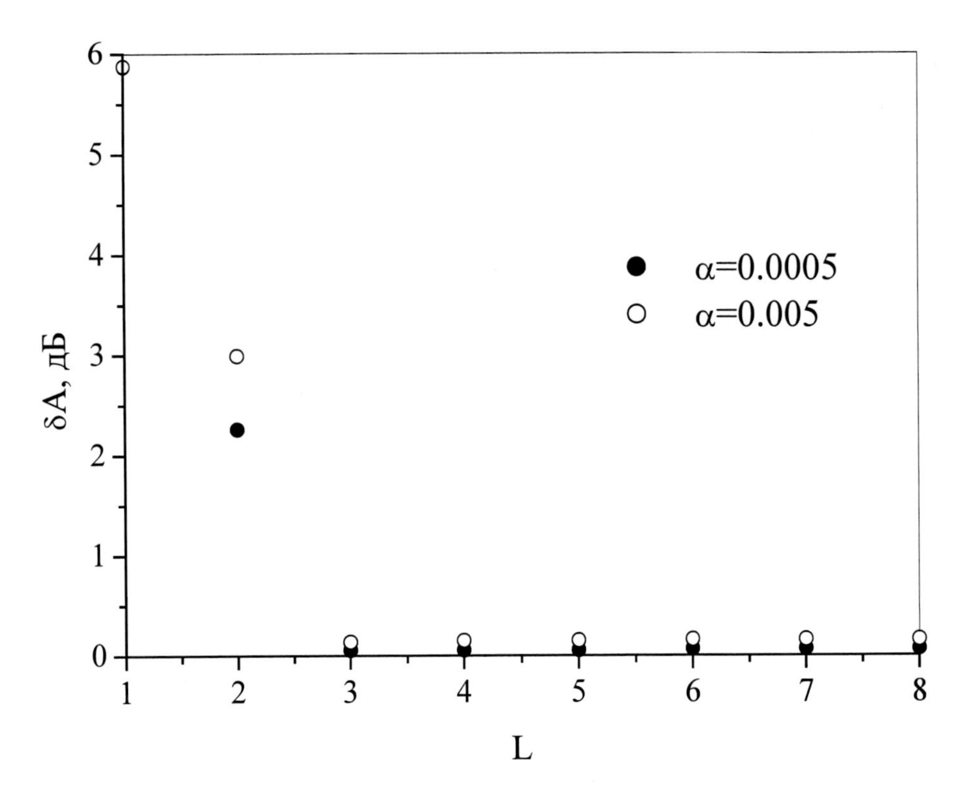

Figure 11. The relative error in the probing signal amplitudes after wavelet filtering using the FWT algorithm versus the number of decomposition levels. The parameters: FWT basis – sym8; hard thresholding; signal-to-noise ratio – 16.5 dB; $F_0 = 0.125$.

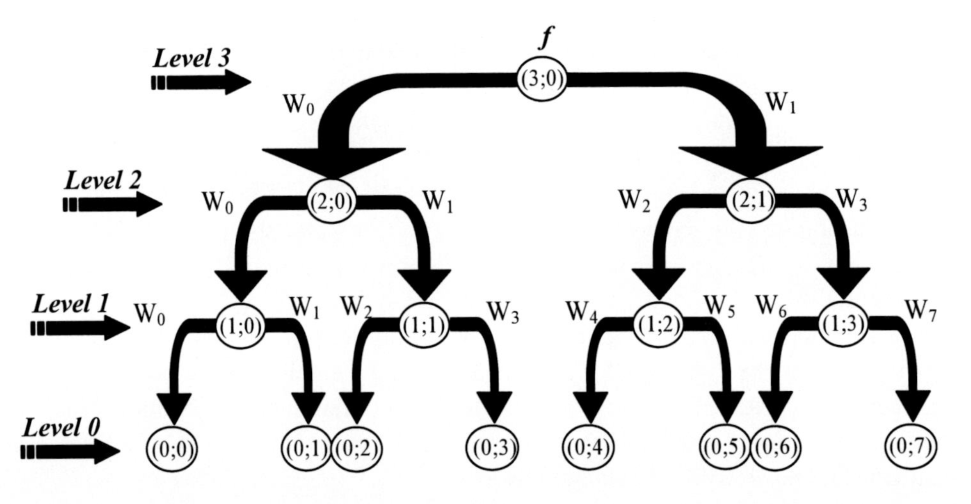

Figure 12. The wavelet packet decomposition tree illustrating the process of decomposition of an analyzed signal on four decomposition levels.

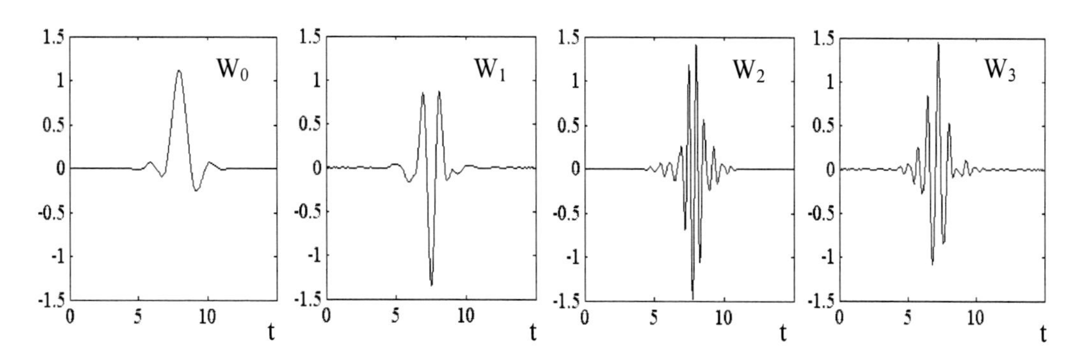

Figure 13. Four wavelet packet functions obtained from the sym8 scaling function and wavelet.

Filtering with the Use of Special Types of the Discrete Wavelet Transform: Stationary Wavelet Transform and Wavelet Packets

De-noising with the FWT sometimes exhibits visual artifacts; as an example of these, it is Gibbs phenomena in the neighborhood of discontinuities resulting from the lack of translation invariance of the wavelet basis. One method to suppress such artifacts, suggested by Coifman [12], is to average some slightly different FWT, called ε-decimated FWT, to define the stationary wavelet transform (SWT). This procedure allows to average out the translation dependence and considerably suppress the spikes arisen due to the Gibbs phenomenon. Furthermore, the SWT of a translated version of an analyzed signal is the translated version of SWT of the signal. It is reasonable that the inverse stationary wavelet transform (ISWT) can be fulfilled similarly to the IFWT. The SWT algorithm uses the same wavelet decomposition tree as the FWT but its computational complexity is proportional to $N \log_2 N$ instead of N as with the FWT. Furthermore, there is a restriction: the SWT can be only defined for signals of length divisible by $2^{i_{\max}-1}$.

The wavelet packet (WP) method is a generalization of the FWT that offers a richer range of possibilities for signal analysis [1, 2, 13]. In FWT or SWT, a signal is split into scaling and wavelet coefficients. The approximation is then itself split again into scaling and wavelet coefficients, and this process can be repeated over and over again. But in WP analysis, the wavelet coefficients as well as the scaling coefficients can be split. Thus the wavelet packet decomposition tree, which is shown in Figure 12, is more complicated than the wavelet decomposition tree corresponding to the FWT and SWT algorithms.

The computation scheme for wavelet packets generation is the easiest when using an orthogonal wavelet like a Daubechies wavelet or symlet. Then, to build a set of wavelet packet bases, we have to use two FIR filters with impulse response sequences h and g corresponding to the wavelet. These filters are the same which have been defined above; see, for example, Eqs. (7) and (8). Now by induction we can define the following sequence of functions by [2, 13]:

$$W_{2n}(t) = \sqrt{2} \sum_k h_k W_n(2t-k) \quad , \tag{17}$$

$$W_{2n+1}(t) = \sqrt{2} \sum_k g_k W_n(2t-k) \quad , \tag{18}$$

where $W_0(t) = \varphi(t)$ is the scaling function and $W_1(t) = \psi(t)$ is the wavelet; *n* denotes a node number at each level of the WP decomposition tree as it is shown in Figure 12. Here each node is defined as $(i;n)$ where *i* is the index of a decomposition level.

Using the functions defined by (17) and (18), we can obtain the set of orthonormal bases in the orthonormal space $L^2(\mathbf{R})$ in the same manner as the orthonormal bases $\{\varphi_{ij}\}$ and $\{\psi_{ij}\}$ have been obtained: see Eqs. (2) and (3). In Figure 13 are shown four functions obtained from the sym8 scaling function and wavelet using Eqs. (17) and (18). They correspond to the wavelet packet decomposition tree presented in Figure 12.

The wavelet packet decomposition gives a lot of bases from which one can look for the best representation with respect to a design objective. This can be done by finding the "best tree" based on an entropy criterion. For this purpose the Shannon entropy [2, 13] can be used that is defined by formula:

$$E_{(i;n)} = -\sum_j \left(w_j^{(i;n)}\right)^2 \ln\left(w_j^{(i;n)}\right)^2 , \tag{19}$$

where $w_j^{(i;n)}$ is the WP decomposition coefficient corresponding to the *n*th node on the *i*th decomposition level.

The optimal wavelet packet decomposition tree, from the minimal entropy criterion point of view, may be found as follows. Let us consider a node on the *(i+1)*th decomposition level. It has a pair of children nodes on the *i*th decomposition level. If sum of entropies for two children nodes is more than the entropy of the parent node, then this branch of the tree has to be truncated on the *(i+1)*th decomposition level. Such a parent node is named as the terminated node.

In the wavelet packet framework, denoising ideas are basically the same as those developed in the wavelet framework. The wavelet packet de-noising or compression procedure involves the next steps: 1) WP decomposition; 2) computation of the best tree; 3) thresholding of wavelet packet coefficients; 4) WP reconstruction which can be realized using the inverse algorithm that may be conceived as the motion along the tree shown in Figure 12, from its bottom to top; it is quite analogously to the IFWT or ISWT.

Thresholding of wavelet packet coefficients is rather different from this procedure corresponding to FWT or SWT. In that case, the threshold levels for the corresponding vectors of WP decomposition coefficients can be found in accordance with the expression [14]

$$T = \pm\,\mathrm{med}\left(\left|\mathbf{w}_{\cup}\right|\right) \frac{\sqrt{2\ln(N\log_2 N)}}{0.6745} \quad , \tag{20}$$

where $\mathbf{w}_{\cup}$ is the union of all the WP decomposition coefficient vectors with the exception of $w_j^{(i_0;0)}$.

The computational complexity of the WP decomposition is proportional to $N\log_2 N$, as for the SWT. But supplementary computing is necessary to estimate the entropy for each node of the WP decomposition tree and select the best tree; it requires about $N\log_2 N$ operations more.

Let us estimate the relative errors in the probing signal amplitudes after wavelet filtering using the SWT and WP algorithms. The results obtained are shown in Figs. 14 and 15.

Now we can compare the errors estimated for all the algorithms: FWT, SWT and WP. The appropriate curves, corresponding to the middle probing signal, are demonstrated in Figure 16. It follows from Figure 16a that positions of local minima of the curves are conditioned on the topology of appropriate decomposition trees corresponding to the appropriate type of discrete wavelet transform.

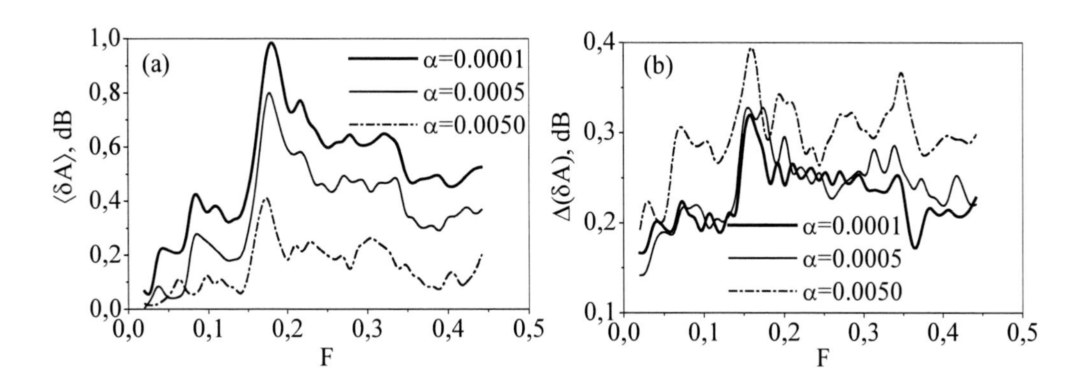

Figure 14. The relative error in the probing signal amplitudes after wavelet filtering using the SWT algorithm: (a) mean value; (b) standard deviation. The parameters: L = 4; SWT basis – sym8; hard thresholding; signal-to-noise ratio – 16.5 dB.

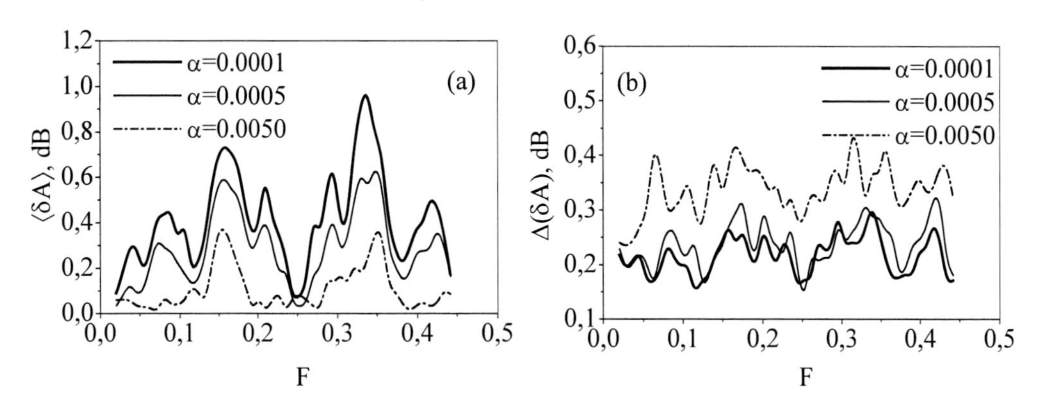

Figure 15. The relative error in the probing signal amplitudes after wavelet filtering using the WP algorithm: (a) mean value; (b) standard deviation. The parameters: L = 4; WP basis – sym8; hard thresholding; signal-to-noise ratio – 16.5 dB.

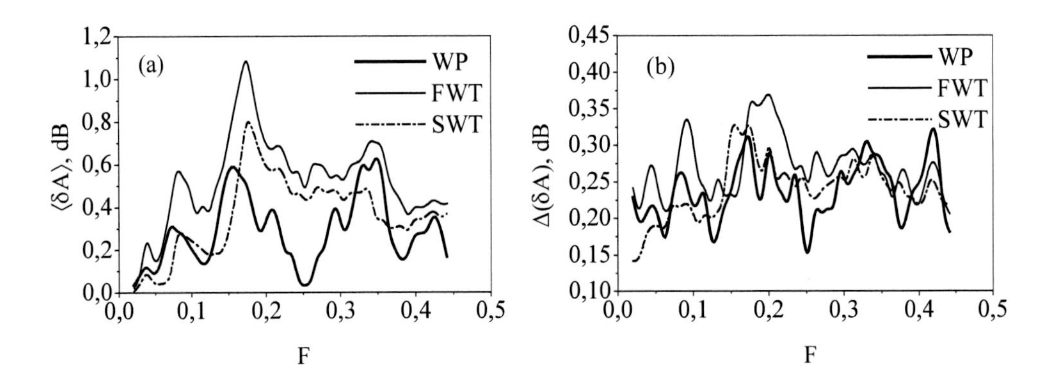

Figure 16. The comparison of the relative errors in the middle probing signal amplitude after wavelet filtering using all the algorithms: (a) mean value; (b) standard deviation. The parameters: $L = 4$; wavelet basis – sym8; hard thresholding; signal-to-noise ratio – 16.5 dB; $\alpha = 0.0005$.

The quality of the wavelet filtering algorithms can be also characterized by normalized RMS distance between the signals s_{out} and s_0, which is defined by the expression

$$\|D\| = \frac{\sum_j \left(s_{out\ j} - s_{0\ j}\right)^2}{\sum_j s_{0\ j}^2} \quad . \tag{21}$$

Because $\|D\|$ is a random number, it may be characterized by its mean value $< D >$ as well as standard deviation ΔD which are obtained by averaging over a hundred independent samplings of random numbers for each normalized frequency.

For the middle probing signal, the dependencies obtained by using Eq. (21) are shown in Figure 17. Comparing Figure 16 with Figure 17 we can conclude that the local minima of all the curves, obtained with the use of either Eq. (16) or Eq. (21), approximately correspond to the dyadic frequencies which have been defined above.

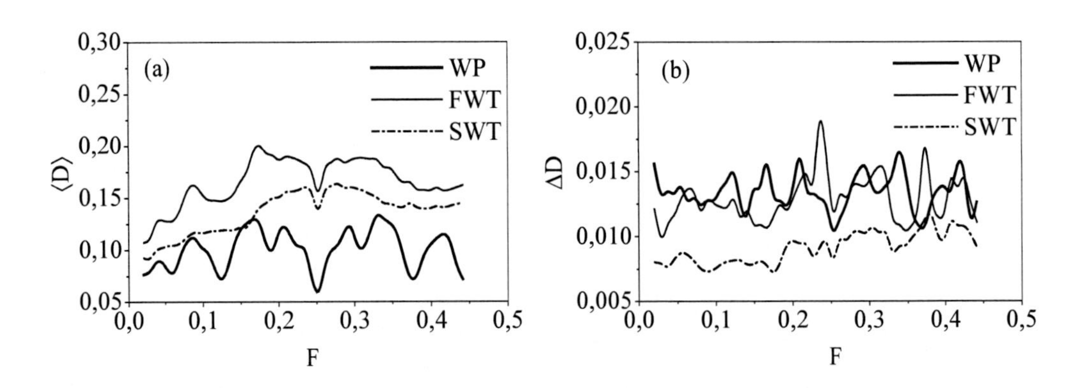

Figure 17. The RMS distance between the signal sout and middle probing signal s0 estimated using all the algorithms: (a) mean value; (b) standard deviation. The parameters: $L = 4$; wavelet basis – sym8; hard thresholding; signal-to-noise ratio – 16.5 dB; $\alpha = 0.0005$.

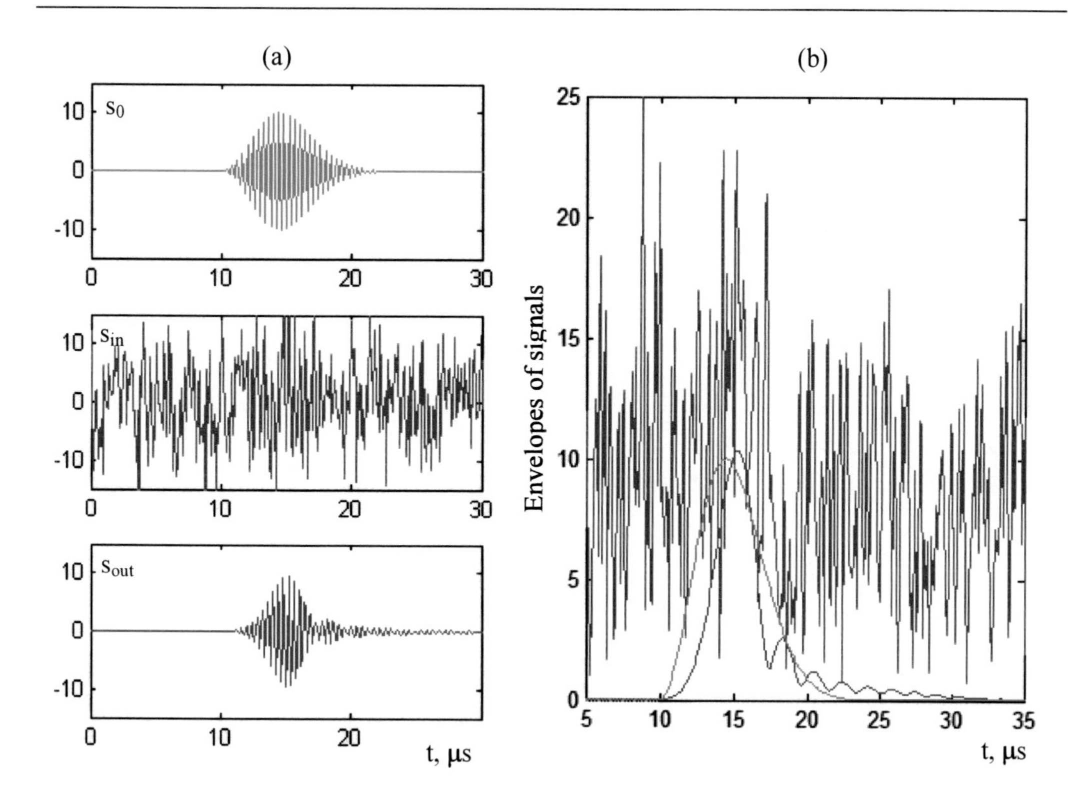

Figure 18. The separation of the middle probing signal from white Gaussian noise using the WP algorithm: (a) the signals s0, sin and sout; (b) the envelopes of the signals. The colors correspond to the signals: s0 (green), sin, (red) and sout (blue). The parameters: L = 4; WP basis – db30; hard thresholding; signal-to-noise ratio – 4.5 dB; $\alpha = 0.0005$; $F_0 = 0.5$.

In Figure 16 one can see that the error is minimal if the WP algorithm is used, because it is able to adapt itself to each analyzed signal. The outstanding ability of the WP algorithm clearly illustrates Figure 18 where the middle probing signal is well separated from tremendous noise.

Two-dimensional Fast Wavelet Transform

In many problems it is necessary to process two-dimensional data arrays. Therefore, it is natural to formulate the problem of generalization of the concept of the discrete wavelet analysis to include the case of the function of two variables.

The very simple and widespread method for solving this problem consists in the choice of the father scaling functions and mother wavelets of the form [1, 2]

$$\varphi_{imn}(t) = 2^i \varphi\left(2^i t - m\right) \cdot \varphi\left(2^i t - n\right) \quad , \tag{22}$$

$$\psi_{imn}^{H}(t) = 2^i \varphi\left(2^i t - m\right) \cdot \psi\left(2^i t - n\right) \quad , \tag{23}$$

$$\psi^{V}_{imn}(t) = 2^{i}\psi(2^{i}t - m)\cdot\varphi(2^{i}t - n) \quad , \tag{24}$$

$$\psi^{D}_{imn}(t) = 2^{i}\psi(2^{i}t - m)\cdot\psi(2^{i}t - n) \quad , \tag{25}$$

where H, V and D denote the horizontal, vertical and diagonal two-dimensional mother wavelets, respectively; and $\varphi(t)$ and $\psi(t)$ are the one-dimensional father scaling functions and mother wavelets; *i*, *j* and *m* are integers.

The two-dimensional FWT and IFWT algorithms can be realized by using the corresponding one-dimensional algorithms, that have been described above, with respect to the rows and columns of the number arrays in the form of the next procedures [1, 2].

At the *i*th wavelet-decomposition level the two-dimensional FWT is carried as follows:

1. Two-dimensional array of the wavelet coefficients $\mathbf{v}_{i+1}$ is supplied along the rows to the inputs of the QMFs with impulse response sequences *h** and *g**;
2. At the outputs of the filters the columns of both arrays are decimated by a factor of two;
3. Each array obtained is once again supplied along the columns to the input of the QMFs with impulse response sequences *h** and *g**;
4. At the outputs of the filters the rows of the four arrays are decimated by a factor of two.

Thus, from the array $\mathbf{v}_{i+1}$ we obtain the following two-dimensional number arrays:

1. After the twofold use of the filter *g** we obtain the scaling coefficient array $\mathbf{v}_i$;
2. After the consecutive use of the filters *g** and *h** we obtain the horizontal wavelet coefficient array $\mathbf{w}_i^H$;
3. After the consecutive use of the filters *h** and *g** we obtain the vertical wavelet coefficient array $\mathbf{w}_i^V$;
4. After the twofold use of the filter *h** we obtain the diagonal wavelet coefficient array $\mathbf{w}_i^D$.

If the array $\mathbf{v}_{i+1}$ has $N{\times}M$ size, then the arrays $\mathbf{v}_i$, $\mathbf{w}_i^H$, $\mathbf{w}_i^V$ and $\mathbf{w}_i^D$ have $(N/2){\times}(M/2)$ size.

Realization of the two-dimensional FWT algorithm in accordance with formulas (11) and (12) can be schematically presented as the wavelet decomposition tree one level of which is shown in Figure 19.

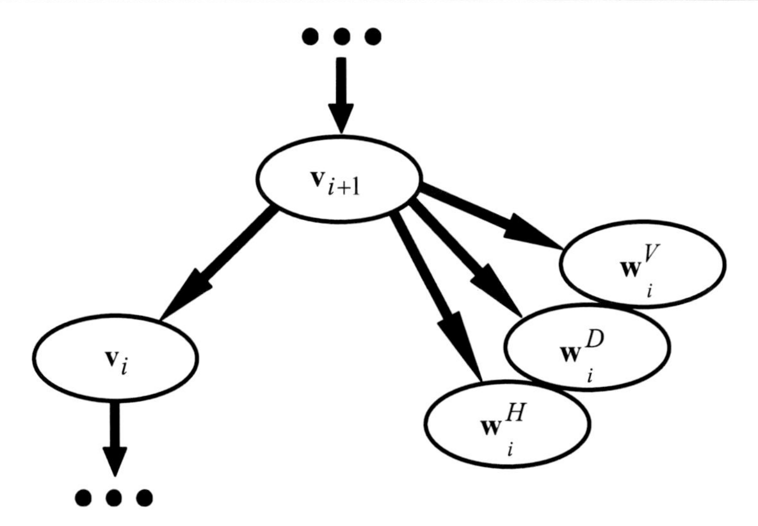

Figure 19. One level of the wavelet decomposition tree illustrating the process of an array expansion into scaling and wavelet coefficients using the two-dimensional FWT algorithm.

At the *i*th wavelet-reconstruction level the two-dimensional IFWT is carried as follows:

1. The length of rows of the two-dimensional arrays $\mathbf{v}_i$, $\mathbf{w}_i^H$, $\mathbf{w}_i^V$ and $\mathbf{w}_i^D$ is increased twice due to insertion of zero values between the elements of rows of the input arrays;
2. The arrays $\mathbf{v}_i$, $\mathbf{w}_i^H$ and $\mathbf{w}_i^V$, $\mathbf{w}_i^D$ completed by the zeros are supplied in pairs to the inputs of the QMFs with impulse response sequences *h* and *g*;
3. At the inputs of the filters the arrays are also summed in pairs. The height of the columns of the two arrays obtained is increased twice due to insertion of zero values between the elements of the columns of the input arrays;
4. The array obtained from $\mathbf{v}_i$ and $\mathbf{w}_i^H$, is supplied to the low-frequency QMF input and the array obtained from $\mathbf{w}_i^V$ and $\mathbf{w}_i^D$ to the high-frequency QMF input.

Thus, from the arrays $\mathbf{v}_i$, $\mathbf{w}_i^H$, $\mathbf{w}_i^V$ and $\mathbf{w}_i^D$ of size $(N/2)\times(M/2)$ we obtain the scaling coefficient array $\mathbf{v}_{i+1}$ of size $N\times M$.

For an array of size $N\times N$ the number of computational operations which are necessary to realize the two-dimensional FWT and IFWT is proportional to N^2 instead of $N^2 \log_2 N$ for the two-dimensional fast Fourier transform [1, 2, 5].

The WSD procedure may be applied to a two-dimensional data array in the same way as it has been described above for one-dimensional signals. Appropriate formula for the fixed-form thresholding strategy can be written as follows:

$$T_i = \pm \operatorname{med}\left(\left|\mathbf{w}_i\right|\right)\frac{\sqrt{2\ln(MN)}}{0.6745}, \qquad (26)$$

where i is the decomposition level number; $\mathbf{w}_i = \mathbf{w}_i^H \cup \mathbf{w}_i^V \cup \mathbf{w}_i^D$; N and M are the dimension sizes of an analyzed array. In principle, thresholding can be applied to any one array $\mathbf{w}_i^H$ or $\mathbf{w}_i^V$ or $\mathbf{w}_i^D$ if you want to single out the appropriate peculiarities of an image [2].

Evolution of contemporary acoustic experimental methods has been mainly developed on the basis of comprehensive investigation of the structure of acoustic fields. One of the most promising methods of investigation of acoustic fields is based on using the Doppler laser interferometer. It features a set of unique properties including high spatial and temporal resolution and the capacity to detect ultrasonic signals of different types and with different frequencies and the possibility to visualize the spatio-temporal structure of field.

As an example we consider the experimental assembly using OFV-3001/OFV-302 Doppler laser interferometer produced by Polytec [16, 17, 18]. A coherent optical beam produced by a helium–neon laser is split into two parts. One part is passed through an acousto-optical modulator (Bragg cell); the second is guided to the object under study through a system of optical lenses. The beam reflected from the object is applied to a photodetector, where interference with the reference beam, which has passed through the Bragg cell, is observed.

If the reflecting surface of the object is illuminated by an ultrasonic wave, components shifted in frequency due to the Doppler effect appear in the spectrum of the reflected signal. A 1-m/s velocity of the object produces a 3.17 MHz Doppler shift. In a harmonic elastic wave, the particle velocity is proportional to the amplitude of the displacement. Therefore, the amplitude distribution of the elastic displacement can be obtained by scanning the object's surface and measuring the particle velocity.

The output signal of the interferometer is recorded and converted into a code by a digital oscilloscope. A scanner moved the sample in the vertical and horizontal directions. The typical scanner increment is 0.125 or 0.25 mm. The equipment is capable of accumulating the signal in order to improve the signal-to-noise ratio. At each point, the signal is averaged over 50 to 300 measurements. Results of measurements were stored by a computer program based on the LabView 5.0 software. The time dependence of the response amplitude at each point of the scanned field, i.e., the A scan, contained 1000 discrete signal samples. The scanned area was 14×14 mm. After the scanning, information about the spatial distribution of the particle velocity on the surface of specimen can be presented as [19]: (1) B scan, i.e., a two-dimensional diagram that combines type A scans along a chosen spatial direction; (2) C scan, i.e., an image which demonstrates the two-dimensional spatial distribution of the maxima of A scans within the selected time gate. The B scans can be used to visualize the shape of the wave front. It can also be useful to optimally choose the time interval for obtaining the C scan.

An example of two-dimensional wavelet filtering applied to the B scan image is shown in Figure 20. The B scan is obtained for the specimen representing a single crystal in the form of a cylinder 20 mm in diameter and 12 mm long made of 60N21 steel [20]. The end edge surfaces are parallel to the crystallographic plane [111]. A longitudinal elastic wave excited by a piezoelectric transducer with a resonance frequency of 5 MHz propagates along the third-order axis of symmetry of the crystal which is orthogonal to the plane [111]. In the image of the B scan, the numbers on the abscissa axis are the sample indexes corresponding to various instants of time and the numbers on the ordinate axis correspond to the various values of one of the space coordinates. In this case the other space coordinate is fixed. In the upper part of Figure 20a we can see horizontal strips induced by the presence of high-frequency noises which make difficult the visual image sensing. Such noise can appear, for example, when the laser beam falls on a high-roughness surface section. After application of the wavelet filtering procedure using the sym4 wavelet with four decomposition levels we obtained the result shown in Figure 20b. Clearly, the image is much convenient for visual sensing.

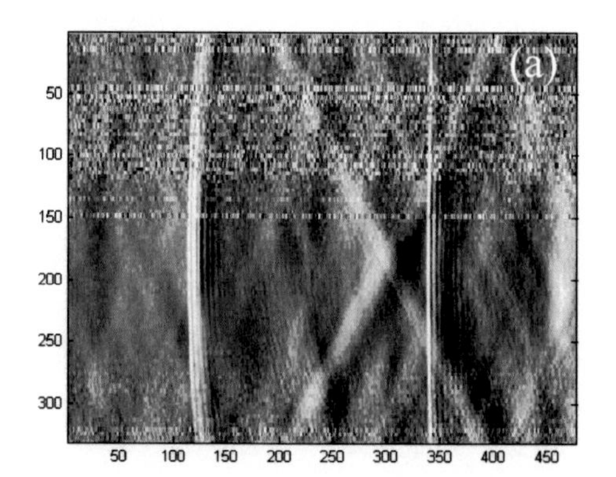

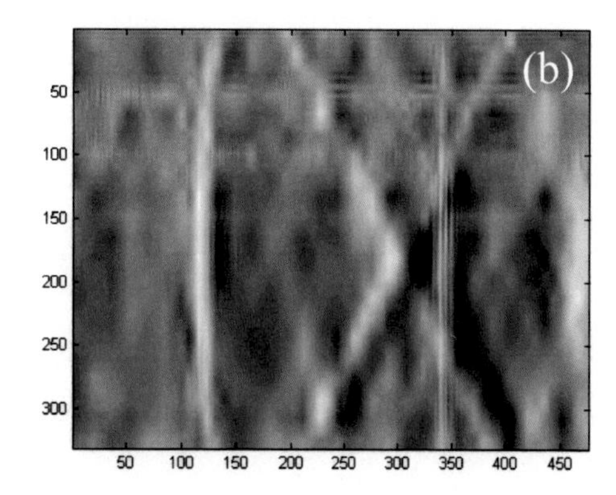

Figure 20. Two-dimensional wavelet filtering of the B scan image: (a) original image; (b) denoised image.
The parameters: $L = 4$; FWT basis – sym4; soft thresholding.

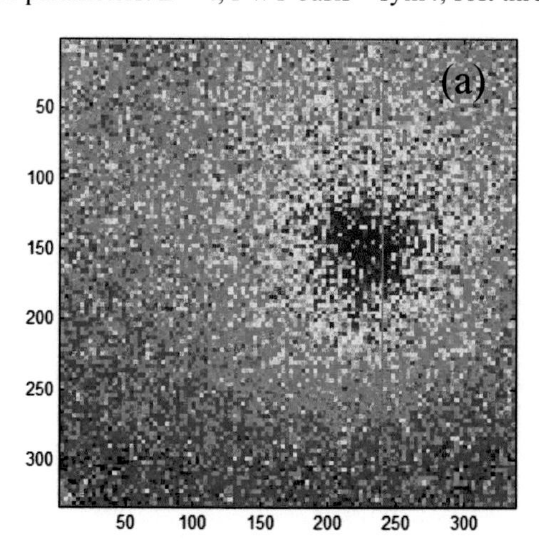

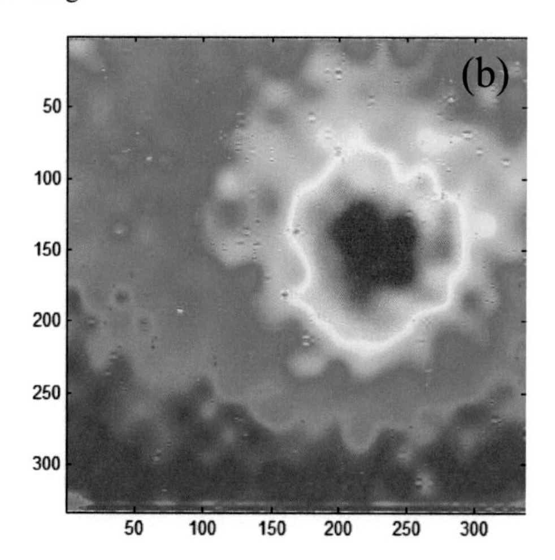

Figure 21. Two-dimensional wavelet filtering of the C scan image: (a) original image; (b) denoised image.
The parameters: $L = 4$; FWT basis – sym8; soft thresholding.

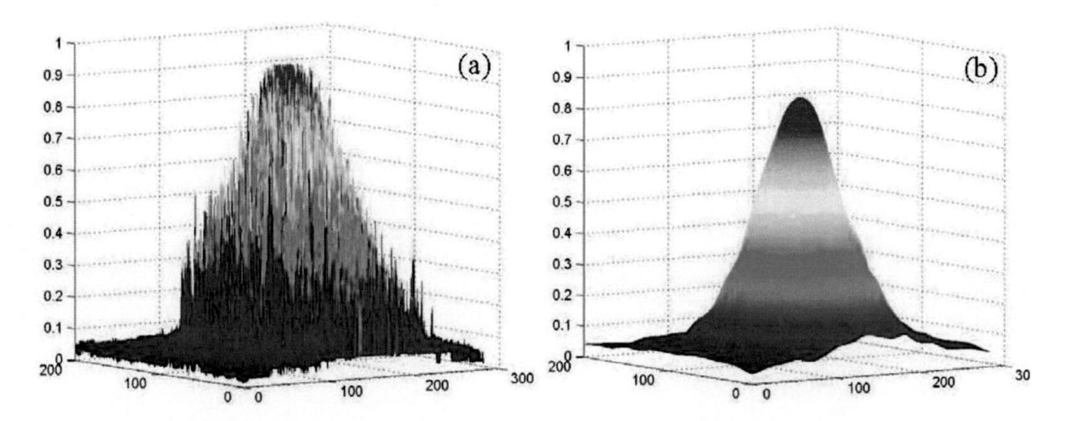

Figure 22. Two-dimensional wavelet filtering of the C scan image presented as three-dimensional surface plot: (a) original image; (b) denoised image. The parameters: $L = 4$; FWT basis – sym8; soft thresholding.

The images of the C scan obtained with the use of the same specimen [16] before and after two-dimensional wavelet filtering are shown in Figure 21. The numbers on the axes correspond to the various values of two spatial coordinates.

One more similar example is shown in Figure 22. Here we can see the C scan images presented as three-dimensional surface plots [18]. The numbers on the abscissa and ordinate axes correspond to the various values of two spatial coordinates.

It must be emphasized that it is preferably to use the soft thresholding strategy to process two-dimensional data arrays because, as it has been shown in [16], the hard thresholding produces a large number of clearly outlined local artifacts which affect the visual perception of the filtered image. However, filtering of one-dimensional signals has to be fulfilled using the hard thresholding because it allows estimate the amplitude of analyzed signal more precisely.

The SWT and WP algorithms can also be generalized for two-dimensional arrays [2].

Continuous Wavelet Analysis

Basic Principles

Let us consider the space $L^2(R)$ of functions $f(t)$ defined on the entire real axis $R(-\infty;+\infty)$ and having a finite norm $\|f\|$. Then the energy, i.e. square of the norm, has the form:

$$E = \|f\|^2 = \int_{-\infty}^{+\infty} |f(t)|^2 \, dt < \infty \quad , \tag{27}$$

and also is finite. The Fourier transform of this signal determines its spectrum which can be found as follows [21]:

$$F(\omega)=\int_{-\infty}^{+\infty} f(t)\exp(-i\omega t)\,dt \quad , \tag{28}$$

where ω is the angular frequency and $i=\sqrt{-1}$.

The spectral analysis based on using complex exponential $\exp(-i\omega t)=\cos(\omega t)-i\sin(\omega t)$ as a kernel of integral operator (28) makes it possible efficiently to analyze samplings of the data whose frequency composition does not change over the realization length. When using the Fourier transform (28) for analyzing nonstationary time dependent processes, we cannot obtain information concerning time variation of the spectral composition of the signals since integration is carried out over the entire time-realization length.

Clearly, the fact that we cannot trace time variations of the spectrum is due to using basis functions unbounded in time. The simplest method for distinguishing the domain in which the spectral analysis is carried out is to use the windowed (or short time) Fourier transform. In this case the spectrum of a signal $f(t)$ multiplied by the window function $z(t)$ which is nonzero on a finite time interval has the form [21]:

$$F_W(\omega,\theta)=\int_{t\in \mathrm{T}} f(t)\, z(t-\theta)\exp(-i\omega t)\,dt \quad , \tag{29}$$

where θ is the translation parameter and T is the value area of the time t, in which the integrand is nonzero. The rectangular window, i.e.

$$z(t)=\begin{cases}1 & t\in[-\tau/2;\tau/2]\\ 0 & t\in(-\infty;-\tau/2)\cup(\tau/2;+\infty)\end{cases} ,$$

is very simple and often used in practice. Here, τ is the window length.

The window Fourier transform has the following significant limitation. In the spectrum $F_W(\omega,\theta)$ defined by expression (29) we cannot separate contributions induced by the analyzing signal $f(t)$ and the window function $z(t)$. The spectral component $F(\omega_0)$ of the signal $f(t)$ corresponding to a certain frequency ω_0 will also be present in the spectrum $F_W(\omega,\theta)$ without distortions only if the condition $\tau >> 2\pi/\omega_0$ is fulfilled. However, in this case the time resolution of the windowed transform (29) can sharply deteriorate. This gives no way of detecting local changes in an analyzed process.

Consequently, using the window Fourier transform, for example, to clarify time-dependent variations of the signal against the background of broadband noises is quite difficult, since in this situation, using the information concerning $f(t)$ and $F(\omega)$, we cannot choose window parameters which are optimum from the standpoint of both minimization of the analyzing signal spectrum distortions and achievement of satisfactory time resolution. This problem is even more difficult for solving when the real time analysis of signals is fulfilled.

Thus, the problem is to find such a functional transform which would have the advantage of the windowed Fourier analysis, namely, the capability to localize finite intervals of the realization considered together with the possibility of scaling the basis functions. In this case conservation of self-similarity of the basis over the entire interval of scale variations is of particular importance since only in this case the results obtained by using different-scale basis functions can be compared between themselves correctly. The continuous wavelet analysis [1, 2] completely satisfies all these conditions.

The continuous wavelet transform (CWT) was firstly suggested by Grossmann and Morlet [22]. In this case the basis of the functional space $L^2(\mathbf{R})$ can be constructed using scaled and shifted versions of a single function, namely, the mother wavelet $\psi(t)$.

The wavelet spectrum, of the function $f(t) \in L^2(\mathbf{R})$ satisfying condition (27) can be determined from the relation [1, 2]:

$$W(s,\theta) = \frac{1}{\sqrt{|s|}} \int_{-\infty}^{+\infty} f(t)\psi^*\left(\frac{t-\theta}{s}\right) dt = \int_{-\infty}^{+\infty} f(t)\psi^*_{s\theta}(t)\, dt \quad , \tag{30}$$

where $\psi_{s\theta}(t) = \frac{1}{\sqrt{|s|}}\psi\left(\frac{t-\theta}{s}\right)$ are the basis functions of the CWT, s is the scaling coefficient, θ is the translation parameter; s and θ are real numbers. Here and in what follows, the asterisk denotes the complex conjugate. The multiplier $1/\sqrt{|s|}$ is introduced in order for all the wavelet functions of a single family to have the unit norm for any value of s. The scaling coefficient s shows the degree of increase ($s > 1$) or decrease ($s < 1$) of the length of a basis function on the time axis as compared with the mother wavelet $\psi(t)$. The translation parameter θ is equal to time at which the basis function is shifted along the time axis from the origin.

The expression for the wavelet spectrum can also be written in terms of the Fourier transforms of the analyzed function $f(t)$ and the wavelet $\psi(t)$ in the form [2]:

$$W(s,\theta) = \frac{\sqrt{|s|}}{2\pi} \int_{-\infty}^{+\infty} F(\omega)\Psi^*(s\omega)\exp(i\omega\theta)\, d\omega \quad , \tag{31}$$

where

$$\Psi(\omega) = \int_{-\infty}^{+\infty} \psi(t)\exp(-i\omega t)\, dt .$$

For the basis parameters s and θ, the inverse continuous wavelet transform (ICWT) can be written using the same basis $\psi_{s\theta}(t)$ as the CWT [2]:

$$f(t) = \frac{1}{C_\psi} \int_{-\infty}^{+\infty} \int_{-\infty}^{+\infty} W(s,\theta) \psi_{s\theta}(t) \frac{ds\, d\theta}{s^2} \quad , \tag{32}$$

where C_ψ is the normalization factor:

$$C_\psi = \int_{-\infty}^{+\infty} \frac{|\Psi(\omega)|^2}{|\omega|} d\omega = 2 \int_{0}^{+\infty} \frac{|\Psi(\omega)|^2}{\omega} d\omega \quad . \tag{33}$$

Here, $|\Psi(\omega)|^2 = \Psi(\omega) \cdot \Psi^*(\omega)$ is the power spectrum of the wavelet $\psi(t)$. It is well known [21] that the power Fourier spectrum always is an even function with respect to the frequency: $|\Psi(\omega)|^2 = |\Psi(-\omega)|^2$. Hence it follows the equality of integrals in formula (33).

We now enumerate the basic properties of the CWT [2].

1. Linearity. If we can represent a function $f(t)$ in the form of a superposition of functions $f_i(t)$ with constant coefficients α_i, i.e., $f(t) = \sum_i \alpha_i f_i(t)$, then its wavelet spectrum must obey the relation:

$$W[f(t)] = W\left[\sum_i \alpha_i f_i(t)\right] = \sum_i \alpha_i W[f_i(t)] = \sum_i \alpha_i W_i(s,\theta) \quad .$$

2. Invariance with respect to time shift: $W[f(t - t_0)] = W(s, \theta - t_0)$. In particular, hence it follows the commutative property of the wavelet transform with respect to differentiation, i.e.,

$$\frac{dW[f(t)]}{dt} = W\left[\frac{df(t)}{dt}\right] \quad .$$

3. Invariance with respect to change in the time scale, i.e., time expansion (or compression):

$$W\left[f\left(\frac{t}{s_0}\right)\right] = \frac{1}{s_0} W\left(\frac{s}{s_0}, \frac{\theta}{s_0}\right) \quad .$$

4. Differentiation of the wavelet transform:

$$W\left[\frac{d^m f(t)}{dt^m}\right] = (-1)^m \int_{-\infty}^{+\infty} f(t) \frac{d^m \psi_{s\theta}^*(t)}{dt^m} dt \quad .$$

5. Parseval equality:

$$\int_{-\infty}^{+\infty} f_1(t) f_2^*(t)\, dt = \frac{1}{C_\psi} \int_{-\infty}^{+\infty} \int_{-\infty}^{+\infty} W_1(s,\theta) W_2^*(s,\theta) \frac{ds\, d\theta}{s^2} \quad .$$

We will here restrict ourselves to examination of one real basis only that is constructed from the second derivative of the Gaussian function [2]. This is so-called MHAT wavelet, or the "Mexican hat", which is determined by the expression:

$$\psi(t) = \frac{2}{\sqrt{3\sqrt{\pi}}} \left(1 - t^2\right) \exp\left(-\frac{1}{2} t^2\right) \quad . \tag{34}$$

The Fourier spectrum of wavelet (34) is

$$\Psi(\omega) = \sqrt{\frac{8\sqrt{\pi}}{3}}\, \omega^2 \exp\left(-\frac{1}{2}\omega^2\right) \quad . \tag{35}$$

The MHAT wavelet is well-localized in both time and frequency domains; it has two zero moments (zeroth and first), see Eq. (10). This wavelet is well suited to analyze complex signals.

In Figure 23 we plot basis functions $\psi_{s\theta}(t)$ of a MHAT wavelet constructed for various values of the scaling coefficient s at $\theta = 0$ and the corresponding Fourier spectra obtained using formulas (34) and (35).

The normalizing coefficient C_ψ for the MHAT wavelet can be founded by substituting Eq. (35) in Eq. (33):

$$C_\psi = \frac{16\sqrt{\pi}}{3} \int_0^{+\infty} \omega^3 \exp\left(-\omega^2\right) d\omega = \frac{8\sqrt{\pi}}{3} \quad .$$

The wavelet spectrum $W(s,\theta)$ of a one-dimensional signal represents a surface in three-dimensional space. There are different methods for visualizing such surfaces. Instead of presenting these surfaces themselves, one often resorts to presenting their projections on the plane (s,θ) by drawing isolines or colored images which enable the changes in the wavelet amplitude intensity to be followed as functions of scaling coefficient and translation parameter.

Let us show some examples of wavelet spectra corresponding to the signals which can be often watched in acoustics experimentally. In Figure 24 the wavelet spectrum of unmodulated cosine wave with 0.5 MHz carrier frequency is shown. It has been obtained using the MHAT wavelet. The variables t and θ are equivalent each other.

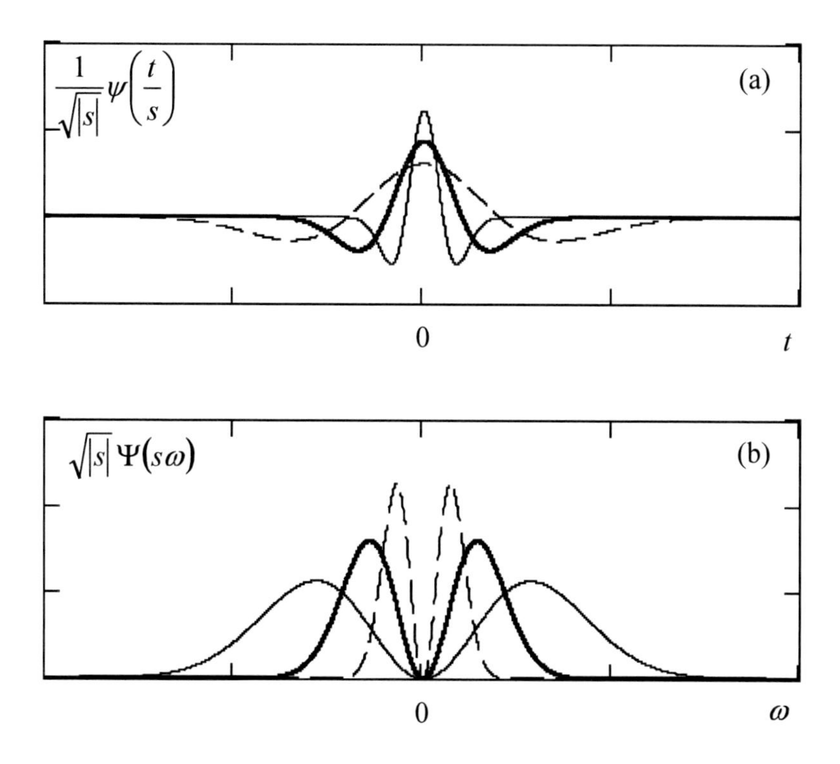

Figure 23. Basis functions of the MHAT wavelet (a) and corresponding Fourier spectra (b) for various values of the scaling coefficient: the thick solid curves correspond to $s = 1$, the thin solid ones to 2; and the dashed ones to 0.5.

Two other examples of wavelet spectra are presented in Figs. 25 and 26. First, it is the harmonic beat signal, i.e., the interference of two cosine waves of equal amplitude with 0.5 and 1 MHz carrier frequencies. Second, it is a chirp signal with the linear-frequency modulation; its frequency varies from 0.5 to 2.09 MHz.

It is important that we can derive the analytical expressions for wavelet spectra corresponding to some types of signals. They may be used, for example, if it is necessary to verify results obtained numerically.

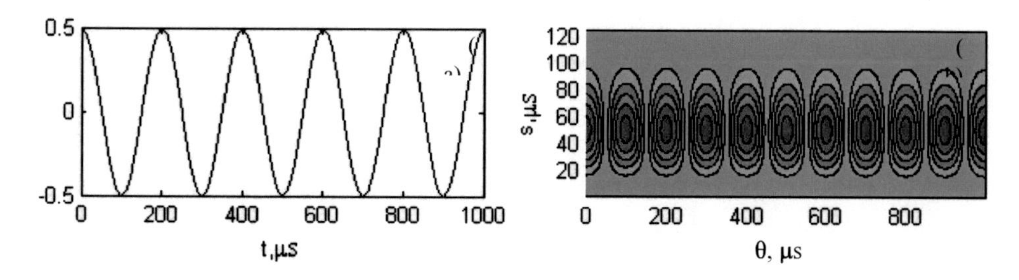

Figure 24. The unmodulated cosine wave (a) and modulus of its wavelet spectrum (b). The parameters: $f_0 = 0.5$ MHz; CWT wavelet – MHAT.

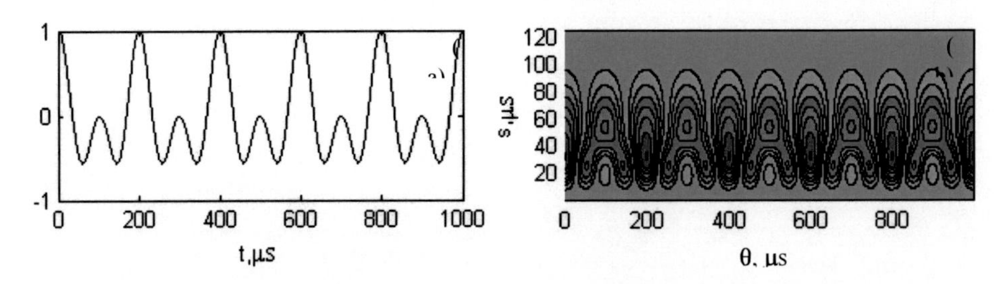

Figure 25. The harmonic beat signal (a) and modulus of its wavelet spectrum (b). The parameters: the carrier frequencies are 0.5 and 1 MHz; CWT wavelet – MHAT.

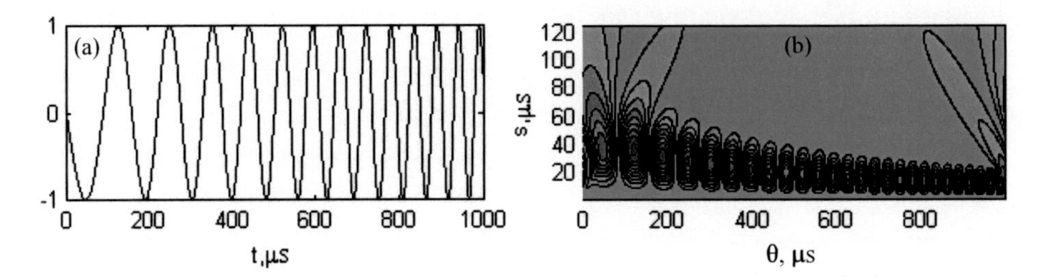

Figure 26. The chirp signal with the linear-frequency modulation (a) and modulus of its wavelet spectrum (b). The parameters: the range of frequencies is 0.5 – 2.09 MHz; CWT wavelet – MHAT.

First, let us consider the rectangular video pulse of length τ and amplitude A:

$$f(t)=\begin{cases} A & t\in[-\tau/2;\tau/2] \\ 0 & t\in(-\infty;-\tau/2)\cup(\tau/2;+\infty) \end{cases} \quad ; \tag{36}$$

and its Fourier spectrum is

$$F(\omega)=\frac{2A}{\omega}\sin\frac{\omega\tau}{2} \quad . \tag{37}$$

Using Eqs. (31), (35) and (37), we obtain the expression

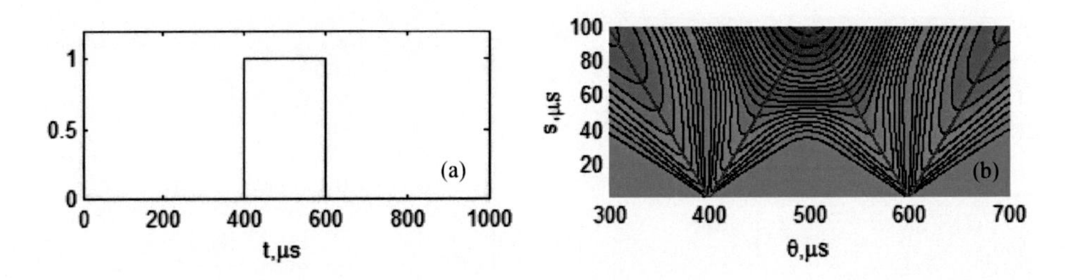

Figure 27. The rectangular video pulse (a) and modulus of its wavelet spectrum (b). The parameters: t_0 = 500 μs; τ = 200 μs; CWT wavelet – MHAT. The red lines correspond to the local maxima of the wavelet spectrum.

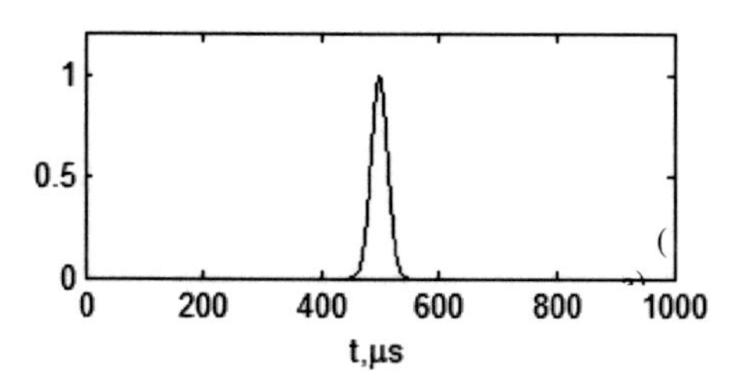

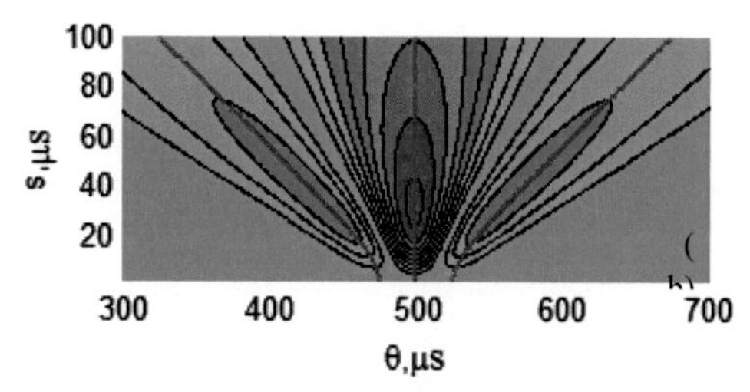

Figure 28. The Gaussian video pulse (a) and modulus of its wavelet spectrum (b). The parameters: t_0 = 500 μs; $\alpha = 5\cdot10^4$ s^{-1}; CWT wavelet – MHAT. The red curves correspond to the local maxima of the wavelet spectrum.

$$W(s,\theta)=\frac{8A}{\sqrt{3\sqrt{\pi}}}\cdot\frac{1}{\sqrt{s}}\cdot\left[\left(\theta+\frac{\tau}{2}\right)\exp\left(-\frac{\left(\theta+\frac{\tau}{2}\right)^2}{2s^2}\right)-\left(\theta-\frac{\tau}{2}\right)\exp\left(-\frac{\left(\theta-\frac{\tau}{2}\right)^2}{2s^2}\right)\right] \quad (38)$$

The signal (36) and modulus of its wavelet spectrum (38) are shown in Figure 27. The time shift of signal: t_0 = 500 μs and τ is equal to 200 μs. The red lines in Figure 27b correspond to the local maxima of wavelet spectrum, which are determined by four equations:

$$s=-\left(\theta-t_0-\frac{\tau}{2}\right);\ s=\left(\theta-t_0-\frac{\tau}{2}\right);\ s=-\left(\theta-t_0+\frac{\tau}{2}\right);\ s=\left(\theta-t_0+\frac{\tau}{2}\right) \quad (40)$$

The Gaussian video pulse signal is

$$f(t)=A\exp\left(-\alpha^2t^2\right)\ , \quad (41)$$

where α is the constant governing the duration of the signal. The Fourier spectrum of this signal is

$$F(\omega)=\frac{A\sqrt{\pi}}{\alpha}\exp\left(-\frac{\omega^2}{4\alpha^2}\right)\ . \quad (42)$$

Substitute Eqs. (35) and (42) into Eq. (31), we obtain the wavelet spectrum

$$W(s,\theta)=\sqrt{\frac{8\sqrt{\pi}}{3}\frac{As^2\sqrt{|s|}}{8\alpha}}\exp\left(-\frac{1}{2}\frac{\theta^2}{s^2+\frac{1}{2\alpha^2}}\right)\cdot\left[\frac{1}{2}\left(s^2+\frac{1}{2\alpha^2}\right)\right]^{-\frac{3}{2}}\cdot\left[1-\frac{\theta^2}{s^2+\frac{1}{2\alpha^2}}\right] \quad (43)$$

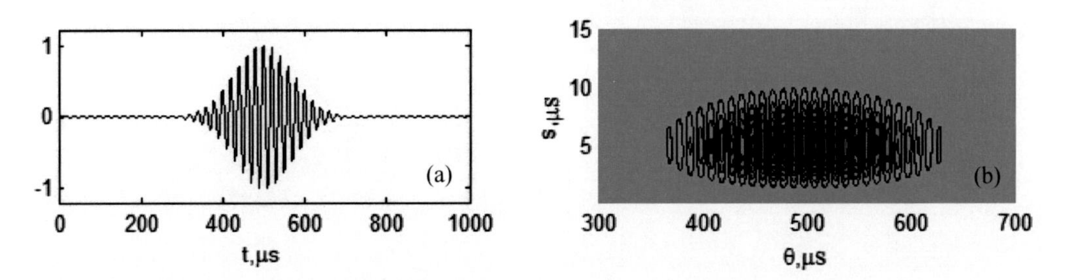

Figure 29. The Gaussian radio-frequency pulse (a) and modulus of its wavelet spectrum (b). The parameters: $t_0 = 500$ μs; $f_0 = 0.5$ MHz; $\alpha = 1{\cdot}10^4$ s^{-1}; CWT wavelet – MHAT.

The signal (41) and modulus of its wavelet spectrum (43) are shown in Figure 28. The red curves in Figure 28b correspond to the local maxima of wavelet spectrum, which are determined by three equations:

$$\theta = t_0;\ s = -\sqrt{\frac{(\theta - t_0)^2}{3} - \frac{1}{2\alpha^2}};\ s = \sqrt{\frac{(\theta - t_0)^2}{3} - \frac{1}{2\alpha^2}} \quad . \tag{45}$$

At last we consider the Gaussian radio-frequency pulse signal with the angular carrier frequency ω_0 :

$$f(t) = A\exp\left(-\alpha^2 t^2\right)\cos(\omega_0 t) \quad , \tag{46}$$

The Fourier spectrum of this signal is

$$F(\omega) = \frac{A\sqrt{\pi}}{2\alpha}\left[\exp\left(-\frac{(\omega - \omega_0)^2}{4\alpha^2}\right) + \exp\left(-\frac{(\omega + \omega_0)^2}{4\alpha^2}\right)\right] \quad . \tag{47}$$

Substitute Eqs. (35) and (47) into Eq. (31), we obtain the expression for wavelet spectrum

$$W(s,\theta) = \sqrt{\frac{8\sqrt{\pi}}{3}}\frac{As^2\sqrt{|s|}}{8\alpha}\exp\left(-\frac{1}{2}\frac{\frac{\omega_0^2}{2\alpha^2} + \theta^2}{s^2 + \frac{1}{2\alpha^2}}\right)\cdot\left[\frac{1}{2}\left(s^2 + \frac{1}{2\alpha^2}\right)\right]^{-\frac{5}{2}} \times$$

$$\times\left[\left(s^2 - \theta^2 + \frac{1}{2\alpha^2}\left(1 + \frac{\omega_0^2}{2\alpha^2}\right)\right)\cos\left(\frac{\omega_0\theta}{1 + 2\alpha^2 s^2}\right) - \frac{\omega_0\theta}{\alpha^2}\sin\left(\frac{\omega_0\theta}{1 + 2\alpha^2 s^2}\right)\right] \tag{48}$$

The signal (46) and modulus of its wavelet spectrum (48) are shown in Figure 29.

An important characteristic is the energy distribution of the analyzed signal over time scales usually called scalogram or wavelet variance [2] which can be written in the form:

$$E_W(s) = \int_{-\infty}^{+\infty} \frac{|W(s,\theta)|^2}{|s|} d\theta \quad . \tag{49}$$

Clearly, the scalogram represents the signal energy density $|W(s,\theta)|^2$ divided by the absolute value of the time scale $|s|$ averaged over all possible values of the translation parameter θ.

Substituting (31) into (49), we obtain one further formula for the scalogram:

$$E_W(s) = \frac{1}{2\pi} \int_{-\infty}^{+\infty} |F(\omega)|^2 \, |\Psi(s\omega)|^2 \, d\omega \quad . \tag{50}$$

Hence it follows that the scalogram is determined by the power spectrum of the analyzed signal $|F(\omega)|^2$ which is smoothed over each time scale by the power spectrum of the analyzing wavelet $|\Psi(s\omega)|^2$. Therefore, the scalogram represents essentially more smooth curve than the signal power Fourier spectrum. It will be illustrated below.

Since the analyzed function must satisfy condition (27) and the zero mean condition is fulfilled for any wavelet, then the function $E_W(s)$ tends to zero both for very small and for very large scale factors and has at least one maximum. The location of such maxima of the Fourier spectrum corresponds to the characteristic frequencies of the analyzed signal carrying its basic energy. The scalogram maxima have a similar physical meaning, they correspond to the characteristic time scales of the process.

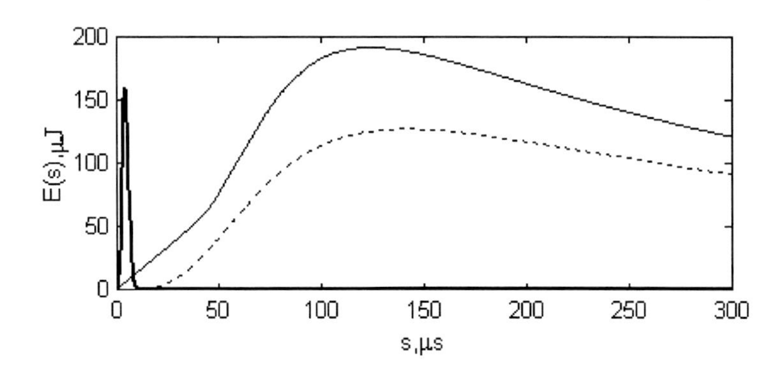

Figure 30. The scalograms corresponding to the next signals: rectangular video pulse ($\tau = 200$ μs) – solid thin line; Gaussian video pulse ($\alpha = 5 \cdot 10^4$ s^{-1}) – dotted line; Gaussian radio-frequency pulse ($f_0 = 0.5$ MHz; $\alpha = 1 \cdot 10^4$ s^{-1}) – thick solid line. The CWT wavelet – MHAT.

Substituting Eqs. (35), (37), (42) and (47) into Eq. (50), we obtain the next analytical expressions for scalograms:

$$E_W(s) = \frac{4}{3}A^2 s\left[1-\left(1-\frac{\tau^2}{2s^2}\right)\cdot\exp\left(-\frac{\tau^2}{4s^2}\right)\right] \quad ; \tag{51}$$

$$E_W(s) = 4\sqrt{2}\pi A^2\alpha^3 s^4\left(1+2\alpha^2 s^2\right)^{-5/2} \quad ; \tag{52}$$

$$E_W(s) = 2\sqrt{2}\pi A^2\alpha^3 s^4\left(1+2\alpha^2 s^2\right)^{-5/2}\exp\left(-\frac{\omega_0^2}{2\alpha^2}\right)\times$$

$$\times\left[1+\exp\left(\frac{\omega_0^2}{2\alpha^2(1+2\alpha^2 s^2)}\right)\cdot\left(1+\frac{2\omega_0^2}{\alpha^2(1+2\alpha^2 s^2)}+\frac{\omega_0^4}{3\alpha^4(1+2\alpha^2 s^2)^2}\right)\right] \tag{53}$$

These formulas correspond to the signals (36), (41) and (46) respectively. The scalograms plotted with the use of Eqs. (51), (52) and (53) are shown in Figure 30.

Relationships between the Scaling Coefficient and Frequency

One of the most important problems is to find relationship between the scaling coefficient s and frequency ω. It is necessary to interpret the results of spectral analysis carried out with the use of the CWT correctly.

From the beginning we consider a radio-frequency pulse signal, namely, a segment of sinusoid with the angular frequency ω_0, corresponding to the period $T_0 = 2\pi/\omega_0$, and a duration τ for which the condition $\tau >> T_0$ must be satisfied. The Fourier spectrum of this signal differs only slightly from the infinity sinusoid spectrum which has the form [21] $F(\omega)=\pi i[\delta(\omega+\omega_0)-\delta(\omega-\omega_0)]$, where $\delta(\omega)$ is the Dirac delta function. Substituting this formula in Eq. (31), we obtain the corresponding expression for the wavelet spectrum:

$$W(s,\theta)=\frac{i}{2}\sqrt{|s|}\left[\Psi^*(s\omega_0)\exp(i\omega_0\theta)-\Psi^*(-s\omega_0)\exp(-i\omega_0\theta)\right] \quad . \tag{54}$$

Substituting Eq. (35) into Eq. (51) and then substituting the result into Eq. (31) and solving the equation

$$\frac{d|W(s,\theta)|}{ds} = (s\omega_0)^{\frac{5}{2}}\exp\left(-\frac{(s\omega_0)^2}{2}\right) = 0 \quad ,$$

which is the necessary condition for a local extremum of the wavelet spectrum absolute value, we obtain the relationship $\omega_0 s = \sqrt{5/2}$. It establishes the interrelation between the angular frequency of the monochromatic sinusoidal signal and scaling coefficient for the wavelet spectrum when the MHAT wavelet is used.

Doing similarly, it may be demonstrated that the analogous relationship for the scalogram, which can be obtained using Eqs. (35), (50) and (54), is $\omega_0 s = \sqrt{2}$.

Consequently, when the MHAT wavelet is used for the CWT analysis, we can present wavelet spectra and scalograms as functions of the frequency f (or period $T = 1/f$) rather than the scaling coefficient s in accordance with the follows expressions:

$$f = \frac{1}{\pi s}\sqrt{\frac{5}{8}}$$

for wavelet spectrum; and

$$f = \frac{1}{\sqrt{2}\pi s} \tag{55}$$

for scalogram. It follows from these formulas that a wavelet spectrum and scalogram can be presented as $W(f,\theta)$ and $E_W(f)$ instead of $W(s,\theta)$ and $E_W(s)$.

In practice, however, an analyzed signal has limited duration. To obtain the formula interconnecting ω and s in that case, we expand the function $|\Psi(\omega)|^2$ into the Taylor series in the neighborhood of the point ω_0:

$$|\Psi(s\omega)|^2 = \sum_{n=0}^{\infty} \frac{1}{n!} \frac{d^n |\Psi(s\omega)|^2}{d\omega^n}\Bigg|_{\omega=\omega_0} (\omega - \omega_0)^n \quad , \tag{56}$$

assuming that the width of the Fourier spectrum of analyzed signal is considerably less than the width of the Fourier spectrum of wavelet when $\omega_0 s \sim 1$.

Substituting (56) into (50), we get the representation of scalogram in the form of the series

$$E_W(s) = \sum_{n=0}^{\infty} \frac{M_n}{n!} \frac{d^n |\Psi(s\omega)|^2}{d\omega^n}\Bigg|_{\omega=\omega_0} \quad , \tag{57}$$

where $M_n = \frac{1}{2\pi} \int_{-\infty}^{+\infty} (\omega - \omega_0)^n |F(\omega)|^2 \, d\omega$ is the nth central moment of the power spectrum of analyzed signal.

Using formulas (35) and (57), we define the scalogram of signal (46) as it was done in the work [23]. Let us consider only the five initial terms of the series. Substituting (47) into the expression for M_n, we obtain the next values of the moments: $M_0 = \frac{\sqrt{2\pi}}{8\alpha} A^2$; $M_1 = M_3 = 0$; $M_2 = \frac{\sqrt{2\pi}}{8} \alpha A^2$; $M_4 = \frac{3\sqrt{2\pi}}{8} \alpha^3 A^2$. Then it follows from (57) that

$$E_W(s) = \frac{8\sqrt{2}}{3}\frac{\pi}{\alpha} s^4 \omega_0^4 \exp\left(-s^2\omega_0^2\right)\left[1 + \left(\frac{\alpha}{\omega_0}\right)^2 \left(6 - 9s^2\omega_0^2 + 2s^4\omega_0^4\right) + \right.$$

$$\left. + \frac{1}{2}\left(\frac{\alpha}{\omega_0}\right)^4 \left(6 - 84s^2\omega_0^2 + 123s^4\omega_0^4 - 44s^6\omega_0^6 + 4s^8\omega_0^8\right)\right] . \tag{58}$$

Then differentiating (58) with respect to s and equating the result to zero as well as disregarding the terms proportional to the highest degrees of $s\omega_0$, we obtain the relation

$$(s\omega_0)^2 = 2 - \frac{-1 + 11(\alpha/\omega_0)^2 - 107(\alpha/\omega_0)^4}{10\,(\alpha/\omega_0)^2\left(1 - 17(\alpha/\omega_0)^2\right)} -$$

$$- \frac{\sqrt{1 - 22(\alpha/\omega_0)^2 + 375(\alpha/\omega_0)^4 - 3194(\alpha/\omega_0)^6 + 14169(\alpha/\omega_0)^8}}{10(\alpha/\omega_0)^2\left(1 - 17(\alpha/\omega_0)^2\right)} . \tag{59}$$

The results of calculations with the formula (59) are presented in Figure 31. For comparison, the analogous relationship is shown in the same figure, which has been calculated numerically using the exact analytical expression (53).

As is well known, the RMS width of a spectrum can be determined as [19]: $\Delta\omega = \sqrt{M_2/M_0}$. For the signal of form (46), it equals α. Consequently, the ratio α/ω_0 is the relative width of spectrum of the radio-frequency pulse signal with the Gaussian envelope. From Figure 31, it follows that the simple relation $\omega_0 s = \sqrt{2}$ is obey quite enough for the narrow bandwidth signals only, when an analyzed function has a lot of periods of main frequency oscillations.

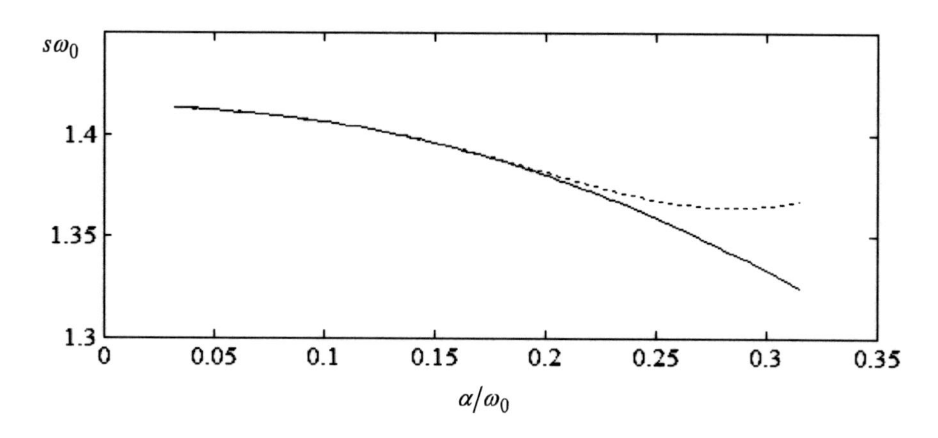

Figure 31. The relation between the product of the angular frequency into the scaling coefficient and relative bandwidth of spectrum of the Gaussian radio-frequency pulse, obtaining with the use of the exact (solid line) and approximate (dotted line) expressions for scalogram.

Now we consider the signal with power spectrum which is constant, $\left|F(\omega)\right|^2 = \left(\dfrac{F_0}{2\Omega}\right)^2$, in the bounded frequency range: $\omega \in \left[-\Omega\,;\,\Omega\right]$. It can be interpreted as the model of white noise. From Eq. (50) it follows that

$$E_W(s) = \frac{2F_0^2}{3\Omega^2\sqrt{\pi}} s^4 \int_0^{\Omega} \omega^4 e^{-\omega^2 s^2}\, d\omega = \frac{2F_0^2}{3\Omega^2\sqrt{\pi}} \frac{1}{s} \int_0^{s\Omega} \xi^4 e^{-\xi^2}\, d\xi \quad . \tag{60}$$

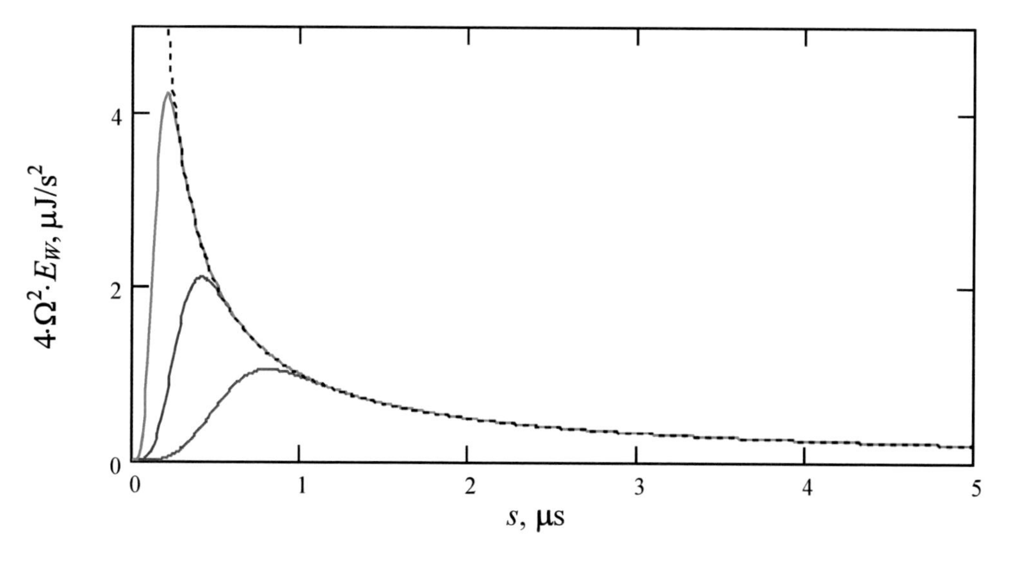

Figure 32. The scalograms corresponding to the model of white noise. The parameters: Ω is equal to 2.5 rad/μs (red), 5 rad/μs (blue), 10 rad/μs (red); Ω tends to infinity (dotted black line); CWT wavelet – MHAT.

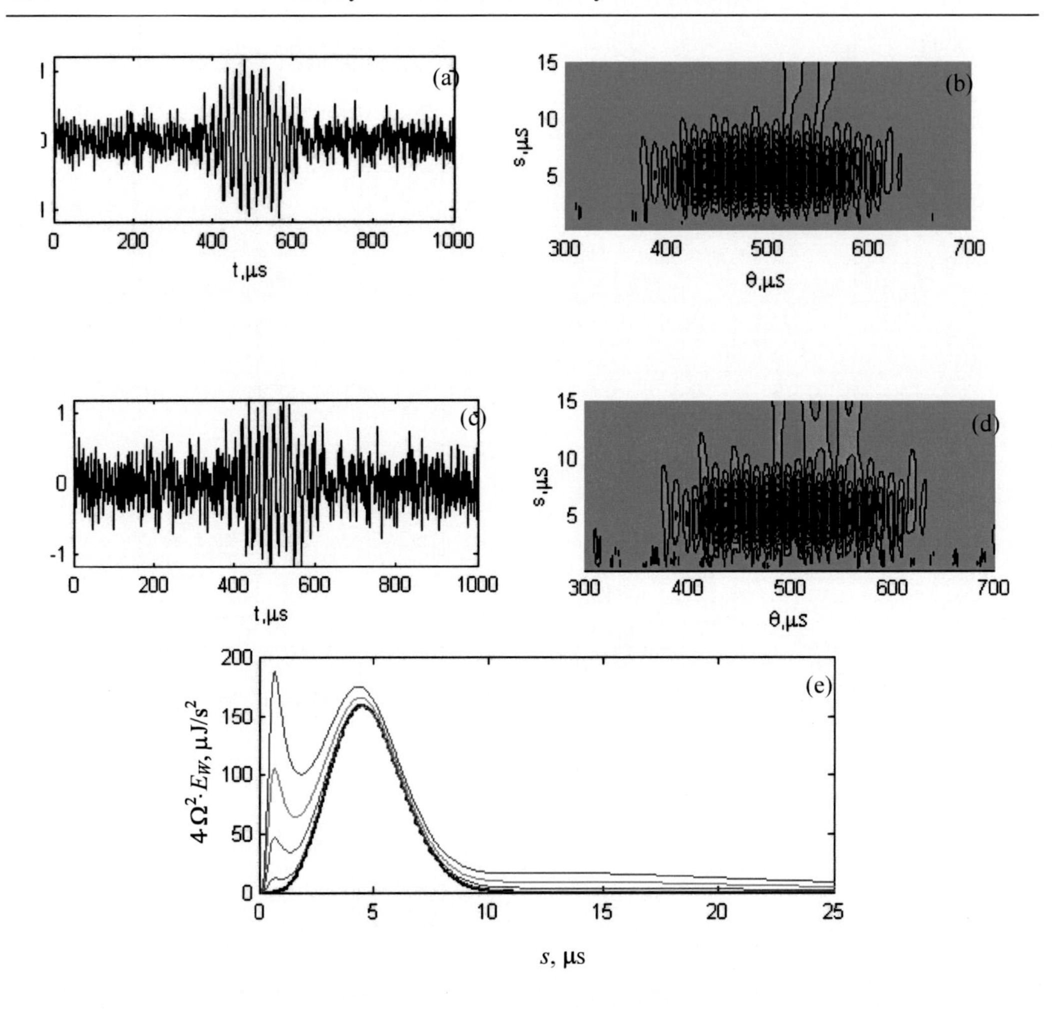

Figure 33. The sum of the Gaussian radio-frequency pulse signal and white normally distributed noise (a, c), modulus of its wavelet spectrum (b, d) and the family of scalograms (e). The signal-to-noise ratio is equal to 14 dB (a, b) and 10.5 dB (c, d). The colors of scalograms correspond to the signal-to-noise ratio of 20 dB (red), 14 dB (magenta), 10.5 dB (green) and 8 dB (blue); the black color is used here when the signal-to-noise ratio tends to infinity. The parameters: $t_0 = 500$ μs; $f_0 = 0.5$ MHz; $\alpha = 1\cdot10^4$ s^{-1}; CWT wavelet – MHAT.

If the frequency bandwidth Ω tends to infinity, the integral in Eq. (60) can be founded analytically: $\int_0^\infty \xi^4 e^{-\xi^2} d\xi = \frac{3\sqrt{\pi}}{8}$; in which case $E_W(s) = \frac{F_0^2}{4\Omega^2 s}$. The family of scalograms (60) corresponding to different values of Ω is shown in Figure 32.

There is an asymptotic expansion for the integral in Eq. (60). Using it, we obtain the expression for the scalogram:

$$E_W(s) \approx \frac{2F_0^2}{3\Omega^2\sqrt{\pi}}\frac{1}{s}\left\{1-\frac{4}{3\sqrt{\pi}}s^3\Omega^3\exp\left(-s^2\Omega^2\right)\cdot\left[1+\frac{3}{4s^4\Omega^4}\left(1+2s^2\Omega^2+R(s)\right)\right]\right\}, \quad (61)$$

where
$$R(s)=\sum_{k=1}^{\infty}\frac{(-1)^k(2k-1)!!}{2^k s^{2k}\Omega^{2k}} \quad .$$

On putting the highest terms of series (61) equal to zero, i.e. $R(s)=0$, we can determine the position of maximum of $E_W(s)$ as a root of the transcendental equation

$$\exp\left(-s^2\Omega^2\right)\left(4s^6\Omega^6+2s^4\Omega^4+3s^2\Omega^2+3\right)-\frac{3\sqrt{\pi}}{2}s\Omega=0 \quad ;$$

it is $s\Omega \approx 2.0244$.

The CWT is a very stable algorithm with respect to the effect of noise [9]. Let us consider the sum of the Gaussian radio-frequency pulse signal and white normally distributed noise. The parameters of the signal are f_0 = 0.5 MHz; $\alpha = 1\cdot10^4$ s^{-1}; Ω = 5 rad/μs. Changing the signal-to-noise ratio, we obtain the results shown in Figure 33. Comparing Figure 33 with Figure 29 anyone can see how the white normally distributed noise affects the shape of wavelet spectra and scalograms.

For comparison, we demonstrate the power Fourier spectrum and scalogram of noised Gaussian radio-frequency pulse signal in Figure 34. These both curves are normalized to their maximal values for the sake of convenience.

Further examples of using the CWT for analyzing acoustic signals can particularly be found in the works [11, 15, 17, 24, 25].

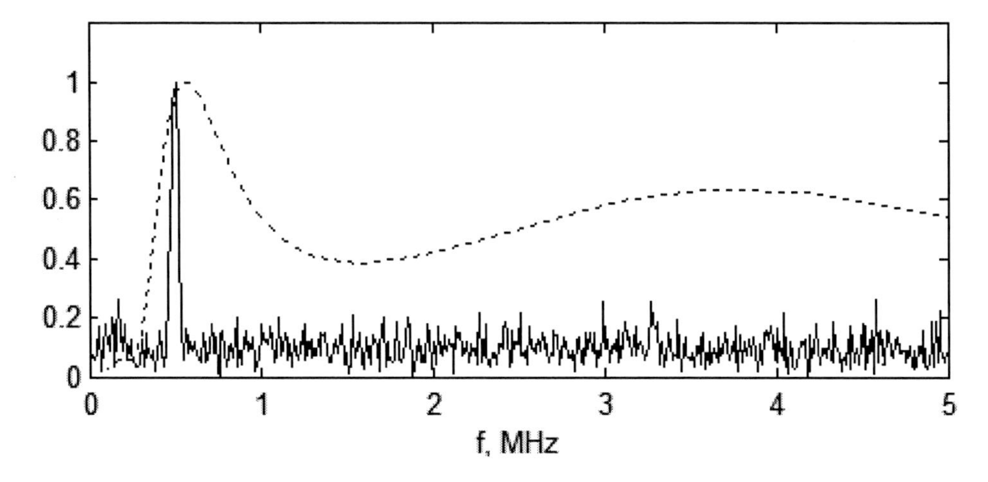

Figure 34. The normalized power Fourier spectrum (solid) and scalogram (dashed) corresponding to the sum of the Gaussian radio-frequency pulse signal and white normally distributed noise. The parameters: t_0 = 500 μs; f_0 = 0.5 MHz; $\alpha = 1\cdot10^4$ s^{-1}; signal-to-noise ratio – 10.5 dB; CWT wavelet – MHAT.

Conclusion

In this chapter we have considered some problems which are related to applying different types of wavelet transform for analyzing acoustic fields and signals in elastic media. Now we briefly outline the main results which have been obtained.

The wavelet filtering procedure can be successfully used for denoising acoustic signals and images. However it is necessary to choose the parameters of the algorithm correctly. An important point in the wave filtering procedure is the correct selection of the order of the wavelet used in filtering and the number of decomposition levels in the FWT scheme. For filtering radio-frequency pulses, we recommend to use the wavelets of the sym or db family with the number of zeroth moments ranging from 8 to 15. At the same time, the number of decomposition levels for FWT algorithm equals to four is sufficient.

An important result of the reported research is obtaining the dependence of the relative error in the signal amplitude after wavelet filtering on the central frequency of acoustic transducer, signal duration, sampling period and number of elements in the sampling. The simple formula given in the paper enables the operator to select the optimal dimension of the sampling to minimize the error.

Comparison of the different types of discrete wavelet transform from the viewpoint of wavelet filtering demonstrates that the denoising quality is the best if the wavelet packets algorithm is used because it is able to adapt itself to each analyzed signal. However, the fast wavelet transform algorithm needs the least computing time.

Furthermore, it has been shown that the method of fixed-form thresholding is appropriate for denoising acoustic fields and signals. It must be emphasized, however, that it is preferably to use the soft thresholding strategy to process two-dimensional data arrays because the hard thresholding produces a large number of clearly outlined local artifacts which affect the visual perception of the filtered image. On the other hand, filtering of one-dimensional signals has to be fulfilled using the hard thresholding because it allows to estimate the amplitude of analyzed signal more precisely.

The continuous wavelet transform can be successfully applied for the purpose of spectral analyzing of nonstationary acoustic signals. It makes possible to estimate the instantaneous frequency of signal, which dominates at different times, search for modulations and relations between oscillations with different periods.

It is important that we have derived the analytical expressions for wavelet spectra and scalograms corresponding to some types of signals which are widespread in acoustics. They may be used, for example, if it is necessary to verify results obtained numerically.

One of the most important problems of wavelet analysis is to find relationship between the scaling coefficient and frequency. It is necessary to interpret the results of spectral analysis carried out with the use of the continuous wavelet transform correctly. In this chapter we have obtained such relationships for wavelet spectrum and scalogram by the example of the MHAT wavelet and Gaussian radio-frequency pulse.

We have also shown that the continuous wavelet transform is a very stable algorithm with respect to the effect of noise. Furthermore, it has been demonstrated how the white noise affects the wavelet spectrum and scalogram.

References

[1] Daubechies, I. *Ten lectures on wavelets;* CBMS–NSF Regional Conference Series in Applied Mathematics; Society for Industrial and Applied Mathematics: Philadelphia, PA, 1992; Vol. 61.

[2] Mallat, S. A *wavelet tour of signal processing;* Academic Press: San Diego, CA, 1998.

[3] Mallat, S. *IEEE Trans. Pattern Anal. Mach. Intell.* 1989, 11, 674-693.

[4] Mallat, S. *Trans. Am. Math. Soc.* 1989, 315, 69-87.

[5] Rabiner, L.; Gold, B. *Theory and applications of digital signal processing*; Prentice-Hall: Paramus, NJ, 1986.

[6] Donoho, D. L.; Johnstone, I. M. *Biometrika* 1994, 81, 425-455.

[7] Donoho, D. *IEEE Trans. Inf. Theory*, 1995, 41, 613-627.

[8] Perov, D. V.; Rinkevich, A. B.; Smorodinskii, Ya. G. *Russian Journal of Nondestructive Testing* 2002, 38, 869–882.

[9] Perov, D. V.; Rinkevich, A. B.; Nemytova, O. V. *Russian Journal of Nondestructive Testing* 2007, 43, 369–377.

[10] Perov, D. V., & Rinkevich, A. B. (2003). Estimation of accuracy of wavelet filtration applied to narrow-band ultrasonic signals. http://www.akin.ru/Docs/Rao/Ses13/U7.pdf

[11] Perov, D. V.; Rinkevich, A. B. *Acoustical Physics* 2005, 51, 443–448.

[12] Coifman, R. R.; Donoho, D. L. *Lecture Notes in Statistics* 1995, 103, 125-150.

[13] Coifman, R.; Wickerhauser, M. V. *IEEE Trans. Inf. Theory*, 1992, 38, 713-718.

[14] Donoho, D. L.; Johnstone, I. M. *CRAS Paris Ser I* 1994, 319, 1317–1322.

[15] Perov, D. V.; Rinkevich, A. B.; Smorodinskii, Ya. G.; Keler B. *Russian Journal of Nondestructive Testing* 2001, 37, 889–899.

[16] Perov, D. V.; Rinkevich, A. B. *Acoustical Physics* 2004, 50, 86–90.

[17] Rinkevich, A. B.; Perov, D. V. *Russian Journal of Nondestructive Testing* 2005, 41, 93–101.

[18] Zhitluhina J. V; Perov, D. V.; Rinkevich, A. B.; Smorodinsky, Y. G.; Kröning M.; Permikin V. S. *Insight* 2007, 49, 267–271.

[19] Kino, G. S. *Acoustic waves: devices, imaging, and analog signal processing;* Prentice-Hall: Englewood Cliffs, NJ, 1986.

[20] Perov, D. V.; Rinkevich, A. B. *Russian Journal of Nondestructive Testing* 2002, 38, 288–305.

[21] Franks, L. E. *Signal theory;* Prentice-Hall: Englewood Cliffs, NJ, 1969.

[22] Grossmann, A.; Morlet, J. *SIAM J. Math. Anal.* 1984, 15, 723–736.

[23] Perov, D. V., & Rinkevich, A. B. (2005). The features of the analysis of wave processes in elastic media with the use of wavelet transform. http://www.akin.ru/Docs/Rao/Ses16/D02.pdf

[24] Perov, D. V.; Rinkevich, A. B. *Russian Journal of Nondestructive Testing* 2001, 37, 879–888.

[25] Permikin, V. S.; Perov, D. V.; Rinkevich, A. B. *Russian Journal of Nondestructive Testing* 2004, 40, 9–18.

In: New Research on Acoustics
Editor: Benjamin N. Weiss, pp. 111-140

ISBN: 978-1-60456-403-7

Chapter 2

MEASUREMENT OF PARAMETERS OF THE ACOUSTIC ANTENNA ARISING AT BRAKING AND STOPPING OF THE PROTON BEAM IN WATER AND RESEARCH OF CHARACTERISTICS OF CREATED FIELD

***V.B. Bychkov*[1], *V.S. Demidov*[2] *and E.V. Demidova*[2]**
[1]the State centre of science the All-Russia scientific research institute of physicotechnical and radio engineering measurements, Mendeleevo, the Moscow region, Russia
[2]the State centre of science Institute of theoretical and experimental physics, Moscow, Russia

Abstract

The purpose of this paper is the experimental research of properties for acoustic antenna arising during braking of an intensive beam of accelerated protons in a water environment. Research was conducted in a near-field zone that had allowed us to allocate signals from separate elements of the antenna and to carry out the analysis of such parameters of signals, as amplitude, width and time of their propagation.

As a source of protons the external beam of the Institute of Theoretical and Experimental Physics (ITEP, Moscow) accelerator with energy of 200 MeV and the time duration of 70 ns has been used. The beam intensity was supported at the level of $4 \cdot 10^{10}$ protons per pulse and supervised by the current transformer. The experiment was carried out in the parallelepiped plexiglas basin of a square section 95 cm in length and with the volume of 250 liters filled 85% with water. Input of the proton beam inside the volume was realized through a pipe with the diameter of 59 mm, 46 cm length and wall thickness of 1.5 mm inserted into a lateral side of the basin and closed by a plug made from organic glass with a thickness of 2 mm. The average ionizing range of protons in water was 25.2 cm. So, the sizes of the basin and the applied equipment have allowed us to study the un deformed structure of a hydroacoustic field induced by the proton beam.

Measurements of an acoustic field were made by means of a relocatable hydrophone in two mutual-perpendicular directions. Along the beam axis hydrophone movement was carried out with the step of 8.9 mm at the distance of 3.5 cm from the beam axis. In the cross-cut direction the trace passed in the horizontal plane passing through the beam axis at the distance of 35.6 cm from the point of the entrance of the proton beam into the water. In this case the scanning step was equal to 4.45 mm.

According to a thermoacoustic model in the area of beam action for time, comparable with the action time, an acoustic antenna arises. In the present work the problem of reconstruction of the form of the antenna using the experimental results is being solved. The technique of calculation of the hydrophone response to the radiation of separate elements of the acoustic antenna has been developed. The dependences of amplitude of the signals and their time parameters on the relative position of the antenna and the hydrophone have been obtained. The angular distribution of the field created by the terminal area of the radiation zone has been obtained. This characteristic, generally speaking, is similar to the directional diagram of an audio antenna.

To test the experimental results, the full-scale simulation of set-up geometry and the physical processes accompanying the propagation of protons in water had been carried out using GEANT-3.21 package. The simulation of the generation process of an acoustic signal was performed as a first approximation in the assumption of proportionality of the signal intensity to the energy which is generated at the ionization of water atoms by a proton without taking into account heat conductivity and the elastic properties of environment, leading to relaxation. The model calculations confirm the qualitative conclusions and the results obtained at the processing of experimental data.

1. Introduction

Experimental and theoretical research on radiation acoustics has been carried out during the last few decades [1-2]. It was established that the intensive fluxes of ionizing radiation create in the substance, lengthy acoustic antenna (AA), the sizes and form of which are defined by distribution in the medium of the thermal field induced by radiation. General characteristics of the radiation acoustic waves arising in the process of propagation of ionizing particles through the substance [3-9] are well studied: the proportional dependence of the response of acoustic receiving systems on the proton beam intensity, the dependence of the signal duration on the beam diameter, temperature dependence of intensity of the acoustic signal arising in liquids, etc. These characteristics are well described by the thermoacoustic model of the occurrence of mechanical oscillations during the propagation of ionizing radiation through the substance. However, many problems of radiation acoustics remain unsolved. The properties of a radiation acoustic antenna were studied mainly with reference to hadron-electromagnetic showers (HES), created by cosmic rays of ultrahigh energy in the water environment, with the hope of being able to measure their energy. It was shown theoretically and by the simulation that acoustic antenna as well as showers have the form described by smooth function with a maximum in the middle of a shower. On the other hand, to specify mechanisms of ultrasonic generation experiments on the study of braking of intensive monochromatic beams of low-energy protons to their stoppage can be interesting. In this case the distribution of energy loss has another shape: almost uniform distribution along all range of protons comes to the end with a sharp maximum in a zone of so-called Bragg peak in the area of ionization range. In papers [10,11] the space-time picture of a field arising in water during the stoppage of a proton beam has been registered for the first time. Measurements have been fulfilled in the points located on trajectories, parallel to beam direction. Three sources of radiation of AA - from the area nearest to the receiver, from Bragg peak and from the point of beam entrance in water have been identified.

The purpose of the present work is the restoration of the form of acoustic antenna arising at the braking of an intensive beam of accelerated protons in the water environment as well as the research of properties of radiation - dependences of pressure and frequency of radiation on co-ordinates of a point of reception of a signal and on the observation angle.

2. Methodical Aspects

2.1. The Arrangement of the Experiment

The experiment was carried out at the external proton beam of the ITEP accelerator similar to that which was used in papers [7,9-11]. The proton beam energy was equal to (200±0.4) MeV, duration of a bunch - 70 ns. The average beam intensity was supported at the level of about $4 \cdot 10^{10}$ protons per pulse and was supervised by the current transformer. The spatial form of beam in the cross-section direction was quasi-gauss with root-mean square deviation equal to 1.5 cm. The beam was limited by the lead collimator with the diameter of 6 cm and the length of 5 cm. The acoustic oscillations were generated in a water basin (see Figure 1).

The box of the basin (1) is made of organic glass and has the length of 94.5 cm and the cross-section of 50.8×52.3cm (in vertical direction). The input of the proton beam into the centre of the measuring volume was carried out through the duralumin pipe (2) with the diameter of 59 mm, length 46 cm and thickness of a wall of 1.5 mm, inserted into the lateral side of the basin and closed by the Teflon end-cap (3) with the thickness of 2 mm.

Figure 1. Photo of the experimental basin.

The basin has been filled by salty water up to 85 % of its volume. Concentration of sea salt was about 3 %. The water temperature was equal to18.5°C and did not change during the experiment.

The research was carried out by the method of a scanning hydrophone [10-11]. The electromechanical scanner (4) with manual remote control allowed us to place the hydrophone (5) discretely with step s_1 =8.9 mm or s_2 = 4.45 mm within the linear aperture with the length 40 cm.

2.2. The Run of the Experiment

The measurements of an acoustic field were made in two mutual-perpendicular directions. The scheme of the arrangement of points in which the measurements in coordinates Z (along beam) - X (in horizontal direction) were carried out is shown in Figure 2.

The cross-section of the area of beam action is represented as a contour of two-dimensional distribution of average loss of energy by protons in water in which 67 % of the energy deposited by protons concentrate approximately. Scanning traces are marked by symbols I, II.

Trace I passed transversely to the beam direction at Z=35.6 cm, trace II in parallel with beam axis at X=3.4 cm.

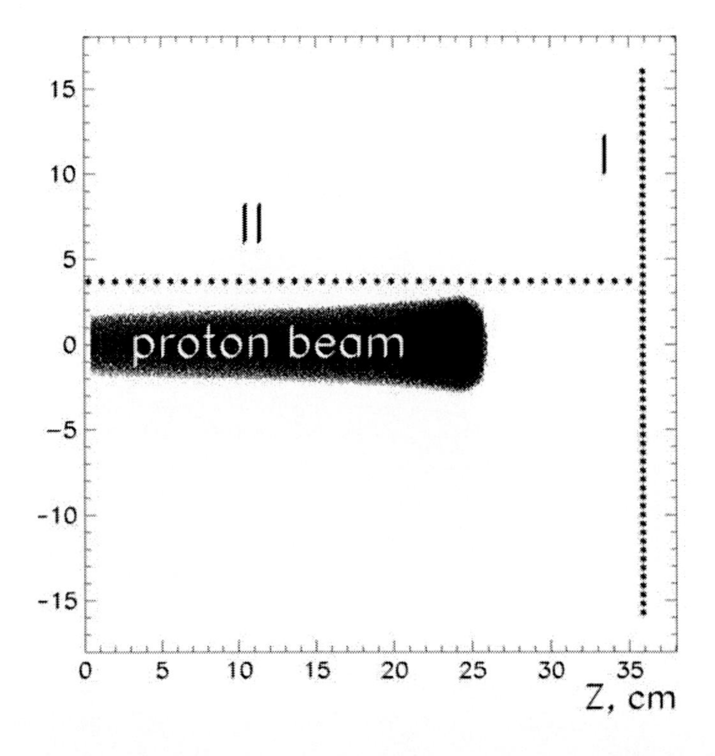

Figure 2. The schematic picture of beam (simulation) and the arrangement of points in which measurements were made.

Along trace I measurements have been carried out at 74 points, located from each other by step size s_2, along trace II - in 39 points with step s_1. In the experiments two different hydrophones and various operating modes of reading devices were applied.

2.3. Hydrophones, Amplification Equipment and Readout Technique

Transducing of acoustic signals to the electric ones was carried out by means of specially designed hydrophones based on piezoceramics ZnTiPb of spherical (trace I) and cylindrical shapes (trace II). In the latter case piezoceramics has been tangentially polarised. The cylindrical hydrophone had the built-in preamplifier, at the trace I the charge amplifier 2635 produced by Bruel&Kjaer was applied for signal amplification. In the table the parameters of hydrophones, amplifiers, readout equipment and their operating modes are presented.

The quantization of signals from the exit of preamplifiers was made by 2-beam digital oscilloscope TEKTRONIX TDS 3032 connected to a personal computer by means of interface GPIB. The information in volume of 10^4 points was recorded to the computer disk in the format *.sht. Except the hydrophone response, the signal from the current transformer (CT) measured the proton beam intensity has been recorded. Electronics of the CT worked in integrating mode, so the amplitude of its signal was proportional to the proton beam current. In Figure 3 the example of the oscillogram obtained on trace I at the point № 32 is presented. The exit of the acoustic signal amplifier was connected to the first channel of the oscilloscope. The signal is shown for time evolvement of 200 μs per a square at the scale of 200 mV per a cell. The signal of current transformer is presented on the scale of 20 μs x 500 mV.

Parameters	**A series of measurements *I***	**A series of measurements *II***
The shape and size of a piezoelement	Sphere ∅ = 1.5 cm	Cylinder ∅= 4 mm, h=6 mm
Average sensitivity of a piezoelement, mcv/Pa	2500	1500
Range of uniformity the amplitude-frequency characteristics in the limits ± 5 dB, Hz	$2.10^2 \div 5.10^4$	$10^2 \div 10^5$
Signal / noise	> 10 dB	> 12 dB
The amplifier, the amplification coefficient	Bruel&Kjaer – 2635, k=400	Special design, k = 200
Frequency of numbering of signal, MHz	5	2. 5
Own noise of the amplifier, nV/SQRT (Hz)	-	<2

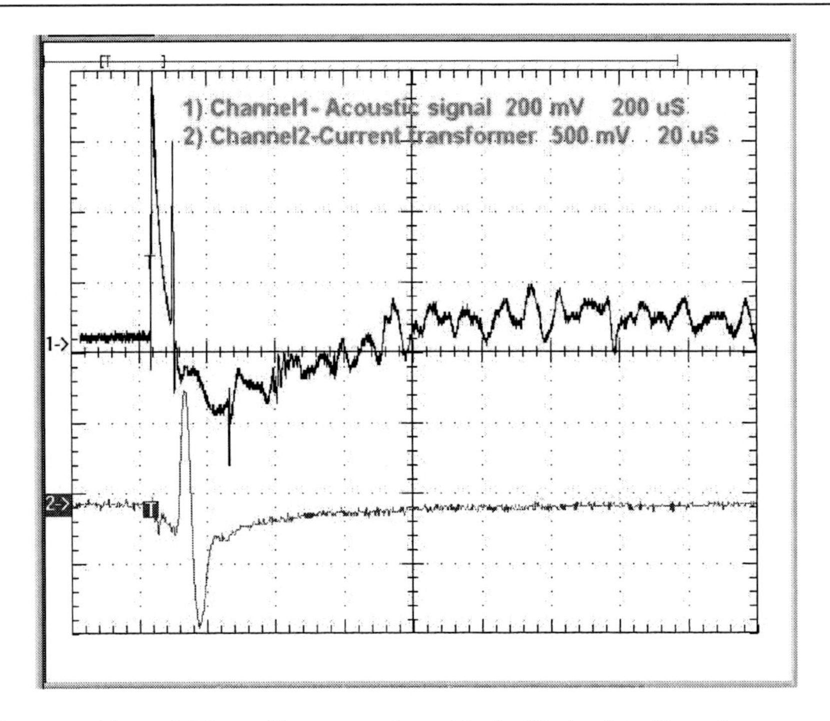

Figure 3. The oscillogram registered in the 32nd point of trace I.

The moment of arrival of the trigger starting the oscilloscope, is marked by symbol T in small black square at the bottom curve.

The intensity of the proton beam was not constant and could deviate from the average value by 30-50 %. However, it had no adverse effect on the accuracy of results because during the analysis of acoustic signals they were normalized on 1V amplitudes of the current transformer, which corresponds to 1.5 10^{10} protons with the precision not more than 5%.

2.4. Mathematical Processing of Signals

Basically, the radiation acoustic field can be described mathematically as the family of functions depending at the constant beam intensity at least on two parameters: the distance from sound source and the direction of emission of signals by the antenna. However, the analytical description of the space-time picture (STP) as a whole represents a very difficult mathematical problem. In the present work, graphic representation is used in general. Analytically only some separate signals are described in the first approximation. Below (in sections 3.3 and 4.3 of this chapter) the results of the description by five-parameter analytical function of signal $\alpha\gamma$ on the trace I and of signal α on the trace II are presented. Because of the complexity of the registered signals and the big number of background impulses the stroboscopic method was applied for separation of useful signals. The gate interval was selected taking into account the conditions of the concrete problem.

Mathematical processing was carried out, in general, using the PAW (Physics Analysis Workstation) package in the LINUX system. Transformation of the information from TEKTRONIX internal code (files *.sht) to electronic table EXEL format was made by means of program WAVE STAR applied to the oscilloscope. Further, standard procedure PAW has transformed the information to three vectors with dimension 10^4 which contain the information on time, an acoustic signal and intensity of proton beam. Mathematical test of quality of measurements, averaging, fit and the statistical analysis were carried out by means of special PAW macro - programs.

Some results have been obtained using ORIGINE program. Algorithms of selection of signals and calculation of their parameters are described below in the corresponding sections.

2.5. Simulation

Simulation has been carried out within the frame of GEANT-3.21 package [12] which is the tool for the description of experimental installations and simulation of all known physical processes occurring at the passage of microparticles in substance.

The purpose of simulation was the reproduction of the basic properties of an acoustic field in the conditions of the experiment. The passage of 200 MeV proton beam through a water medium has been simulated. The value of energy was selected as a random number from Gauss distribution with the standard deviation equal to the experimental energy dispersion of protons. In transversal plane the beam was limited by lead cylinder collimator with the diameter of 6 cm and thickness of 10 cm. Distribution of protons in this plane was simulated by two-dimensional normal distribution with the standard deviation equal to 1.5cm. The shape of model distribution is presented in Figure 4. The electromagnetic and strong interactions were taken into account during the simulation. For generation of acoustic fluctuations, it is important to reproduce the distribution of energy which was deposited in the substance by the proton beam.

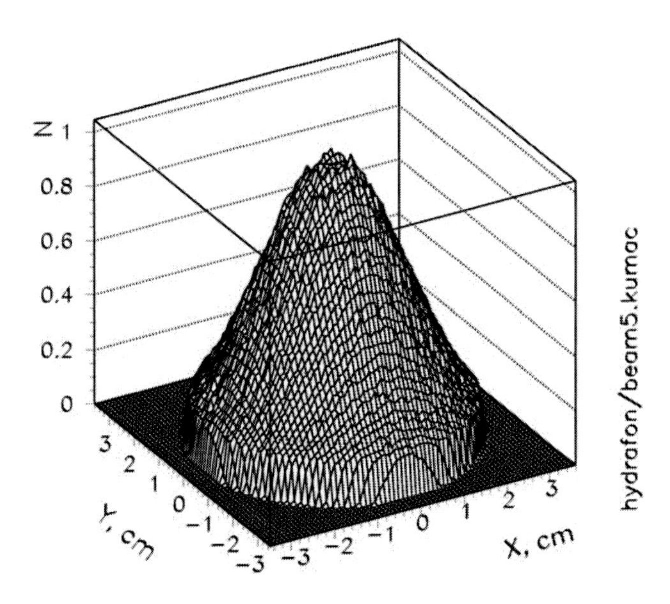

Figure 4. Model distribution of beam protons in a plane perpendicular to the beam axis.

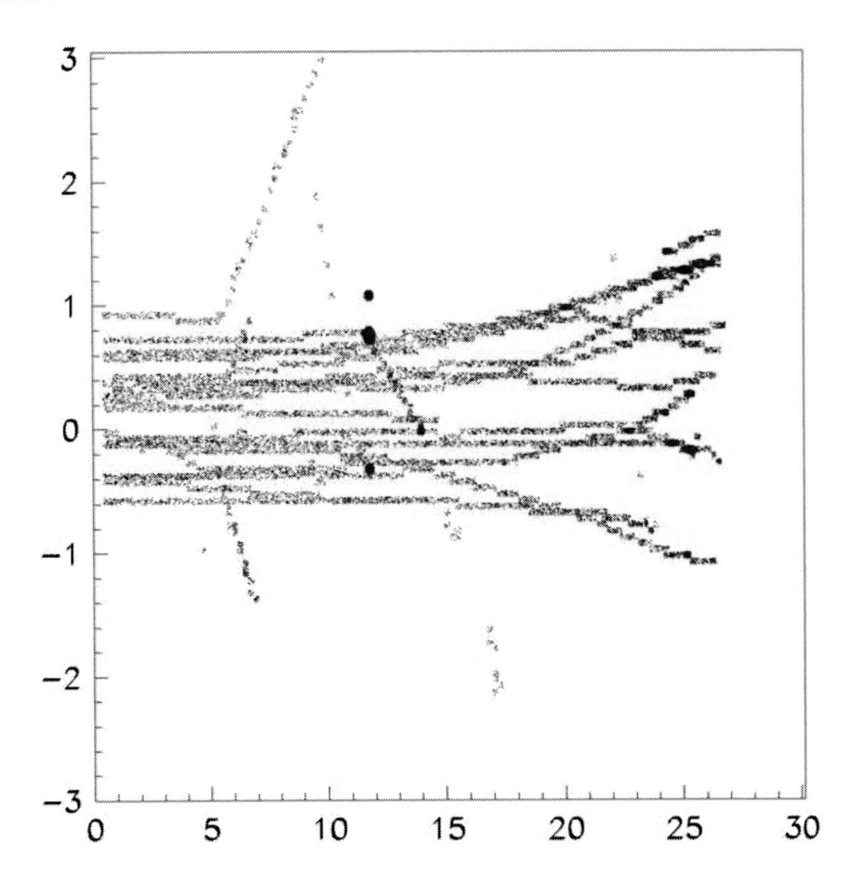

Figure 5. Simulated trajectories of 20 protons in water. Abscissa – coordinate Z, ordinate – X. All in centimeters.

In our conditions the transmission of energy basically is carried out due to ionization of the medium by protons that leads to quick emission of heat in the area of beam action and, as a consequence, to expansion of water and the generation of ultrasonic oscillations. Following the thermoacoustic model of this phenomenon it is supposed, that the intensity of fluctuations is proportional to the energy deposited in matter.

Spatial distribution of the deposited energy is influenced by four factors: distribution of protons in the beam, the dependence of ionization losses on the range of protons, the multiple scattering of protons on atoms of water and by strong interaction with atomic nuclei of hydrogen and oxygen.

For the example 20 trajectories of protons in water are shown in Figure 5.

Owing to multiple scattering and strong interaction, the particles deviate from the initial direction and the area of beam influence extends in transversal direction depending on the passed distance. Simulation of the medium under investigation has been carried out for parallelepiped filled with water.

The size of the parallelepiped in the direction of beam propagation was equal to 44 cm, in transversal direction - 50 x 50 cm^2. The energy $\boldsymbol{\varepsilon}$ deposited by a proton beam in water, and the model of acoustic signals in points of lines I and II in which measurements were made were adopted as a subject of study.

The distributions of energy $\boldsymbol{\varepsilon}$ in basin space were analyzed in the form of two histograms: one-dimensional H_1, representing the distribution along Z axis and two-dimensional histogram H_2 in the plane XZ. The width of channels of histograms along Z was equal to 2.2 mm, along X - 1 mm.

Models of acoustic signals were analyzed in the form of histograms h_i in a time interval up to 400 μs with the accuracy of 1 μs . The signals were reproduced in 70 points of the trace I ($0 < i < 71$) and in 40 points of the trace II ($70 < i < 111$). Let's define coordinates of these points as (X_i, Y_i, Z_i). All the histograms were filled according to the following algorithm. On each step $\boldsymbol{k}$ of GEANT program the value of energy $\boldsymbol{\varepsilon_k}$ deposited in water in a small area and the coordinates of this area (x_k, y_k, z_k) are known. Histograms H_1 and H_2 were filled with weights equal to $\boldsymbol{\varepsilon_k}$. In Figure 6 the distribution H_2 of the energy $\boldsymbol{\varepsilon_k}$ is presented in the projection to the plane containing the beam axis in the form of equidistant contours. On the abscissa axis coordinate Z along beam, on the axis of ordinates - coordinate X are laid off. It is seen from the figure that the maximum of energy deposition concentrates in the area of Bragg peak in the end of the ionization range of protons.

The models of acoustic signals were created in the form of histograms h_i with the following assumptions: 1) the area of energy liberation is considered_a point source of mechanical oscillations of medium, 2) energy of oscillations is proportional to $\boldsymbol{\varepsilon_k}$, 3) there is no mutual influence of adjacent points on the oscillations at the observation point, 4) the response of the acoustic system is proportional to the sound pressure in the observation point. Assumptions 1) and 2) seem to be obvious.

Assumption 3) probably is fulfilled only partly. As to assumption 4) it is, strictly speaking, not fulfilled, since every component of the reception acoustic antenna (hydrophone, the electronic amplifier, cables) has its own amplitude-frequency response which is not taken into account in this experiment. The last assumption leads to the fact that the model calculations can differ from the experimental data outside the ranges of the hydrophones amplitude-frequency characteristics, presented in the table above. One more essential approximation is that the model does not take into account the elastic properties of environment and the properties of heat distribution in water. Therefore the shape of model signals does not describe real signals in detail.

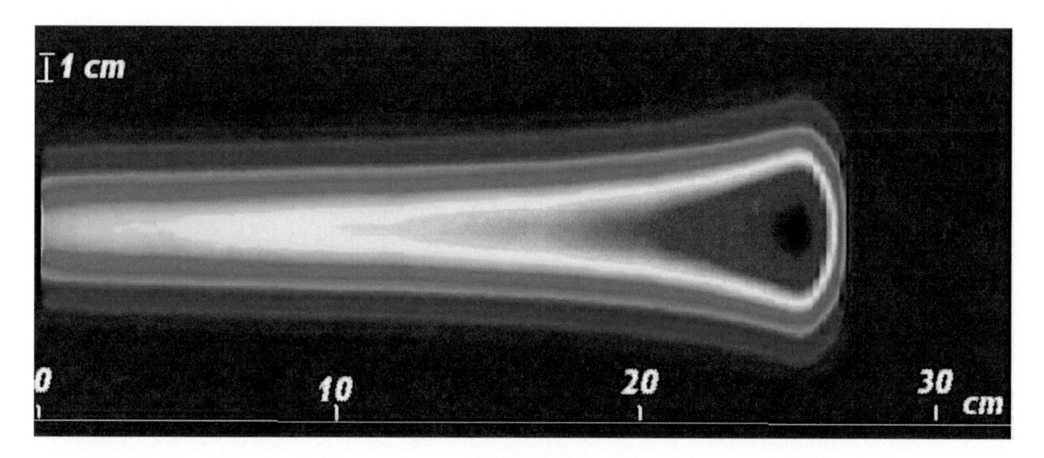

Figure 6. The distribution of simulated energy $\boldsymbol{\varepsilon}$ in coordinates Z (abscissa) and X (ordinate).

At the filling of histograms the time of signal propagation from k-th source to i-th point along the trace was calculated as $\boldsymbol{t_{ik}} = \boldsymbol{R_{ik}}\ \boldsymbol{C}$, where $\boldsymbol{R_{ik}}$ - distance between the points with the co-ordinates (x_k, y_k, z_k) and (X_i, Y_i, Z_i). The sound velocity was adopted to be equal to the value $\boldsymbol{C}$ = 1505 km/s, obtained in the present experiment (see section 3.2 in this chapter). Histograms were filled with the weight W ($\boldsymbol{R_{ik}}$) = $\boldsymbol{\varepsilon_k}$ / F ($\boldsymbol{R_{ik}}$), where F ($\boldsymbol{R_{ik}}$) is the function describing the dependence of signal attenuation. At trace II (in a near zone) it was accepted $F_{II} = \boldsymbol{R_{ik}}^{0.5}$, and at trace I where the signal from the area with sizes much less than $\boldsymbol{R_{ik}}$ is observed, the function $F_I = \boldsymbol{R_{ik}}^{2}$ was used. The analysis of models of signals and comparison with experiment are presented below in this chapter.

3. Trace I

3.1. General Characteristic of the Experimental and Simulated Signals

Trace I (Figure 2) is located in horizontal plane XZ transversely to the beam axis and crosses the beam axis at the distance 36 cm from the point of the beam entrance into water. Measurements were made from the point at X = -15cm to X = 17.93cm. In each point one measurement has been made. As a result of the analysis of quality the measurements of nine last points have been rejected because of failure of the mechanical scanner and of the point i=8 because of electronics failure. In Figure 7 (the upper window) the response of the acoustic system registered in the point i=32 at the distance X =10 cm from the end of the protons range is shown. Time t in μs is laid off as an abscissa, value t = 0 corresponds to the moment of arrival of sync pulse. The positive polarity of the impulse corresponds to increase of pressure of water, negative - to fall. The response is normalized to 1V of the current transformer. In the bottom window the example of oscillograph trace of the electric noise accompanying the work of accelerator is shown. It was obtained in the conditions when the proton beam stopped in front of the basin by the lead moderator with the thickness of 5cm. The noise had an effect during the entire time of the experiment at the trace I. The amplitude and the shape of noise were not stable and changed within 50 % from shot to shot that excluded the separation of a useful signal by a method of background subtraction.

Coming back to the top picture, we can notice, that the registered waves have a complex spectral structure and represent a number of signals of various form and duration, the majority of which do not represent the acoustic response to the passage of the proton beam. The useful signals were analyzed by the stroboscopic method, which is widely used in hydrolocation. The essence of this method is the selection of some interval T_0 on a time axis to increase the probability to reveal the useful signal against noise. The interval T_0 (50 μs < t <300 μs) in which signals from the investigated acoustic antenna are expected, is marked in the figure by a horizontal line with vertical arrows. The interval limits have been chosen taking into account a priori information on the location of trace I with respect to the proton beam. The left limit is a little bit less than the distance to the area of the antenna nearest to the hydrophone - Bragg peak, the right overlaps the sizes of the antenna. In this interval two signals rise above the electric noise.

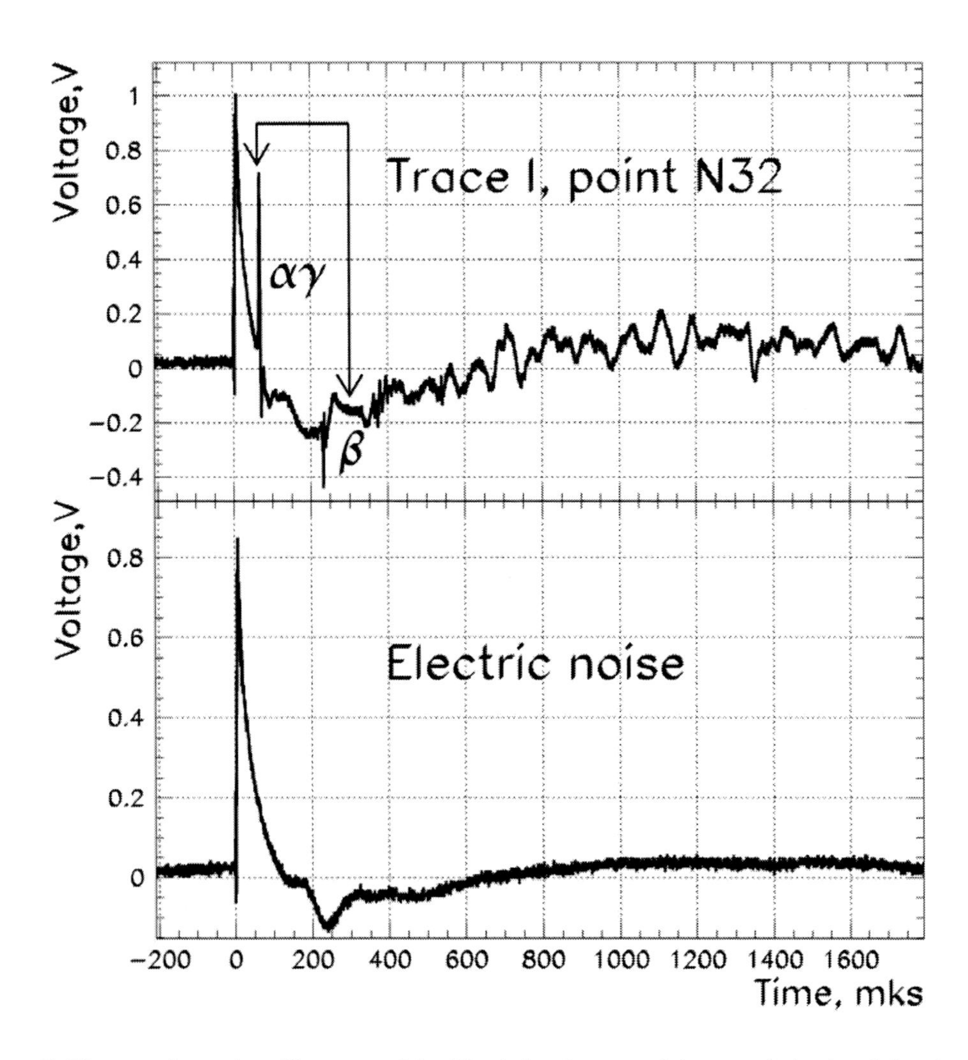

Figure 7. The experimental oscillograms of the 32-nd signal at trace I (upper window) and the electric noise (lower window)

The positive signal designated according to terminology, offered in works [10,11] by symbols αγ comes from Bragg peak which is the area of the antenna nearest to the receiver simultaneously. The source of a negative signal β is the area of plug, located at the point of beam entrance in water (3 in Figure 1). Quasi-harmonic fluctuations outside the interval represent the reflections from the walls of the basin and do not distort the investigated field since the nearest reflected signal arrives130 μs later then the signal β. Only the interval T_0 is analyzed in the present work.

In Figure 8 (above) the response of the acoustic system in the same point № 32 in the range of 50-270̰̃ μs is presented.

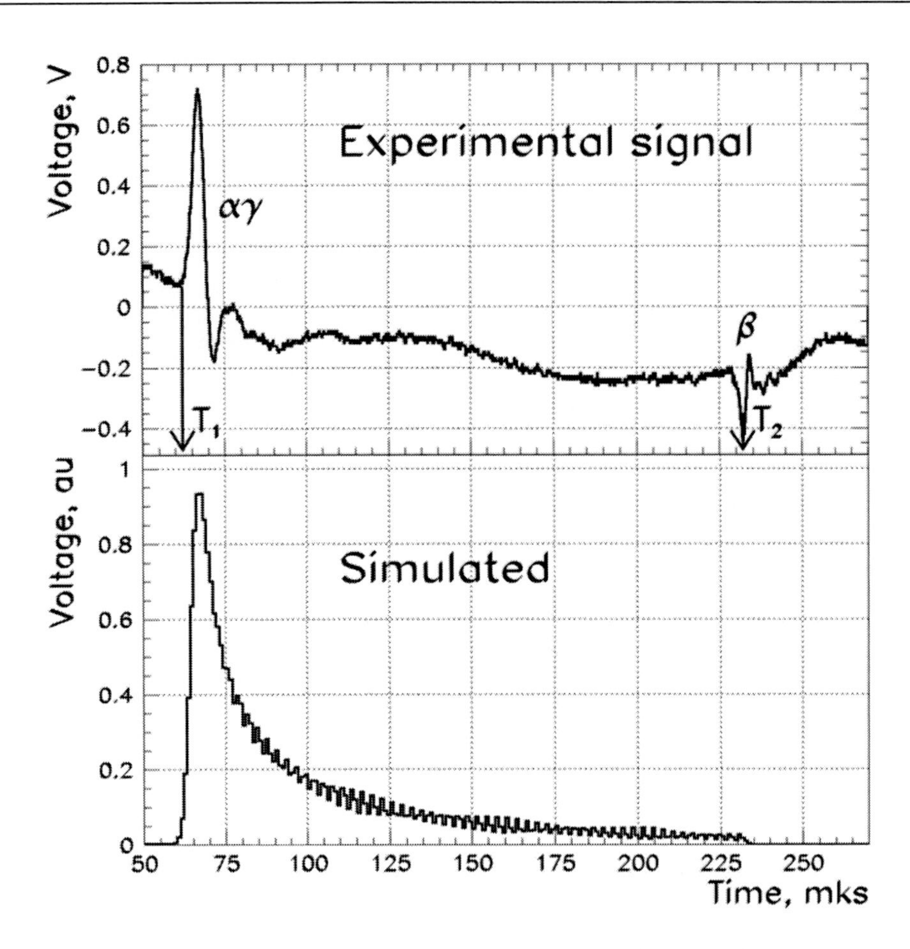

Figure 8. The response of the acoustic system at the point № 32 and simulated signals in this point in the strobing interval 50-270 μs.

The signal $\alpha\gamma$ from the Bragg peak at region T_1 consists of several periods of oscillations with the fading amplitude and variable frequency. The subject of the analysis of the present work is the first positive half-period. The time T_2 of arrival of signal β corresponds to the distance from the point of the proton beam entrance to the hydrophone equal to 352.5 mm. For comparison in the bottom picture in Figure 8 it is shown the model response in the same point - the histogram h_{35} in the same interval of time. Since the model does not take into account some properties of water essential to the formation and propagation of acoustic signals, the model signals do not reproduce the form of real signals. However, they allow us to understand the basic features of the appearing acoustic field. The simple form of model signals is explained by the approximation with which the simulation was fulfilled. The model signal $\alpha\gamma$ begins with the increase to maximum at $t=T_1$ and terminates at $t=T_2$. The signal behaviour between T_1 and T_2 is caused by the sound coming from the lengthy part of the antenna. The coincidence of time and polarity of experimental and model signals β testifies that this signal, partially, can be explained by border effect at the entrance of the proton beam in water. Another possible reason for the appearance of this signal is the oscillation of the plug separating water volume from air. The detailed explanation of source of the signal β is not the subject of the present work.

3.2. The Space-time Picture of an Acoustic Field at Trace I

The space-time picture of the radiation-acoustic field has been reproduced for the first time in papers [10,11] under the experimental conditions close to the present experiment except using a proton beam with another geometrical parameters. In mentioned works the measurements were carried out along three longitudinal traces. In the present experiment the technique of space-time pattern reconstruction, developed in these papers is applied.

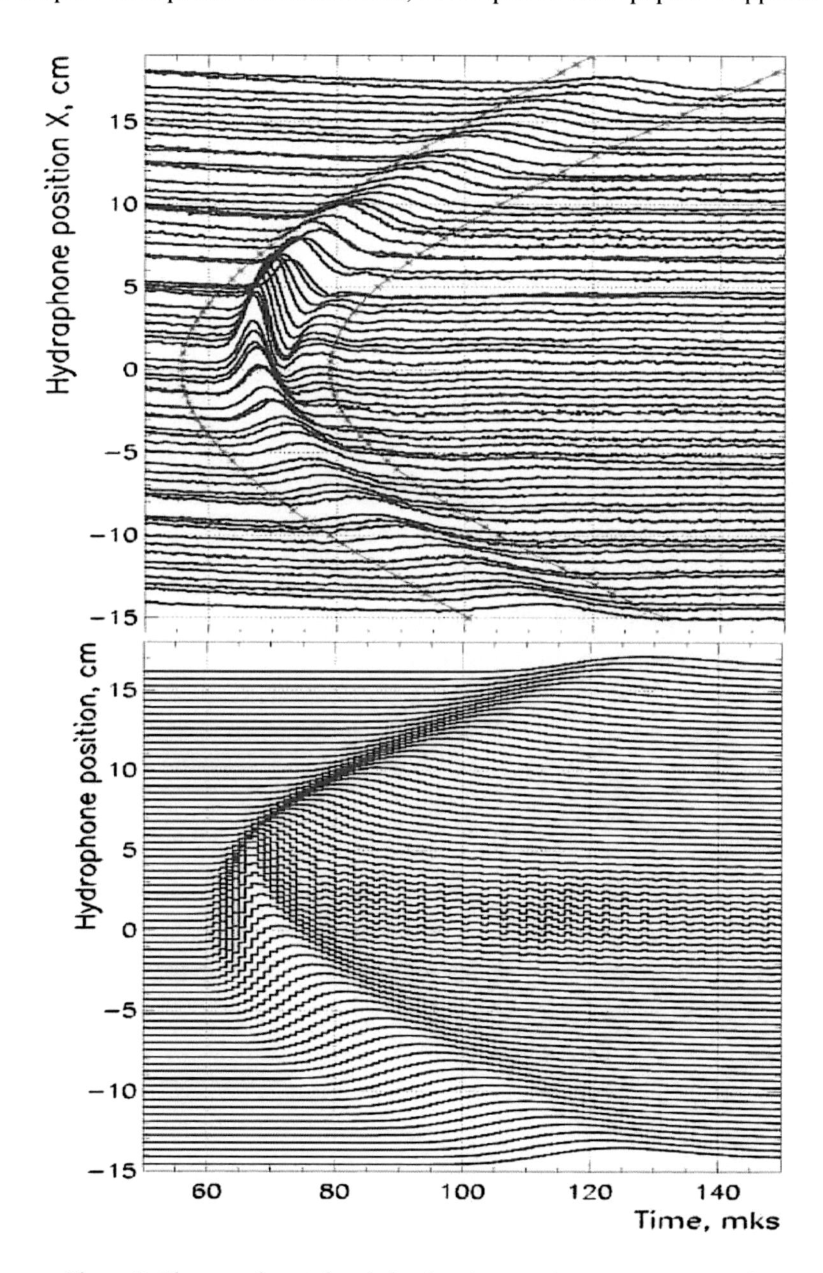

Figure 9. The experimental and simulated space-time pattern at trace I.

In Figure 9 the space-time pictures of the acoustic field registered at trace I are graphically represented: experimental (above) and model (below). Pictures are shown in the form of three-dimensional histograms representing the diagrams of responses of the acoustic system to the sound of the radiation antenna. Along the abscissa the time in microseconds is laid off. For the upper picture this is the time of the oscilloscope beam sweep, the model pictures are represented in the same scale. Time interval is chosen in such a way that the conditions of observation of the signal $\alpha\gamma$ would be optimal. Along the vertical axis the coordinate X of the hydrophone at the moment of registration is laid off. For the experimental histogram along applicate axis the response of the acoustic system is laid off, normalized by 1Volt of the current transformer. For the bottom histogram the applicate axis represents the model form of signal in relative units. Identical trajectories of a signal from Bragg peak are observed in both pictures in the form of a parabola which reflects the time of propagation of a signal from the peak area to the receiver. Some general qualitative conclusions can be made from these pictures concerning the signal $\alpha\gamma$ evolution. In the region $t \cong 70$ μs and $X \cong 0$ the durations of signals are minimal and increase in the process of moving the receiver away from the source. As to signal amplitude, it is minimal at the distant point and increases with the decrease of $|X|$. The identical behavior of signals in the top and bottom pictures justifies that the model correctly reproduces the basic features of the field.

3.3. The Analytical Description of the Signal $\alpha\gamma$ and Calculation of Its Characteristics

Quantitatively signals are characterized by three parameters: amplitude A, time τ at which the signal reaches the maximum value and duration. For the third parameter the standard deviation σ was accepted. For each experimental signal, these parameters were calculated by the method of least squares. The parameters of the approximating function consisting of a linear background and normal distribution, describing *i*-th (i=1, ... 65) acoustic signal $U_i = b_i + c_i t + A_i \exp(-0.5\,((t - \tau_i)/\sigma_i)^2)$, where U_i - the signal value in volts in *i* th measurement; t-time, b_i, c_i, A_i, τ_i and σ_i - parameters, were evaluated. The errors in each bin of the experimental histogram have been chosen as 0.1 from the value U, but not less than 10 mV. The intervals of the strobing time for which fit was carried out are shown in the top picture in figure 9 by asterisks connected by a thin line. The intervals were selected in such a way that they contained the investigated signal with small time segments of the background for the condition of the description of trajectories of borders of intervals by a simple analytical function. In this case the parabola was such a function.

In Figure 10 the results of approximation of fragments of the experimental oscillograms for two signals i=1 (the green histogram) and i=32 (the dark blue histogram) are shown. The approximation of the noise is shown by a black line, by red and yellow lines - the results of approximation of the response U of acoustic system by Gauss distribution. It is seen from the figure that in a distant point (i=1), the response is described well by the approximation function. The χ^2 value on degree of freedom for this point is equal to 0.25. As to the diagrams, registered in the points located near the radiation area, the reliability of the descriptions of experimental signal in the specified intervals is not high - the χ^2 value for i=32 is 25.5. The matter is, that the signal has multipolar form, and the first positive half-period is asymmetrical on time - the forward front has considerably smaller duration, than the back one, and therefore it is not described well by the symmetric Gauss function (Figure 10). Apparently, the form of the acoustic signal reflects the form of radiation area in the given direction - sharp edge effect in the end of the antenna and slower falling of intensity toward the middle of it and further. In the present work the first half-period is analyzed only. For this interval it was accepted that the estimations of parameters A, τ and σ , are close to parameters of the real signal. The χ^2 value on degree of freedom, averaged on all points, was equal to 1.5.

The dependence of the χ^2 value on X coordinate is shown in Figure 11 a).

It follows from the figure, that the signal keeps the Gaussian form in the interval |X |> 2cm, and gradually deforms outside the interval getting in the nearest point the multipolar shape with flat back front. In pictures 11 b-d) the circles represent the dependences of signal parameters on the hydrophone position at trace I. All three parameters have an extremum in point X =0. The values calculated by the least-squares technique are marked by symbols. The pictures confirm numerically the qualitative conclusions concerning the dependence of the parameters A, τ and σ on X from space-time pictures.

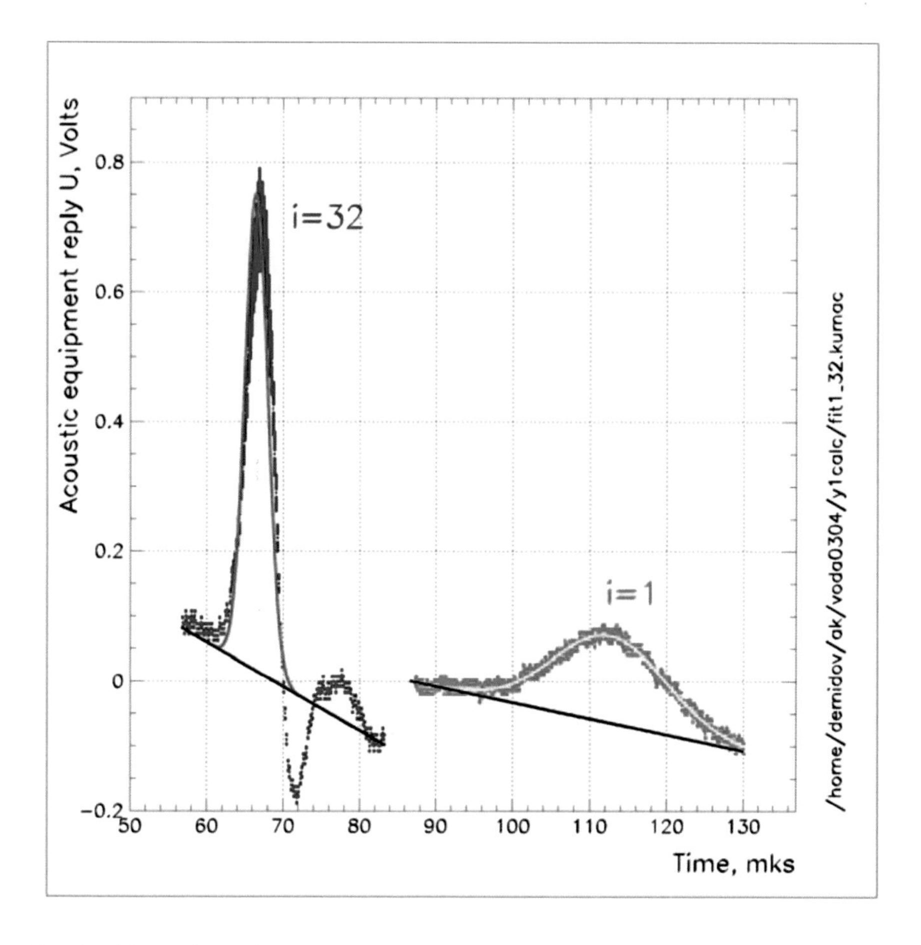

Figure 10. The illustration of two signals approximation by the Gauss distribution and the linear background.

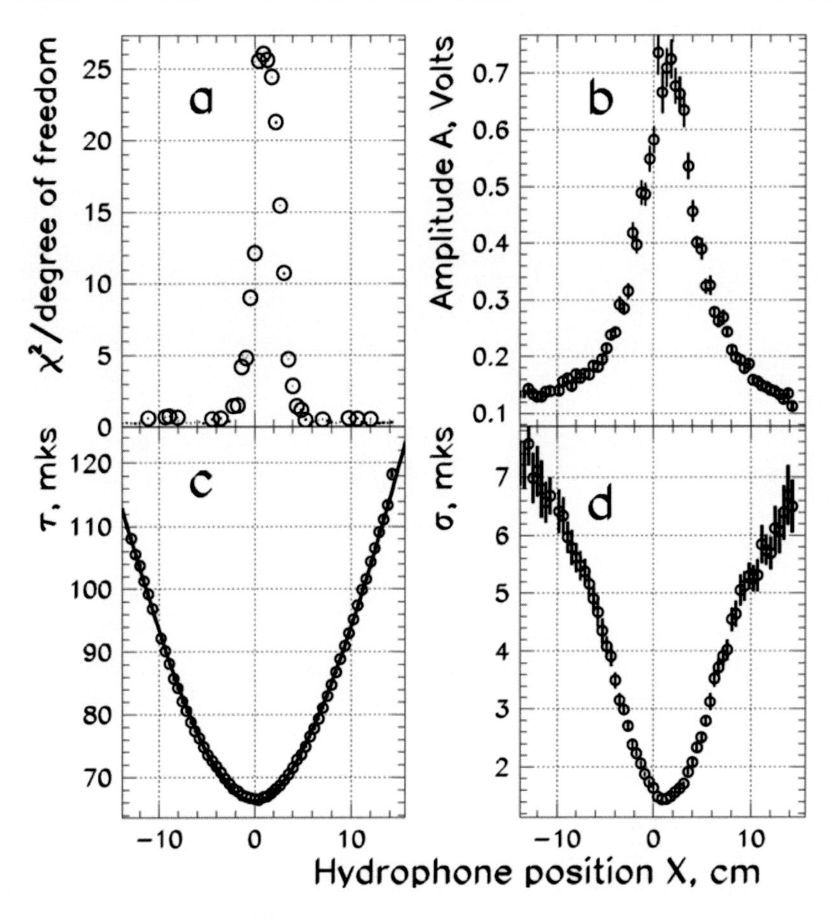

Figure 11. Dependence on the hydrophone position of the values: a) the χ^2/N value of the fitting procedure, b – d) the calculated parameters of the signals. The smooth curve in the window c) – the approximation function described in the text.

Let's notice, that minimal time of signal arrival (Figure 11c) does not coincide with the amplitude maximum (11 b) and the minimum of the signal width (Figure 11d). It can be explained by the fact that the real maximum of the proton beam is shifted along **X** by ~1cm from the centre of the pipe 2 (see Figure 1). Concerning the accuracy of reconstruction of the shape of distribution, we will make the important note. The propagation time of the proton beam does not cause appreciable distortions in the created acoustic field because of small beam duration. During the beam impact, the acoustic wave extends at the distance of an order of 0.1 mm, that is much less the beam size and the dimension of the relative acoustic antenna.

Considering τ to be the time of a sound propagation from the point of the antenna with maximum radiation to the hydrophone, and taking into account that the minimal distance between them is L_1=10 cm, it is possible to define the sound velocity C_0. It has been calculated by fitting of the dependence 11 c) by the function $\tau = C_0^{-1}\sqrt{L_1^2 + (X+l)^2}$ using the method of least squares.

Here l is the possible deviation of an extremum of function from **X** =0 which was calculated by fitting. Graphically the fit results are presented in Figure 11c by the smooth curve. The following numerical values of the parameters have been obtained:

$C_0 = (1505 \pm 8)$ m/s, $l = (0.1 \pm 0.1)$ cm. At that the χ^2 value on one degree of freedom was equal to 2.2. From the obtained value l the accuracy of mechanical adjustment of the measuring system can be estimated as ~ 1mm. As to sound speed its numerical value depends on the way of definition and can be used only within the frame of the present experiment and model.

It is interesting to compare these dependences with model results. The absence of electric noise and the asymmetric form of model signal in the time scale (see for example Figure 8 (lower window) has led to the necessity of working out the other methods of determination of signal characteristics different from those applied to the experimental oscillograms. The peculiarity of model signals is that the investigated signal $\alpha\gamma$ is maximal in all the areas of observation and the algorithm of its program search is based on this fact. For every histogram h_i the value A_i and the position τ_i of the maximum have been found.

The standard deviation σ_i of the signal forward front assuming that it is described by Gauss distribution was accepted as valuation of signal width. How to obtain this value is shown in Figure 12a by the example of histogram h_{35} which is given in the interval 50-90 μs. The histogram is normalized by 1 upon the signal maximum.

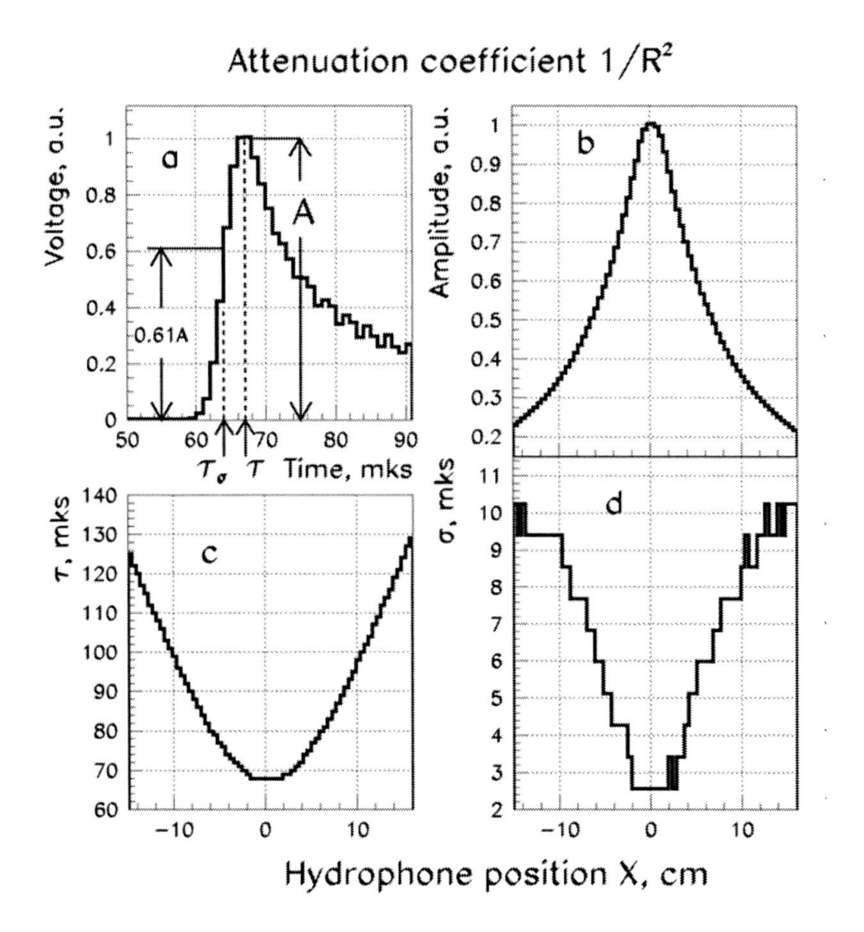

Figure 12. a) Explanation of the calculation procedure of simulated signal parameters at trace I; b-d) dependence of amplitude, time and standard deviation of the signals on the hydrophone position.

The position τ of the maximum A is marked on the abscissa axis by the right arrow. To the left of the maximum there is the bin τ_σ of the histogram with contents equal to 0.61 from the maximum value A within the accuracy of bin width equal to 1 μs. The position of τ_σ is marked on the abscissa axis by the left arrow. Then the value $\sigma = \tau - \tau_\sigma$ was accepted as an estimation of the standard deviation of the signal. Dependences of parameters A_i, τ_i and σ_i on hydrophone coordinate X are shown in windows b)-d) of the figure 12 in the form of histograms. Let us compare these dependences with experimental ones in figure 11. All model histograms, unlike the experimental ones, have an extremum at X=0 (the beam maximum in the model was located in the centre of the basin). The signal amplitude A increases (picture b) in the experiment wore rapidly during the movement of the hydrophone to the beam center. The ratio of the maximum amplitude to minimum one is equal to 6.1 and 4.5 for experiment and model, respectively. It means, that in the experiment the sound attenuation coefficient is slightly higher than in the model where it was accepted, that attenuation is inversely proportional to the square of the distance between the radiation point and the receiver. The time τ, corresponding to the signal amplitude has similar dependence upon X in figures 11 and 12. The small distinction can be explained by the fact that the position of maximum of the approximating function in the experiment not always coincides with the maximum position in the oscillogram. As to the signal width (pictures 11d and 12d) it practically coincides if we take into account that the accuracy of σ evaluation in the model is 1 μs. The first half-period of the response of acoustic system is characterized by frequencies ~ 4.7 σ, that makes ~30 and ~150 kHz for the distant and near point respectively.

Such dependence of the signal frequency on the registration direction reflects the geometrical sizes of the antenna in the given direction - large frequencies correspond to the smaller sizes of radiation area.

3.4. The Directivity Pattern

Calculated in the previous section 405 parameters of the experimental response of acoustic system and model signals allow us to carry out the analysis of dependence of the acoustic field pressure upon the radiation angle, i.e. to calculate experimental and model diagrams analogous to the directivity pattern of a sonic and radio aerials.

The directivity pattern for experimental data can be represented with the dependence 11b presented in polar co-ordinates. The value A_i was taken as a radius-vector for i-th point and the value $\theta = \text{asin}\,(X_i / C_0\tau_i)$ as the angle between the beam axis and the radiation direction. The diagram for model results is plotted in the same way. Both diagrams are presented in Figure 13.

As it was mentioned above, the model diagram is constructed taking into account the basic experimental conditions, including the factor of signal attenuation because of the limited solid angle which was accepted inversely proportional to R^2. The fact, that the experimental diagram has more narrow form, than the model one, can indicate that except the geometrical factor the signal attenuation is influenced by other factors. The establishment of these factors can become as a subject of special research.

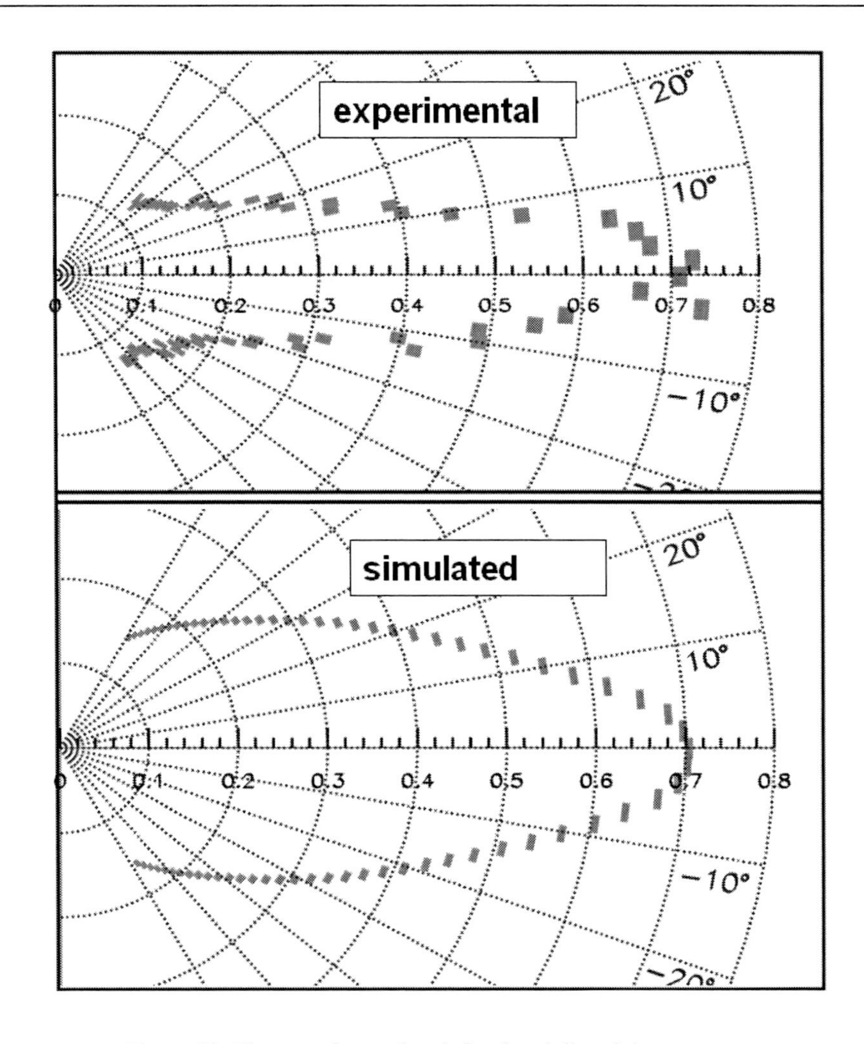

Figure 13. The experimental and simulated directivity patterns.

3.5. Hydroacoustic Location of Most Intensively Radiating Area of the Antenna

The experimental data can be used to define the shape and to calculate coordinates of the most intensively emitting area of the radiation-acoustic antenna, applying the usual methods used in hydroacoustic location for the detection of sources of acoustic radiation. The response to the acoustic signal arising in this area is the impulse marked as $\alpha\gamma$ on the registered oscillograms. Accepting that the signal maximum corresponds to the most intensively radiating area of the antenna and that its width characterizes the width of this area in the measured direction, it is possible to build the contours of radiating area.

The algorithm of building was the following. The contours were plotted in the plane XZ. For each value of coordinate X_i of the hydrophone position time τ_i is known. It follows, that the source is located at a sphere surface of radius $R_i = C_0\tau_i$. In a planar case the source is placed at a circle of the specified radius. If the hydrophone in all points of trace received at the time τ_i a sound from the same source, the arches of all circles plotted from different points

X_i with the different radius R_i (i = 1, …., 65) should cross in one point and coordinates of this point should coincide with the coordinates of the radiation spot. In Figure 14 (the upper picture) the arches of circles of the radius R_i in the proton beam coordinate system are displayed. As a reference point the centre of the end cup of the pipe through which the beam enters into the measuring volume is accepted. Along the abscissa axis the coordinate Z in cm is placed, along the vertical axis coordinate X along which the hydrophone moved is laid off. The first arch (i = 1) is drawn from the point with coordinates X =-15 cm, Z=36 cm, the last (i = 65) has the centre at the point X=13.9cm, Z = 36 cm. Not all the arches intersect in one point which can be explained by the following reasons. First, the present experiment doesn't have the sufficient precision because of the inaccuracy of mechanical adjustment of measuring system, the instability of position of the proton beam in space and the inaccuracy of measurement of time τ_i which corresponds to the signal maximum.

Secondly, at different angles different points of the most intensively radiating area can be registered, and the plain approach used during the mathematical processing, does not allow us to restore the radiation area in three-dimensional space. Nevertheless, such method of an acoustic location is justified: the majority of cross points are located in the expected area of Bragg peak and the centre of this area is established within the accuracy of several millimetres.

Let's notice, that it is not the ultimate accuracy for the given method, at more perfect performance of the experiment the accuracy of location can be raised by several times.

In the bottom window of Figure 14 except the restored "point" of radiation the results of reconstruction of a cross-section of area of the most intensive source of sound are presented in plane XZ.

The cross-section is carried out at the level "plus - minus" of one standard deviation from the radiation maximum. Two series of arches of circles are presented. Dark blue colour marks the arches of circles with radius $R_i = C_0 ((\tau_i+\sigma_i)$, green - $R_i = C_0 ((\tau_i-\sigma_i)$. The arches plotted by a dotted pink line, repeat a picture represented in the top drawing.

It is seen from the figure that the size of the investigated area makes an order of two centimetres in a transversal direction that corresponds to the sizes of proton beam and about 0.5 cm along beam.

Therefore, the developed method of hydroacoustic location allows us to define basically the position and the sizes of Bragg peak arising in liquid at the end of a range of protons within the accuracy of several millimetres. The accuracy of measurement can be improved first of all at the expense of improving the algorithm to calculate the parameters of separate acoustic signals.

4. Trace II

Trace II (see Figure 2) passed along the beam axis in a near zone at the distance of 3.4 cm from the axis, comparable with the diameter of the proton beam. The measurements were carried out with the step s_2 starting from the point with coordinates (X= 3.4,Y=Z=0.) cm. In each point m ($0 < m < 40$) from 3 up to 32 measurements have been fulfilled. In total 334 measurements have been made. At the first stage of mathematical processing the quality of measurements has been tested. The oscillograms without acoustic signals or without a signal in the oscilloscope channel connected with the current transformer were rejected. The number of such oscillograms was 27, at the same time for each point not less than three good measurements remained. At every point the oscillograms were subjected to statistical processing. As a result each histogram was normalised on 1 volt of the current transformer amplitude; average histograms of signals were calculated and histograms of the standard deviation were created for use at some stages of the mathematical processing.

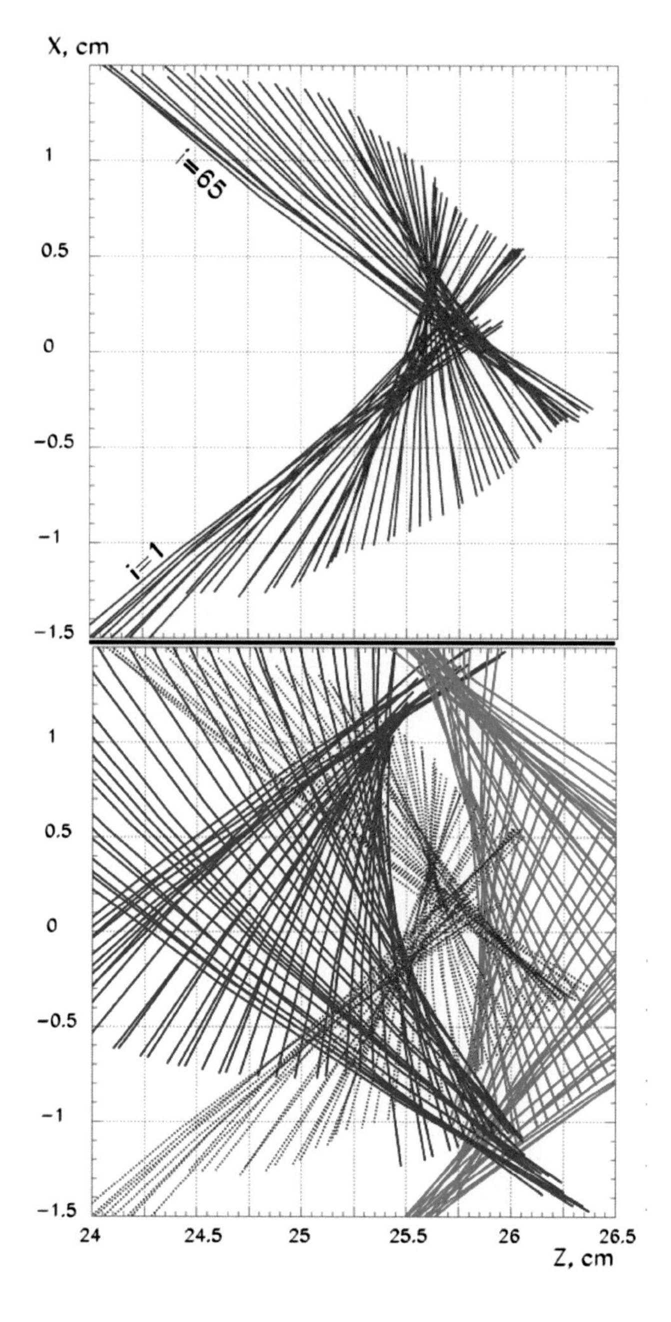

Figure 14. Reconstruction of coordinates of the most loudly sounding area.

4.1. The Structure of the Registered and Model Signals

In Figure 15 two signals - experimental and model in the range of time up to 150 μs are presented. This interval is sufficient for the registration of a sound from the whole acoustic antenna. On the other hand, the reflected waves from the basin walls appear later and do not get into the considered time interval. The signal displayed in the top window, represents the fragment of the experimental oscillogram registered at the point $m = 9$, being located from the point of proton beam entrance to the basin at 7.12 cm along Z axis. At this point 12 measurements were carried out.

In the picture except the average acoustic signal the shaded histogram representing the statistical error at level of one standard deviation, calculated using these measurements is shown. The coordinate t =0 corresponds to the moment of generation of sync pulse.

The response of the acoustic system to the proton beam passage has a complex configuration at this trace and represents a number of waves of various shape and duration. Here, as well as at trace I, the polarity of an electric impulse corresponds to polarity of water pressure. Following the classification introduced in the paper [10] let's consider three signals α,β and γ. The signal α which reaches maximum at t=22.1 μs, corresponds to the time of sound transmission from the centre of the proton beam to the hydrophone. It is formed under the influence of the acoustic field from the nearest to hydrophone area of the cylindrical antenna.

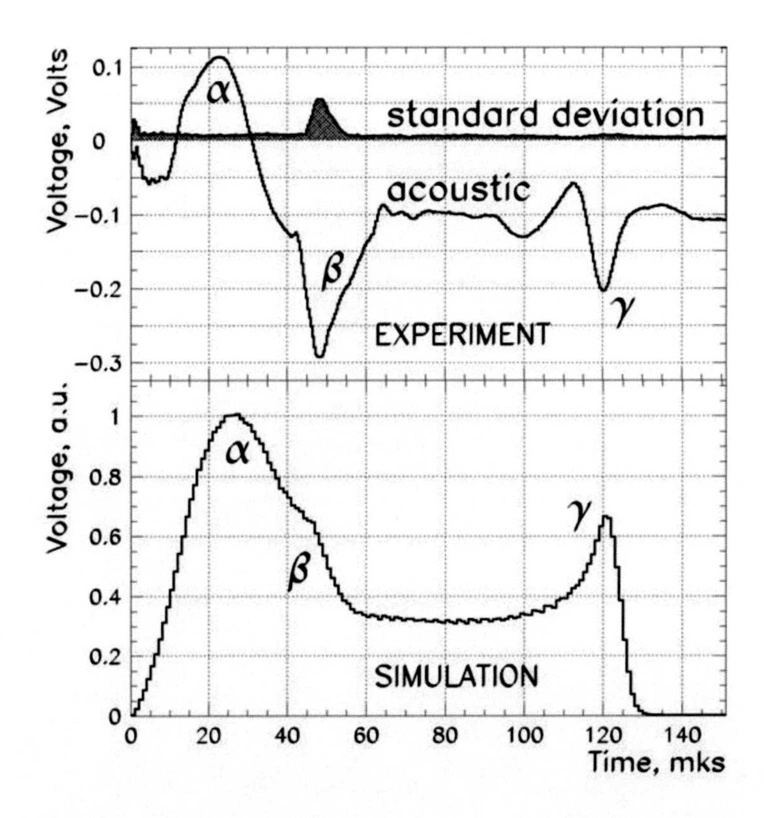

Figure 15. Experimental and model signals at 9-th point at the trace II.

The plug at the entrance point of the proton beam is a source of the signal β of the negative polarity. The signal has a minimum at the time t=48 μs, that corresponds to the distance from the plug to the hydrophone. The hydrophone response to the acoustic impulse, arising in Bragg peak, is the signal γ. The correspondent time of arrival of this signal and the distance to ultrasonic source are equal to 120 μs and 18 cm.

As to model distributions (see lower picture in Figure 15) they have the bimodal form - one maximum is formed by the signal α, another – by the signal γ which has the positive polarity. In the majority of distributions the maximum of the signal α is higher, than that of the signal γ, but there are also opposite cases. Signal β looks like a kink at the back front of the signal α.

4.2. The Space-time Picture of the Acoustic Field Measured on the Trace II

In Figure 16 dark blue lines represent the images of profiles of signals in space "time-coordinate-response" for the experimental data (the top window) and model calculations. The gating of the signal has been fulfilled in the range of $0 < \tau < 200$ μs. The picture was constructed in the same way for measurements on the trace I. Along the axis of ordinates distance Z at which the receiver was located from the origin of coordinates, along the abscissa axis - the fragments of distribution of signals in strobe interval is laid off. The interval overlaps the sizes of the acoustic antenna, but on the other hand there is no noise caused by reflexion of signals from the walls of the basin. The trajectories which are formed by the signals α, β and γ from three sources mentioned above can be seen in the pictures.

The acoustic signal α from the point of the radiating antenna nearest to the hydrophone (the first source) corresponds to the trajectory represented in the form of lengthy crests (half waves of higher pressure) in the direction A-C. Letter B marks the area of beam stoppage and the end of acoustic antenna accordingly. At the initial region AB trace II passes in parallel with the axis of the cylindrical antenna. At that the time of propagation of a signal to a hydrophone remains constant. At this area the cylindrical wave dispersing, according to the theory [2], from the antenna is registered and the trajectory passes in parallel with Z axis.

Futher, the hydrophone, continuing to move along a straight line, moves away from the antenna, time of the signal arrival increases and the signal trajectory on the region B-C changes its direction. The other source, the signal γ, which is presented in the picture as the trajectory DC, is the area of Bragg peak (at the end of the proton range) where the density of the proton beam energy emission sharply increases. When the hydrophone is located at the beginning of the trace II, near to the point of beam entrance into water, the area of Bragg peak is located near the mark D at the distance about 25 cm from the hydrophone that corresponds to the time of τ =160 μs.

At the movement of the hydrophone along the beam, Bragg peak comes nearer the hydrophone and the corresponding branch of trajectory DB goes from right to left. Approaching point B at 20<Z<25cm the interaction of signals α and γ and their conjunction at the site BC is observed. It is important to note, that the signal from this source becomes apparent at site DB in general as a rarefaction wave, and further (at the segment BC) the signal changes polarity, in this direction the wave of the higher pressure propagates.

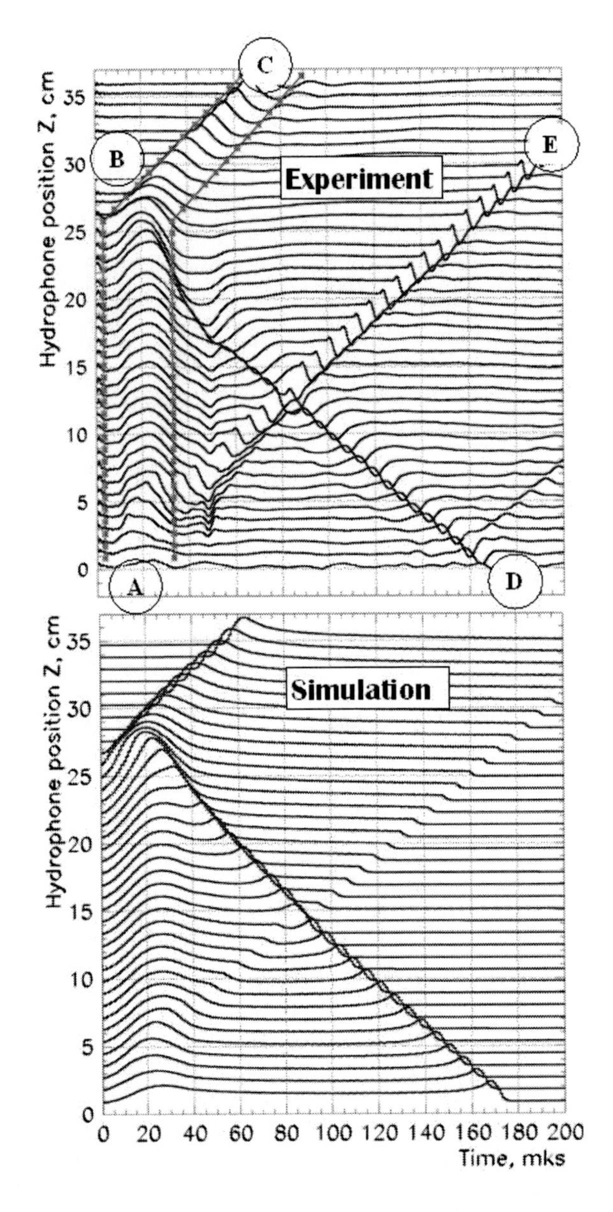

Figure 16. The space-time picture of an acoustic field reconstructed using the results of measurement on trace II and simulated replies.

The trajectory AE of the signal β from the third source goes from the left upwards to the right. It means that at the initial point the hydrophone was located close to a sound source, but in the process of moving the hydrophone along the beam direction, the distance to the source increased. The only area possessing such a property is the area of entrance of the beam into water, and, apparently, the pipe cap separating the water target from the environment served as the source of this signal. The polarity of the signal coincides with the polarity of a signal from Bragg peak at the line DB. The trajectory curvature at the beginning of the trace at $Z<8$ cm occurs owing to the interaction of the given signal (β) with the signal α from the first source.

The space-time picture constructed from model distributions (the bottom window of Figure 16) adequately reproduces the basic characteristics of the experiment: a general view of the picture, peculiarities of trajectories of signals and the behaviour of signals in areas of their interaction.

4.3. Calculation of Parameters of Signal α at the Trace II Using Experimental Data and Model Evaluation

The calculations are fulfilled using the method described in section 3.3. The values of the parameters A,τ and σ of approximating Gauss function were accepted as the estimation of signal amplitude, its center position and duration. The region in which the fit of the experimental signal was made, is shown in Figure 16 by red lines. The average value of χ^2/N_{df} (degrees of freedom) which defines the quality of approximation is equal to 1.1. The dependence of this parameter on Z coordinate of the hydrophone position is shown in Figure 17a. There are three areas in which the parameter has the increased value. Two areas at $Z < 10$ cm and $20 < Z < 25$ cm represent areas of interaction of signals αβ and αγ, correspondingly. The increase of the value of parameter χ^2/N_{df} in the third area, at the end of trace, is connected, as well as in section 3.3, with the approximation of a multipolar signal αγ by the unipolar Gauss function.

The distribution of the radiation intensity along the antenna is shown in Figure 17b. In this case as a numerical estimation of the intensity, the amplitude A (in Volts) of the function, approximating the response of acoustic system, is accepted. The flat part of distribution at the level 0.19 - 0.21 V corresponds to phonation of the cylindrical part of the antenna. In this case the hydrophone accepts the acoustic waves equally from the right and from the left.

The left maximum reaching the value of 0.25 V is formed by signals from the interaction area αβ. The falling of the ultrasonic field intensity at $Z < 5$ cm is connected with boundary conditions: near the point of beam entrance into water the signal arrives to hydrophone only from the right. At the increase of the hydrophone coordinate the importance of the left signal increases and at $7 < Z < 20$ cm the influence of the left and right sides of the antenna becomes equal. The reasons of occurrence of 0.32 V maximum are the increased deposition of energy by a proton beam in the region of Bragg peak and the amplitude enhancement due to the interaction of signals α and γ. The amplitude fall at the region $23 < Z < 27$cm is connected with boundary effect in the end of proton range.

The flat part at $Z > 27$ cm needs the detailed analysis in further experiments. Apparently, it can be explained by a methodical error at mathematical processing, since the fall of signal amplitude should be observed here.

The dependence of the parameter τ, which is the estimate of the position of the maximum of the investigated signal on the position of the registration point is shown in Figure 17c. The constancy of this parameter, observed over the total length of acoustic antenna, confirms that trace II goes in parallel with the beam axis. The small deviation at the initial part of the dependence is explained by αβ - interaction mentioned above. The linear growth of τ at high values of Z reflects the removal of hydrophone from the acoustic antenna as a whole and from Bragg peak in particular.

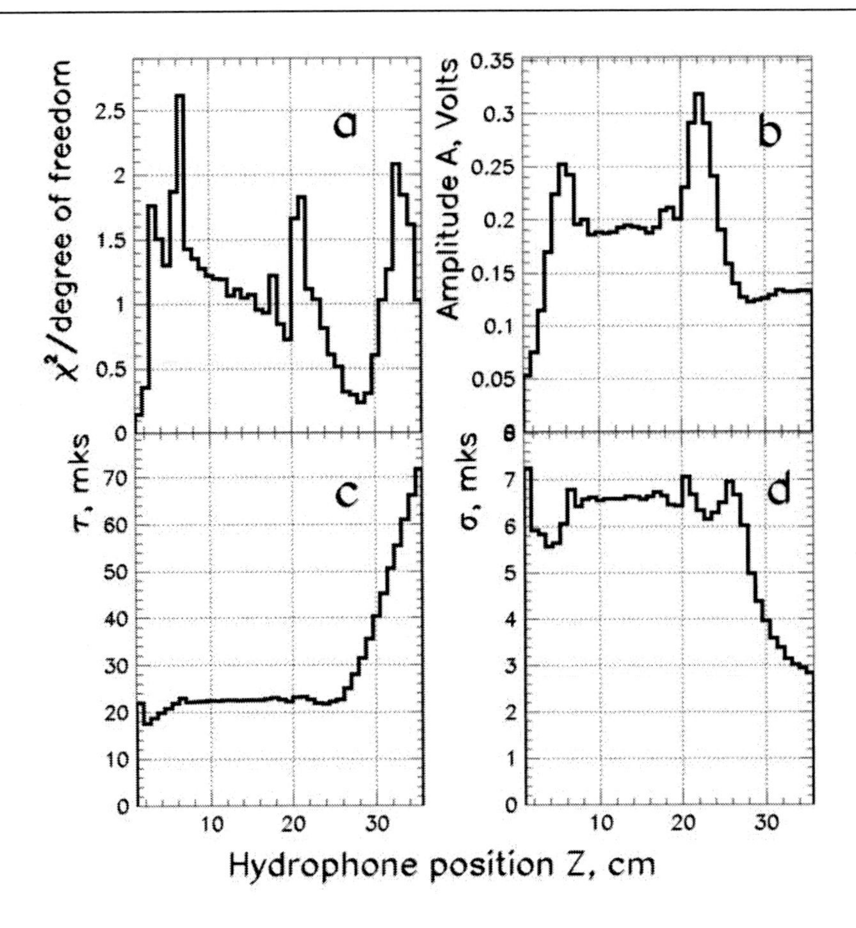

Figure 17. Dependence on Z coordinate of a hydrophone at trace II: a) of value χ^2/N_{df}; b-d) of the parameters of the function, approximating the experimental signal α.

As to the duration of signal, displayed in windows d), it is constant within 10 % practically in the whole antenna area. The behaviour of the dependence at Z> 27cm needs to be confirmed in additional experiments or by mathematical processing by means of other algorithms.

Similar dependences for model signals are presented in Figure 18. The algorithm of an estimation of amplitude A, signal duration and time τ of their propagation from a source to a hydrophone was the same as at the trace I. The additional problem at the trace II had appeared because the distribution of model signals is bimodal. As a result the obtained maximum sometimes relates not to the signal from the nearest point to the hydrophone but to the signal from Bragg peak. In Figure 18a such case, when the procedure of extremum search finds a false maximum, is shown. Analogous situation arises in three points near the area of interaction of signals α and γ. Accordingly, in this area at $19 < X < 22$cm in Figures 18c and 18d false peaks are observed. In the field of interaction αβ the signal from the plug does not prevent the search of the useful signal since the last one has inverse polarity. The other features of behaviour of the signals, except the mentioned bursts in values of time (pictures c) and duration (pictures d) in cylindrical part of the AA, are adequately described by the model. These bursts can be related to the errors of the method of signal processing.

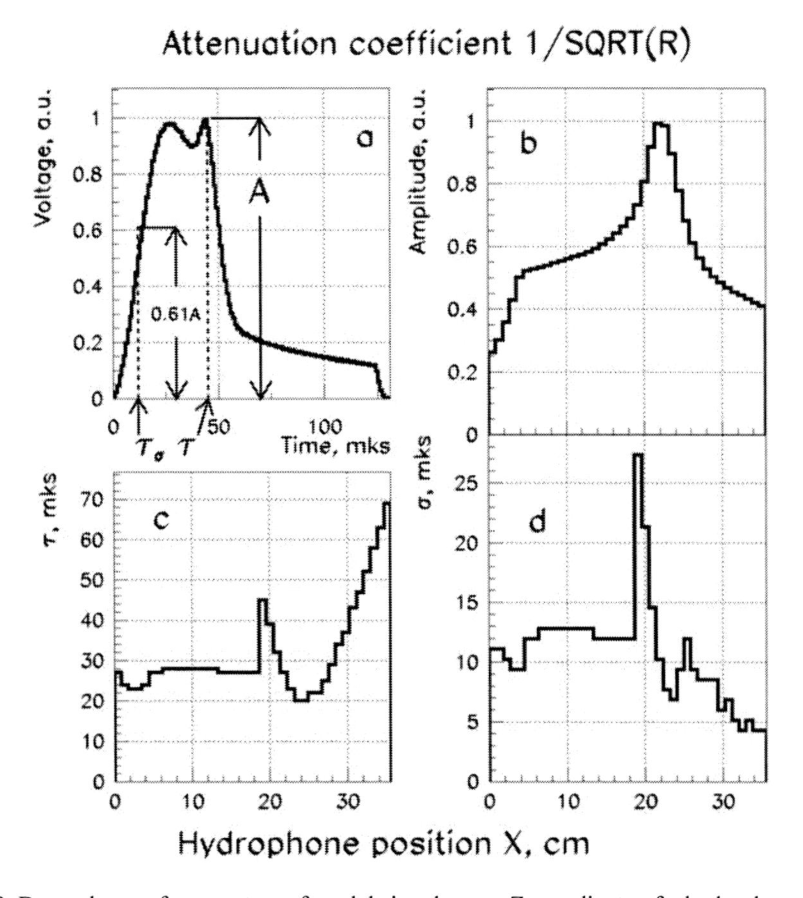

Figure 18. Dependence of parameters of model signals α on Z coordinate of a hydrophone at trace II.

4.4. Reconstruction of the Longitudinal Form of the Proton Beam

Using the obtained parameters of an acoustic signal it is possible to reconstruct the spatial proton beam form. The following hypothesis, that the accepted signal is «the acoustic image» of beam cross-section in the direction of signal propagation and the time of arrival of its maximum displays the most intensively radiating point located on the beam axis, makes it possible to reconstruct the trajectory of this point in plane ZX which passes through the beam axis and the centre of a hydrophone. In Figure 19 the part of this plane including the area of the beam exposure and trace II is represented. The scanning line is drawn by red symbols located in the points of measurement. Also as in the section 3.5 of this chapter, on this plane with the origin of coordinates in the point where the beam axis crosses the plug plane from each point of measurement arches of circles with the radius corresponding to values τ_m have been drawn. These arches are shown by green colour. In general the X coordinate of the most loudly sounding point coincides with the beam axis. The deviation by 0.8 cm upwards is observed only near the edge of the acoustic antenna and, apparently, is connected with the interaction of a signal from the antenna with the sound, arising in the plug.

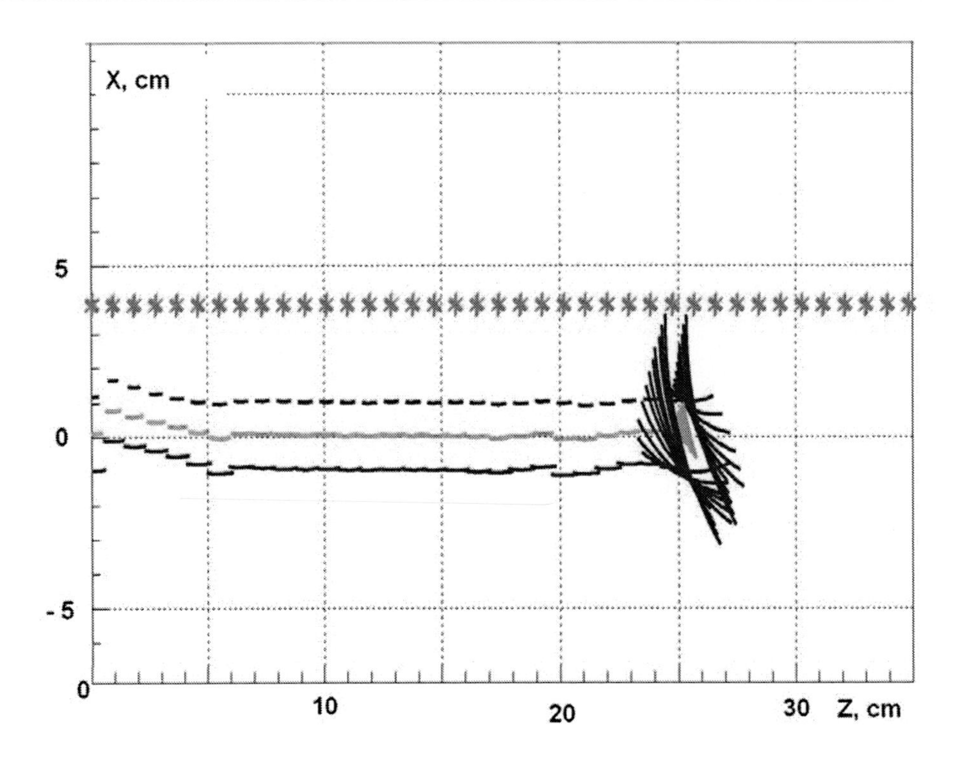

Figure 19. The results of reconstruction of the form of the acoustic antenna created by a proton beam using the experimental data obtained at trace II. Hydrophone positions on the trace at the moment of measurement are designated by red symbols.

The arches represented by a dark blue colour have the radius which are distant from the green trajectory up and down by the value of one standard deviation. It was estimated that between dark blue arches about 68% of protons is concentrated on the average. Unfortunately, we do not possess data about the measurement of the spatial beam distribution obtained by other methods and we can not assert, as far as our conclusions are reliable.

5. Conclusion

By the method of a scanning hydrophone it was shown, that at the braking of an intensive proton beam with energy of 200 MeV in water the area of the beam exposure turns into an acoustic antenna of the cylindrical form. Scanning was carried out along two traces. One of them passed transversely to the beam axes behind the exposure zone at the distance of 10 cm from the end of the beam, another - parallel to the beam direction at the distance close to the beam radius. The analysis of strobes of the registered responses of the acoustic system in three-dimensional space CTVR («Coordinate-Time-Value of Response») had allowed us not only to identify three separate acoustic objects: a signal α from the area of the antenna nearest to the hydrophone, a signal β from area of entrance of a proton beam into water and γ - a signal from the most loudly sounding area in the end of proton range, but to observe their interaction as well.

As a result of the optimal choice of values of the time gating intervals and their trajectories in CTVR space, the parameters of acoustic signals have been calculated first. For the numerical description of the signal parameters a simple function in which the useful signal is described by the normal distribution and the background by a first-degree polynomial has been used. As a result the estimations of the key parameters have been obtained: amplitudes and durations of signals, as well as the time of the signal maximums. The research of correlations between the parameters has allowed us to restore the acoustic image of the proton beam at the level of "plus-minus" of one standard deviation. It represents a lengthy cylindrical part deformed at the ends of the antenna by boundary conditions. It was established, that antenna borders are the sources of the separate signals β and γ, connected with the limitation of area of the beam exposure. The marginal effect at the end of the protons range (signal γ) amplifies due to a sharp increase in energy losses by the proton beam in a small volume of water. The sizes of this area have been measured for the first time.

By the analysis of results of measurement of the signal amplitude and the time of propagation in polar coordinates the angular distribution of the acoustic field - the analogue of the directivity pattern has been constructed. Sharp elongation of distribution along beam direction has not been explained by the simulation procedure results which takes into account the geometry of experiment and attenuation of the signal, inversely proportional to the square of the distance from sound source to the hydrophone.

A full-scale Monte-Carlo simulation of the experiment, in which all kinds of interaction of protons in the water environment, has been performed. The model hydrophone response is calculated in the assumption of its proportionality to the energy emitted by a proton beam at the ionization of atoms of water. The model adequately describes the characteristics of experimentally registered signals, except some details.

Acknowledgments

Work supported by the Russian Agency of the Atomic Energy and Russian Found for Basic Research (grant 02-02-17148-a). The help of V.E.Luckjashin is gratefully acknowledged.

References

[1] Askarijan, G.A. *AE* 1957,3 (N8), 152-153.
[2] Lyamshev, L.M. *Radiation acoustics;* ISBN 5-02-014626-042; Moscow, *Science, Physmatlit*, 1996; pp 1-304.
[3] White, R.M. *J Appl Phys* 1963, 34, 3559-3567.
[4] Sulak, L.; Armstrong, T.; Baranger, H.; et al. *NIM* 1979, 161, 203-217.
[5] Askariyan, G.A.; Dolgoshein, B.A.; Kalinovsky, A.N.; Mokhov, N.V. *NIM* 1979, 164, 267-278.
[6] Learned, J.G. *Phys Rev D* 1979, 19 (³11), 3293-3307.
[7] Albul, V.I.; Buchkov, V.B.; et al. *NET* 2001, №3, 327-334.
[8] Hunter, S.D.; Jones, W.V.; Malbrough, O.J. *JASAm* 1981, 69 (№ 6), 1557-1562.
[9] Albul, V.I.; Buchkov, V.B.; et al. *PTE* 2004, №4, 94-99.
[10] Albul, V.I.; Buchkov, V.B.; et al. *PTE* 2004, №4, 89-93.

[11] Albul, V.I.; Buchkov, V.B.; et al. *AJ* 2005, 51 (^{3}1), 47-51.

[12] GEANT Detector Description and Simulation Tool (Long Writeup W5013) *CERN Program Library,* CERN Geneva, Switzerland, 1995.

In: New Research on Acoustics
Editor: Benjamin N. Weiss, pp. 141-157
ISBN: 978-1-60456-403-7

Chapter 3

TRANSFER EFFECTS ON AUDITORY SKILLS FROM PLAYING VIRTUAL THREE-DIMENSIONAL AUDITORY DISPLAY GAMES

Akio Honda*[1, *]*, Hiroshi Shibata*[2]*, Souta Hidaka*[2]*,
***Yukio Iwaya*[3]*, Jiro Gyoba*[2] *and Yôiti Suzuki*[3]**
[1]Department of Psychology, Faculty of Humanities,
Iwaki Meisei University, 5-5-1 Chuodai Iino, Iwaki City, Fukushima, 970-8551 Japan
[2]Department of Psychology, Graduate School of Arts and Letters,
Tohoku University, 27-1, Kawauchi, Aoba-ku, Sendai, 980-8576 Japan
[3]Research Institute of Electrical Communication and Graduate School of Information Sciences, Tohoku University, 2-1-1, Katahira, Aoba-ku, Sendai, 980-8577 Japan

Abstract

Along with the advancement of virtual auditory displays (VAD), which render three-dimensional auditory perceptual space by controlling sound paths drawn from a sound source to a listener's ears, auditory games with VAD have been attracting increasing interest. The VAD games offer advantages over visual action-video-games in that both sighted persons and visually impaired people can enjoy them. Although some previous studies have attempted to apply auditory virtual reality games to the auditory education of visually impaired people, few studies have investigated the transfer effects of playing virtual auditory games. In this chapter, we present VAD games as an effective training tool for skills related to auditory information processing in daily life situations. Our studies investigated transfer effects on various auditory skills from playing VAD games. We particularly confirmed transfer effects in the following human aspects: sound localization performance, communication behaviors in face-to-face situations, and avoidance behavior from approaching objects. Finally, we propose new perspectives and future applications of auditory games which use the VAD system.

* Corresponding author. Tel: +81 246-29-5111; Fax: +81 246-29-5105 E-mail address: honda@iwakimu.ac.jp

1. Introduction

A virtual auditory display (VAD) is a system for generating spatialized sounds and conveying them to a listener. The VAD techniques for creating three-dimensional virtual sound sources are typically based on convolving head-related transfer function (HRTF) to a sound source signal [1]. In fact, HRTF is related to the sound propagation path from a sound source to a listener's ear and is dependent on the sound source position. For that reason, HRTFs to the listener's two ears comprehensively involve perceptual cues to localize the sound source position [2]. Langendijka and Bronkhorst [3] have reported on how VAD techniques have applications in various fields such as avionics [4–6] and music reproduction [2]. Recently, this technology has been applied to some virtual auditory games. In this chapter, we introduce the findings of our studies [7, 8] and demonstrate that VAD games can be an effective training tool for skills related to auditory information processing in daily life situations. We also describe new perspectives and propose future applications for auditory games with VAD systems.

Previous studies have examined the transfer effects of playing action video games [9–12]. For example, Green and Bavelier [10] showed that action video game playing can alter the range of visual skills and modify visual selective attention. In addition, a recent study showed that playing action video games can alter fundamental characteristics of the visual system such as the spatial resolution of visual processing across the visual field [11]. Moreover, Fery and Ponserre [12] reported a positive transfer of golf video game playing to actual putting skills. Results of those studies underscore that transfer effects are based on perceptual motor-learning in visual action video game playing. Transfer effects are commonly observed for various motor as well as verbal learning tasks. Notwithstanding, few studies have investigated transfer effects of playing auditory games. Actually, VAD games offer advantages over visual action-video-games in that both sighted persons and visually impaired people can enjoy auditory games. Some researchers have tested various newly developed VAD games. For example, Röber and Masuch [13] designed three simple action games and one story-based adventure. Results of their study showed that the majority of participants had only very brief or no experiences with audio games, but nearly everyone liked the idea and found the concept interesting as well as challenging [13]. In another study, Winberg and Hellström [14] proposed an auditory version of the game "Towers of Hanoi". In the game, a player is presented with three towers and three or more discs, which are stacked on the left-most tower. The object of the game is to move the discs from the left to the right-most tower. The player can place a disc on any of the three towers as long as he or she does not move a larger disc on top of a smaller one. All participants reported that the game was entertaining despite its complexity [14]. Recent studies have used psychological and behavioral testing to investigate the use of VAD games as auditory training tools. For example, Ohuchi *et al.* [15] proposed the use of VAD techniques in auditory skills education for visually impaired people. In this study, a three-dimensional sound game called "Hoy-Pippi" was developed with the VAD technique. The characteristic feature of this game was a "touching sound," namely, positioning a hand at the position of an imaginary creature called "Hoy-Pippi", which was rendered as a virtual point sound source. Training with the game, which was conducted for 10 days, involved 10 listeners who were sighted persons with normal hearing: 5 were assigned to the trained group, and 5 were assigned to the non-trained group. All participants conducted all tasks while blindfolded. The ability to identify the sound source position of each listener was measured before and after training. A comparison of results revealed that the ability to identify the sound source position improved only in the trained group, suggesting that the VAD game is an effective training tool for sound localization skills supporting visual impaired people [15]. Afonso *et al.* [16] also examined spatial cognition in an immersive virtual audio environment. Participants in their study were asked to acquire knowledge of a scene either through a purely verbal description or active exploration of the virtual scene. The results indicated that active exploration of a virtual-reality environment enhances absolute positioning of sound sources compared to that obtained from learning using verbal descriptions [16]. These findings suggest that VAD technologies can provide effective auditory training tools for visually impaired people. However, few studies have investigated that auditory training using a VAD game transferred various skills related to auditory information processing in daily life situations. Our previous studies examined thoroughly whether the VAD game can be an effective training tool for various auditory skills using behavioral and psychological testing [7, 8]. In particular, we investigated transfer effects related to sound localization performance [7], communication behaviors in face-to-face situations [8], and avoidance behavior from approaching objects [8]. For example, vision-impaired people are known to have better auditory perception capabilities than sighted people, especially those capabilities associated with sound localization. In fact, Lessard *et al.* [17] found that people who became sightless at an early age can localize sound sources better than sighted people. Nevertheless, that same study showed that vision-impaired individuals with residential peripheral vision localized sounds less precisely than either sighted or entirely sightless people. Moreover, Gougoux *et al.* [18] found that people who became sightless in infancy have better listening skills than those who became sightless later. These findings indicate that sensory compensation varies according to the etiology and extent of vision impairment [17] and according to the age at which blindness occurs [18]. Therefore, early support to improve sound localization skills has important meaning for improving the quality of life for people with low vision or late blindness. For that reason, we examined the effects of playing the VAD game transfer to sound localization performance [7]. In addition, we considered the transfer effects of communication skills in a face-to-face situation because we anticipated that effective communication skills would be related to sound localization skills [8]. Eye contact in face-to-face situations has a regulatory function [19]. The use of eye contact in social interaction with sighted people is a critical component of rewarding social exchanges [20]; for instance, the breaking off of listener eye contact indicates disagreement or dissatisfaction with another's speech [21]. Results of these studies revealed that normal sighted people consider that eye/face-contact plays important roles in social interaction. In contrast, visually impaired people used more non-visual cues in social interaction [22]. The difference of communication cues affects impressions of visually impaired people. In fact, Raver [23] found that visually impaired children were perceived by sighted adults as possessing less positive social and intellectual attributes when they did not manifest gaze direction than when gaze direction was manifested. Several researchers attempted to find effective training methods for communication skills education supporting visually impaired people [24–26]. For example, Sanders and Goldberg [25] proposed a training program for increasing the rate of eye contact in social interactions. This method used auditory feedback for the correct eye/face contact. They reported that the eye contact of clients (almost totally blind men) was increased to over 80% and the effect remained at the 74% level after 10 months [25]. The findings of that study suggest that visually impaired people can be taught more effective communication skills by training sound localization skills through perceptual-motor learning. We predicted that the sound localization training with the VAD game would be transferred to the communication behaviors in social interaction [8]. Moreover, the trained sound localization skills might be transferred also to avoidance behaviors in relation to approaching objects [8]. When avoiding such looming objects, the generation of an appropriate response includes five tasks: detection of a looming stimulus, localization of the stimulus position, computation of the direction of the stimulus movement, determination of an escape direction, and selection of a proper motor action [27]. Moreover, the ability to avoid unwanted collisions is a critical aspect of adaptive behavior [28–30]. People with normal vision are able to use not only auditory information but also visual information when they judge the object's approaching course or time to arrival. Previous studies have revealed that visual information tends to be used more efficiently than auditory information in accuracy of judging time to arrival [30]. However, visually impaired people are obviously restricted in their use of visual cues. Consequently, for them it is very important to formulate and execute perform avoidance behaviors using acoustical information. For performing collision avoidance behaviors, visually impaired people must perceive a correct location of an approaching object (sound source) using auditory information. Furthermore, when the object approaches on a collision course, they must avoid it with minimum distance from their own position because avoidance with greater distance might cause another collision with surrounding obstacles, which might be very difficult to detect from auditory information. Therefore, when visually impaired people try to conduct appropriate avoidance behaviors, it is crucial that they are able to perceive sound source positions accurately. Appropriate avoidance behaviors relate to sound localization skills. For that reason, we predicted that sound localization training using VAD games would be transferred to avoidance behaviors in response to approaching objects. Previous studies have demonstrated that VAD games are effective in improving the sound localization skill [15]. Results of other studies suggest that higher sound localization skill is important to conduct better communication and practice more appropriate avoidance behaviors based on auditory information. However, few studies have investigated transfer effects on various auditory skills from playing a VAD game. In our previous studies [7, 8], we confirmed transfer effects in the following human aspects: sound localization performance, communication behaviors in face-to-face situations, and avoidance behavior from approaching objects. In this chapter, we describe the transfer effects of playing VAD games on skills related to auditory information processing in daily life situations. We propose new perspectives and future applications of VAD games.

2. The VAD System and the Newly Developed VAD Game

2.1. The VAD System Used in Our Studies

This section will introduce the functions of the VAD used in our studies. Our VAD system (Simulative environment for three-dimensional Acoustic Software: SifASo) was installed on a PC (Pentium 4, 3.2 GHz) running on Windows XP (Microsoft Corp.). The development environment consisted of Microsoft Corp. Visual C++.NET□Microsoft Corp. Direct X 9.0 for graphic application program interface (API) and sound APT, and MAUDIO's DELTA Audiophile 2496 with audio stream in/out (ASIO) used as a sound card. We set the sampling frequency at 32 kHz and the quantization bit rate at 16 bits. The VAD engine had a multithread construction. Specifically, a convolution process generates convolution threads when a convolution method is required. Once the rendered sound has been computed, the data are buffered in the ring buffer and are mixed with other sounds in the mixer thread; the sound is finally passed to the Direct Sound interface or ASIO to output device. In our VAD system, impulse responses corresponding to HRTFs were measured using a spherical speaker array (see Figure 1) installed in an anechoic room of the Research Institute of Electrical Communication, Tohoku University.

The HRTFs for sound sources located 1.5 m from the center of the spherical array were measured using an equal interval angle of 5° in the horizontal plane and 10° in the medium plane. The pulse signals to measure HRTFs were Optimized Aoshima's Time Stretched Pulse (OATSP) [31] with sampling frequency of 48 kHz and length of 8.192 points. For one direction, OATSP was generated four times for synchronized summation to improve the signal-to-noise ratio. The response to OATSP was processed to determine the head-related impulse response (HRIR), which is the IFFT of HRTF, to the direction. Then, calculated HRIRs were downsampled to 32 kHz and truncated to 128 points adequately using a rectangle window to lessen the VAD engine calculation load. The HRTFs or HRIRs for a direction that was not measured were calculated through interpolation between the nearest four HRTF (HRIR) values. To achieve stable sound localization against a listener's head movement, a three-dimensional gyro-sensor (MDP-A3U9S; NEC-Tokin Corp.) was attached to the top of the listener's headphones to monitor the head angle data and appropriate HRTFs convolved in real time. The head angle data were transmitted from the gyro-sensor at 1/125 s (8.0 ms) intervals. The head angles were temporally interpolated for the 8.0 ms span based on the method proposed by Watanabe *et al.* [32] to compensate for any discontinuity that might have occurred because of large/fast head movements between intervals. The VAD engine itself had a system latency of 16.4 ms. Data from the gyro-sensor on the headphones were transmitted at 8.0 ms intervals, as described above. Therefore, a maximum delay of 16 ms long was anticipated because of asynchronous transmission via the USB interface for the gyro-sensor as well as temporal interpolation. In all, the maximum value of the total system latency (TSL) including the latency of the gyro-sensor is 32.4 ms. To render a virtual sound source, Yairi *et al.* [33] measured the average absolute and discrimination thresholds for the latency of VADs because of the listener's head movements. The result was some 75 ms in average [33]. Consequently, the maximum total system latency of 32.4 ms of our VAD system seems well shorter than the absolute and discrimination thresholds for the latency of VADs. The VAD engine that we developed can render several acoustic effects including Doppler effects, early reflections, late reverberation, distance decay, and air absorption. Our research group has used this VAD engine not only for VAD games introduced in this paper, but also in relation to the content of cognitive mapping training methods that have been studied for their influence on human cognition and behavior through psychological experiments [7][8][34]□We compare the characteristics of our VAD engine to those of other VAD engines that were proposed previously such as Miller's DIVA [35] and Lokki's SLAB [36]. The SifASo that we developed was used with SLAB, which was developed as a Windows application. The total system latency of SLAB is approximately 24 ms; it is possible to perform dynamic switchovers of HRTFs because of head movements and rendering of Doppler effects□However, SLAB does not render acoustic effects such as air absorption and late reverberation. In contrast, DIVA was developed on Unix and can render early reflections, late reverberation, air absorption, and some other acoustic effects. A disadvantage is the total system latency of DIVA, which is larger than the detection threshold: approximately 100 m. Head movement data are updated only 20 times per second. Although these two software systems are highly sophisticated, neither fully fits the applications that the authors would like to have developed. For that reason, we developed our own VAD engine, which has high and well-balanced performance, to develop application systems for visually impaired people.

Figure 1. Spherical speaker array.

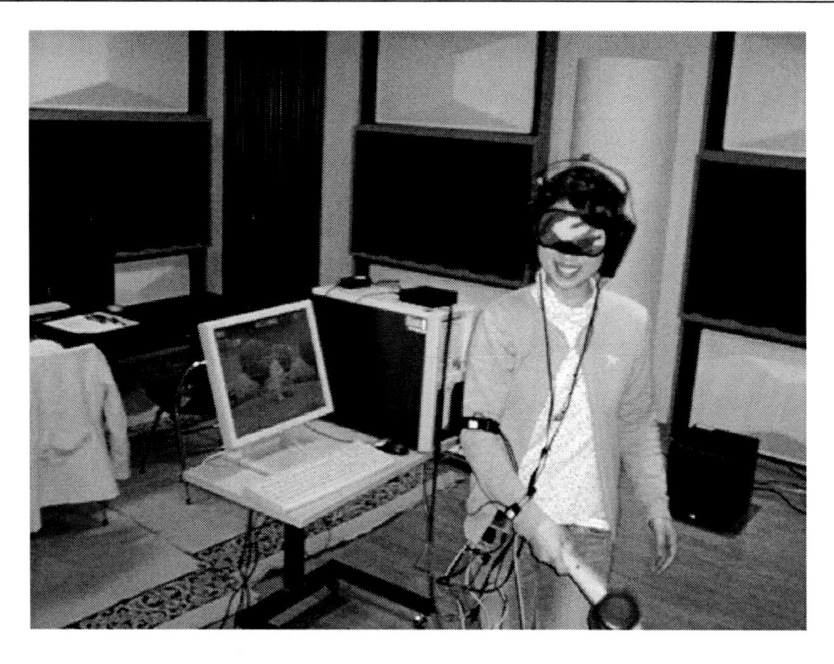

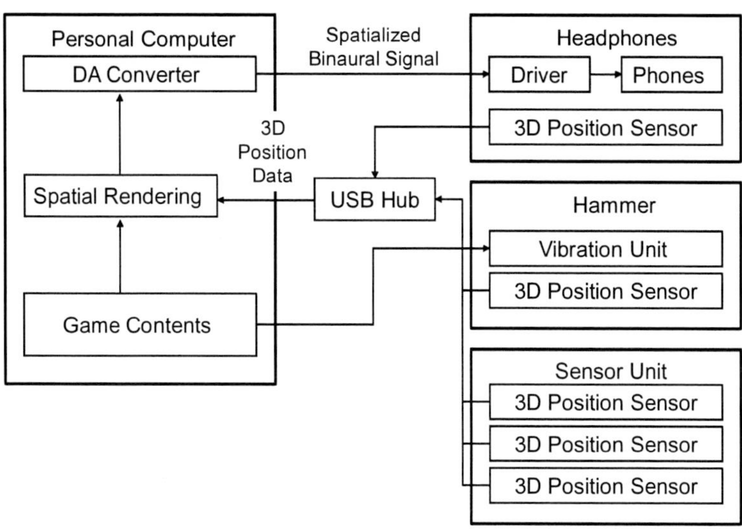

Figure 2. Auditory training with the VAD game and the architecture of this VAD system.

2.2. Our Newly Developed VAD Game

For our studies [7, 8], we used a newly developed VAD game called "*Bee Bee Beat*". The VAD game is an action game that is intended to train sound localization skills of its players. This game resembles the popular whack-a-mole game, but the player must hit flying bees instead of moles. A sphere of 0.6 m radius is considered as the play area. The center of both shoulders of the participant is treated as the sphere's center. In playing the game, a honeybee, represented by its buzz sound, appears at an arbitrary position on the sphere surface. Because HRTFs are measured from 1.5 m distance, the sound pressure level is adjusted to the level from 0.6 m with a decrease by 6 dB for doubling the distance. The buzz sound consists of an AM triangle wave (200 Hz) modulated with a 25-Hz sinusoidal envelope, with a sampling frequency of 32 kHz. The bees' positions of appearance are determined automatically and randomly. The game player has a plastic hammer that is used to hit the honeybee. A three-dimensional gyro-sensor (MDP-A3U9S; NEC-Tokin Corp.) and a vibration unit are installed in the hammer. Another three-dimensional gyro-sensor is attached to the headphones' headband, as described above, to sense the orientation of the listener so that the honeybee's buzzing can be localized steadily in real-world coordinates through convolution of HRTFs appropriately relative to the position of the honeybee according to the listener's movement. This real-time signal processing to reflect the influence of the listener's dynamic motion is remarkably effective for realizing correct sound localization [37].

The player's task can be summarized as follows: First, the player listens attentively to the sound of the bee's appearance (target stimulus) in the background music. Secondly, the player must detect the target sound as rapidly as possible. At this time, the player frequently uses head movements to detect the target sound. Thirdly, the player turns to face a location of the presented target and looks at the sound location. Finally, the player localizes the target sound position using the hammer. The player receives an immediate vibration feedback signal from the hammer; another honeybee is given to the player when the sound localization is correct (see Figure 2). In this game, the player is asked to localize the honeybee position and hit it with the hammer as quickly and correctly as possible. The radius of the hammer's hit-zone is 5 cm. A hit is judged whether the bee is within the hit-zone; if it is, then the player receives one point. Consequently, although the sound bee position is defined two-dimensionally (azimuth, elevation, and fixed distance), the player is required to hit the sound position correctly in three-dimensional space.

3. Transfer Effects on Various Auditory Skills from Playing the VAD Game

Previous studies have not investigated that a VAD game can be an effective training tool for various auditory skills related to auditory information processing in daily life situations. Our studies examined transfer effects on various auditory skills from playing the VAD game [7, 8]. We examined the performances of 40 participants: 20 women and 20 men, who are university graduate and undergraduate students with ages of 19–25 years [7][8]. All participants reported that they had normal hearing, normal or corrected normal vision, and no prior experience of conducting psychoacoustic tasks. They were separated into two groups, maintaining the same ratio of men to women in each group. The training group of 20 participants participated in the VAD game. For 10 of them, their own HRTFs, which were measured with the blocked ear canal technique [38], were used for rendering with the VAD; for the other 10, fitted HRTFs from 16 pairs of those of other listeners with the tournament-style listening test were used [39, 40]. The control group of 20 participants did not play the VAD game. One participant of the training group was excluded from the experiment because of a health problem. All participants performed the three tasks (sound localization task, face-contact task, and collision avoidance task) on the first day (pre-test). They were asked to perform the same tasks two weeks later (post-test). The two training groups were asked to play the game for seven days (30 min/day) within a 2-week period, but the participants of the control group did not play the game within that period. All participants conducted all these tasks blindfolded. They provided informed consent and their rights as participants were protected. Moreover, the ethics review committee of the Research Institute of Electrical Communication, Tohoku University, approved the protocol for the study.

3.1. Transfer Effects of Sound Localization Performance

All blindfolded participants conducted a sound localization task for real sound sources. This task was conducted in a soundproof room with a speaker array. The participant sat on a chair placed in the center of the speaker array. The speaker array was equipped with 36 loudspeakers. The loudspeakers were positioned at 30° intervals and were elevated to 0°, ±30°on the speaker array with a 1.2 m radius. The speaker array frame was covered with soft material. The experimenter adjusted the chair height such that the participant's ear level was positioned at 140 cm elevation. The target stimulus was delivered through a randomly selected loudspeaker three times for each position. The target sound consisted of the same frequency spectrum as the target stimulus of honeybee buzz in the VAD game. The sound pressure level (SPL) was adjusted to 70 dB; its duration was 1 s. In addition, the background sound was presented to the participant using four loudspeakers that were spaced equally around the speaker array during the task. The background sound was uncorrelated pink noise and the SPL was maintained at 60 dB. In the task, the participants were asked to localize the target sound position using a rod (100 cm long, 60 g). All participants performed the sound localization task on the first day (pre-test). They were asked to perform the same task 2 weeks later (post-test). The two training groups were asked to play the game for seven days (30 min/day) within a 2-week period, but the participants of the control group did not play the game during that period.

The results revealed that by playing the VAD game, the hit scores increased while misses decreased with training, irrespective of the type of HRTF: own or fitted. In addition, the sound localization performance of participants for actual sound sources using fitted HRTF was similar to their performance using their own HRTF. The VAD game-playing results revealed that: (1) the hit rate of the sound localization task for real sound sources increased approximately 20%; (2) the vertical and horizontal localization error (wrong azimuth angle, correct elevation) decreased remarkably. Furthermore, we conducted a follow-up test to investigate the persistence of transfer effects.

In this follow-up test, we asked eight participants of the training group to join. Their hit rates had increased more than 20% by playing the VAD game. Seven participants agreed to participate in the follow-up study. These participants performed the same sound localization task for real sound sources 1 month after they finished the post-test. The participants were told not to play the VAD game within that 1-month period. The follow-up tests revealed that transfer effects persisted more than 1 month, suggesting that the effects of playing the VAD game transfer to sound localization performance (see Figure 3). For the follow-up test, we asked participants to report their subjective estimation of their sound localization skill. They felt that their sound localization skills had increased from playing the VAD game and that they had decreased 1 month later. Therefore, their consciousness about the transfer effects did not reflect the actual persistence of the effects.

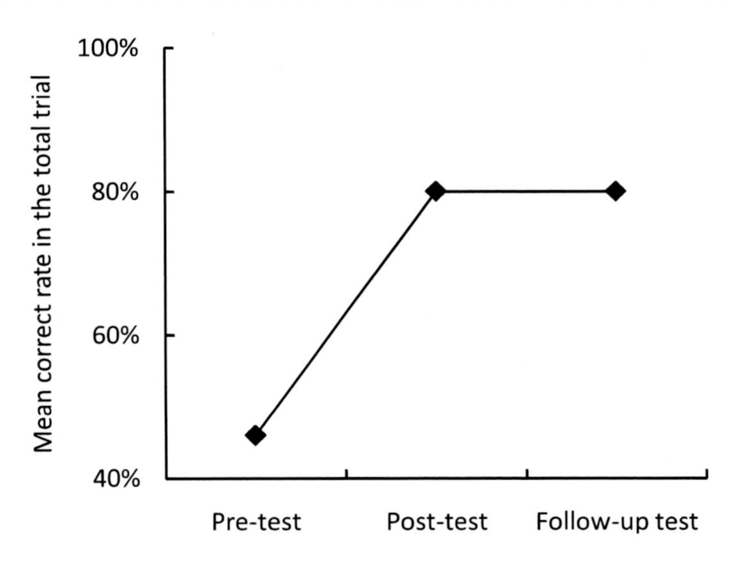

Figure 3. Persistence of real-sound localization performance from playing the VAD game.

3.2. Transfer Effects of Face-contact Performances

In this task, all blindfolded participants conducted a communication task in a soundproof room. The participants sat on a chair that allows free head and body movement. They were asked to talk about some topics with two interviewers (pre-test). Two trained female experimenters assisted as interviewers to ensure uniformity of the task. These interviewers had never met the participants. In the communication task, the interviewers sat on chairs that were positioned at 30° left and right and facing the participant. The distance between the participant and the interviewers was 2.5 m. Two video cameras recorded the scene from these positions. The interviewers provided questions for the participants according to a script (see Figure 4). For example, this script included weather topics, health problems, and opinions concerning this experiment. In all, 36 questions were used. The face-contact behaviors were defined as the participant's face turning to a location of the interviewer when they listened to each question from the interviewer or tried to answer each question. Each interviewer was asked to check whether the participants showed face-contact behaviors on each topic with four phases (start-phase of listening, end-phase of listening, start-phase of speaking, and end-phase of speaking). Another interviewer confirmed whether the participants showed face-contact behavior to the querying interviewer when the one interviewer provided the question and the participant answered it. Reliability obtained using the corresponding rate between the interviewers was 83%. They were resolved through verification using recorded videotapes and discussion of the two interviewers if mismatches of checked face-contact behaviors existed. After the task, the participants were asked to rate the subjective levels of tension during the communication task ("Did you feel tense during the communication task?"). Tension was rated on a 7-point Likert scale (1 = *Not at all* to 7 = *Very much*). All participants were asked to perform the same task two weeks later (post-test). In the post-test, several topics were altered and the position of the interviewers was exchanged.

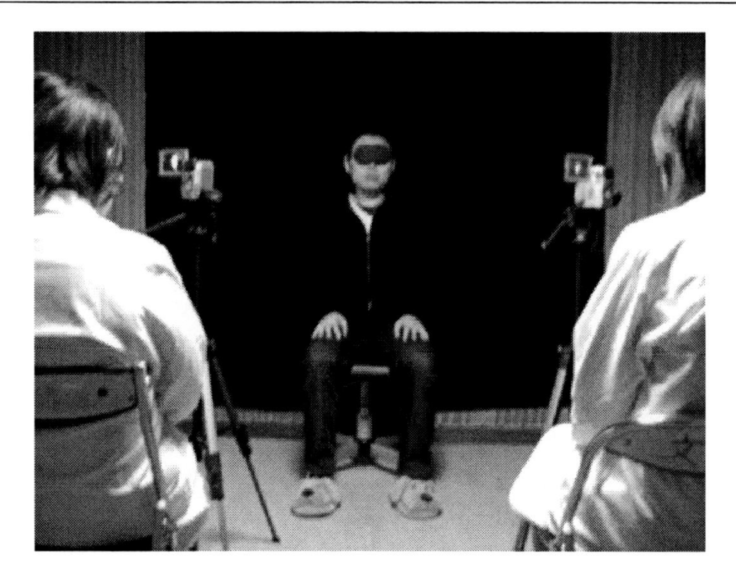

Figure 4. Communication task in our study.

We compared the communication performances of the training group (including own-HRTF and fitted-HRTF groups) and the control group. The results of the sound localization task [7] revealing the sound localization performance for the real sound sources using fitted HRTFs was similar to their performance using their own HRTFs. Results in the communication task revealed that, by playing the VAD game, (1) the face-contact in social interaction increased remarkably, and (2) subjective rated levels of tension during the situation did not change [8]. These results demonstrated that playing the VAD game transferred participants' communication skills in the social interaction. The results suggest that the transfer effects are based on automatic perceptual-motor learning through playing the VAD game because no difference was detectable in the subjectively rated levels of tension during the communication task between the two conditions [8].

3.3. Transfer Effects of Collision Avoidance Performance

All blindfolded participants performed a collision avoidance task in an experiment room. They were asked to avoid an approaching object when they felt it was moving on a collision course (relevant path). Furthermore, they were asked to perform avoidance behaviors with minimum moving distance from their position. They were instructed not to avoid an approaching object when they felt it was moving on an irrelevant path. The distances between the relevant path and two irrelevant paths were 80 cm. The colliding object was a toy car (width = 30 cm, weight = 2.5 kg). The front of the toy car was covered with soft material. The participant confirmed the safety of the approaching object before initiating the task. Furthermore, one experimenter stood behind the participant to eliminate the possibility of a participant's falling. The approaching stimulus was presented randomly and repeatedly to participants from either relevant or irrelevant paths. The SPL of the approaching sound was 75 dB and the background noise level of the experiment room was 35 dBA. The toy car was placed at 50 cm height on lanes (initial velocity = 0 m/s) and slid along the lane slopes (2 m/s). Three lanes were used. The center lane was for the relevant (collision) path and the lanes of both sides were for irrelevant paths. The distance between the participant and the lanes was 4.0 m. The participants were required to perceive a correct sound localization of the approaching object based on auditory information and to decide their behaviors within 2 s. The trials numbered 36 in all. The approaching object was sent on a randomly selected lane 12 times for each course. The body direction of the participant was changed for each trial; consequently, the toy car approached from the front, back, left, or right direction (each direction = total nine trials)(see Figure 5). The experimenter then checked whether the participant had done the avoidance behaviors for each trial. In addition, they measured the distance from the participants' start position to the end point of their actions. Two video cameras recorded the scene. After the task, the participants were asked to rate the subjective levels of difficulty of the collision avoidance task ("Did you feel that the collision avoidance task was difficult?"). The difficulty was rated on a 7-point Likert scale (1 = *Not at all* to 7 = *Very difficult*). Additionally, we asked participants to judge the subjective levels of threat related to the collision ("Did you feel a threat that you might bump against the moving object?"). The threat levels were rated on a 7-point Likert scale for each direction (1 = *Not at all* to 7 = *Very fearful*). All participants were asked to perform the same task two weeks later (post-test). In pre-tests and post-tests, they were given a short practice period to adapt themselves to this task.We compared the performances of the training group (including own-HRTF and fitted-HRTF groups) and the control group. The results [8] showed that the avoidance distance by playing the VAD game decreased when the object was approached from the irrelevant path. In contrast, no difference was detectable between the training group and the control group regarding the number of collisions with approaching objects in the avoidance task. In addition, the results showed that their subjective estimations of difficulty or threat about the task did not change by playing the VAD game. These findings indicate that auditory training using the VAD game modified the executed avoidance behaviors because the appropriate avoidance behaviors relate to sound localization skills. In addition, the results suggest that these transfer effects are based on automatic perceptual-motor learning by playing the VAD game because training did not alter their subjective awareness [8].

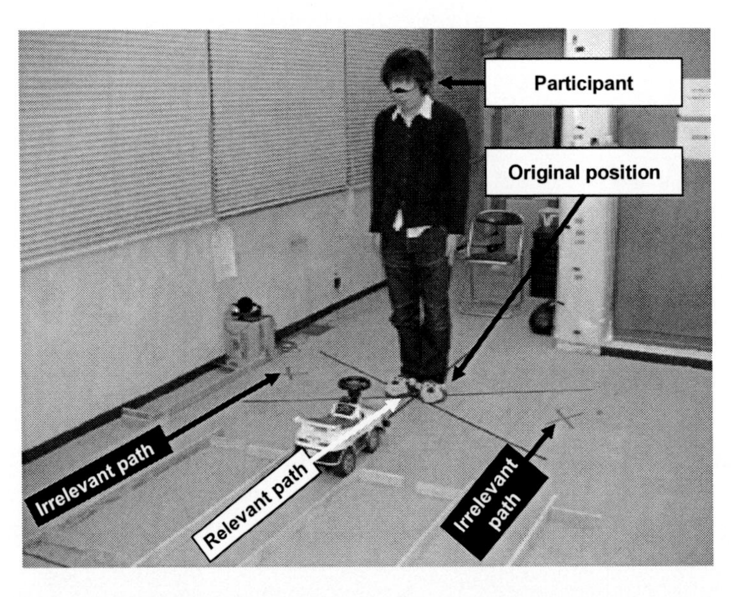

Figure 5. Collision avoidance task in our study.

4. Conclusion

This chapter introduced our findings from our investigations of the transfer effects of playing an auditory game installed on a virtual auditory display (VAD). Although prior research had not explored how playing VAD games would impart transfer effects on various auditory skills, our studies [7, 8] revealed that, by playing the VAD game, (1) transfer effects are visible not only in sound localization performance but also in communication and collision avoidance behaviors, and (2) these transfer effects must be based on automatic perceptual motor-learning because the subjective estimations of participants did not reflect their actual performances. Furthermore, results showed that transfer effects on actual sound localization were not diminished, even after one month. These findings underscore that VAD games can be an effective training tool for skills related to auditory information processing in daily life situations. Moreover, our data suggest that, was with playing visual video action games, playing VAD games can have positive transfer effects on various human behaviors in a short period. In fact, several researchers have recently developed VAD games which are designed to improve sound localization skills of visually impaired people as well as application software systems with VAD which are intended to improve spatial cognition. For example, Ohuchi *et al.* [41] have proposed a new assistive technology for sightless people based on VAD. While providing a safe virtual sound environment in which to navigate, their system is intended to encourage users to improve their map-forming skills and use them more efficiently. The system virtually realizes a sound maze through which a user navigates using a game controller instead of traveling physically. The experimental results indicate that the use of this system promotes formation of cognitive maps by users. However, few studies have examined the influence of such education systems with VAD on auditory training or on the everyday behavior of visually impaired people. Future experimental studies should specifically address development and testing of application software systems' efficacy in helping visually impaired people to lead safer and more comfortable lives. For example, in our experiment, several participants reported that they felt their bodies stiffen because of feelings of anxiety or fear in initial trials of the collision task [8], which clarifies that future studies should be undertaken to design VAD games that can better train individuals in safe navigation or otherwise reduce their way-finding anxiety in real environments. The findings of these studies can contribute to improvement of the quality of life for visually impaired people. Previous studies have suggested that action video games have better transfer effects than other games because of a greater sense of immersion [9–12]. Both VAD games and visual action video games appear to offer great potential for applications to many different kinds of skill learning. The important point is the control of transfer effects through skill learning using these kinds of games. The findings of Honda *et al.* [7] indicate that transfer effects derived from playing the VAD game persist for at least one month after play stops. Recently, video games which require physical movement to play are gaining in popularity. VAD games and visual action video games are similar in that they also require physical movement and provide instant feedback. Compared to regular television games, we can surmise that there are greater transfer effects to everyday behavior because these games demand more physical movements which resemble those used in real life. Some evidence suggests that a great deal of effort and time is required to eliminate behavior that might be transferred inappropriately through playing these games. Both VAD games and visual action video games share this problem of controlling transfer effects. Future studies should (1) investigate what kinds of game contents and devices transfer to specific behavior, and (2) clarify and organize previous experimental findings. Explaining the determinant factors underlying these kinds of transfer effects can engender development of concrete methods for eliminating inappropriate behaviors learned from playing action video games. Such findings would also provide insight into how action video games affect human behavior and suggest ways that these games can be used to promote skill learning. Moreover, future studies should investigate how to optimize the transfer effects of VAD games. Because the acquisition of new skills in a short period of time requires effective and appropriate learning, it will be necessary to explain the features of VAD games and the particular effects of these features to further develop and improve transfer effects. For example, in VAD games requiring sound localization, play involves hearing a sound stimulus and then positioning the sound source in response through head and body movements. By attaching audio and physical feedback when a play is executed, the player's behavior can be rewarded or corrected. Nevertheless, few studies have examined how the learning of actual sound source positioning is affected by head and body movements when a sound stimulus is presented or by the feedback given in response to a listener's sound localization performance. Our research group partially addressed this issue by conducting experiments to determine what role physical movement and kinetic feedback play in the process of sound localization learning [42]. Our results indicate that effective sound localization learning involves active head movement and that kinetic feedback has the function of correcting head movements and encouraging skill learning in the early stages [42]. These findings suggest that, to optimize learning, VAD games that are designed to improve the ability to locate actual sound positions should include (1) contents that encourage active head movements by the user and (2) optional modes in the early stages of sound localization learning that gives concrete feedback to users related to the way in which their sound source positioning was misaligned. Understanding how various elements in VAD games contribute to effective skill learning will provide important insights into the conditions necessary to optimize transfer effects. Moreover, the possibility exists that transfer effects are influenced by the motivation or goals of participants involved in skill learning through VAD games. In fact, an earlier study demonstrated that transfer effects are influenced by motivation levels or goal setting of visual action video game players [12], which suggests that any effort to optimize transfer effects should also incorporate a psychological approach to understanding the emotions and cognitive state of skill learners. These studies will certainly also contribute to fundamental research in fields such as engineering, psychology, and education. The VAD technology is being applied currently to various auditory skills learning for visually impaired people. Results of previous studies show that VAD games have positive transfer effects to various auditory skills on daily life situation. Future research must identify exactly what factors contribute to transfer effects in VAD training, and to design VAD systems that engender the best possible transfer effects. These studies will benefit not only applications of VAD technology, but also applications of the transfer effects of action video games. Meaningful research will require the contributions and cooperation of researchers from diverse fields.

Acknowledgements

An initial phase of this study was supported by Consortium R&D Projects for Regional Revitalization (15G2025), Ministry of Economy, Trade and Industry. Later phases of this work were supported by a Grant-in-Aid for Young Scientists (Start-up) No. 18830064 from the Japan Society for the Promotion of Science.

References

[1] Shinn-Cunningham, B. G., Lehnert, H., Kramer, G., Wenzel, E. M., and Durlach, N. I. (1998). "Auditory Displays," In R. Gilkey and T. Anderson (Eds.), *Binaural and Spatial Hearing in Real and Virtual Environments*. Mahwah, NJ: Lawrence Erlbaum, 611–664.

[2] Blauert, J. (1996). "*Spatial Hearing: The Psychophysics of Human Sound Localization (revised edition)*," Cambridge, The MIT Press.

[3] Langendijka, E. H. A., and Bronkhorst, A. W. (2000). "Fidelity of three-dimensional-sound reproduction using a virtual auditory display," *The Journal of the Acoustic Society of America*, **107**, 528–537.

[4] Sorkin, R. D., Wightman, F. L., Kistler, D. J., and Elvers, G. C. (1989). "An exploratory study of the use of movement-correlated cues in an auditory head-up display," *Human Factors*, **31**, 161–166.

[5] McKinley, R. L., Erickson, M. A., and D'Angelo, W. R. (1994). "3-dimensional auditory displays: Development, applications, and performance," *Aviation, space, and environmental medicine*, **65**, 31–38.

[6] Bronkhorst, A. W., Veltman, J. A., and Breda, L. (1996). "Application of a three-dimensional auditory display in a flight task," *Human Factors*, **38**, 23–33.

[7] Honda, A., Shibata, H., Gyoba, J., Saitou, K., Iwaya, Y., and Suzuki, Y. (2007). "Transfer effects on sound localization performance from playing a virtual three-dimensional auditory game," *Applied Acoustics*, **68**, 885–896.

[8] Honda, A., Shibata, H., Gyoba, J., Iwaya, Y., and Suzuki, Y. "Transfer effects on communication and collision avoidance behavior from playing a three-dimensional auditory game based on a virtual auditory display," *Applied Acoustics*, submitted.

[9] Castel, A. D., Pratt, J., and Drummond, E. (2005). "The effects of action video game experience on the time course of inhibition of return and the efficiency of visual search," *Acta Psychologica*, **119**, 217–230.

[10] Green, C. S., and Bavelier, D. (2003). "Action video game modifies visual selective attention," *Nature*, **423**, 534–537.

[11] Green, C. S., and Bavelier, D. (2007). "Action-video-game experience alters the spatial resolution of vision," *Psychological Science*, **18**, 88–94.

[12] Fery, Y. A., and Ponserre, S. (2001). "Enhancing the control of force in putting by video game training," *Ergonomics*, **44**, 1025–1037.

[13] Röber, N., and Masuch, M. (2005). "Leaving the screen: New perspectives in audio-only gaming", *Proc. of ICAD 05-Eleventh Meeting of the International Conference on Auditory Display*, 92–98.

[14] Winberg, F., and Hellström, S. O. (2001). “Qualitative aspects of auditory direct manipulation: A case study of the towers of Hanoi,” *Proc. of the 2001 International Conference on Auditory Display*, 16–20.
[15] Ohuchi, M., Iwaya, Y., Suzuki, Y., and Munekata, T. (2005). “Training effect of a virtual auditory game on sound localization ability of the visually impaired,” *Proc. of ICAD 05-Eleventh Meeting of the International Conference on Auditory Display*, 1–4.
[16] Afonso, A. A., Katz, B. F. G., Blum, A., Jacquemin, C., and Denis, M. (2005). “A study of spatial cognition in an immersive virtual audio environment: Comparing blind and blindfolded individual,” *Proc. of ICAD 05-Eleventh Meeting of the International Conference on Auditory Display*, 228–235.
[17] Lessard, N., Pare, M., Lepore, F., and Lassonde, M. (1998). “Early-blind human subjects localize sound sources better than sighted subjects,” *Nature*, **395**, 278–280.
[18] Gougoux, F., Lepore, F., Lassonde, M., Voss, P., Zatorre, R. J., and Belin, P. (2004). “Pitch discrimination in the early blind: People blinded in infancy have sharper listening skills than those who lost their sight later,” *Nature*, **430**, 309.
[19] Kendon, A. (1967). “Some functions of gaze-direction in social interaction,” *Acta Psychologica,* **26**, 22–63.
[20] Ellsworth, P. C., and Ludwig, L. M. (1972). “Visual behavior in social interaction,” *Journal of Communication,* **22**, 375–403.
[21] Wiener, M., Devoe, S., Rubinow, S., and Geller, J. (1972). “Nonverbal behavior and nonverbal communication,” *Psychological Review,* **79**, 185–214.
[22] Fichten, C. S. Judd, D., Tagalakis, V., Amsel, R., and Robillard, K. (1991). “Communication cues used by people with and without visual impairments in daily conversations and dating,” *Journal of Visual Impairment & Blindness,* **85**, 371–378.
[23] Raver, S. A. (1987). “Training gaze direction in blind children: Attitude effects on the sighted,” *Remedial & Special Education,* **8**, 40–45.
[24] Erin, J. N., Dignan, K., and Brown, P. A. (1991). “Are social skills teachable? A review of the literature,” *Journal of Visual Impairment & Blindness*, **85**, 58–61.
[25] Sanders, R. M., and Goldberg, S. G. (1977). “Eye contacts: Increasing their rate in social interactions,” *Journal of Visual Impairment & Blindness,* **71**, 265–267.
[26] Raver, S. A. (1987). Training blind children to employ appropriate gaze direction and sitting behavior during conversation. *Education & Treatment of Children,* **10**(3), 237–246.
[27] Liaw, J. S., and Arbib, M. A. (1993). “Neural mechanisms underlying direction-selective avoidance behavior,” *Adaptive Behavior*, **1**, 227–261.
[28] Li, F. X., and Laurent, M. (2001). “Dodging a ball approaching on a collision path: Effects of eccentricity and velocity,” *Ecological Psychology,* **13**, 31–47.
[29] Schiff, W., and Detwiler, M. L. (1979). “Information used in judging impending collision,” *Perception,* **8**, 647–658.
[30] Schiff, W., and Oldak, R. (1990). “Accuracy of judging time to arrival: Effects of modality, trajectory, and gender,” *Journal of Experimental Psychology: Human Perception and Performance*, **16**, 303–316.
[31] Suzuki, Y., Asano, F., Kim, H. Y., and Sone, T. (1995). “An optimum computer-generated pulse signal suitable for measurement of very long impulse responses,” *Journal of the Acoustic Society of America*, **97**, 1119–1123.

[32] Watanabe, K., Takane, S., and Suzuki, Y. (2003). "Interpolation of head-related transfer functions based on the common-acoustical –pole and residue model," *Acoustical Science and Technology*, **24**, 335–337.

[33] Yairi, S., Iwaya, Y., and Suzuki, Y. (2006). "Investigation of system latency detection threshold of virtual auditory display," *Proc. of the 12th International Conference on Auditory Display*, 217–222.

[34] Ohuchi, M., Iwaya, Y., Suzuki, Y., and Munekata, T. (2006). "Cognitive-map forming of the blind in virtual sound environment," *Proc. of the 12th International Conference on Auditory Display*, 1–7.

[35] Miller, J. D. (2001). "Slab: A software-based real-time virtual acoustic environment rending system" *Proc. of the 2001 International Conference on Auditory Display*, 1–2.

[36] Lokki, T. (2002). "Physically-Based Auralization: Design, Implementation, and Evaluation" *Ph.D. Thesis, Helsinki University of Technology.*

[37] Kawaura, J., Suzuki, Y., Asano, F., and Sone, T. (1989). "Sound localization in headphone reproduction by simulating transfer functions from the sound source to the external ear," *Journal of Acoustical Society of Japan* **45**, 756–766.

[38] Moller, H. (1992). "Fundamental of binaural technology," *Applied Acoustics*, **36**, 171–218.

[39] Iwaya, Y. (2006). "Individualization of head-related transfer functions with tournament-style listening test: Listening with other's ears," *Acoustical Science and Technology*, **27**, 340–343.

[40] Iwaya, Y., Yairi, S., and Suzuki, Y. "Individualization of head-related transfer functions in virtual auditory display with listening-test," *The Japan-China Joint Conference of Acoustics 2007*, 6 pages in CDROM.

[41] Ohuchi, M., Iwaya, Y., Suzuki, Y., and Munekata, T. "Cognitive-map forming of the blind in virtual sound environment," *Proc. of the 12th International Conference on Auditory Display*, 1–7.

[42] Honda, A., Shibata, H., Hidaka, S., Gyoba, J., Iwaya, Y., and Suzuki, Y. *"The effects of head movement and kinetic feedback on sound localization learning"*, in preparation.

In: New Research on Acoustics
Editor: Benjamin N. Weiss, pp. 159-196
ISBN: 978-1-60456-403-7

Chapter 4

RECENT STUDIES OF CAR DISC BRAKE SQUEAL

Abd Rahim Abu-Bakar
Department of Automotive Engineering, Faculty of Mechanical Engineering
Universiti Teknologi Malaysia, 81310 UTM Skudai, Johor, Malaysia
Huajiang Ouyang
Department of Engineering, University of Liverpool
Brownlow Street, Liverpool L69 3GH, U.K.

Abstract

Friction-induced vibration and noise emanating from car disc brakes is a source of considerable discomfort and leads to customer dissatisfaction. The high frequency noise above 1 kHz, known as squeal, is very annoying and very difficult to eliminate. There are typically two methods available to study car disc brake squeal, namely complex eigenvalue analysis and dynamic transient analysis. Although complex eigenvalue analysis is the standard methodology used in the brake research community, transient analysis is gradually gaining popularity. In contrast with complex eigenvalues analysis for assessing only the stability of a system, transient analysis is capable of determining the vibration level and in theory may cover the influence of the temperature distribution due to heat transfer between brake components and into the environment, and other time-variant physical processes, and nonlinearities. Wear is another distinct aspect of a brake system that influences squeal generation and itself is affected by the surface roughness of the components in sliding contact.

This chapter reports recent research into car disc brake squeal conducted at the University of Liverpool. The detailed and refined finite element model of a real disc brake considers the surface roughness of brake pads and allows the investigation into the contact pressure distribution affected by the surface roughness and wear. It also includes transient analysis of heat transfer and its influence on the contact pressure distribution. Finally transient analysis of the vibration of the brake with the thermal effect is presented. These studies represent recent advances in the numerical studies of car brake squeal.

Introduction

Passenger cars are one main means of transportations for people travelling from one place to another. Indeed, vehicle quietness and passenger comfort issues are a major concern. One of the vehicle components that occasionally generate unwanted vibration and unpleasant noise is the brake system. As a result, carmakers, brake and friction material suppliers face challenging tasks to reduce high warranty payouts. Akay (2002) estimated that the warranty claims due to the noise, vibration and harshness (NVH) issues including brake squeal in North America alone were up to one billion US dollars a year. Similarly, Abendroth and Wernitz (2000) noted that many friction material suppliers had to spend up to 50 percent of their engineering budgets on NVH issues.

In a recent review, Kinkaid *et al* (2003) listed a wide array of brake noise and vibration phenomena. Squeal, creep-groan, moan, chatter, judder, hum, and squeak are among the names that can be found in the open literature. Of these noises, squeal is the most troublesome and irritant one to both car passengers and the environment, and is expensive to the brakes and car manufacturers in terms of warranty costs (Crolla and Lang, 1991). It is well accepted that brakes squeal is due to friction-induced vibration or self-excited vibration via a rotating disc (Chen *et al*, 2003a). Brake squeal frequently occurs at frequency above 1 kHz (Lang and Smales, 1983) and is described as sound pressure level above 78 dB (Eriksson, 2000) or usually at least 20dB above ambient noise level in the automotive industry.

Brake squeal has been studied since 1930's by many researchers through experimental, analytical and numerical methods in an attempt to understand, predict and prevent squeal occurrence. Although experiments used to be the more credible way of studying disc brake squeal (Nishiwaki *et al*, 1990; Yang and Gibson, 1997), there has been a great advancement in the numerical analysis methodology in recent years. More specifically, the finite element (FE) method is the preferred method in studying brake squeal. The popularity of finite element analysis (FEA) is due to the inadequacy of experimental methods in predicting squeal at early stage in the design process. Moreover, FEA can potentially simulate any changes made on the disc brake components much faster and easier than experimental methods. A recent review (Ouyang *et al*, 2005) stated that experimental methods are expensive due to hardware costs and long turnaround time for design iterations. In addition, discoveries made on a particular type of brake are not always transferable to other types of brake and quite often product developments are based on a trial-and-error basis. Furthermore, a stability margin is frequently not found experimentally.

With the refinements on the methodology and analysis of the disc brake squeal using the finite element method being progressively reported in the open literature, it is thought that refinement in the disc brake model should also be made in parallel. This can be seen from the previous works of Liles (1989); Ripin (1995) and Lee *et al* (1998) where a number of linear spring elements are employed at the friction interface. The introduction of a number of spring elements at the disc/pads interface is necessary to generate friction-coupling terms (asymmetric stiffness matrix) that lead to the complex eigenvalues, i.e., unstable behaviour in which the positive real parts indicate the likelihood of the squeal occurrence. With the contributions of many researchers (for example, Yuan, 1996,

Blaschke *et al*, 2000) and the initiative of a finite element software company (ABAQUS, Inc., 2003), linear spring elements are no longer required as friction-coupling terms can now be directly implemented into the stiffness matrix. As a result the effect of non-uniform contact pressure and residual stresses can be incorporated in the complex eigenvalue analysis (Bajer *et al*, 2003, Kung *et al*, 2003). Another advantage of the current approach is that the surfaces in contact do not need to have matching meshes and essentially it can reduce data-preparation time. In contrast, the former approach required nodes on the two contacting surfaces to coincide and hence similar meshes.

Some previous studies assumed full contact at the pads and disc interfaces (Liles, 1989, Ouyang *et al*, 2000). Other works (Samie and Sheridan, 1990; Tirovic and Day, 1991; Ghesquiere and Castel, 1992; Ripin, 1995; Lee *et al*, 1998; Hohmann *et al*, 1999; Tamari *et al*, 2000, Ioannidis *et al*, 2003; Abu-Bakar and Ouyang, 2004 and Abu-Bakar *et al*, 2005b) have shown that the contact pressure distributions at the disc/pads interfaces are not uniform and there exists partial contact over the disc surfaces. Traditionally, contact at the disc/pads interface was simulated using either linear spring elements (Nack, 2000) or non-linear node-to-node gap elements (Samie and Sheridan, 1990; Tirovic and Day, 1991; Ripin, 1995; Lee *et al*, 1998, Tamari *et al*, 2000). Recent contact analyses (Hohmann *et al*, 1999; Ouyang *et al*, 2003a, Ioannidis *et al*, 2003) no longer use such elements, where a surface based element can provide more realistic and accurate representation of contact pressure distribution (Bajer *et al*, 2003, Kung *et al*, 2003). Incidentally, the contact pressure distribution is believed to be very influential on squeal generation. Fieldhouse (2000) experimentally and Abu-Bakar *et al* (2005a) numerically demonstrated that different pressure contact distributions could promote or inhibit squeal occurrence.

When developing a finite element model, it is important to validate it in order for the model to correctly represent the actual structure in terms of the geometry and its material properties. A validated model should be able to predict squeal sufficiently accurately. Liles (1989) was the first researcher who conducted and expounded the complex eigenvalue analysis with a large finite element model and used modal analysis to compare natural frequencies and the mode shapes for each of disc brake components. This component validation later became a standard practice (Ripin, 1995; Lee *et al*, 1998; Guan and Jiang, 1998; Liu and Pfeifer, 2000; Kung *et al*, 2000, Ioannidis *et al*, 2003). Even though precise representations of brake components are possible, results obtained by the complex eigenvalue analysis did not always correspond to the experimental squeal results. Realising this shortcoming Richmond *et al* (1996), Dom *et al* (2003), Ouyang *et al* (2003a) and Goto *et al* (2004) used a systematic procedure to correlate and update the FE model at both the component and assembly levels. A tuning process was performed to reduce relative errors in natural frequencies between predicted and experimental results whereby material properties and spring stiffness were adjusted in the tuning process.

As progressive refinements on the methodology and analysis stated above are made, it is thought that a more realistic finite element model should be developed or, in other words, refinement on the modelling aspects of disc brakes should be made. It is already known that contact geometry between the disc and friction material interface has a significant contribution towards squeal generation (Tarter, 1983; Ripin, 1995; Eriksson *et al*, 1999; Ibrahim *et al*, 2000; Soom *et al*, 2003; Sherif, 2004; Hammerström & Jacobson 2006; Trichês *et al*, 2008; Fieldhouse *et al*, 2008). They believed that squeal could be generated at particular conditions of pads topography. Another fact is that the friction material is more prone to wear. Furthermore, friction material has a much more irregular or corrugated surface than the disc. It was found that none of the existing finite element models considered friction material surface topography. All of the models assumed that the friction material had a smooth and flat interface whereas in reality it was rough.

The complex eigenvalue analysis was the method preferred by the industry to study their squeal noise issues. This method is largely dependent on the results of contact analysis, which can determine instability in the disc brake assembly. Determination of dynamic contact pressure through experimental methods remains impossible. However there are methods to obtain static contact pressure, when the disc is stationary. It is believed that static contact pressure distribution and its magnitude can be used as a validation tool where correlation between calculated and measured results can be made. This validation level can enhance one's confidence in the developed model as well as provide better prediction of squeal occurrence. Although complex eigenvalue analysis is the standard methodology used in the brake research community, transient analysis is gradually gaining popularity. In contrast with complex eigenvalues analysis for assessing only the stability of the system, transient analysis is capable of determining the vibration level and in theory may cover the influence of time-variant physical processes and nonlinearities.

This chapter reports recent advances in the numerical studies of car brake squeal conducted at the University of Liverpool. The major contributions are a more realistic model of the disc and pads interface (including surface roughness and wear), inclusion of the thermal effect on the interfacial pressure distribution and implementation of transient analysis of the vibration of the brake with the above thermal effect. Results have been drawn from the PhD thesis of the first author, an MPhil thesis (Li, 2007) and most importantly a number of recently published papers of the authors and their colleagues in the University of Liverpool. For the recent advances in the studies of disc brake squeal worldwide, refer to a book written by some established experts working in this area (Chen *et al*, 2005)

Recent Numerical Methodology

As mentioned before, most of the FE models of disc brakes are either validated at only component level or a combination of the components and assembly levels. In the open literature, it was found that the complex eigenvalue analysis and dynamic transient analysis were typically employed by the brake research community to study their squeal noise issues. These two methods require the results of the contact pressure for the subsequent analysis. Therefore, it would be worthwhile to verify the static pressure distributions between measured data and simulated results. These three validation stages (Abu-Bakar and Ouyang, 2008) have formed an improved methodology recently developed by the authors. Figure 1 shows the authors' recent numerical methodology to study car disc brake squeal.

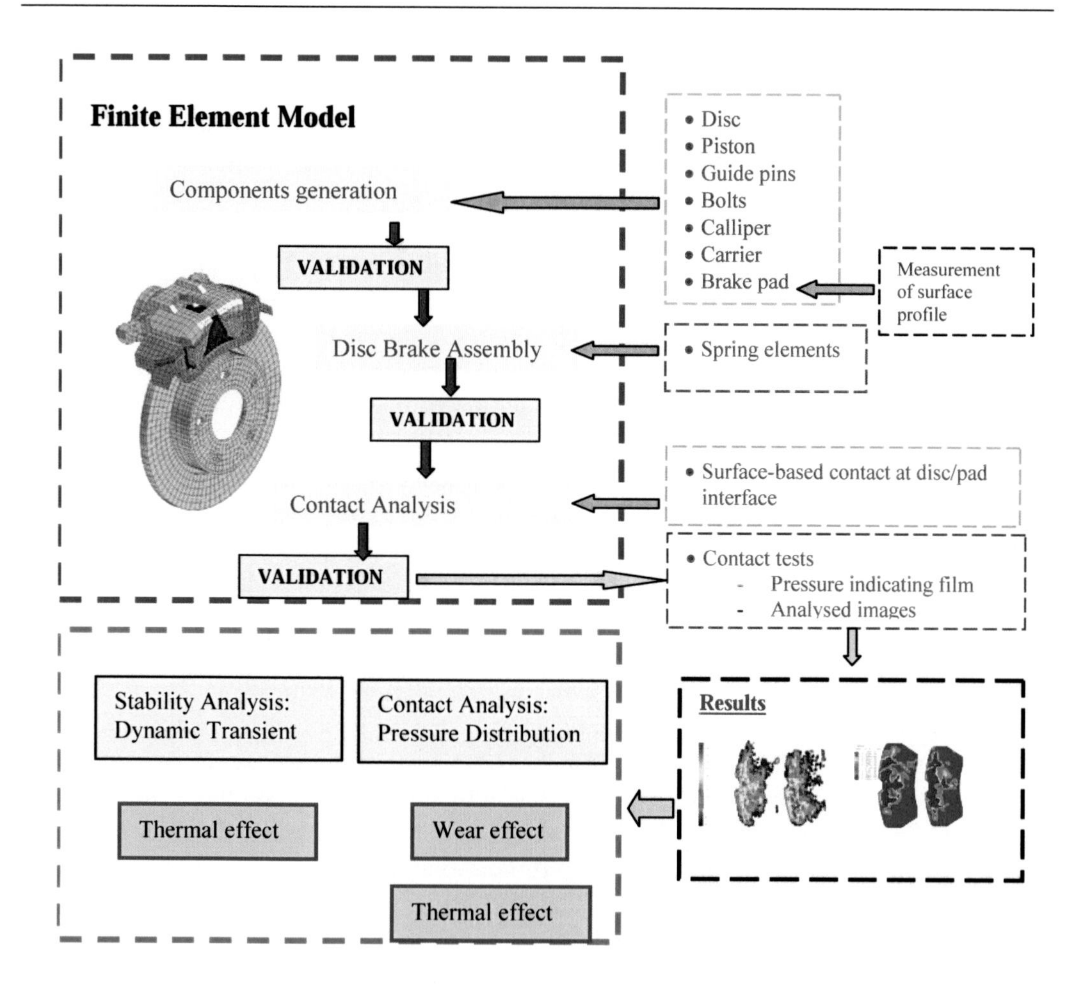

Figure 1. Overall simulation scheme.

Development and Validation Process of an FE Model

This section will describe in detail the development and validation process of a disc brake model throughout this research. The first validation stage is using modal analysis, in which the natural frequencies and mode shapes will be compared with the experimental results. It has two levels of validation, namely, components and assembly. The second validation stage uses contact analysis in which static contact pressure distributions at piston and finger pads, between simulated and measured are compared. There are a number of considerations that should be taken into account in order to model disc brake components and assembly, and simulate contact. These considerations are given in the subsequent sections. Chen *et al* (2003b) put forward a systematic way of carrying out testing and validation from an experimentalist's point of view.

Construction of a Disc Brake Model

Two commercial software packages are employed in order to generate the disc brake model and to simulate different mechanical behaviour. A software package called MSC PATRAN R2001 is utilised to generate elements and nodes of the disc brake. The advantages of using this software include the ease of changing one element type to another, e.g., a linear 8-node solid element (C3D8) to a quadratic 20-node solid element (C3D20), specifying contact surfaces between the disc and pads, connecting components at interfaces with spring elements, and simplifying geometry modifications. The software is also capable of generating an input file that is compatible with ABAQUS v6.4, which will then be used to perform subsequent analyses such as modal analysis, contact analysis and the complex eigenvalue analysis.

A detailed 3-dimensional finite element (FE) model of a Mercedes solid disc brake assembly is developed. Figure 2(a) and 2(b) show the Liverpool rig of a real disc brake of floating calliper design and its FE model respectively. The FE model consists of a disc, a piston, a calliper, a carrier, piston and finger pads, two bolts and two guide pins. A rubber seal (attached to the piston) and two rubber washers (attached to the guide pins) are not included in the FE model. Damping shims are also not present since they have also been removed in the squeal experiments. The FE model uses up to 8350 solid elements and approximately 37,100 degrees of freedom (DOFs). This figure excludes the spring elements that have been used to connect disc brake components other than between the disc and pads.

The disc, brake pads, piston, guide pins and bolts are developed using a combination of 8-node (C3D8) and 6-node (C3D6) linear solid elements while the other components are developed using a combination of 8-node (C3D8), 6-node (C3D6) and 4-node (C3D4) linear solid elements. Element types in the brackets show the notation in ABAQUS nomenclature. Details for each of the components are given in Table 1.

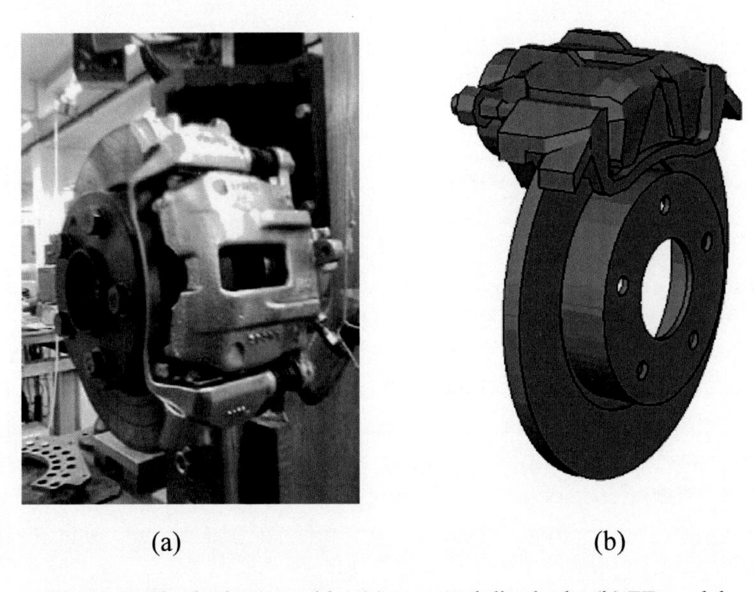

(a) (b)

Figure 2. Disc brake assembly; (a) an actual disc brake (b) FE model.

Since the contact between the disc and the friction material surfaces is crucial, realistic representation of those interfaces should be made. The friction material has a rougher surface and is low in Young's modulus than the disc, which has quite a smooth and flat surface, and is less prone to wear. Therefore in this work, actual surfaces at macroscopic scale of piston and finger pads are considered and measured. A Mitutoyo linear gauge LG-1030E and digital scale indicator are used to measure and provide reading of the surface respectively as shown in Figure 3. The linear gauge is able to measure surface height distribution ranging from 0.01mm up to 12 mm.

Table 1. Description of disc brake components

COMPONENTS	***TYPES OF ELEMENT***	***NO. OF ELEMENTS***	***NO. OF NODES***
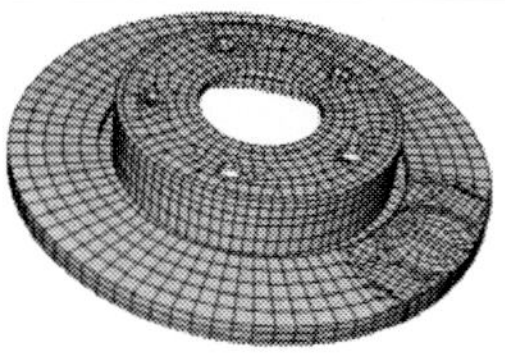 ***Disc***	C3D8 C3D6	3090	4791
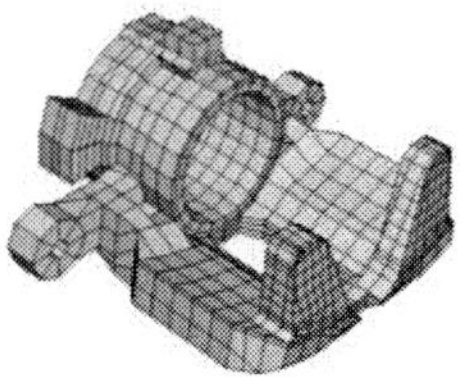 ***Calliper***	C3D8 C3D6 C3D4	1418	2242
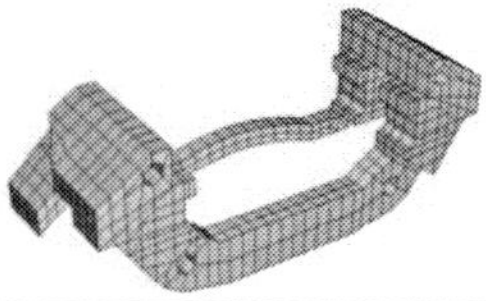 ***Carrier***	C3D8 C3D6 C3D4	862	1431
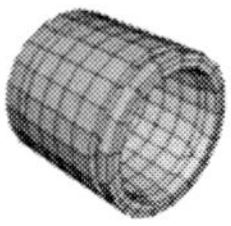 ***Piston***	C3D8 C3D6	416	744
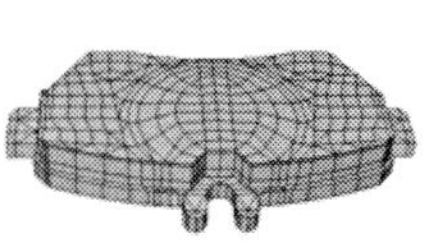 ***Back plate***	C3D8 C3D6	2094	2716
Friction Material	C3D8 C3D6		
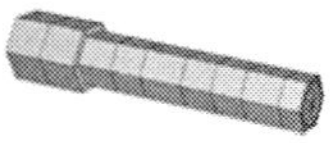 ***Guide pin***	C3D8 C3D6	388	336
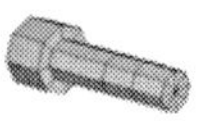 ***Bolt***	C3D8 C3D6	80	110

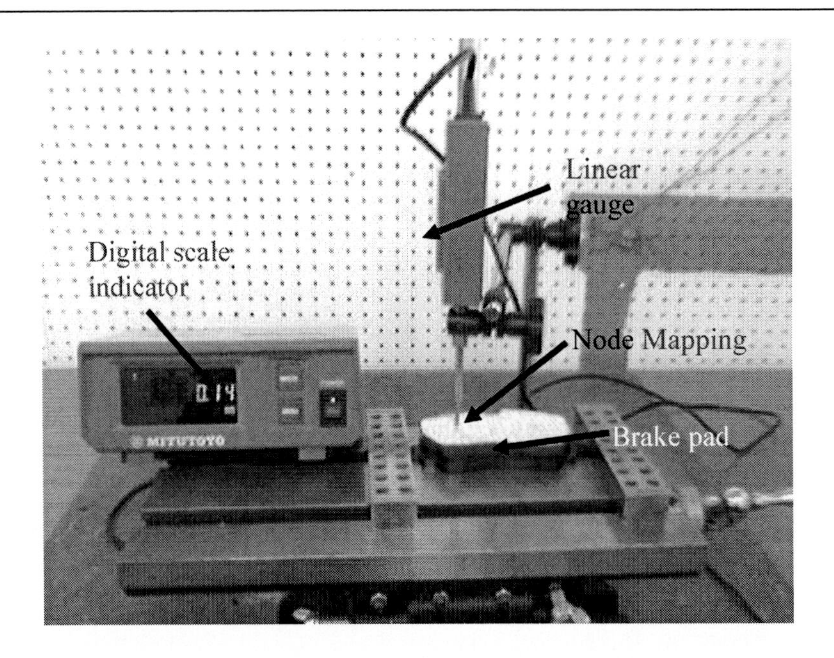

Figure 3. Arrangement of tools for surface measurement.

Prior to the measurement, the back plate must be flat and level. This can be confirmed by taking four measurement points at both pad abutments and the indicator should show similar height position. Node mapping, as shown in Figure 3, is required so that surface measurement can be made at particular positions, which are nodes of the FE model. By doing this, information that is obtained in the measurement can be used to adjust the coordinates of the piston and finger pad nodes in the brake pad interface model. There are about 227 nodes at the piston pad interface and 229 nodes at the finger pad interface. Since measurement is taken manually, it takes about two and half hours to complete this for a single pad. In this work, three pairs of brake pads (6 pieces) are measured, in which one pair of them are worn while the rest are new and unworn. All the brake pads are from the same manufacturer. Thus, it is assumed that the global geometry and material properties of the brake pad are the same. Upon completion of the modelling, all the disc brake components must be brought together to form an assembly model. Contact interaction between disc brake components is represented by linear spring elements (SPRING 2 in the ABAQUS nomenclature) except for the disc/pads interface where surface-to-surface contact elements are employed. This selection is due to the fact that contact pressure distributions at the disc/pad interface are more significant than at the contact interfaces of other components.

Figure 4 shows a schematic diagram of contact interaction that has been used in the disc brake assembly model. A rigid boundary condition is imposed at the boltholes of the disc and of the carrier bracket, where all six degrees of freedom are rigidly constrained in the rig.

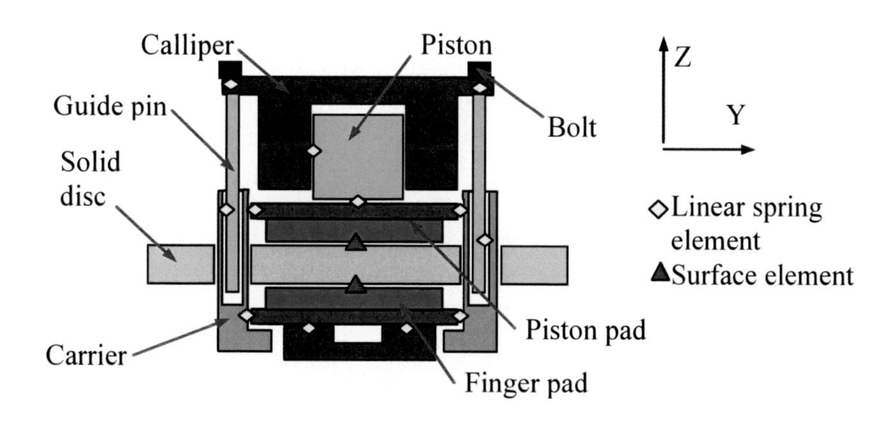

Figure 4. Schematic diagram of contact interaction in a disc brake assembly.

Modal Analysis

The experimental study of structural vibration has made significant contributions to better understanding of vibration phenomenon and for providing countermeasures in controlling such phenomenon in practice. Typically, experimental observations always have two-fold objectives (Ewins, 1984):

- Determining the nature and vibration response levels
- Verifying theoretical models and predictions

The first objective is referred to as a test where vibration forces or responses are measured during a structure's normal service or operation while the second is a test where the structure or component is vibrated with a known excitation. The second test is much more closely carried out under controlled conditions and this type of test is nowadays known as modal testing or experimental modal analysis (EMA). There are two different methods of comparison available to verify a theoretical model over EMA. They are a comparison in terms of response properties and modal properties. Although response properties of a tested structure can directly be produced in EMA, it is less convenient for some finite element software packages when it comes to generate frequency response function (FRF) plots. Furthermore, comparisons of modal properties are perhaps most common and convenient in the current practice where natural frequencies and mode shapes (either graphical or numerical) are used to obtain correlation between predicted and EMA results.

In this work, experimental modal analysis that was conducted by James (2003) was utilised to verify the developed FE model. Natural frequencies and graphical mode shapes obtained in the experiments are used to compare with the finite element results. Comparisons are made at two levels, namely, models of disc brake components and an assembly model. Firstly, FE modal analysis is performed at components level where the dynamic behaviour of disc brake components in the free-free boundary condition is captured. The second stage is to perform FE modal analysis on the disc brake assembly where the disc and the carrier are mounted to the knuckle. A certain level of brake-line pressure is applied to the stationary disc brake. During the analysis a tuning process (also known as model updating) is required in order to reduce relative errors between the predicted and experimental results. Normally, material properties, such as density and spring stiffness, are tuned or adjusted for disc brake components and assembly models respectively in order to bring closer predicted natural frequencies to the experimental data.

All the components are firstly simulated in free-free boundary condition and there are no constraints imposed on the components. Natural frequencies up to 9 kHz are considered since this study takes into account squeal frequencies between 1~ 8 kHz. The finite element model of the disc is validated by the authors and compared with the experimental data while the material data of the other components were validated and provided by an industry source. It is always desirable to validate all the components at once. This has not been done due to limitation in the equipment and tools available in the laboratory. It is thought that laser vibrometers can capture natural frequencies and mode shapes more accurately, as no contact with the components is needed. This can reduce errors in measuring dynamic behaviour of the components, compared with using accelerometers, which need to be attached to the components.

For the free-free boundary condition of the brake disc, a number of modes for up to frequencies of 9 kHz are extracted and captured. There are various mode shapes exhibited in the numerical results. However, only nodal diameter type mode shapes are considered because they were found to be the dominant ones in the observed squeal events of this particular disc brake. The calculated natural frequencies and mode shapes are given in Figure 5, which includes 2ND up to 7ND (nodal diameters). The number of nodal diameters is based on the number of nodes and anti-nodes appearing on the rubbing surfaces of the disc. Using standard material properties for cast iron the predicted frequencies are not well correlated with the experimental results. Hence tuning of the density and Young's modulus is necessary to reduce relative errors between the two sets of results. Having tuned the material properties the relative errors are shown in Table 2 and the maximum relative error is – 0.5%. The new material properties after tuning are given in Table 3.

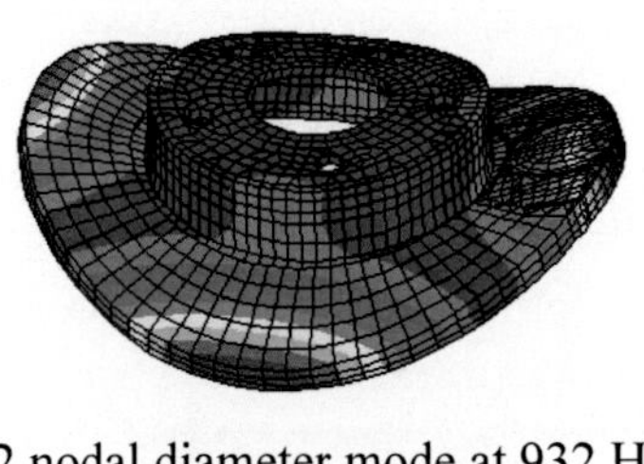

a) 2 nodal diameter mode at 932 Hz

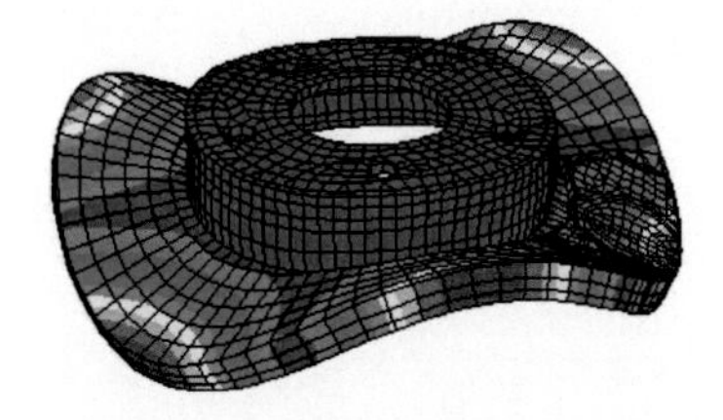

b) 3 nodal diameter mode at 1814 Hz

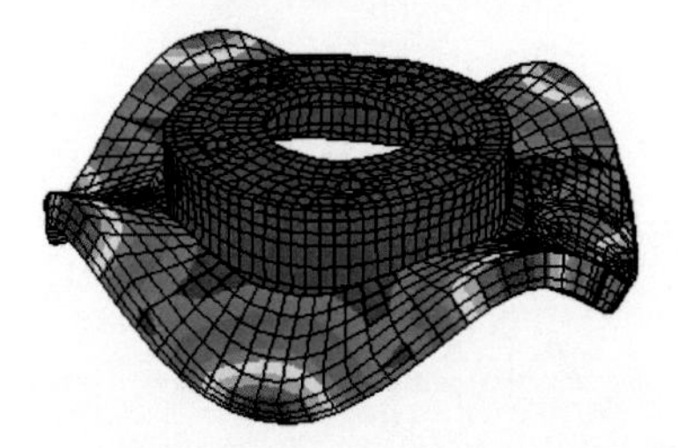

c) 4 nodal diameter mode at 2940 Hz

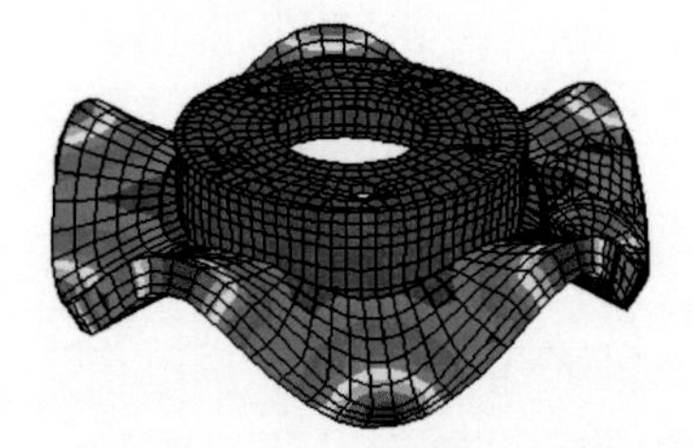

d) 5 nodal diameter mode at 4369 Hz

Figure 5. Continued on next page.

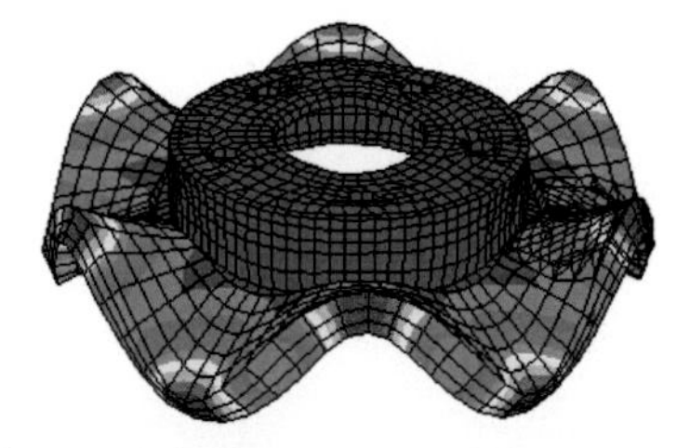

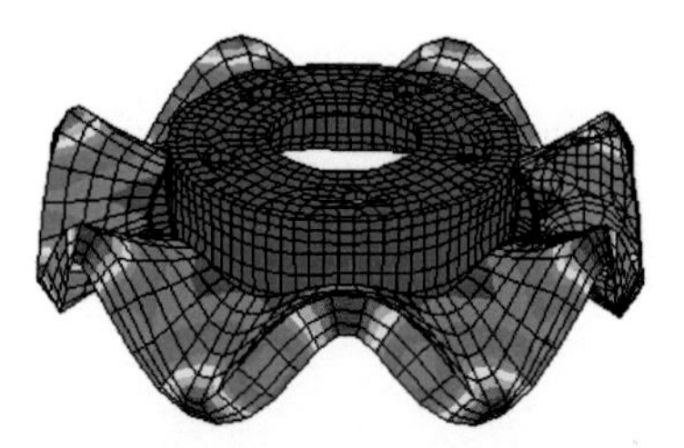

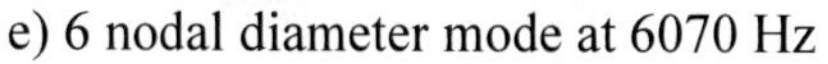

e) 6 nodal diameter mode at 6070 Hz f) 7 nodal diameter mode at 7979 Hz

Figure 5. Mode shapes of the disc at free-free boundary condition.

Table 2. Modal results of the disc at free-free boundary condition

MODE	2ND	3ND	4ND	5ND	6ND	7ND
Test (Hz)	937	1809	2942	4371	6064	7961
FE (Hz)	932	1814	2940	4369	6070	7979
Error (%)	-0.5	0.3	-0.1	0.0	0.1	0.2

Table 3. Material data of disc brake components

	DISC	BACK PLATE	PISTON	CALLIPER	CARRIER	GUIDE PIN	BOLT	FRICTION MATERIAL
Density (kgm^{-3})	7107.6	7850.0	7918.0	7545.0	6997.0	7850.0	9720.0	2798.0
Young's modulus (GPa)	105.3	210.0	210.0	210.0	157.3	700.0	52.0	Orthotropic
Poisson's ratio	0.211	0.3	0.3	0.3	0.3	0.3	0.3	-

The second stage of the methodology is to capture dynamic characteristics of the assembled model. The previous separated disc brake components must be now coupled together to form the assembly model. As discussed earlier in this chapter, a combination of linear spring elements and surface-to-surface contact elements are used to represent contact interaction between disc brake components and disc/pad interface, respectively. Table 4 shows details of disc brake couplings that are employed in the FE assembly model.

In the experimental modal analysis, a brake-line pressure of 1 MPa is imposed to the stationary disc brake assembly. A similar condition is also applied to the FE brake assembly model. In this validation, measurements are taken on the disc as it has a more regular shape than the other components. For the FE assembly model, spring stiffness values are tuned systematically as follows:

- At the interface of any two components that allow sliding between them the tangential spring constant is set at a very low stiffness, e.g., around 0.5 N/m. Example of this is between the guide pin and the carrier as given in Table 4.
- At the interface of any two components that restrict movement in any directions, e.g., between the bolt and the calliper arm, the spring constant is set a very high stiffness, e.g., around 1E+10 N/m.
- Any two interacting components that experience intermittent contact, e.g., the back plate and the piston, the spring stiffness is set around 1E+6 N/m.

Table 4. Disc brake assembly model couplings

No	Connections	DOF	Coordinate System	No. of Spring	Stiffness (N/m)
1	Piston wall-Calliper housing	1	Local	66	1.00E+9
2	Piston- Back plate	1	Global	38	2.80E+6
3	Piston- Back plate	2	Global	38	2.80E+6
4	Piston- Back plate	3	Global	38	4.00E+6
5	Calliper finger- Back plate	1	Global	104	1.02E+6
6	Calliper finger- Back plate	2	Global	104	1.02E+6
7	Calliper finger- Back plate	3	Global	104	1.46E+6
8	Leading abutment- Carrier	1	Global	24	0.50E+0
9	Leading abutment- Carrier	2	Global	24	1.00E+9
10	Trailing abutment- Carrier	1	Global	24	1.00E+9
11	Trailing abutment- Carrier	2	Global	24	1.00E+9
12	Leading bolt- Calliper arm	1	Local	16	3.00E+10
13	Leading bolt- Calliper arm	2	Local	16	3.00E+10
14	Leading bolt- Calliper arm	3	Local	16	3.00E+10
15	Trailing bolt- Calliper arm	1	Local	16	3.00E+10
16	Trailing bolt- Calliper arm	2	Local	16	3.00E+10
17	Trailing bolt- Calliper arm	3	Local	16	3.00E+10
18	Leading guide pin- Carrier	1	Local	18	1.00E+9
19	Leading guide pin- Carrier	3	Local	18	0.50E+0
20	Trailing guide pin- Carrier	1	Local	18	1.00E+9
21	Trailing guide pin- Carrier	3	Local	18	0.50E+0

Once those spring constants are set, modal analysis is performed to obtain natural frequencies of the disc and their associated mode shapes. A comparison is made between predicted and experimental results of the disc. If there are large relative errors, the spring stiffness values for linking two components need to be adjusted or updated. This updating process is continued until the relative errors are reduced to an acceptable level. Since the process is performed based on the trail-and-error process, it takes a lot of time and requires engineering intuition to identify more influential springs and pick up appropriate spring constants.

After a number of attempts, good agreements between predicted and experimental results are achieved. Correlation between the two sets of frequencies that include 2ND up to 7ND of the disc is given in Table 5. From the table, it is found that the maximum relative error is - 5.2%. These predicted results are based on the spring stiffness values given in Table 4. Mode shapes of the FE assembly are described in Figure 6. The simulated FE modal analysis is able to predict two frequencies at 3-nodal diameter as obtained in the experiments, which are generated at 1730.1 Hz and 2151.1 Hz. While in the experiments these frequencies are found at 1750.7 Hz and 2154.9 Hz. The highest relative error is found on a 6-nodal diameter mode, for which the predicted frequency is 5837.1 Hz while the experimental frequency is 6159.0 Hz. The lower relative error is about – 0.1 % on the second 3-nodal diameter mode, for which the frequencies are 2151.1 Hz and 2154.9 Hz in theory and in experiments respectively. In this validation process, static friction coefficient (at pads/disc interface) also plays an important role to reduce the relative errors. It is found that static friction coefficient of $\mu = 0.7$ give better correlation in the assembly model as described in Table 5.

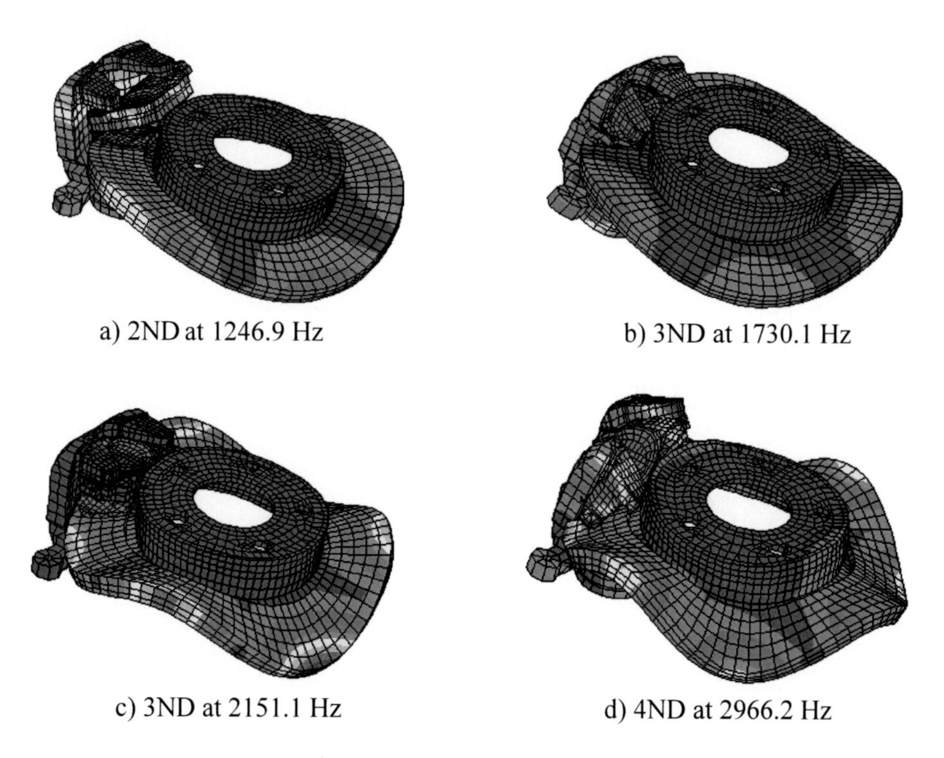

a) 2ND at 1246.9 Hz b) 3ND at 1730.1 Hz

c) 3ND at 2151.1 Hz d) 4ND at 2966.2 Hz

Figure 6. Continued on next page.

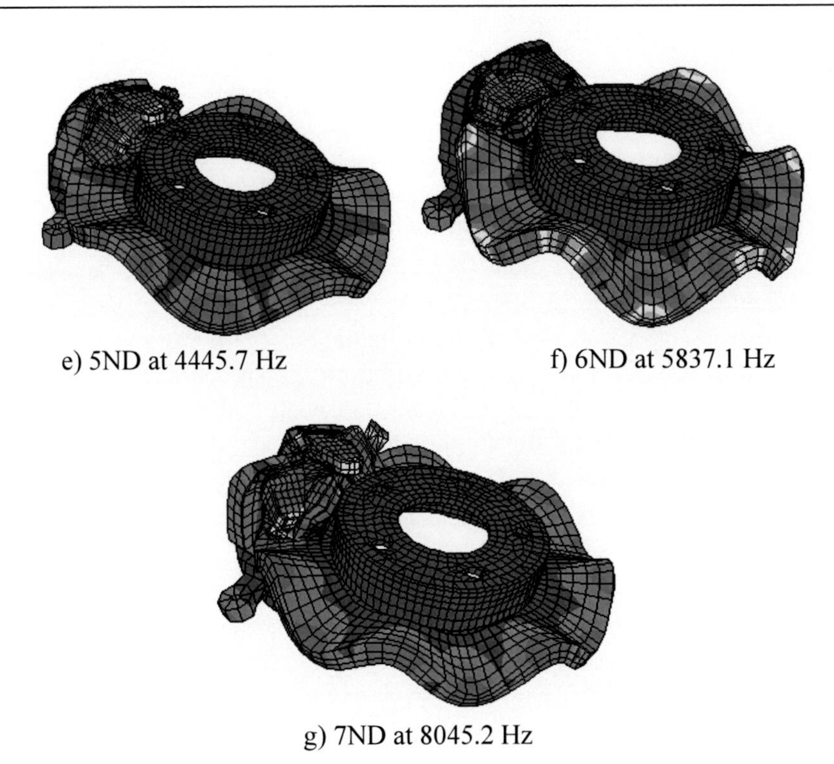

e) 5ND at 4445.7 Hz f) 6ND at 5837.1 Hz

g) 7ND at 8045.2 Hz

Figure 6. Mode shapes of the assembly model.

Contact Analysis

The third and final stage of the proposed methodology is to conduct experiments and simulations of contact pressure distributions under static application of the disc brake (that is, application of brake with no torque to or rotation of the disc). The experimental results will be used to confirm contact pressure distribution predicted in the FE model. In this section, brake pad models with real surface topography illustrated in Figures 7(a) ~ 7(c) are employed. The new and unworn pad pairs are used in order to confirm the measurements taken from the linear gauge and also to show the accuracy and reliability of the available tool.

Table 5. Modal results of the assembly measured on the disc

MODE	2ND	3ND	3ND	4ND	5ND	6ND	7ND
Test (Hz)	1287.2	1750.7	2154.9	2980.4	4543.7	6159.0	7970.0
FE (Hz)	1246.9	1730.1	2151.1	2966.2	4445.7	5837.1	8045.2
Error (%)	-3.1	-1.1	-0.1	-0.4	-2.1	-5.2	0.9

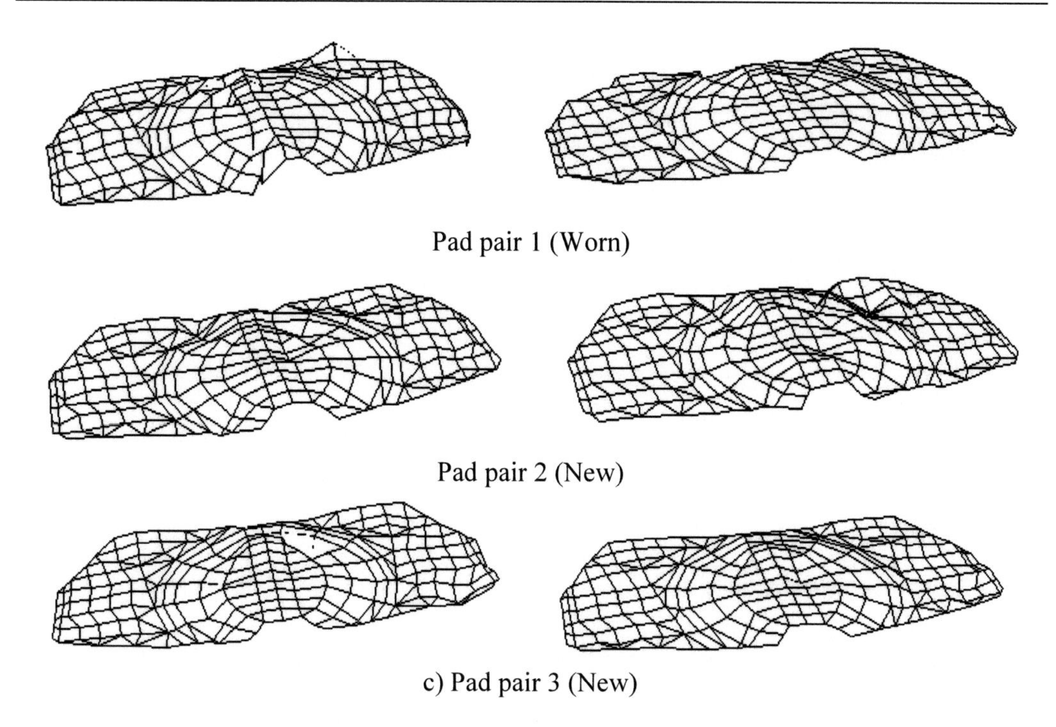

c) Pad pair 3 (New)

Figure 7. Surface topography at the piston pad (left) and finger pad (right).

In order to capture static contact pressure (stationary disc), Pressurex® Super Low (SL) pressure-indicating film, which can accommodate contact pressure in the range of between 0.5 ~ 2.8 MPa, is used. Pressure-indicating film is widely used to measure contact pressure distribution or surface roughness in the automotive industry. Tests conducted before and after a brake application often showed a noticeable difference between the measured pressure distribution at the disc and pads interface (Chen *et al*, 2003b).

In the current investigation, the films are tested under certain brake-line pressures for 30 seconds and then removed from the disc/pad interfaces. Figure 8 shows an example of pressure-indicating film before and after the contact testing. From the figure, the tested film only provides stress marks without revealing its magnitude. Topaq® Pressure Analysis system that can interpret the stress marks is then used. Configurations of the tested pad pairs are given in Table 6.

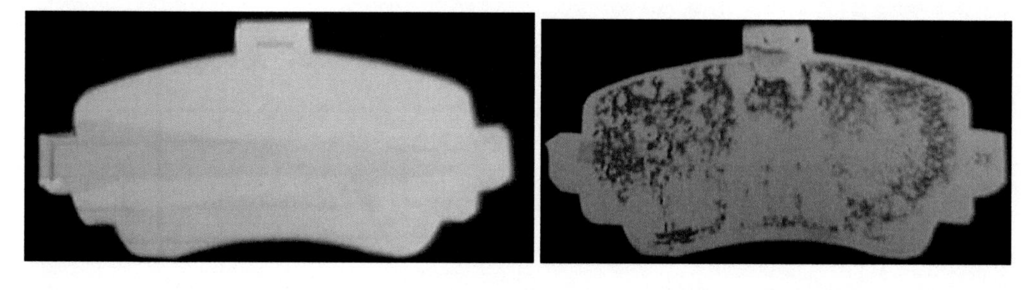

Figure 8. Pressure-indicating films before (left) and after (right) the static contact pressure testing.

Table 6. Configurations of tested pad

Identification	Pad conditions	Damping shim	Brake-line pressure (MPa)
Pad pair 1	Worn	No	2.5
Pad pair 2	New	No	2.5
Pad pair 3a	New	No	2.5
Pad pair 3b	New	No	1.5

It is shown that contact pressure distributions for the worn pad (Pad pair 1) seem to be concentrated (red colour) at the outer border region of the pads, while zero pressure exists at the inner border region of the pads. It is also shown that contact pressure distributions of both the piston and finger pads are asymmetric. This might be due to irregularities in the surface topography of the friction material. Contact pressure distributions of the worn pad are shown in Figure 9(a). The red colour shows the highest contact pressure. Areas in contact for the piston and finger pads are 1.436e-3m^2 and 1.484e-3m^2 respectively.

For the new and unworn pads, i.e., Pad pair 2 and Pad pair 3a that come from the same box and the same manufacturer, it is seen that they have different contact pressure distributions both at the piston and finger pad surfaces as shown in Figures 9(b) and 9(c). These variations are due to the surface topography as shown in Figure 7. It is also seen from figure 9(b) that contact pressures of Pad pair 2 are distributed more evenly than Pad pair 3a. There is contact at the trailing edge for Pad pair 2. But there seems to be a loss of contact in that region for Pad pair 3a. The areas of contact for the piston and the finger pads are 1.361e-3 m^2 and 1.069e-3 m^2 respectively. From Figure 9(c), the contact pressure seems to be zero at the centre of the pads. Most of the highest contact pressures appear at the outer border of the pads. Areas in contact for Pad pair 3a are 8.090e-4m^3 and 9.230e-4m^2 for the piston and the finger pads respectively.

By applying different levels of brake-line pressure, the higher the pressure the bigger the contact areas should be generated. This is illustrated in Figure 9(d) where the areas of highest pressure are reduced significantly in comparison with Figure 9(c). It is also confirmed that the areas of contact for the piston and the finger pads are reduced to 6.370e-4 m^2 and 6.857e-4 m^2 respectively. This means a reduction of about 21% and 26% for the piston and the finger pads respectively.

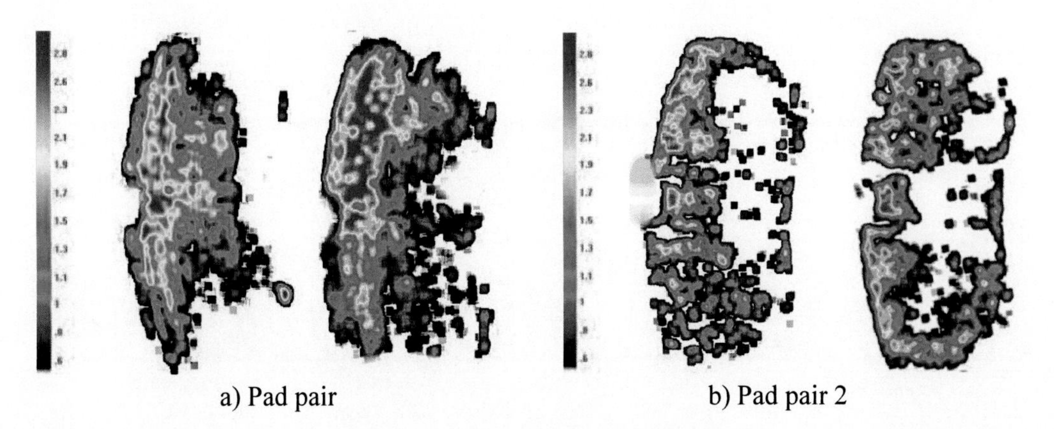

a) Pad pair b) Pad pair 2

Figure 9. Continued on next page.

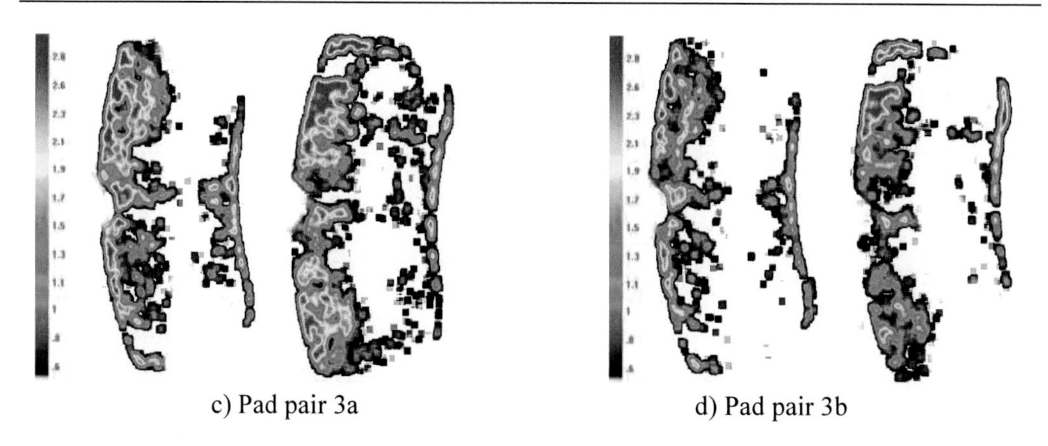

c) Pad pair 3a

d) Pad pair 3b

Figure 9. Analysed images of the tested pads: piston pad (left) and finger pad (right) in MPa. Top of the images are the leading edge.

In the FE contact analysis, the brake pad models are similar to those used in the contact tests. Now the real surface profile of the brake pads are considered in the sense that the surface profile information is incorporated in the FE model of the brake pad surface by adjusting its surface coordinates in the normal direction. Similar configurations of the test are also adopted in order to make comparison between the two sets of results, predicted versus experimental.

The first contact simulation is performed on the worn pad or Pad pair 1 at a brake-line pressure of 2.5 MPa. It can be seen in Figure 10(a) that the areas in contact are almost the same as those found in the experiment. Predicted contact areas in the contact analysis are 1.441e-3m^2 and 1.784e-3m^2 for the piston and the finger pads respectively. The results suggest that there is fairly good agreement between the two as illustrated in Figure 11. The contact area of the piston pad seems closer to the experimental one, compared with the finger pad.

The second contact simulation is done for Pad pair 2, which is subjected to the same brake-line pressure. From Figure 10(b) the contact pressure seems to be biased towards the outer radius of the pads. These patterns are most likely to be the same for those obtained in Figure 9(b). In the simulation it is found that the contact areas for the piston pad are 8.476e-4m^2 and for the finger pad is 8.131e-4 m^2. These contact areas are smaller than those measured in the experiments. However, quite reasonable correlations against the experimental results are obtained especially at the finger pad as described in Figure 11.

The third contact analysis is simulated for Pad pair 3a, subjected to the same brake-line pressure. Predicted areas of the highest contact pressure are in good agreement with the experimental results. Contact pressure distribution of Pad pair 3a is illustrated in Figure 10(c). For Pad pair 3a, predicted contact areas are 1.046e-3 m^2 and 1.020e-3 m^2 for the piston and the finger pads respectively. It can be seen from Figure 11 that the difference in the finger pad is small while there is a quite large difference at the piston pad. However, overall, fairly good agreement is achieved between predicted and experimental results.

The last contact analysis is similar to the third except under a different brake-line pressure of 1.5 MPa applied to the assembly model. The predicted areas in contact should be smaller than those predicted in the third analysis and are shown in Figure 10(d). Once again, good correlations are achieved between predicted and experimental results in terms of areas of the highest contact pressure. The locations of different levels of the contact pressure distribution are almost identical to the experimental one. In the contact simulation, it is predicted that the contact areas of the piston pad and the finger pad are 5.943e-4 m^2 and 6.860e-4 m^2. These values are nearly the same as those measured in the experiment. Figure 11 shows that there are small differences in the contact area for both the piston and the finger pads.

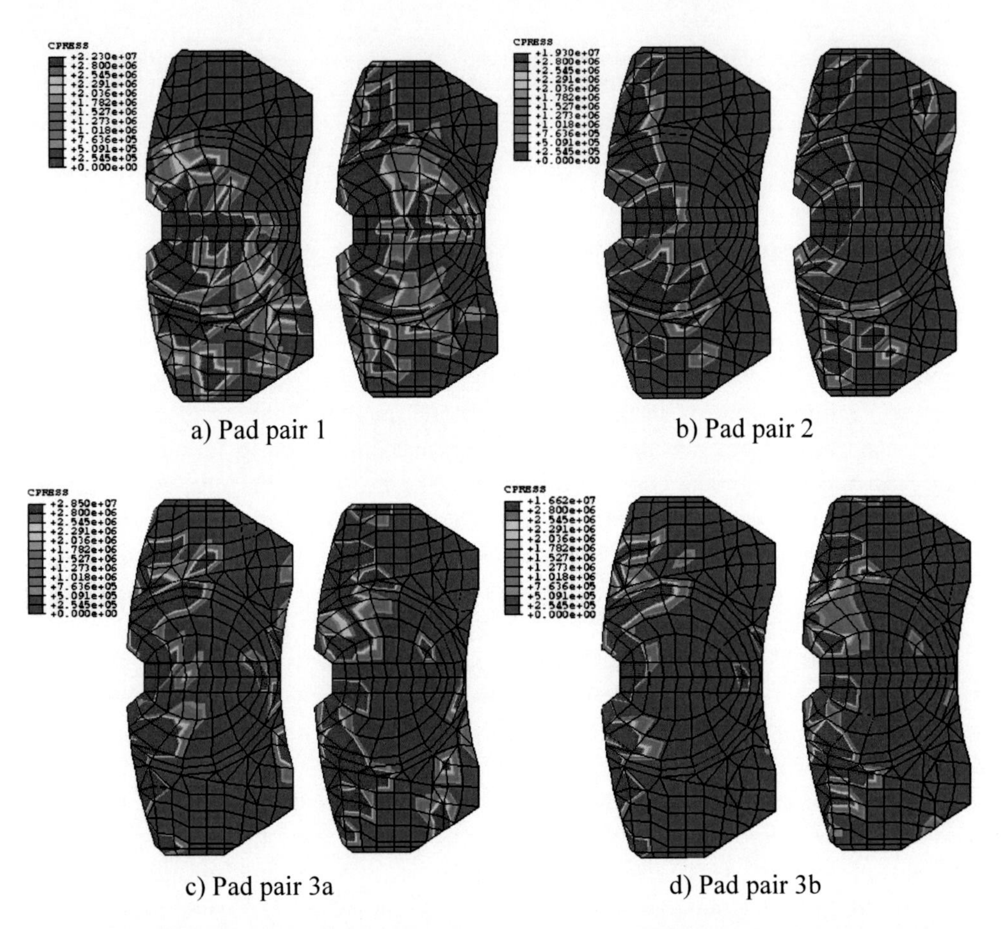

Figure 10. Predicted contact pressure distribution: piston pad (left) and finger pad (right) in Pascal. Top of the diagrams are the leading edge.

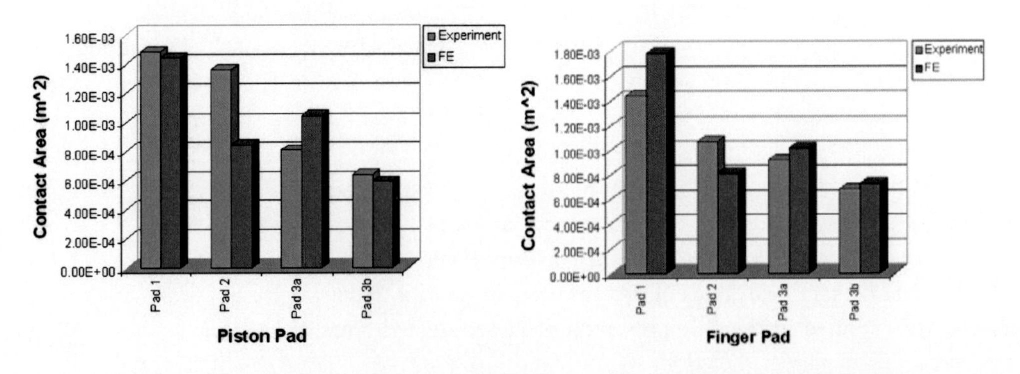

Figure 11. Comparison between experiment and FE analysis in the contact area.

Contact Pressure Distributions

At the contact interface of disc and brake pads, friction induces wear and heat during braking applications. Wear is one of the distinct aspects of brake systems that influence the contact pressure distribution, and itself is affected by the surface roughness of the components in sliding contact. Temperature is another distinct aspect and it normally rises up at some local area and thermal deformation of these areas occurs. Due to thermal deformation, the pressure distribution is also affected. Thermal and mechanical deformations affect each other strongly and simultaneously. This section looks into the effects of wear and temperature on contact pressure distributions.

Wear Effects

When two solid bodies are rubbed together they experience material removal, i.e., wear. In engineering applications the wear depth is a function of normal pressure, sliding distance and specific wear coefficient and other factors. Rhee (1970) in his study showed that the wear rate of most friction materials could be given as follows:

$$\Delta W = kF^a v^b t^c \tag{1}$$

where ΔW is the wear volume, F is the contact force, v is the sliding speed, t is the time and k is the wear constant which is a function of the material and temperature. a, b and c are constants that should be determined experimentally and c is usually close to unity. This original formula however cannot be used in the present investigation. Since mass loss due to wear is directly related to the displacements that occur on the rubbing surface in the normal direction, Rhee's wear formula is then modified as:

$$\Delta h = kP^a (\Omega r)^b t^c \tag{2}$$

where Δh is the wear displacement, P is the normal contact pressure, Ω is the rotational disc speed (rad/s), r is the pad mean radius (m) and a, b and c are all constants which remain to be determined. In incorporating wear into the FE model, the methodology that was proposes by Podra *et al* (1999) and Kim *et al* (2005) is adopted in this work. Bajer *et al* (2004) also performed wear simulation particularly for a disc brake. They used ABAQUS v6.5 and adopted a very simple wear model, i.e., a function of wear rate coefficient and contact pressure. On the other hand, Abu-Bakar *et al* (2005a) simulated wear progress over time using Equation (2) and assumed all constants were unity. In simulating wear, contact analysis is firstly performed in order to determine the normal contact pressure generated at the piston and the finger pads interface. Using Equation (2), wear displacements/depths are calculated based on the following parameters:

1. Predicted contact pressure generated in the contact analysis, P
2. Sliding time, t
3. Specific wear rate coefficient, k

4. Pad effective mean radius, r
5. Rotational speed, Ω

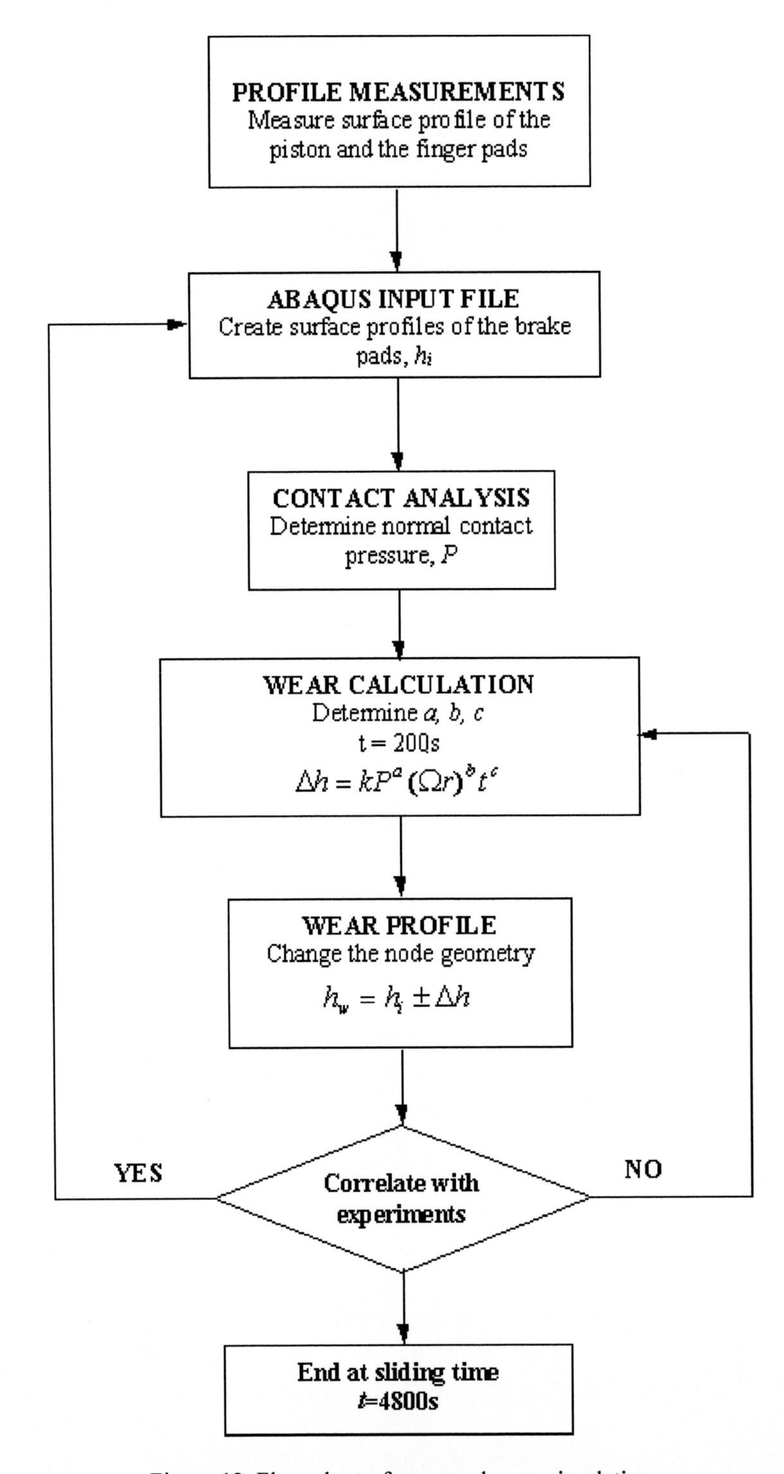

Figure 12. Flow chart of proposed wear simulation.

As a case study, Pad pair 2 in Table 6 is used to observe wear evolution on the brake pad interface. In the wear analysis, a rotational speed is maintained at 6 rad/s and the total braking time is set to 4800s (8 minutes). The seemingly short duration of wear tests is due to a numerical consideration. In the wear formula of Equation (2), the duration of wear, t, must be specified. The longer the duration of wear, the more the dimensional loss and the greater change of the contact pressure. However, if t is too big, there will be numerical difficulties in an ABAQUS run. It has been found through trial-and-error that t = 200 s gives reasonably good results and good efficiency. Consequently a simulation of 80-minute wear means twenty-four ABAQUS runs. In line with this numerical consideration, wear tests have not lasted for numerous hours as normally done in a proper wear test or a squeal test. In theory, however, numerical simulations of wear may cover an arbitrary length of time. A constant specific wear rate coefficient is assumed for all braking applications and is set to k = 1.78e-13m^3/Nm (Jang *et al*, 2004) and the effective pad radius is r = 0.11m. Then, based on the calculated wear displacements at steady state, a new surface profile for the piston and the finger pads is created. Figure 12 shows the overall procedure of wear simulation that has been used by the authors.

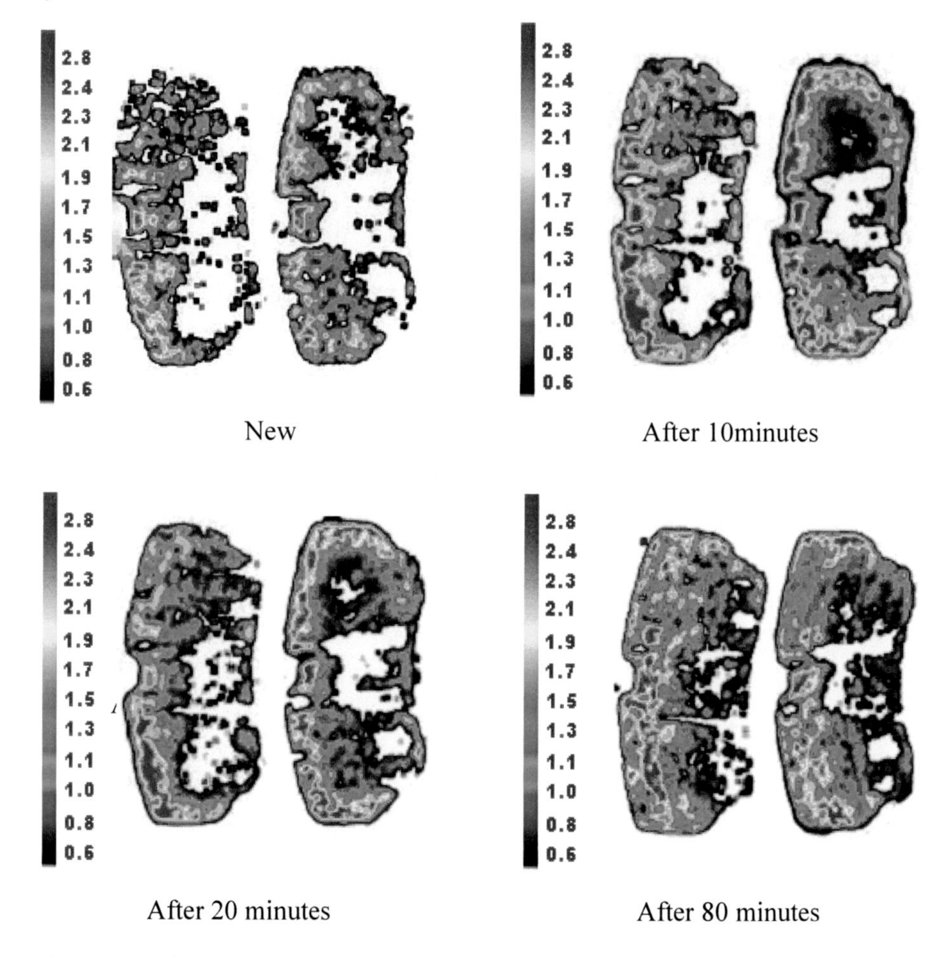

Figure 13. Measured contact pressure distribution: piston pad (left) and finger pad (right) in MPa. Top of the diagrams are the leading edge.

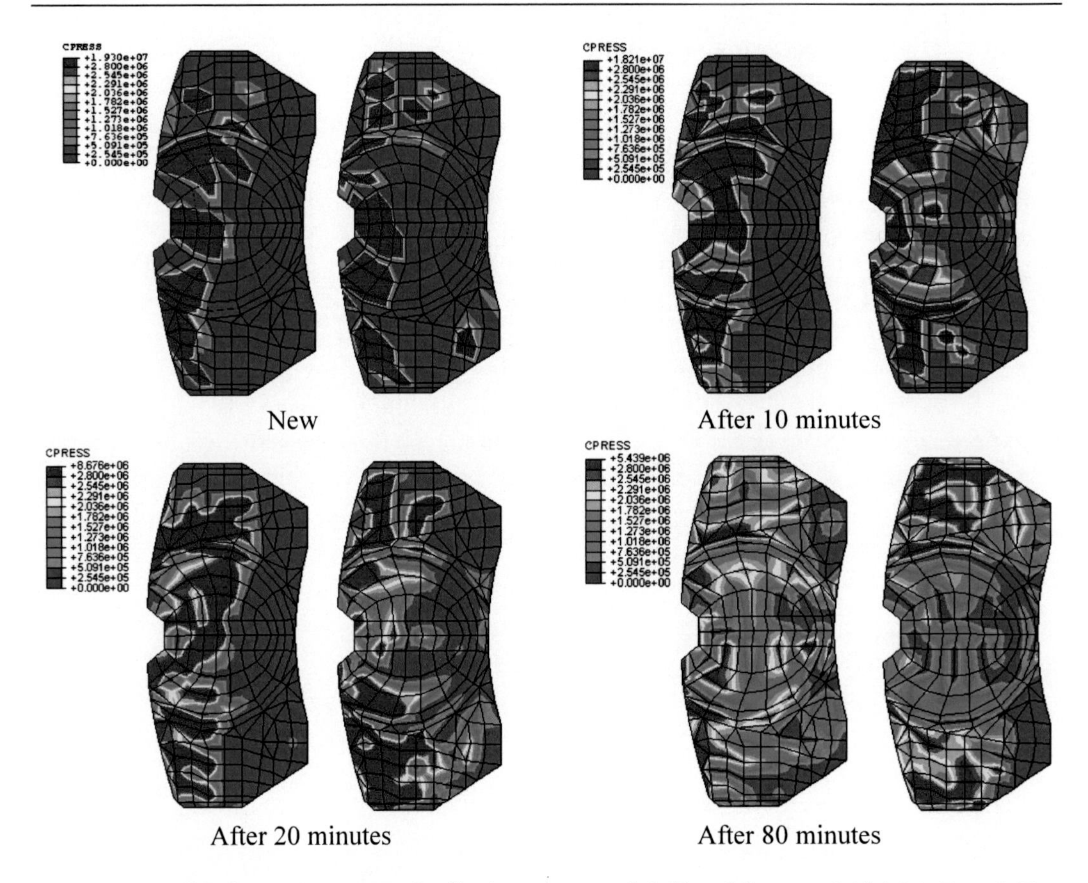

Figure 14. Predicted contact pressure distribution: piston pad (left) and finger pad (right) in Pascal. Top of the diagrams are the leading edge.

During this wear calculation all constants in Equation (2) need to be determined. Having simulated for various values of constants *a*, *b* and *c*, it is found that the wear formula below gives reasonably good results.

$$\Delta h = k_0 (\frac{P}{P'})^{0.9} \Omega r t \tag{3}$$

In Equation (3), P' is the maximum allowable braking pressure (8MPa for a passenger car) and $k_0 = 2.9 \times 10^{-7} \mathrm{m}^3 / \mathrm{Nm}$. Figures 13 and 14 show measured and predicted static contact pressure distributions at the piston and finger pads, respectively. It can be seen that most locations of the highest contact pressure (in red colour) predicted are at the outer region of the brake pads and these are almost identical to the measured data shown in Figure 13. It is also seen that areas in contact increase as braking duration approaches 80 minutes described in Figure 15. From the figure, the initial contact areas are predicted as about 7.0e-4 m^2 for both pads and then are predicted as much as 2.9e-3 m^2 in the final stage of braking duration. This is an increase by more than four folds.

Due to wear progress, it is found from the FE analysis that the surface profile of the brake pads becomes smoother after 80 minutes of wear, as shown in Figures 16(a) and 16(b). As a result, greater areas of the brake pads come into contact with the disc surface, as also illustrated in Figures 13 and 14. Graphs in Figure 16 are axial coordinates of the nodes that form the circumferential centre lines of the pads. It is shown that at the early stage of braking application, i.e., within 10 – 20 minutes, the axial coordinates change slightly overall in comparison with the new (unworn) brake pads. Having completed 80 minutes of braking application the surface height seems to level off and hence implies that some initial rough patches have been worn out to form smoother ones.

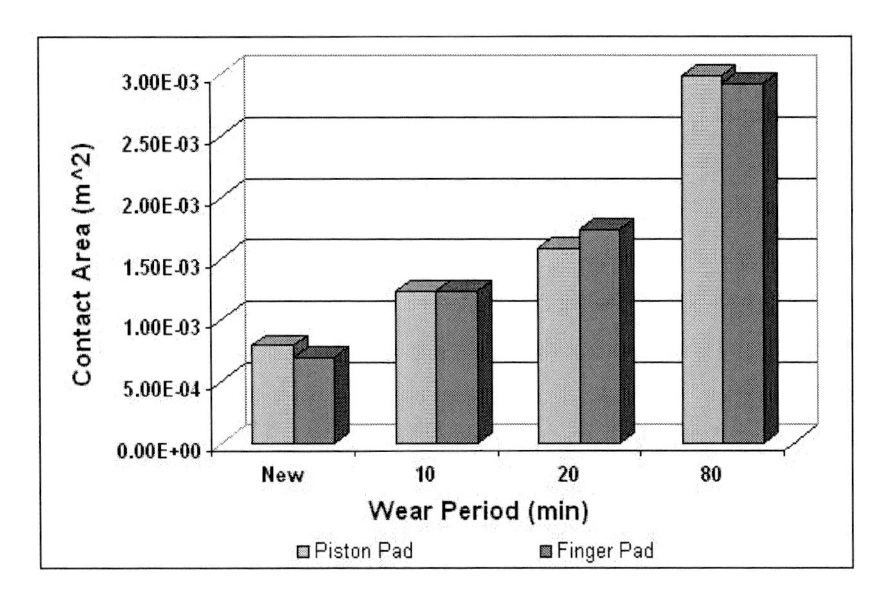

Figure 15. Predicted contact area for different braking application.

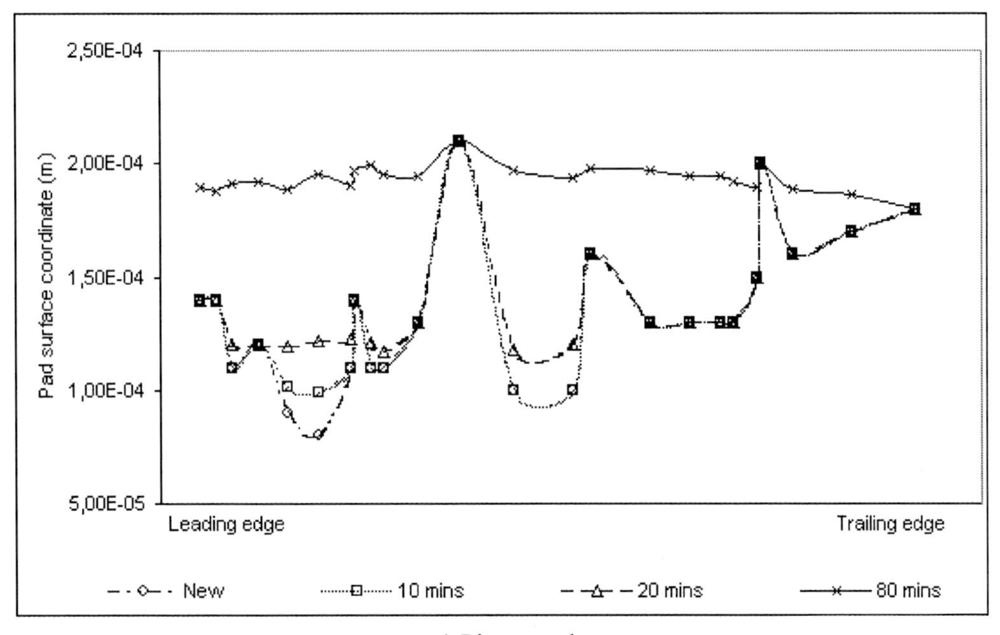

a) Piston pad

Figure 16. Continued on next page.

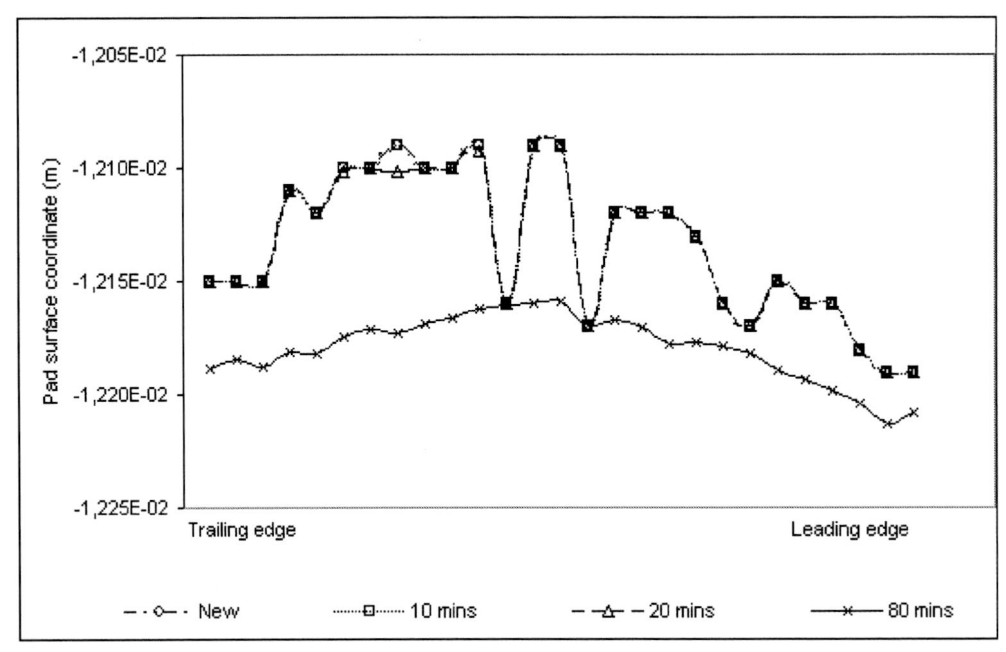

b) Finger pad.

Figure 16. Predicted surface profile due to wear.

Thermal Effects

In a disc brake system, the brake pads are pressed against the disc in order to generate friction and therefore to slow down the vehicle. Once friction occurs, it induces a large amount of heat in the system. Therefore, thermal effects should be one of the most important aspects, which is likely to affect squeal generation in a disc brake system. Due to the complex phenomenon of heat transfer and the difficulty of numerical modelling, thermal effects have largely been ignored in the past research into disc brake squeal. However, there were a few researchers who studied thermal characteristics in brakes for purposes other than studying brake squeal in the past. Day and Newcomb (1984) investigated the friction-generated heat energy dissipated from the contact interface. Brooks *et al* (1993) looked into the brake judder phenomena by using thermo-mechanical finite element model. Kao *et al* (1994) studied thermo-elastic instability of disc brake system. Qi *et al* (2004) investigated temperature distributions at the friction interface. Recently, Trichěs *et al* (2008) and Hassan *et al* (2008) incorporated thermal effects in complex eigenvalue analysis to investigate instability of the disc brake assembly.

There are two aspects of thermal effects, namely, thermal deformation and temperature dependence of material properties. Take contact as an example. If thermal deformation is considered, then the contact area changes and pressure distribution also becomes different. As mentioned before, thermal deformation effects are considered in the present work and thermal analysis is implemented in the baseline model. Therefore, it requires the use of elements with both temperature and displacement degrees of freedom. The elements of disc and brake pads are meshed with C3D6 (solid 3-dimensional 6-nodes element) and C3D8 (solid 3-dimensional

8-nodes element) in the baseline disc brake finite element model. These elements are now replaced by C3D6T and C3D8T, which include the temperature degree of freedom. However, there is a limitation of ABAQUS software package regarding the element types. Specifically C3D8T is not available in ABAQUS/Standard version but is available in ABAQUS/Explicit version. There are two ways to deal with this problem: either creating a new FE model using those elements available in an ABAQUS/Standard version that allows thermal analysis or using ABAQUS/Explicit version instead. The former approach is considered a very difficult task and time-consuming. On the other hand, heat transfer is a transient process and as a result temperature varies with time. Therefore, dynamic transient analysis in ABAQUS/Explicit version is considered a more suitable analysis method to simulate the squeal generation under thermal loading and therefore the latter approach is adopted.

In order to determine the temperature distribution in a medium, it is necessary to solve the appropriate form of heat transfer equation. However, such a solution depends on the physical conditions existing at the boundaries of the medium and on conditions existing in the medium at some initial time. To express the heat transfer in the disc brake model, several thermal boundary conditions and initial condition need to be defined. As shown in Figure 17, at the interface between the disc and brake pads heat is generated due to sliding friction, which is shown in blue colour. In this work, it is assumed that all the mechanical energy is converted into thermal energy. Al-Bahkali and Barber (2006) noted that the heat flux due to friction could be expressed as

$$q = \mu V p \tag{4}$$

where μ is the friction coefficient, V is the sliding velocity of the disc and p represents the contact pressure at the interface.

For the exposed region of the disc and brake pads, it is assumed that heat is exchanged with the environment through convection. Therefore, convection surface boundary condition is applied there (shown in red colour in Figure17). This can be expressed as

$$-k \left.\frac{\partial T}{\partial x}\right|_{x=0} = h[T_\infty - T(0,t)] \tag{5}$$

where h is convection heat transfer coefficient, k is thermal conductivity, and T_∞ is atmosphere temperature and $T(0,t)$ is the temperature at that boundary denoted by $x = 0$.

Finally, at the surface of the back plate, adiabatic or insulated surface boundary condition is used and shown in black colour in Figure 17. This can be expressed as

$$\left.\frac{\partial T}{\partial x}\right|_{x=0} = 0 \tag{6}$$

which means there is no heat transfer through the back plate into other disc brake components. This simplification removes the need to define the convection surface boundary condition of the exposed regions of the other components and is mainly a numerical consideration. Lin (2001) and Al-Bahkali and Barber (2006) used the same boundary conditions in their models. This simplification should be sufficient for short braking application where heat can hardly propagate far when squeal may occur already. It should also be noted that Equations (5) and (6) describe one-dimensional heat transfer for the sake of explanations and three-dimensional heat transfer is actually simulated in the authors' research. The initial condition of the model is 20°C at every node of the disc and brake pads. The atmosphere temperature is also 20°C all the time.

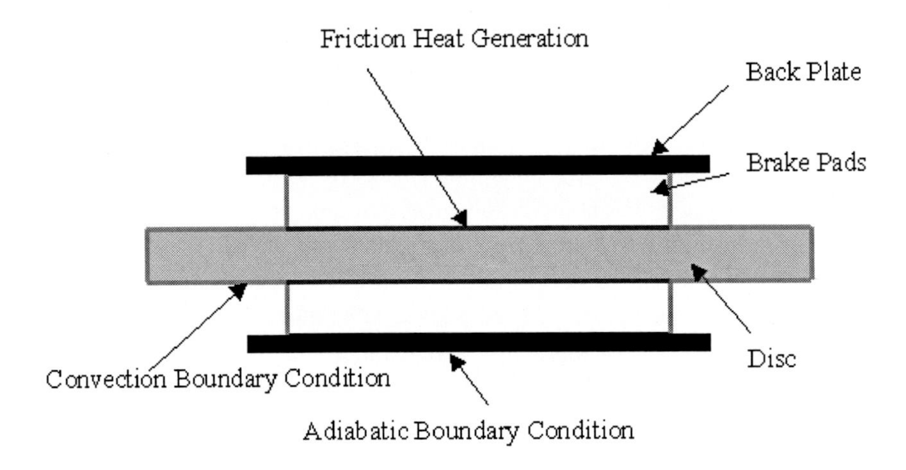

Figure 17. Boundary Condition of Thermal Analysis.

Table 7 lists the thermal properties of the disc and brake pads and all these data are from Lin (2001). However, it turns out that using those appropriate values of thermal properties leads to exceedingly long computing time. A typical example of thermal analysis of the disc brake system takes a few weeks to finish. To overcome this problem, Choi and Lee (2003) used a value of specific heat that is much lower than the realistic value and found that much faster convergence to the steady state in the transient thermo elastic analysis could be achieved. Therefore, a much smaller value of specific heat capacity (20 J/kg K) is also adopted here instead.

A comparison is made between the contact pressure distributions with and without thermal effects. Figure 18 is the results of the contact pressure distribution with thermal effects. Figure 19 shows the contact pressure distribution without thermal effects. These results are obtained at Ω = 100 rad/s and P = 1 MPa. Comparing these two figures, the pressure distribution is different. The FE model with thermal effects shows that the contact pressure at the piston pad is spreading towards the leading edge, compared with the trailing edge for the FE model without thermal effects. For the finger pad, it shows that a larger contact area is established at the trailing edge with the thermal effects, compared with the FE model without thermal effects. The contact pressure is also higher in the model with thermal effects, which is 10.67 MPa, than the model without thermal effects, which is 6.83 MPa. Figures 20 and 21 are another example with Ω = 50 rad/s and P = 1 MPa. These two figures also show that the contact pressure at the piston and the finger pad is higher with the inclusion of thermal effects than the model without thermal effects. Distributions of contact pressure are also seen significantly different between the two models.

Table 7. Material thermal properties data of disc and brake pads

	Disc	Brake pad
Thermal Conductivity (W/m K)	46.73	2.06
Specific Heat (J/kg K)	690.8	749
Thermal Expansion Coefficient 10^{-6} (1/K) m^2	6.6	14.3

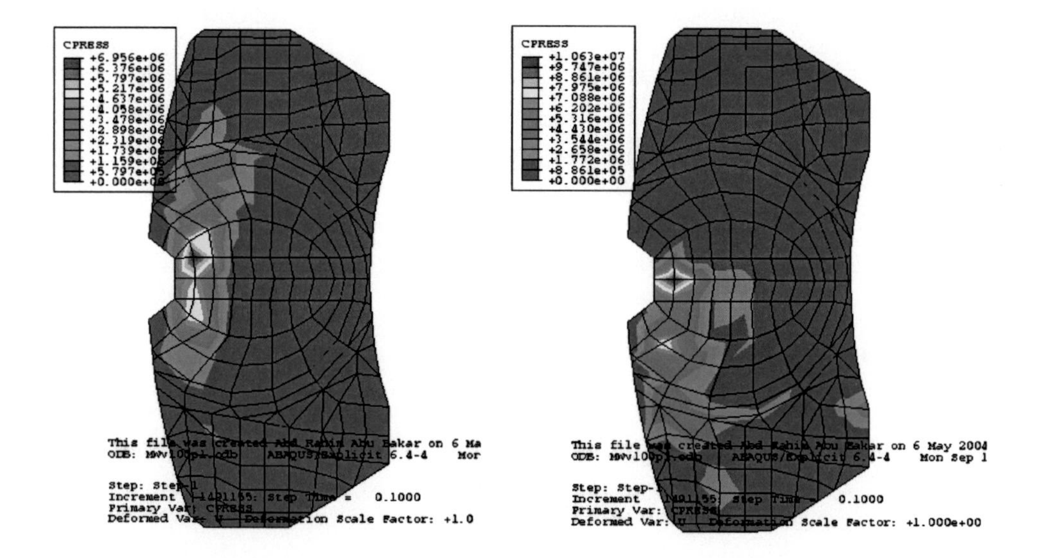

Figure 18. Pressure distribution with thermal effects in 100 rad/s and 1 MPa.

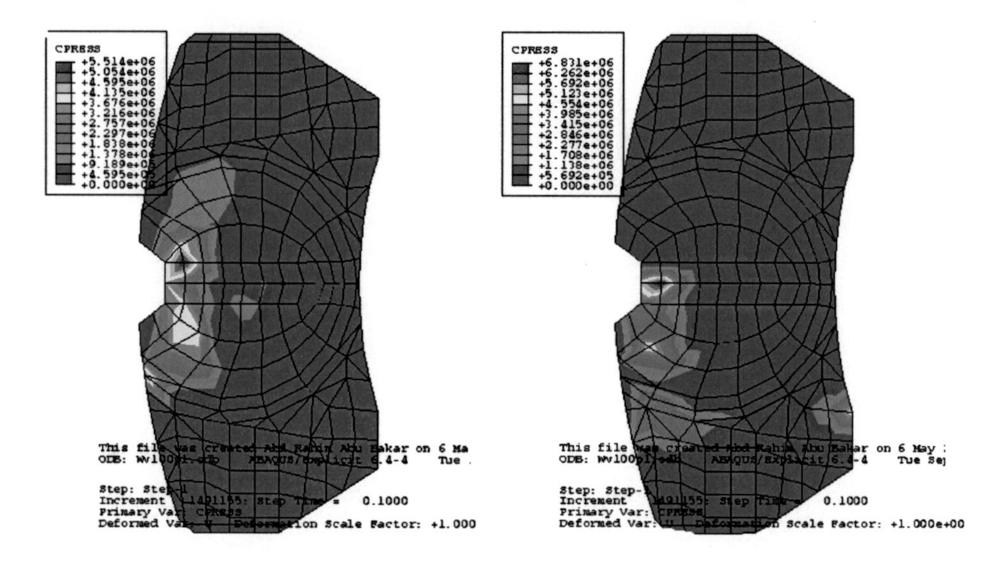

Figure 19. Pressure distribution without thermal effects in 100 rad/s and 1 MPa.

Vibration Analysis

There are two major numerical methods used in the studies of brake noise by researchers, namely, complex eigenvalue analysis and dynamic transient analysis. The advantages and limitations of both methods were commented by Mahajan *et al* (1999) and Ouyang *et al* (2005). In recent years, the dynamic transient analysis is gradually gaining popularity. A number of researchers pioneered this approach in their studies of squeal behaviour (Chargin *et al*, 1997, Hu and Nagy, 1997, Hu *et al*, 1999, Mahajan *et al*, 1999). Massi and Baillet (2005), Abu-Bakar and Ouyang (2006), Massi *et al* (2007) and Abu-Bakar *et al* (2007) furthered this approach. However, none of them considered thermal effects. Dynamic transient analysis in ABAQUS v6.4 is the approach used in this investigation into the vibration of the finite element disc brake model. ABAQUS uses central difference integration rule together with the diagonal lumped mass matrices. The following finite element equation of motion is solved:

$$\mathbf{M}\ddot{\mathbf{x}}^{(t)} = \mathbf{f}_{\text{ex}}{}^{(t)} - \mathbf{f}_{\text{in}}{}^{(t)} \tag{7}$$

At the beginning of the increment, accelerations are computed as follows:

$$\ddot{\mathbf{x}}^{(t)} = \mathbf{M}^{-1}(\mathbf{f}_{\text{ex}}{}^{(t)} - \mathbf{f}_{\text{in}}{}^{(t)}) \tag{8}$$

where $\ddot{\mathbf{x}}$ is the acceleration vector, $\mathbf{M}$ the diagonal lumped mass matrix, $\mathbf{f}_{\text{ex}}$ the applied load vector and $\mathbf{f}_{\text{in}}$ the internal force vector. The superscript t refers to the time increment.

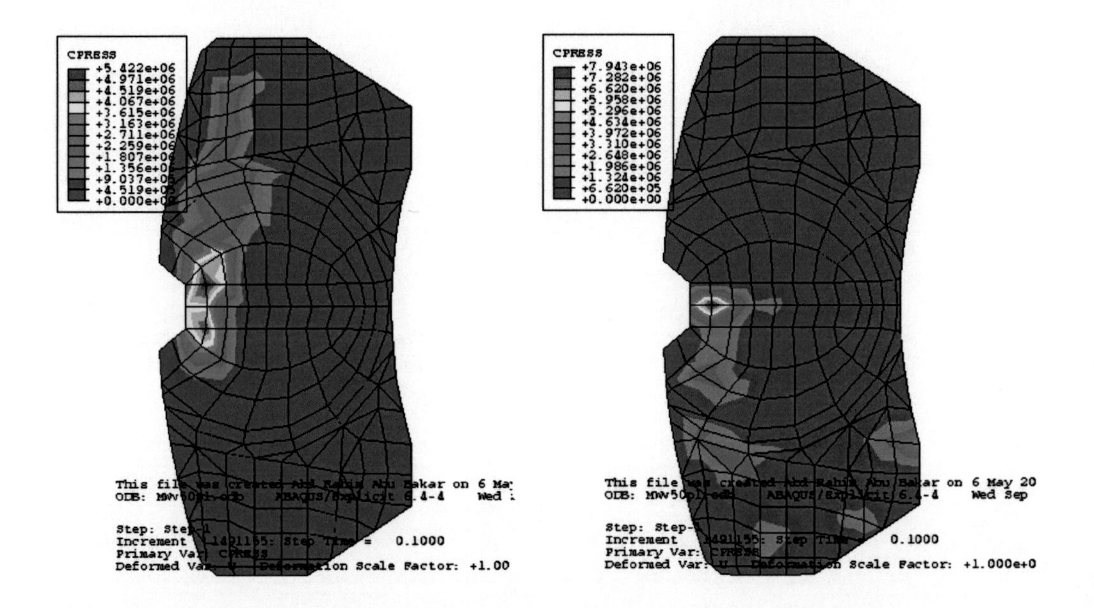

Figure 20. Pressure distribution with thermal effects in 50 rad/s and 1 MPa.

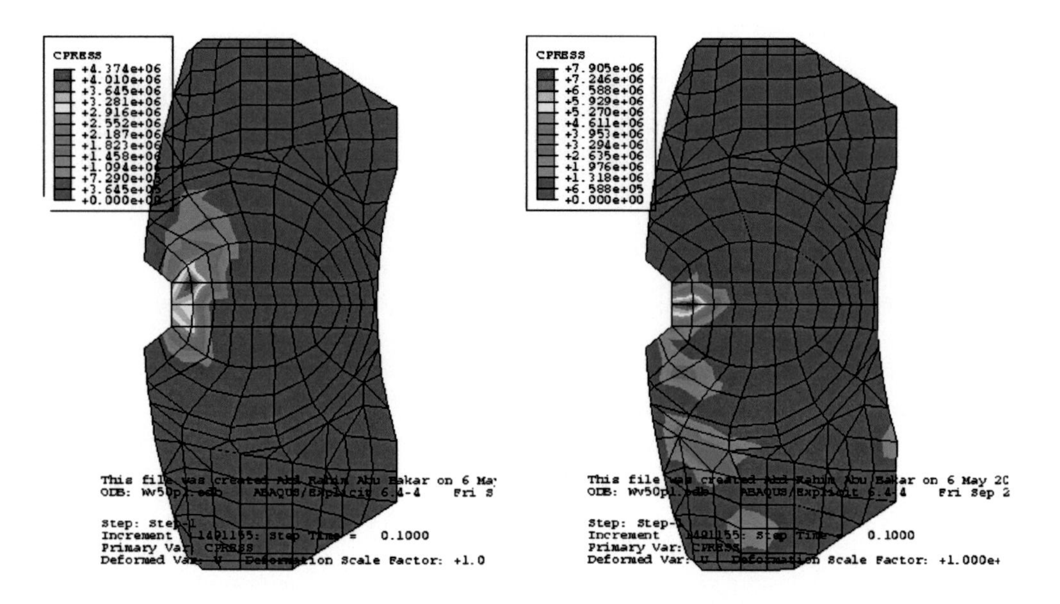

Figure 21. Pressure distribution without thermal effects in 50 rad/s and 1 MPa.

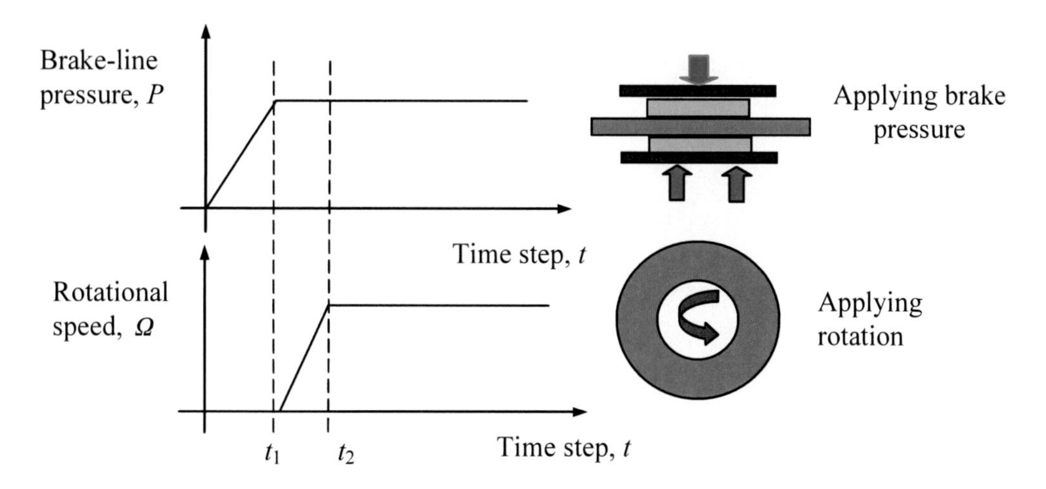

Figure 22. Time history of brake-line pressure and rotational speed.

The velocity and displacement of the body are given in the following equations:

$$\dot{\mathbf{x}}^{(t+0.5\Delta t)} = \dot{\mathbf{x}}^{(t-0.5\Delta t)} + \frac{\Delta t^{(t+\Delta t)} + \Delta t^{(t)}}{2} \ddot{\mathbf{x}}^{(t)} \tag{9}$$

$$\mathbf{x}^{(t+\Delta t)} = \mathbf{x}^{(t)} + \Delta t^{(t+\Delta t)} \dot{\mathbf{x}}^{(t+0.5\Delta t)} \tag{10}$$

where the superscripts $(t - 0.5\Delta t)$ and $(t + 0.5\Delta t)$ refer to mid-increment values. Since the central difference operator is not self-starting because of the mid-increment velocity, the initial values at time $t = 0$ for velocity and acceleration need to be defined. In this case, both values are set to zero as the disc is stationary at time $t = 0$.

The time history of the brake-line pressure and rotational speed are used for describing operating conditions of the disc brake model, as shown in Figure 22. At the first stage, a brake pressure is applied gradually until it reaches t_1 and then it becomes constant. The disc starts to rotate at t_1 and gradually increases up to t_2. Then the rotational speed becomes constant too.

As a case study, two different operating conditions are considered in order to observe squeal behaviour in the disc brake assembly. The objective of this investigation is to reveal how thermal aspects affect squeal behaviour. Thus, a comparison between the disc brake model with and without thermal effects is made in this section. Figures 23 and 24 show the results of disc brake model with thermal effects at $\Omega = 50$ rad/s and $P = 1$ MPa. Figure 25 and Figure 26 show the results from the model without thermal effects. From these figures, it is found that the vibration amplitude for the model with thermal effects is higher than the model without thermal effects. Moreover, the patterns of vibration of both examples are also different. However, the highest frequency components in these examples both are around 1200 Hz. Other examples are shown in Figures 27, 28, 29 and 30. The operational conditions are $\Omega = 50$ rad/s and $P = 0.5$ MPa. The vibration amplitude also increases in the model with thermal effects compared with the results from the model without thermal effects. The highest frequency components both are around 1400 Hz this time. All these examples indicate that thermal effects do affect the vibration level of disc brake system and therefore are very likely to affect the squeal generation. Therefore, it would be worthwhile to include thermal effects in the prediction of disc brake squeal.

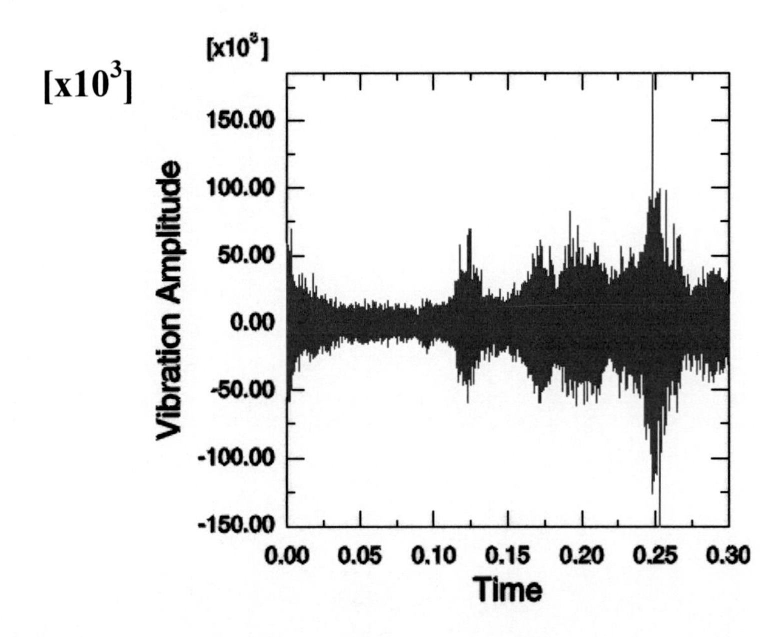

Figure 23. Time history of acceleration at a particular node with thermal effects (50 rad/s and 1 MPa).

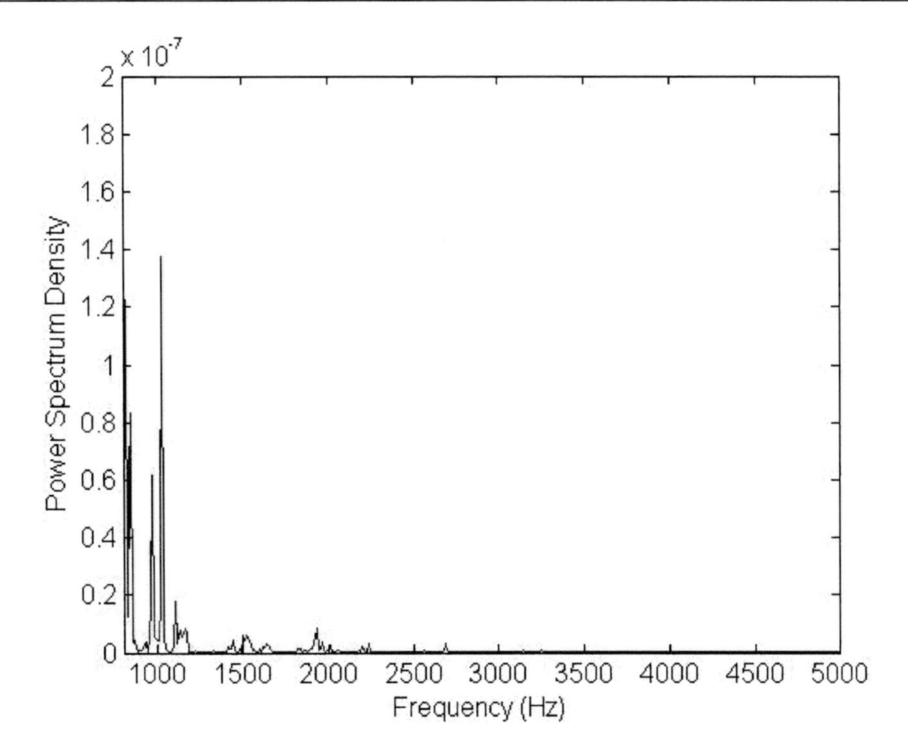

Figure 24. Frequencies after converting from time domain with thermal effects (50 rad/s and 1 MPa).

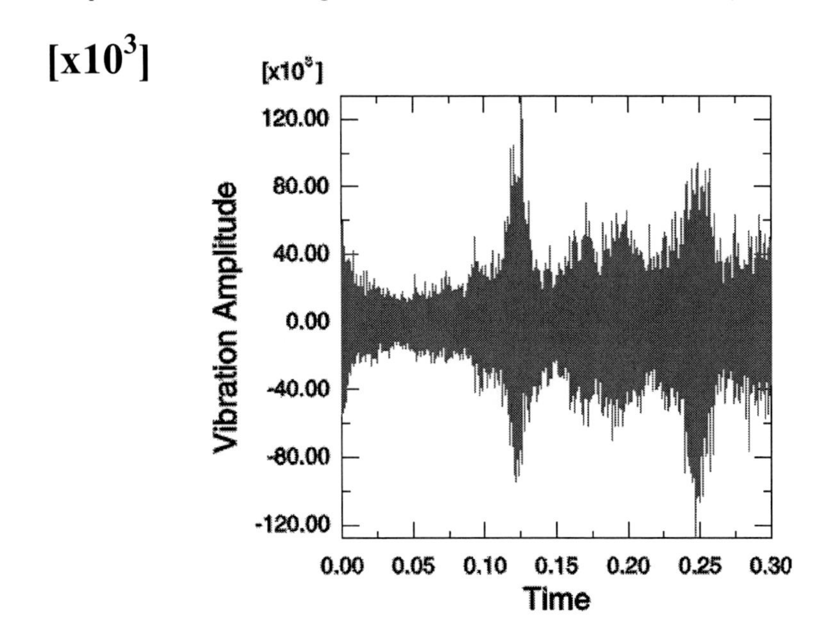

Figure 25. Time history of acceleration at a particular node without thermal effects (50 rad/s and 1 MPa).

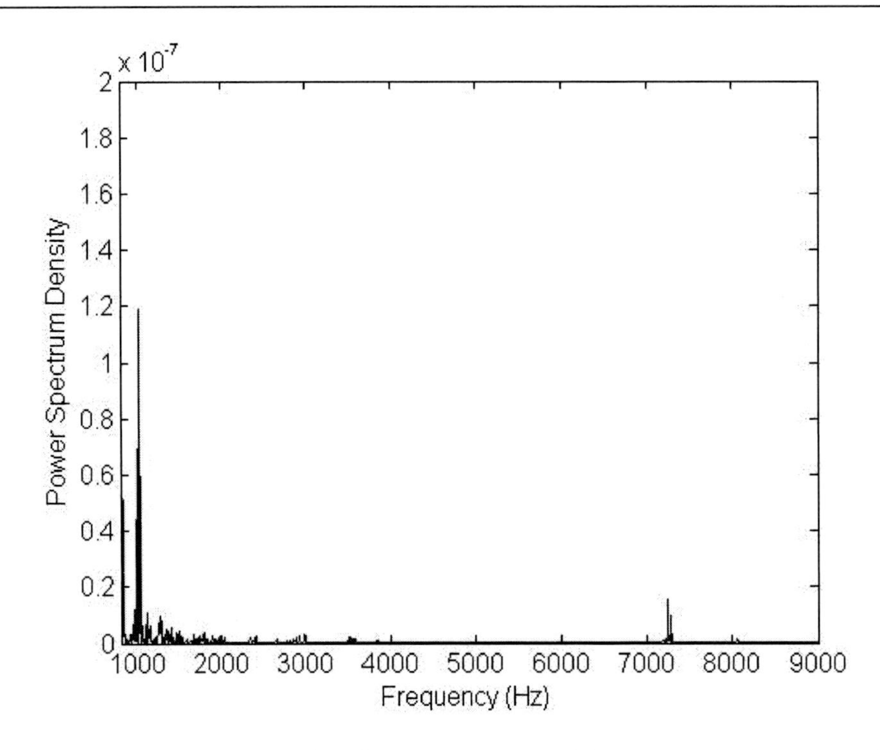

Figure 26. Frequencies after converting from time domain without thermal effects (50 rad/s and 1 MPa).

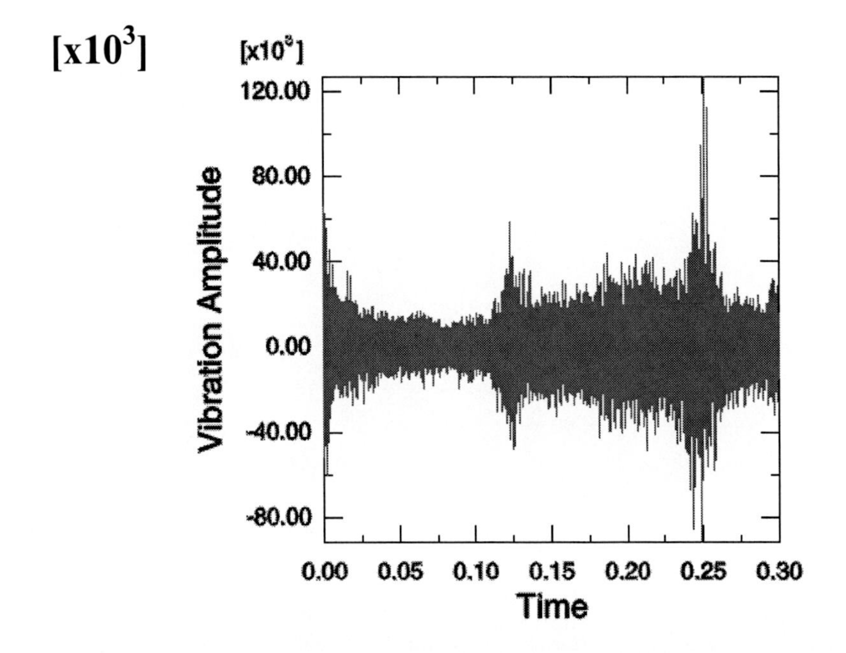

Figure 27. Time history of acceleration at a particular node with thermal effects (50 rad/s and 0.5 MPa).

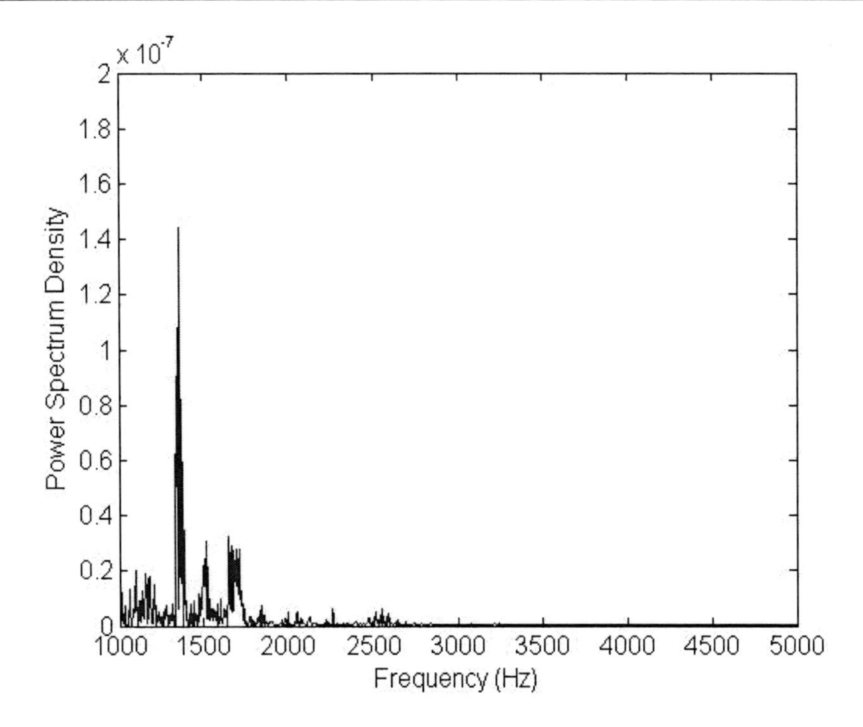

Figure 28. Frequencies after converting from time domain with thermal effects (50 rad/s and 0.5 MPa).

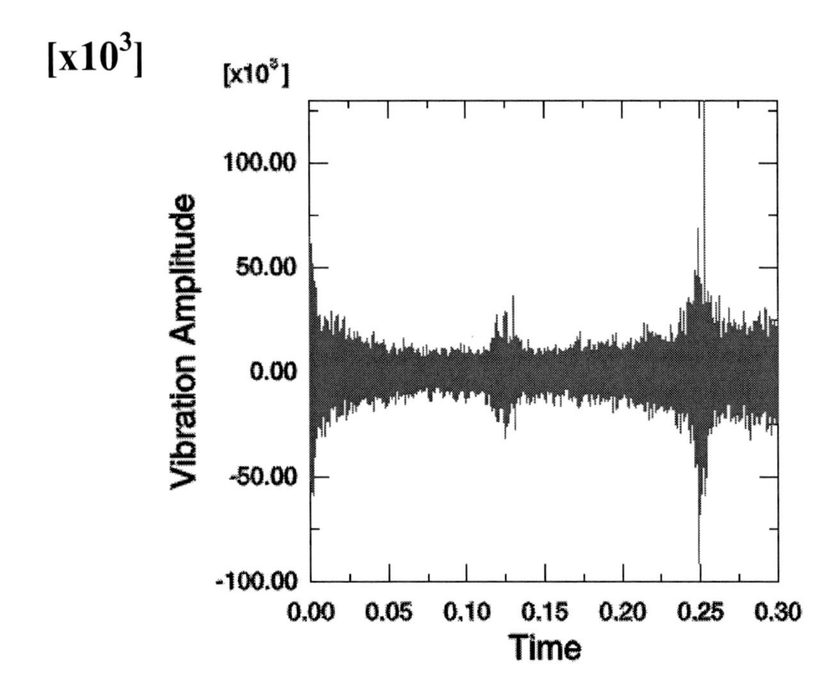

Figure 29. Time history of acceleration at a particular node without thermal effects (50 rad/s and 0.5 MPa)

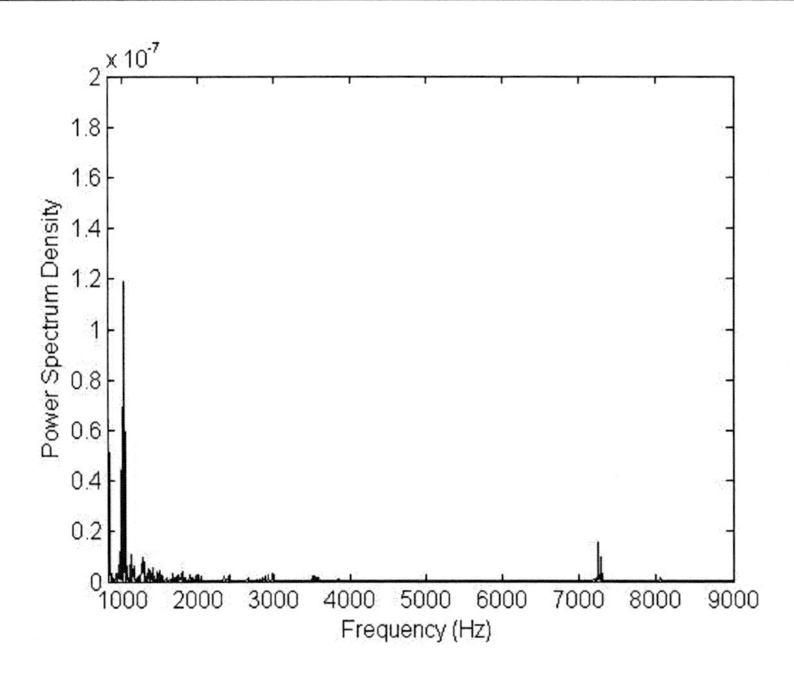

Figure 30. Frequencies after converting from time domain without thermal effects (50 rad/s and 0.5 MPa).

Conclusion

This chapter outlines recent studies into car disc brake squeal conducted at the University of Liverpool since 2004. The focus is on the numerical analysis using the finite element method. The simulation results are supported with measured data in order to verify predictions. An improved numerical methodology is presented by considering three-validation stages, namely, modal analysis at component and assembly levels and verification of contact analysis. Prior to that, a realistic surface roughness of the brake pad at macroscopic level is considered in the finite element model instead of assuming a smooth and perfect surface that has been largely adopted by most previous researchers. These two aspects have brought about significant improvement to the validation as well as analysis. Wear and thermal effects are other distinct aspects of disc brakes that influence contact pressure distributions and squeal generation in a disc brake assembly and they are also included in the current investigation. Transient analysis of disc brake vibration using a large FE model that includes thermal effects is carried out for the first time.

Acknowledgements

Some of the work reported in this chapter has been financially supported by TRW Automotive, Sensor Products LLC and Universiti Teknologi Malaysia. A number of people have helped this work at Liverpool, notably Dr S. James, Dr Q. Cao and Mr H. Tuah. Their contributions are gratefully acknowledged. Dr Tie Li of Ford at Basildon and Dr Frank Chen of Ford at Dearborn have kindly provided some papers.

References

ABAQUS Manual Version 6.4. (2003). Rhode Island, USA: Hibbitt, Karlsson & Sorensen, Inc.

Abendroth, H., & Wernitz, B. (2000). The integrated test concept: dyno-vehicle, performance-noise. *SAE Technical Paper* 2000-01-2774.

Abu-Bakar, A.R., & Ouyang, H. (2004). Contact pressure distribution by simulated structural modifications. In: Barton, D. C., Blackwood, A. *Proceedings of IMechE International Conference*- Braking 2004. UK: Professional Engineering Publishing Ltd; 123-132.

Abu-Bakar, A. R., Ouyang, H., Li, L., & Siegel, J. E. (2005a). Brake pad surface topography Part II: Squeal generation and prevention. *SAE Technical Paper* 2005-01-3935.

Abu-Bakar, A. R., Ouyang, H., & Siegel, J. E. (2005b). Brake pad surface topography Part I: Contact pressure distributions. *SAE Technical Paper* 2005-01-3941.

Abu-Bakar, A. R., & Ouyang, H. (2006). Complex eigenvalue analysis and dynamic transient analysis in predicting disc brake squeal. *International Journal of Vehicle Noise and Vibration,* Vol. 2, No. 2, 143-155.

Abu-Bakar, A. R., Sharif, A., Rashid, M. Z. A., & Ouyang, H. (2007). Brake squeal: Complex eigenvalue versus dynamic transient analysis. *SAE Technical Paper* 2007-01-3964.

Abu-Bakar, A. R., & Ouyang, H (2008). A new prediction methodology of disc brake squeal using complex eigenvalue analysis. *International Journal of Vehicle Design* (in press).

Akay, A. (2002). Acoustics of friction. *Journal Acoustics Society of America,* Vol. 111, No. 4, 1525-1548.

Al-Bahkali, E. A., & Barber, J. R. (2006). Nonlinear steady state solution for a thermoelastic sliding system using finite element method. *Journal of Thermal Stresses,* Vol. 29, 153-168.

Bajer, A., Belsky, V., & Kung, S. (2004). The influence of friction-induced damping and nonlinear effects on brake squeal analysis. *SAE Technical Paper* 2004-01-2794.

Bajer, A., Belsky, V., & Zeng, L. J. (2003). Combining a nonlinear static analysis and complex eigenvalue extraction in brake squeal simulation. *SAE Technical Paper* 2003-01-3349.

Blaschke, P., Tan, M., & Wang, A. (2000). On the analysis of brake squeal propensity using finite element method. *SAE Technical Paper,* 2000-01-2765.

Brooks, P. C., Barton, D., Crolla, D. A., Lang A. M., & Schafer, D. R. (1993). A new approach to disc brake judder using thermomechanical finite element model. *Proceedings of I.Mech.E.,* Paper No. C462/31/064.

Chargin, M. L., Dunne, L. W., & Herting, D. N. (1997). Nonlinear dynamics of brake squeal. *Finite Elements in Analysis and Design,* Vol. 28, 69-82.

Chen, F., Tan, C. A., & Quaglia R. L. (2003a). On automotive disc brake squeal. Part I: mechanisms and causes. *SAE Technical Paper* 2003-01-0683.

Chen, F., Abdelhamid, M. K., Blashke, P., & Swayze, J. (2003b). On automotive disc brake squeal. Part III: test and evaluation. *SAE Technical Paper* 2003-01-1622.

Chen, F., Tan, C. A., & Quaglia, R. L. (Eds.), (2003). *Disc brake squeal: Mechanism, analysis, evaluation, and reduction/prevention.* Warrendale, USA: SAE International.

Choi, J., & Lee, I. (2003). Transient thermoelastic analysis of disk brakes in frictional contact. *Journal of Thermal Stresses,* Vol. 26, 223-244.

Crolla, D. A., & Lang, A. M. (1991). Brake noise and vibration: The state of the art. *Vehicle Tribology Leeds-Lyon Tribology Series,* No. 18, 165-174.

Day, A., & Newcomb, T. P. (1984). The dissipation of friction energy from the interface of annular disc brake. *Proceedings of I.Mech.E.,* Paper No. 69/84.

Dom, S., Riefe, M., & Shi, T. S. (2003). Brake squeal noise testing and analysis correlation. *SAE Technical Paper* 2003-01-1616.

Eriksson, M. (2000). Friction and contact phenomenon of disc brakes related to squeal. *PhD Thesis,* Faculty of Science and Technology, Uppsala University, Sweden.

Eriksson, M., Bergman, F., & Jacobson, S. (1999). Surface characterization of brake pads after running under silent and squealing conditions. *Wear,* Vol. 232, No. 2, 163-167.

Ewins, D. J. (1984). *Modal testing: theory and practice*. Letchworth: Research Studies Press.

Fieldhouse, J. D (2000). A study of the interface pressure distribution between pad and rotor, the coefficient of friction and calliper mounting geometry with regard to brake noise. In: Barton, D. C., Earle S *Proceedings of the International Conference on Brakes* 2000 Automotive Braking – Technologies for 21st Century. UK: Professional Engineering Publishing Ltd; 2000; 3-18.

Fieldhouse, J. D., Ashraf, N., & Talbot, C. (2008). The measurement and analysis of the disc/pad interface dynamic centre of pressure and its influence on brake noise. *SAE Paper* 2008-01-0826.

Ghesquiere, H., & Castel, L. (1991). High frequency vibrational coupling between an automobile brake-disc and pads. *Proceedings of I.Mech.E.,* Paper No. C427/11/021.

Goto, Y., Amago, T., Chiku, K., Matsushima, T., & Ishihara, T. (2004). Experimental identification method for interface contact stiffness of FE model for brake squeal. In: Barton, D. C., Blackwood, A. *Proceedings of IMechE International Conference-* Braking 2004. UK: Professional Engineering Publishing Ltd; 143-155.

Guan, D., & Jiang, D. (1998). A study on disc brake squeal using finite element methods. *SAE Technical Paper* 980597.

Hammerström, L., & Jacobson, S. (2006). Surface modification of brake discs to reduce squeal problems. *Wear*, Vol. 261, 53-57.

Hassan, M. Z., Brooks, P. C., & Barton, D. C. (2008). Fully coupled thermal-mechanical analysis of automotive disc brake squeal. *Automotive Research Conference 2008*, Paper No. 05AARC2008.

Hohmann, C., Schiffner, K., Oerter, K., & Reese, H. (1999). Contact analysis for drum brakes and disk brakes using ADINA. *Computers and Structures*, Vol. 72, 185-198.

Hu, Y., & Nagy, L. I. (1997). Brake squeal analysis by using nonlinear transient finite element method. *SAE Technical Paper* 971510.

Hu, Y., Mahajan, S., & Zhang, K. (1999). Brake squeal DOE using nonlinear transient analysis. *SAE Technical Paper* 1999-01-1738.

Ibrahim, R. A., Madhavan, S., Qiao, S. L., & Chang, W. K. (2000). Experimental investigation of friction-induced noise in disc brake system. *International Journal of Vehicle Design,* Vol. 23, Nos. 3-4, 218-240.

Ioannidis, P., Brooks, P. C., & Barton, D. C. (2003). Drum brake contact analysis and its influence on squeal prediction. *SAE Technical Paper* 2003-01-3348.

James, S. (2003). An experimental study of disc brake squeal. *PhD Thesis,* Department of Engineering, University of Liverpool, UK.

Jang, H., Ko, K., Kim, S. J., Basch, R. H., & Fash, J. W. (2004). The effect of metal fibers on the friction performance of automotive brake friction materials. *Wear,* Vol. 256, 406-414.

Kao, T. K., Richmond, J. W., & Moore, M. W. (1994). The application of predictive techniques to study thermo-elastic instability in braking. *SAE Technical Paper* 942087.

Kim, N. M., Won, D., Burris, D., Holtkamp, B., Gessel, G. R., Swanson, & Sawyer, W. G. (2005). Finite element analysis and experiments of metal/metal wear in oscillatory contacts. *Wear,* Vol. 258, Nos. 11-12, 1787-1793.

Kinkaid, N. M., O'Reilly, O. M., & Papadopolous, P. (2003). Review of automotive disc brake squeal. *Journal of Sound and Vibration,* Vol. 267, 105-166.

Kung, S., Dunlap, K. B., & Ballinger, R. S. (2000). Complex eigenvalue analysis for reducing low frequency brake squeal. *SAE Technical Paper* 2000-01-0444.

Kung, S., Steizer, G., Belsky, V., & Bajer, A. (2003). Brake squeal analysis incorporating contact conditions and other nonlinear effects. *SAE Technical Paper* 2003-01-3343.

Lang, A. M., & Smales, H. (1983). An approach to the solution of disc brake vibration problems. *Proceedings of I.Mech.E.,* Paper No. C37/83.

Lee, Y. S., Brooks, P. C., Barton, D. C., & Crolla, D. A. (1998). A study of disc brake squeal propensity using a parametric finite element model. *Proceedings of I.Mech.E.,* Paper No. C521/009/98, 191-201.

Li, L. (2007). Preliminary investigation of contact pressure and squeal of a disc brake considering thermal effects. *MPhil Thesis,* Department of Engineering, University of Liverpool, UK.

Liles, G. D. (1989). Analysis of disc brake squeal using finite element methods. *SAE Technical Paper* 891150.

Lin, J. (2001). The study of thermal and stress analysis of the disc brake of motorcycle. *MPhil Thesis,* Department of Mechanical Engineering, National Taiwan University, Taiwan.

Liu, W., & Pfeifer, J. (2000). Reducing high frequency disc brake squeal by pad shape optimisation. *SAE Technical Paper* 2000-01-0447.

Mahajan, S. K., Hu, Y., & Zhang, K. (1999). Vehicle disc brake squeal simulations and experiences. *SAE Technical Paper* 1999-01-1738.

Massi, F., & Baillet, L. (2005). Numerical analysis of squeal instability. *International Conference on Emerging Technologies of Noise and Vibration Analysis and Control,* NOVEM2005, 1- 10.

Massi, F., Baillet, L., Giannini, O., & Sestieri, A. (2007). Brake squeal: Linear and nonlinear numerical approaches. *Mechanical Systems and Signal Processing,* Vol. 21, No. 6, 2374-2393.

Nack, W. V. (2000). Brake squeal analysis by finite elements. *Journal of Vehicle Design,* Vol. 23, Nos. 3-4, 263-275.

Nishiwaki, M.R. (1990). Review of study on brake squeal. *JSAE Review,* Vol. 11, No. 4, 48–54.

Ouyang, H., Cao, Q., Mottershead, J. E., & Treyde, T. (2003a). Vibration and squeal of a disc brake: modelling and experimental results. *Proceedings of I.Mech.E.*, Part D: *Journal of Automobile Engineering,* Vol. 217, 867-875.

Ouyang, H., Mottershead, J. E., & Li, W. (2003b). A moving-load model for disc-brake stability analysis. *Journal of Vibration and Acoustics,* Vol. 125, No. 1, 53-58.

Ouyang, H., Mottershead, J. E., Brookfield, D. J., James, S., & Cartmell, M. P. (2000). A methodology for the determination of dynamic instabilities in a car disc brake. *International Journal of Vehicle Design,* Vol. 23, Nos. 3-4, 241-262.

Ouyang, H., Nack, W. V., Yuan, Y., & Chen, F. (2005). Numerical analysis of automotive disc brake squeal: a review. *International Journal Vehicle Noise and Vibrations,* Vol. 1, Nos. 3-4, 207-230.

Podra, P., & Andersson, S. (1999). Simulating sliding wear with finite element method. *Tribology International,* Vol. 32, 71-81.

Qi, H. S., Day A. J., Kuan K. H., & Forsala, G. F. (2004). A contribution towards understanding brake interface temperatures. In: Barton, D. C., Blackwood, A. *Proceedings of I.Mech.E., International Conference-* Braking 2004. UK: Professional Engineering Publishing Ltd; 251-260.

Rhee, S. K. (1970). Wear equation for polymers sliding against metal surfaces. *Wear*, Vol. 16, 431-445.

Richmond, J. W., Smith, A. C., Beckett, P. B., & Hodges, T. (1996). The development of an integrated experimental and theoretical approach to solving brake noise problems. *Advances in Automotive Braking Technology, Design Analysis and Material Developments,* 3-23.

Ripin, Z. B. M. (1995). Analysis of disc brake squeal using the finite element method. *PhD Thesis,* Department of Mechanical Engineering, University of Leeds, UK.

Samie, F., & Sheridan, D. C. (1990). Contact analysis for a passenger car disc brake. *SAE Technical Paper* 900005.

Sherif, H. A. (2004). Investigation on effect of surface topography of pad/disc assembly on squeal generation. *Wear,* Vol. 257, Nos. 7-8, 687-695.

Soom, A., Serpe, C. I., & Dargush, G. F. (2003). High frequency noise generation from components in sliding contact: flutter instabilities including the role of surface roughness and friction. *Tribology and Interface Engineering Series*, Vol. 43, 477-485.

Tamari, J., Doi, K., & Tamasho, T. (2000). Prediction of contact pressure of disc brake pad. *JSAE Review,* Vol. 21, 136-138.

Tarter, J. H. (1983). Disc brake squeal. *SAE Technical Paper* 830530.

Tirovic, M., & Day, A. J. (1991). Disc brake interface pressure distributions. *Proceedings of I.Mech.E.,* Vol. 205, 137-146.

Trichês, M. J., Gerges, S. N. Y., & Jordan, R. (2008). Analysis of brake squeal noise using the finite element method: a parametric study. *Applied Acoustics*, Vol. 68, No. 2, 147-162.

Yang, S., & Gibson, R. F. (1997). Brake vibration and noise: review, comments and proposals. *International Journal of Materials and Product Technology,* Vol. 12, Nos.4–6, 496–513.

Yuan, Y. (1996). An eigenvalue analysis approach to brake squeal problems. *Proceedings of the 29th ISATA Conference Automotive Braking Systems,* Florence, Italy.

In: New Research on Acoustics
Editor: Benjamin N. Weiss, pp. 197-220

ISBN: 978-1-60456-403-7

Chapter 5

DYNAMIC PROPERTIES OF RAILWAY TRACK AND ITS COMPONENTS: A STATE-OF-THE-ART REVIEW

***Sakdirat Kaewunruen*[1,2,*] *and Alex M. Remennikov*[1,**]**

[1]School of Civil, Mining, and Environmental Engineering, Faculty of Engineering
The University of Wollongong, Wollongong 2522 NSW, Australia
[2]Austrak Pty Ltd, Moorooka, Brisbane, 4105 QLD, Australia

Abstract

Recent findings indicate one of major causes of damages, which is attributed to the resonant behaviours, in a railway track and its components. Basically, when a railway track is excited to generalised dynamic loading, the railway track deforms and then vibrates for certain duration. Dynamic responses of the railway track and its components are the key to evaluate the structural capacity of railway track and its components. If a dynamic loading resonates the railway track's dynamic responses, its components tend to have the significant damage from excessive dynamic stresses. For example, a rail vibration could lead to defects in rails or wheels. The track vibrations can cause the crack damage in railway sleepers or fasteners, or even the breakage of ballast support. Therefore, the identification of dynamic properties of railway track and its components is imperative, in order to avoid any train operation that might trigger such resonances. This chapter deals with the vibration measurement techniques and the dynamic behaviours of ballasted railway tracks, and in particular their major components. It describes the concept of vibration measurements and the understanding into the dynamic behaviour of ballasted railtrack sleepers. It discusses briefly on the track structures and track components in order to provide the foundation of understanding ballasted railway tracks. The highlight in this paper is the state-of-the-art review of dynamic properties of railway track and its components. It summarises the non-destructive acoustic methods, the identification processes, and the properties of each rail track element.

Key words: Dynamics; Acoustics; Concrete sleepers; Rail pad; Ballast; Railway track; Non destructive test; Vibration measurement; Property.

* E-mail address: skaewunruen@austrak.com, Tel.: +61 7 3308 7608; fax: +61 7 3308 7881. (Corresponding author.)
** E-mail address: alexrem@uow.edu.au

1. General Background

Structural dynamic problems result in a serious failure and design restriction over a wide range of engineering structures since natural phenomena and human activities often impart time-dependent burden on every accessible structures. Analysis and design of those structures are based on the consideration of dynamic actions. The dynamic parameters of those structures, such as natural frequency, damping constant, and corresponding mode shape, are of substantial importance in the procedures needed for analysis and design of such structures subjected to dynamic loadings. Consequently, it is inevitable to investigate the dynamic and acoustic properties of an object associated with various sources of noise and vibration. In general, the major outcomes of the vibration measurements are the natural frequency, the dashpot value and the mode shape. Apart from explicit properties, the dynamic stiffness can be extracted and estimated from the vibration responses of structures. The principal reason of the dynamic test is that those outcomes allow verifying an analytical model proposed for the system tested. The analytical model, which has been validated, can be used in design and response prediction with much confidence. It forms a new arena in engineering trials as a '*virtual experiment*'. However, the structural modifications in computer simulations could yield further sensitivity analysis whereas either linear or nonlinear behaviour can be identified though the experimentations.

Enormously, railways have taken major part in transporting population, resources, merchandises, etc. over the large continent of any country. Railway industry has significantly grown in the past century and has continuingly developed new suitable technology for its solution to specific requirements. In many countries, the traditional railway system is the ballasted track, which consists of rails, rail pads, and concrete sleepers (ties) laid on ballast and subgrade. Recently, the demand of track usages has greatly increased. The increase in frequency of traffic for passenger trains and the rise in loads of freight trains over recent years have also been a significant factor contributing to the deterioration of the railway track system. The increase in transport capacity has been stimulated by the growing industrial need for long-distance freight conveyancing, especially in large countries with lots of coalmines like Australia or China. It was evident that railway tracks in Australia have been deteriorating due to increased traffic frequency, heavier wheel loads and improper maintenance. However, undetermined ultimate load capacity of track components has been under suspicion in order to specify a reserved performance in each particular track component. Regardless to a nature that railway tracks are subject to impact dynamics induced by wheel/rail interaction or irregularities, railway civil engineers have paid attention to the real capacity of track components under realistic load. This has been a recent attempt to develop the advances in design of such infrastructure.

At present, many research organizations rigorously aims at providing rail track owners/designers the up-to-date knowledge related to the dynamic behaviors of prestressed concrete sleepers, rail pads, and other track components. Based on the fact that the rail firms do run their own business, a number of projects attempts at giving chances to them for making use of their existing railway structures more efficiently and in particular more cost-effectively. However, the limited information on static and dynamic behaviors nowadays still indicates the methods and procedures of over- analysis and design of such rail structures as concrete sleepers and railway track components. In order to either improve those economical analysis and design or exploit the applicable capacity from existing structures, the better realistic insight of static, dynamic, and impact responses must be ascertained.

For instance, the Australian Cooperative Research Centre for Railway Engineering and Technologies (RailCRC) has been constituted based on the awareness of those problems in Australia. Its mission is to become an internationally recognized National Research Centre for the Australian Railway Industry and to deliver decision-making tools, knowledge and technologies necessary to address industry needs for the effective rail management, operation, maintenance and development of the rail industry. To accomplish this mission Rail CRC promotes several research projects in coordination within six industry partners and six universities. A research investigation under a funded project of Rail CRC No.5/23 was "Dynamic analysis of track and the assessment of its capacity with particular reference to concrete sleepers", which had been allocated into a Research Theme # 2 "Innovative Track Maintenance and Upgrading Technologies". The University of Wollongong (UoW)'s contribution to the project had included the integration of experiments and simulations with reference to concrete sleepers and railway track, and the development of a new limit states design philosophy of concrete sleepers, with close relation to computer package for track analysis and factored load analysis done at Queensland University of Technology (QUT).

In this chapter, the technical procedures and testing methodology for dynamic characteristics of such track components as rail bearing pads, prestressed concrete sleepers, and ballast/sub-ballast/subgrade have been reviewed and summarized. The ballasted track system and the well-known modal testing technique have been briefly described. Previous analytical model and experiments of rail track-component vibrations have also been taken into account. Of the most interest, practical test methods in in-situ and in-field conditions are subsequently addressed.

2. Ballasted Railway Tracks

Rail track is a fundamental part of railway infrastructure and its components can be classified into two main categories: superstructure and substructure. The most obvious parts of the track as the rails, rail pads, concrete sleepers, and fastening systems are referred to as the superstructure while the substructure is associated with a geotechnical system consisting of ballast, sub-ballast and subgrade (formation). Both superstructure and substructure are mutually important in ensuring the safety and comfort of passengers and quality of the ride. The typical shape and construction profiles of a ballasted track are illustrated in Figure 1.

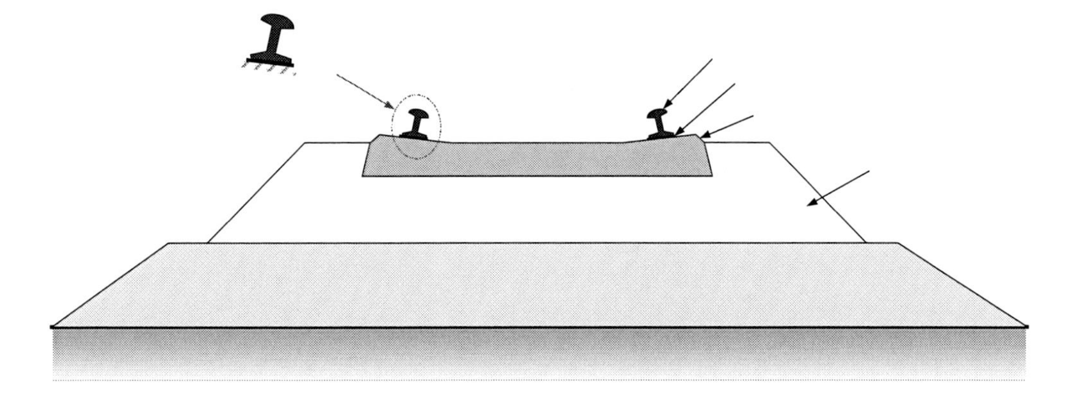

Figure 1. Typical ballasted track profile.

2.1. Rails

Rails are longitudinal steel members that are placed on spaced sleepers to guide the rolling stock. Their strength and stiffness must be sufficient to maintain a steady shape and smooth track configuration (Selig and Waters, 1994), and resist various forces exerted by travelling rolling stock. One of their primary functions is to accommodate and transfer the wheel/axle loads onto the supporting sleepers. Esveld (2001) reported that a modern rail track also conveys signals and acts as a conductor on an electrified line. Adoped from Esveld (2001), Table 2.1 describes typical rail profiles and their applications. The most commonly used profile is flat-bottom rail, also called Vignole rail, and is divided into three parts:

- *rail head*: the top surface that contacts the wheel tire
- *rail web*: the middle part that supports the rail head, like columns
- *rail foot*: the bottom part that distributes the load from the web to the underlying superstructure components.

Table 1. Types of rail profiles and their applications (Esveld, 2001)

Shape	Profile type:	Applications:
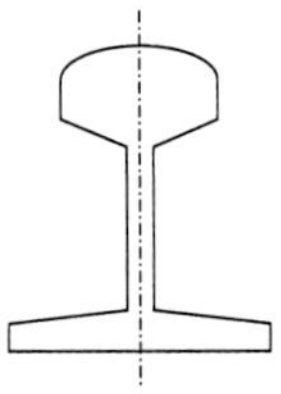	Flat-bottom rail	Standard rail track
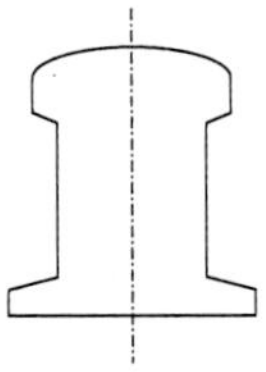	Construction rail	Manufacturing of automobiles and switch parts
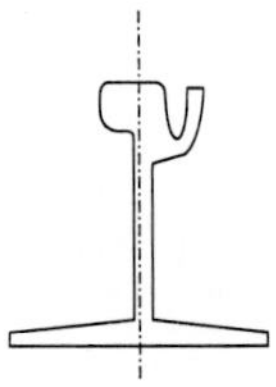	Grooved rail	Railway track embedded in pavements, roads, yards

Table 1. Continued

Shape	Profile type:	Applications:
	Block rail	Railway track used in concrete slab as part of Nikex-structure
	Crane rail	Heavy load hoisting cranes with high wheel loads

2.2. Fastening Systems

The fastening system, or "*fastenings,*" includes every component that connects the rail to the sleeper. Fastenings clamp the rail gauge within acceptable tolerances and then absorb forces from the rails and transfer them to the sleepers. Vibration and impact from various sources e.g. traffics, natural hazards, etc. are also dampened and decelerated by fastenings. Fastenings sometimes act as electrical insulation between the rail and the sleepers. Their primary components are fastener and rail pad. Some tracks might have base plates with or without

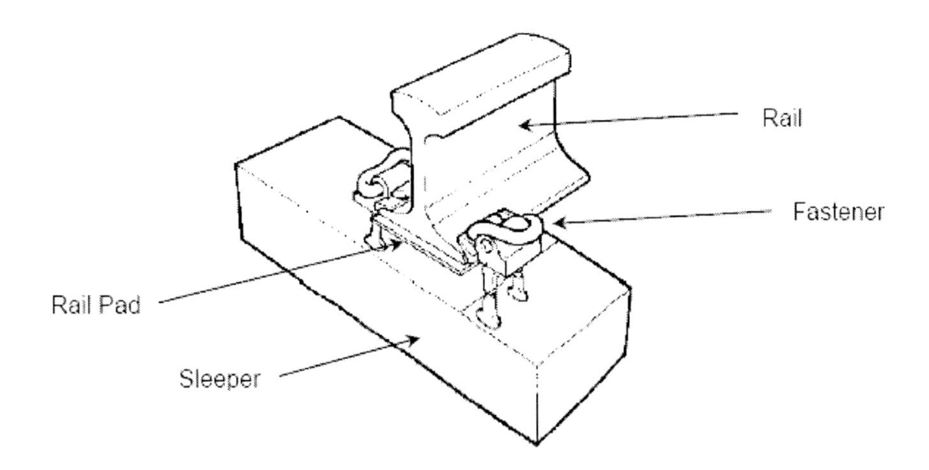

Figure 2. Typical fastening system for concrete sleepers (Steffens, 2005).

pads, which helps the workmen to remove damaged rails without having to untie the fastenings and immediately replace them with new rails. In this case the rail is only connected to the immediate base plate. This system is called *indirect fastenings*, which differs from the usual *direct fastenings* in that the latter device is built-in and holds the rail directly onto the sleepers (Esveld, 2001; Smutny, 2004). A typical concrete fastening system is shown in Figure 2.

2.3. Fasteners

Because there are many different types of fasteners, their application depends on the characteristics, patterns, and structure of the sleepers to be used. The fasteners withstand the vertical, lateral, and longitudinal forces, and overturning moments of the track, as well as keeping the rails in place. They also transfer all the forces caused by the wheels, thermal change, and natural hazards, from the rails to the adjacent sleepers.

2.4. Rail Pads

Rail pads are placed on the rail seat to filter and transfer the dynamic forces from rails and fasteners to the sleepers. The high dashpot value of rail pads reduces excessive high-frequency forces and provides a resiliency between rail and sleeper that helps alleviate rail seat cracking and contact attrition.

2.5. Sleepers

Sleepers are transverse beams resting on ballast and support. Wooden sleepers were used in the paste because timber was readily available in the local area. However, pre-stressed or reinforced concrete sleepers, and to a limited extent steel sleepers, have been adopted in modern railway tracks over the past decades because of their durability and long service life. Esveld (2001) grouped timber sleepers into two types: softwood (e.g. pinewood) and hardwood (e.g. beech, oak, tropical tree). Concrete sleepers are described as either twin-block or mono-block (FIP Commission on Prefabrication, 1987), and are illustrated in Figure 3. Within all these types, concrete sleepers are more widely used because they are not affected very much by either climate or weather.

The important functions of sleepers are:

- To uniformly transfer and distribute loads from the rail foot to the underlying ballast bed;
- To provide an anchorage for the fastening system that holds the rails at their correct gauge and preserves inclination, and
- To support the rail and restrain longitudinal, lateral and vertical movement by embedding itself onto the substructures.

a) monoblock concrete sleeper b) twin-block concrete sleeper

Figure 3. Types of concrete sleepers.

2.6. Ballast

Ballast is a layer of free draining coarse aggregate used as a tensionless elastic support for resting sleepers. This layer comprises graded crushed stone, gravel, and crushed gravel such as granite and basalt which depends on local availability. It not only provides support, it also transfers the load from the track to the sub-ballast and drains water away from the rails and sleepers. For a heavy haul freight line, individual axle loads on rails can be up to 50 tons or around 80 tons. Thus, in addition to the weight of the track, heavy cyclic loading, tamping and impact from rolling stock, ballast provides a static and dynamic stability to the sleepers by distributing a uniform load and reduction over the sub-ballast and subgrade.

The fundamental functions of ballast can be summarised from previous research as follows (Hay, 1982; Selig and Waters, 1994; Esveld, 2001).

- Resist vertical, lateral and longitudinal forces applied to the sleepers, to retain the track in its proper position because the submergence and interlocking of irregularly shaped ballast tends to confine the sleepers, as shown in Figure 4.;
- Absorb shock and impact from the rough interlocking particles acting as a simple spring element with limited action;
- Give resiliency and energy absorption to the sleeper;
- Assist track maintenance in surfacing and lining operations (track geometry adjustment and sleeper replacement) by the ability to manipulate ballast with low energy tamping;
- Reduce bearing stresses from the sleeper to acceptable stress levels for underlying layers;
- Allow optimum global and local settlement;
- Provide some large voids for controlling the storage of internal fouling materials enable small particles to pass through;
- Provide fast drainage of fluid;
- Avoid freezing and thawing problems by being unsusceptible to frost;

- Provide an insulating layer with some electrical resistance;
- Prohibit vegetation growth because of unsuitable cover layer for vegetation
- Absorb airborne noise from travelling traffic, and
- Facilitate reconstruction of the track.

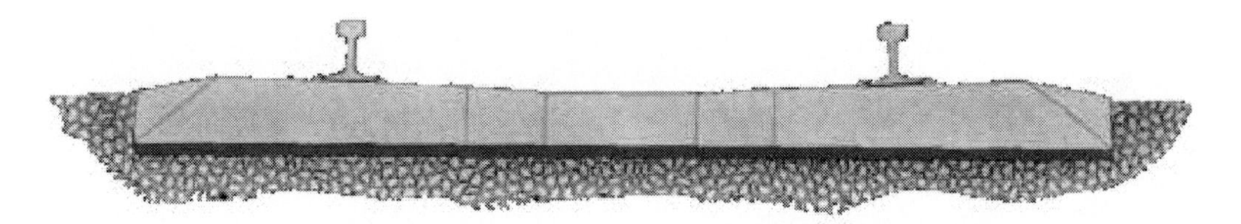

Figure 4. Typical construction of a ballasted track.

2.7. Sub-ballast

Sub-ballast that may be called *the capping layer* is a layer of granular material between the ballast and underlying subgrade. It is generally composed of broadly graded slag or crushed aggregate, although a broadly graded sand-gravel compound is sometimes used. Sub-ballast is usually impervious and therefore its general functions are to:

- Reduce stress at the bottom of the ballast layer to a reasonable level for the top of the subgrade;
- Prevent inter-penetration from the upward migration of fine particles from the layer of subgrade to the upper layer of ballast;
- Provide good drainage that is ascribed to the non-obstructed voids by inter-penetration;
- Also act as a shedding layer to keep water away from subgrade;
- Protect the subgrade from attrition by the hard ballast; and
- Inhibit freezing and thawing problems in the subgrade.

2.8. Subgrade

Subgrade is also referred to as *the formation*. It includes the existing soil and rock, which possess slopes, verges, ditches and other structures or materials within. The subgrade is the last support which bears and distributes the dynamic loading resultant downward along its infinite depth. This deep layer must have sufficient bearing capacity, provide good drainage and yield a tolerably smooth settlement in order to prolong track serviceability. Some synthetic materials such as geo-textile, fabric, etc., have recently been used to upgrade the capacity of the subgrade.

3. Modal Testing

Experimental modal analysis (EMA) or *modal testing* is a non-destructive testing strategy based on the response of structures to vibration. Since the 1940s, modal testing has been widely used to help understand the dynamic behaviour of structures. The original modal testing technique was based on the simple sine dwell method. After some years, innovations based on Fast Fourier Transform (FFT) have been developed and are currently used (Brown, 1982; Allemang and Brown, 1986; Mitchell, 1986; Allemang, 1993; Ewin, 1995; He and Fu, 2001). This vast improvement of modal analysis results in the more precise and accurate vibration measurements. In addition, the integration of analytical models and experimental results has led to the sensitivity analysis of structural behaviour. To solve dynamic problems or analyse and design structures, experimental modal analysis becomes one of the most significant methodologies. It promises civil/mechanical engineers practical procedures and a prompt and reasonable solution to structural dynamic problems.

Numerous researches have been investigated from the modal analysis perspective and have shown substantial efficiency in engineering uses, even if the modal analysis is comparatively young. Assumptions involved in modal testing include the linearity of structures, time-invariant structural parameters, and observability in measurements (Allemang, 1993). It is noteworthy that linear systems mean that the superposition of forces can be applied, although linearity may be inapplicable for some structures (Zaveri, 1985). Besides the time invariant properties mean that structural parameters such as stiffness and damping remain constant, depending on factors excluded in the model, during a time. Observability arises when the input-output measurements have sufficient information to create a characteristic model of the structure, although the forcing and response functions must be measurable for the structure to be observed.

In additional to the generic assumptions, there are five more options to be assumed when one performs an excitation on a structure. They are whether the excitation is impulsive; or white noise; or a step; or a Wiener-Levy signal; and or that there is no excitation (free decay or free response) (Allemang and Brown, 1986). Various methods for performing experimental modal analysis had been categorised by Allemang and Brown (1986) into:

- Forced-normal-mode excitation,
- Frequency response function,
- Damped complex exponential response, and
- Mathematical input-output model

The forced-normal-mode function method is the oldest modal testing approach that uses multiple inputs to approximate the modal parameters but it does not take the complex modes of vibration into account. The frequency response function (FRF) method is currently the most common one to use for assessing modal parameters. It contains spectra computed from the auto-spectrum and cross-spectrum that are measured from the structure. The damped complex exponential response method makes use of the information acquired from the free decay of a system, which is generated by the release of an initial condition. There are two similar approaches, which occur through the development of a damped complex exponential response method. They are the Ibrahim time domain method and Poly reference approach.

- The Ibrahim time domain (ITD) method uses data received from the impulse response function (IRF) to identify the modal parameters of a structure. This approach requires the response measurements simultaneously at a number of locations on the structure. However, a universal position can be chosen for measurements if multiple data collection is unobtainable. Then the displacement and acceleration, or velocity of free response data is used to form the eigenvalue matrix to be solved (He and Fu, 2001).
- The poly reference approach uses FRF data from multi-input multi-output testing to carry out the natural frequency, modal constants and damping loss factors. All data measured simultaneously are used to identify the modal frequencies (Allemang and Brown, 1986).

The mathematical input-output model method considers the input and output responses independently. Applications can be extended to both time and frequency domain models with no limitation in the number of degrees of freedom (dof). The approaches based on this methodology are the auto-regressive moving average approach and the reduced structural matrix approach.

- The auto-regressive moving average approach makes the excitation assumption that the responses obtained at various points are induced by white random noise input to the structure. The two-stage least square method is needed. From statistical results, the confidence factor (coefficients of variation), which are the ratio of the standard deviation of each parameter compared to its real counterpart, can be computed for natural frequencies and damping (Allemang and Brown, 1986).
- The reduced structural matrix approach is based on the condensed matrix obtained from incomplete responses. However, its estimated solution is not unique, and the frequency range is reduced because the matrix is weighted to represent an incomplete model. Besides, the sensitivity needed to reduce the matrix will be decreased due to the limited precision of modal results from experimental errors. As a result, this method is not very successful (Allemang and Brown, 1986).

4. Dynamics of Rail Pads

The standard rail pads used in the Australian track system are usually made from polymeric compound materials. They are installed on rail seats to reduce the dynamic stress from axle loads and wheel impact from both regular and irregular train movements. These pads are highly important because of their dynamic softening between track and sleepers. Inappropriate or inadequate uses of rail pads exacerbate the cracking of sleepers at rail seats. Besides, misuse causes high settlements of global and local tracks, and ballast/subgrade breakage from heavier tamping (Kaewunruen and Remennikov, 2004). This negative effect can be extended to the capacity and integrity of a railway system unless this helps to gain a better understanding of the structural behaviour of them. Based on these grounds Australia has launched a standard referred to as AS1085.19 for testing bearing pads. (Standards Australia, 2001). The dynamic behaviour of rail pads is not well known, although some publications can be found, as presented:

4.1. Linear and Nonlinear Model of Rail Pads

Many investigations have been carried to describe the dynamic behaviour of rail pads mathematically; they are either linear or nonlinear models. Dynamic rail pad models are usually on the fundamental basis of either time or frequency domain. The literature review shows that most are in relation to the frequency domain because this model implicates dynamic properties such as resonant frequencies and damping (Fahey and Pratt, 1998).

A suitable linear '*Fractional derivative model*' was developed by Fenander (1997; 1998) and adopted into a linear track model. In this approach the rail pad model had to be linear. Various damping or viscoelasticity models were included; for example, the viscous model where the loss factor is proportional to frequency. Since the stiffness of a rail pad is nonlinearly dependent on the preload and the fractional derivative model is linear, a different set of parameters, which were fitted to the experimental data, was needed for each preload. This model seems to be a good fit, especially for testing the stiffness of the rail pads.

Sjoberg (2002) developed a time domain model, which uses compressive actions as component forces and accounts for the effect of pre-load, and the frequency and dynamic amplitude dependence. The nonlinear shape factor, neo-Hookean hyper-elastic model, fractional derivative element model, and Coulomb forcing function were included in the proposed model. It is a one dimensional component model using a few parameters whilst giving reasonably measured characteristics. It was proposed initially for rubber isolators but could for other rubber elements.

Since the complex properties of rail pads are frequency dependent, Knothe et al. (2003) developed a frequency domain model which assumed that for each frequency the equivalent complex stiffness can be approximated by a bi-linear function associated with load amplitude and preload. The latter assumption is that the modal parameters rely linearly on the frequency. The estimated values fitted the measured results quite well.

4.2 Dynamic Testing of Rail Pads

The dynamic testing of rail pads has been of interest for many years and many different methods have been developed to determine their dynamic properties. The response of a rail system comprising a concrete sleeper, rail section, and various types of rail pads to vibration were measured (Ford, 1988b) and a series of FRFs was presented to identify the effects that various pads had on the wave signals.

Several types of resilient rail pads with various materials and different surface profiles had been tested in the laboratory and on a track (Grassie, 1989). The average attenuation of the pads was indicated and the results from the laboratory showed that they were conservative. Their dynamic stiffness increases with normal load and is more than or identical to the tangent stiffness from the load deflection curves. Although rail pad damping has almost no affect on the dynamic behaviour of a well compacted ballasted track, more damping causes the pad itself to heat up.

Van't Zand (1993) used FFT technique to measure and assess the dynamic characteristics of pads through impact load tests. The curve fitting method was used to fit an SDOF equation of motion to the experimental results. It was carried out at a specific frequency of 400-2000 Hz. The results was compared with another research and seemed to be in a good agreement. This method was then extended to the urban track structures (Esveld, 1997; Esveld, Kok, and De Man, 1998).

Later, Fenander (1997) studied the vertical stiffness and damping of studded rail pads on a complete track and in a laboratory test rig. Stiffness increased substantially with preload but only weakly with frequency. The fractional derivative model of their dynamic behaviour was presented in this investigation. The 2DOF testing apparatus was developed from research done by Verheij (1982). It consisted of two steel blocks with the resilient element mounted between them. To approximate the dynamic stiffness of the pads, the stiffness of the lower spring supporting the lower block was neglected. The measurements of several new pads indicated that stiffness tends to increase slightly with frequency whilst the effect of the preload is more pronounced.

Thompson et al. (1998) developed an indirect method for measuring the high frequency dynamic stiffness of resilient elements by applying the theory of two-degree-of-freedom (2DOF) system. The resilient element was mounted between two large blocks on a floor, one of which was supported by soft springs beneath it. The resilient element was assumed to be much stiffer than the soft springs. Two exciters were used in the measurement apparatus and some approximations were made to reduce some difficulties with the 2DOF system. This proposed methodology can be extended to elements with low frequency stiffness by rearranging the equations of motion. The results of this case study seem to be reliable and it is noteworthy that this approximated method is being included in international standards.

Another rail pad tester has been constructed based on the SDOF viewpoint, to examine their dynamic properties in controlled conditions in the laboratory (De Man and Esveld, 2000; De Man, 2002). New and worn pads were tested in the laboratory. They are considered to be the only elastic components in the tester, which is an apparatus of tuned masses, preloading springs and elastic supports. In this investigation the 2DOF-based tests on real track had also been done. However, it is noteworthy that this tester gives realistic results and becomes a better way for determining their properties.

There has been a recent investigation by Knothe et al. (2003) dealing with the measurement and modelling of resilient rubber pads. The quasi-static and dynamic measurements were done in the low frequency (0-40 Hz.) and high frequency range (100-2000 Hz.). With dynamic testing two pads are placed between three steel plates which then introduces the secant, tangent and equivalent stiffness. This test showed that the stiffness of the pads are frequency dependent and thus conform to previous researches. They were also hinged on the preload and amplitude of the harmonically varying load. It can be concluded that the equivalent stiffness increases with increasing preload and decreasing amplitude. These tests showed the hyper-elastic action of the rubber in the quasi-static test, its visco-plastic action in the low frequency cyclic load test, and especially its viscoelastic action in the high frequency range test.

At the UoW an alternative method for determining the dynamic properties of rail pads based on measuring the response to SDOF vibration has been proposed. The simplicity and reliability of this method has been shown. It proved to be a fast and non-destructive method for accurately assessing the dynamic stiffness and damping constant of every type of rail pad available in Australia. This strategy enables new types of rail pads to be tested as well as evaluate how ageing affects their dynamic characteristics (Kaewunruen and Remennikov, 2004).

5. Vibrations of Concrete Sleepers

Apart from the rail pads, concrete sleepers have also played a crucial role in this country for many years. Many investigations dealing with the dynamic properties of concrete sleepers have been performed, modal testing in particular (Ford, 1988a). Ballasted track can be divided into five groups of discrete support models (Knothe and Grassie, 1993), as shown in Table 2. To get a better insight into the dynamic strength of sleepers, their response to various dynamic loadings has been studied (Grassie and Cox, 1984; Esveld et al., 1996; De Man, 2002). Standards Australia (2003) developed AS1085.14 for analysing and designing concrete sleepers, including procedures for sampling and testing materials. It also contains the methods for testing static, quasi-static, and dynamic cyclic loading and the allowable limits and serviceability of sleepers.

Modal testing has been very significant as a method for non-destructive dynamic testing of concrete sleepers for over five decades. Ford (1988a) performed a modal analysis on a concrete railway sleeper. It was suspended at each end by soft springs which allowed for "free-free" support. An electro-dynamic shaker was used to excite the sleeper at a single point, an accelerometer was used to measure the response at the remaining points, and then a series of FRFs were obtained. In this test the natural frequencies and corresponding mode shapes were extracted using modal analysis software (SMS modal).

Vincent (2001) carried out modal testing and numerical modelling of a reinforced concrete sleeper. It was laid on a very soft spring support and excitations were done with an impact hammer and a swept sine loading of a shaker with different load amplitudes. Different types of modal parameter extraction methods were used and showed that linear behaviour arises in the low frequency range but is less in the higher frequency range. Also, a 3D linear finite element model was developed by *I-deas* to compare the experimental and numerical results. However, excluding the steel reinforcement, the model only contains sleeper geometry with the material property updating with modal testing results. A comparison of those results shows good agreement. A sensitivity analysis was done by modifying the structural properties.

Various boundary conditions such as free-free, perfectly coupled to the subsoil, and voided sleepers, were investigated for their effects on the dynamic behaviour of concrete sleepers (Plenge and Lammering, 2003). Excitations were given laterally and vertically. FRFs were obtained and became a means to study the effects of various boundary conditions on their response to vibration. A numerical investigation was done and if the sleeper was modelled as a rigid beam the results were not very good but by using the thin layer method, the simulated FRFs coincided with the measured results much better when the frequency was up to 2000 Hz. Additionally, a static load on a sleeper increases the bond with the ballast to stiffen the whole system. These experiments confirmed the results observed by Wu and Thompson (1999).

Table 2. Discrete sleeper support models

Models	Assumptions
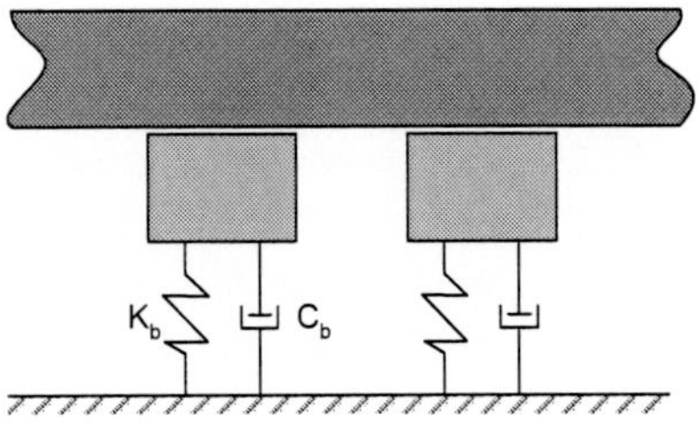 K_b C_b	Type A - discrete sleeper support - ballast: spring and damper
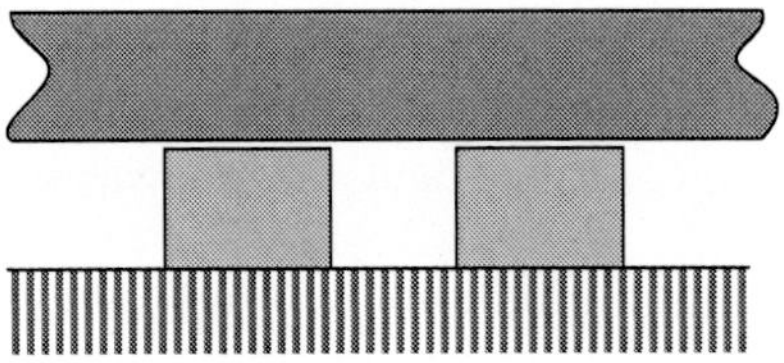	Type B - discrete sleeper support - half-space modeling of ballast and substrate
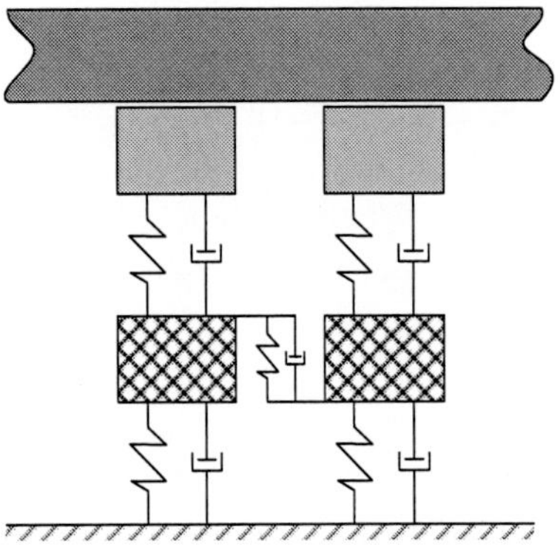	Type C - discrete sleeper support - ballast: spring and damper - interconnected ballasted masses - substrate: spring and damper
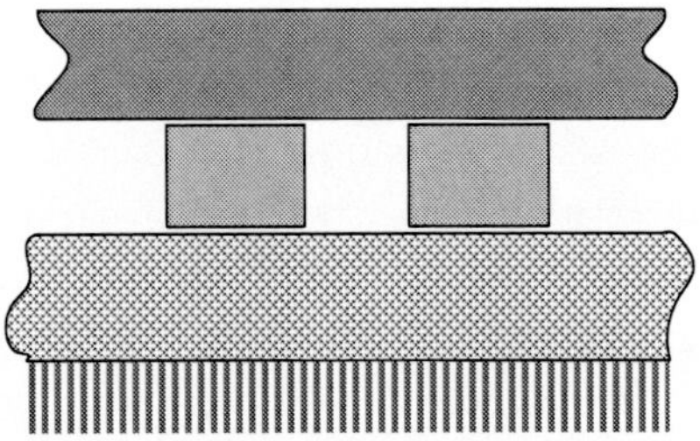	Type D - discrete sleeper support - continuous ballast layer - half-space substrate model

6. Dynamics of Railway Tracks

Timoshenko (1926) was one of the first to model the dynamic behaviour of a railway track. In that model, the rail was considered as an infinite uniform Euler beam, laid on a continuous damped elastic Winkler foundation. Grassie (1982) then found in some experiments that there are only two dominant resonances in the frequency range of interest. The first in-phase mode at about 100 Hz corresponds to the sleeper and rail moving together on the ballast while the second out-of-phase mode at a frequency somewhere between 300-500 Hz depending on the rail pad parameters, corresponds to the opposite vibration of sleepers on ballast and rails on the rail pad. The dynamic moment and impact attenuation were subsequently observed. It was concluded from much research that a rail pad which was resilient at frequencies of several hundred Hertz would substantially reduce the dynamic loads on concrete sleepers. Some rail pads attenuate dynamic bending strains in concrete sleepers by more than 50%. Their stiffness has a greater effect on reducing strain in a sleeper than its depth (Grassie, 1988). The dynamic loading on sleepers was also studied and revealed that when ballast is packed loosely, the dynamic loads become significantly higher and the resilient pads have less effect. Selig and Waters (1994) studied the ground formation of a railway track was and found that the subgrade is the one component of the substructure that has the greatest affect on track stiffness. Raymond (1978) also found that stiffer tracks have smaller differential settlements or unevenness and cause lower impact loading. Coincidentally, Liang and Zhu (2001) found the lower subgrade stiffness leads to greater elastic deformation and less stability of the ballast layer and upper part of the track structure.

In-situ and in-field experiments for determining global ballasted track were mostly done using the instrumented impact hammer technique because an impact hammer is mobile and self-supporting (De Man, 1996) and can be used without damaging or obstructing traffic. The FRF measurements can be obtained through a handy laptop and immediately extracted for such modal properties as resonant frequency, damping constant, and corresponding mode shape. However, these parameters are susceptible to factors such as duration of pad usage, temperature, and preload (Wu and Thompson, 1999). Previous investigations are summarised below.

6.1. Track Behaviour

Railway track modeling has only been developed in recent years to incorporate dynamic load effects. However, a more comprehensive dynamic load modelling to allow more accurate prediction of track degradation and associated railway track maintenance requirements is highly demanding. In practice, the physical railway track has historically been employed as a mechanical model to perform technical analyses. The model is rather simplistic as it could reasonably predict global track responses but the individual component's behaviour.

De Man (2002) classified railway track structures into three types: ballasted track structures, slab track structures and embedded rail structures. This chapter focuses on the ballasted type (see Figure 5) as the behaviour of the track structure depends heavily on the type of structure.

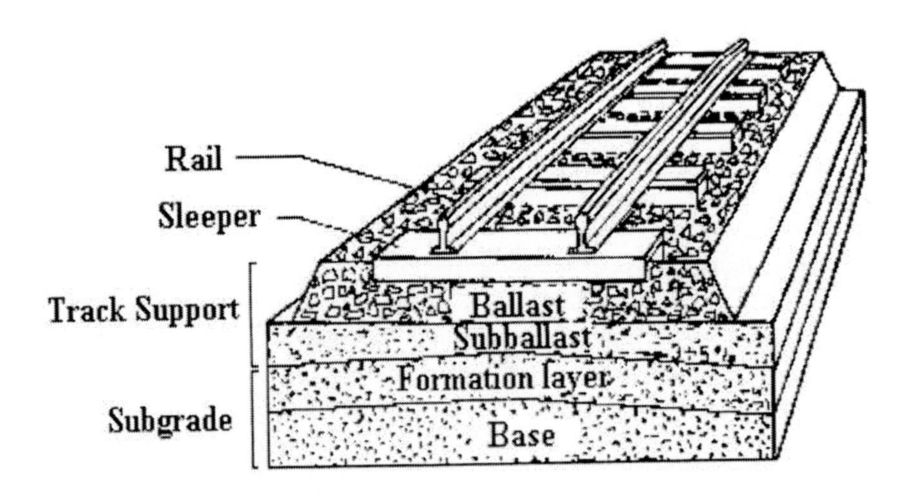

Figure 5. Railway ballasted track components.

In most of mechanical models of railway tracks, the track components have been simulated as a type of elements which are classified according to their properties as follows:

- Components with mass and inertia properties, rail and sleepers;
- Components with elastic properties, railpads; and
- Component with mass, inertia and elastic properties; ballast.

The ballasted track structure dynamic responses vary largely on individual component properties, the contact relationship between components, the physical body of the components, and the dynamic load actions. Rail and sleepers with mass and inertia properties tend to keep the track stable under static longitudinal buckling forces. Rail pads and ballast with elastic properties soften energy transfer and dampen the dynamic force frequency, resulting in a lower and less harmful load action. However, it is important to note that in a well-ballasted track structure, the rail pad does not damp vibration of the sleeper and plays insignificant role in softening transient load action (Kaewunruen and Remennikov, 2008); in contrast, the greater is the rail pad damping constant, the more effectively are dynamic forces transmitted to the sleepers (Grassie and Cox, 1984). It was also found that if the ballast damping is much greater than 100kNs/m, which is typical of well-compacted ballast, the railway track dynamic behaviour is affected little by rail pad damping (Grassie, 1988).

Generally, the vehicular imposed loading on a railway track can be divided into three categories corresponding to the plane of loading: vertical, lateral and longitudinal. The vertical loading on the track structure consists of the static axle weight of the freight vehicle and any additional dynamic augments (e.g. quasi-static, dynamic ride, impact), which are superimposed onto this static load. These dynamic augments are often the impact load caused by a variety of defective factors, such as:

- Irregularities in the geometry of the track structure;
- Irregularities on the surface of the rail;
- Irregularities on the surface of the wheel;

- The vehicle operating speed; and
- The mass of the vehicle suspension characteristics (sprung vs unsprung mass).

Due to the physical nature of track structures, a chain of their structural vibration modes exists depending on the resonant frequencies on the vertical, lateral and longitudinal direction. The lowest possible vertical resonant frequency of the track structure is the full track resonance (ft), which includes both in-phase and out-of-phase vibrations (Kaewunruen and Remennikov, 2007). For ballasted railway tracks in generally good conditions, these full track resonant frequencies are between 40 and 140 Hz, second vibration mode are between 100 and 400 Hz, and third vertical vibration mode are between 250-1500 Hz. Figure 6 shows the typical full track vibrations.

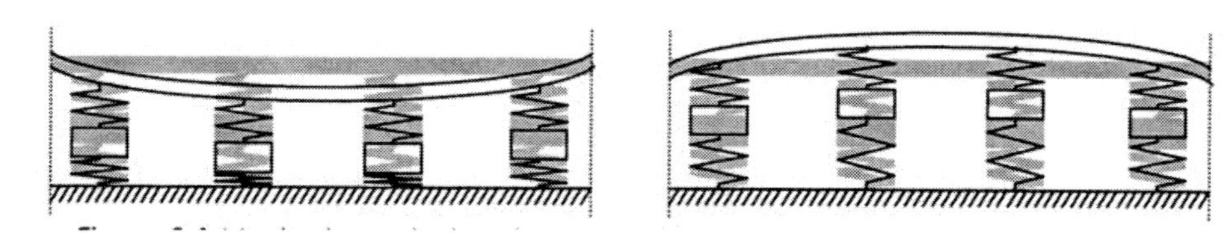

Figure 6. Full track vertical resonant frequency mode shape.

In addition, there exists the rail resonant frequency (fr) whereas the rail vibrating to the supports and is highly dependent on the rail pad properties but is independent to the sleeper and ballast properties, as shown in Figure 7.

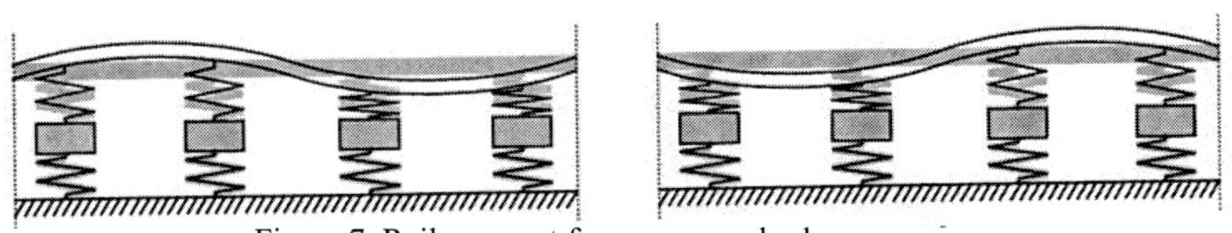

Figure 7. Rail resonant frequency mode shape.

The vibration modes and shapes are dependent on the sleeper support spacing. As a result, the vertical vibrations with less inflection points (see Figure 7) are more likely to develop than others. These vibration modes are so-called '*pin-pin*' resonant modes, corresponding to pin-pin resonant frequencies. The first pin-pin resonant frequencies (see Figure 8) occur between 400-1200 Hz and the second (see Figure 9) at slightly less than 4 times higher frequencies.

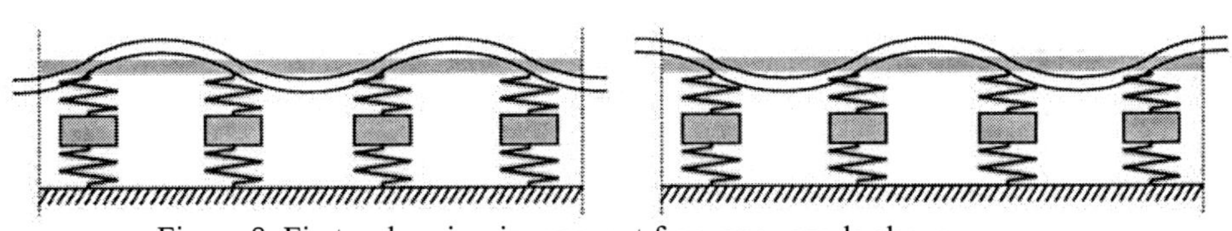

Figure 8. First order pin-pin resonant frequency mode shape.

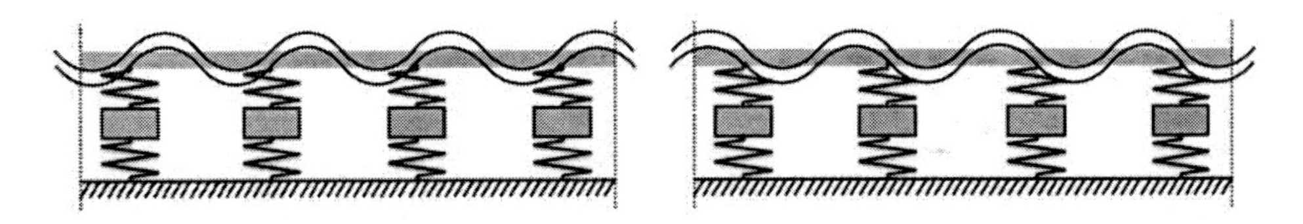

Figure 9. Second order pin-pin resonant frequency mode shape.

Kaewunruen and Remennikov (2007) reported that the frequency range (low, mid and high) has varying effects on the railway track as a structure and on each component. Low frequency range (0-40 Hz) produces damage to the substructure (subgrade, sub-ballast and ballast); mid frequency range (40-400 Hz) produces damage to the superstructure excluding the rail (sleepers, fasteners, and railpads) and high frequency range (400-1500 Hz) produce damage to rails and distortion to fasteners. Also, either rail corrugation, wheel defects (e.g. wheel flats), or combination of both, may generate periodical loading through impact, creating a damaging cycle process (Remennikov and Kaewunruen, 2007; Kaewunruen, 2007).

Lateral instabilities are a function of the vehicle characteristics and speed. The lateral forces on the rail are mostly located at the inner side of the railhead. In addition, lateral torsion modes can be observed at high frequencies. The lateral forces on the track structures are introduced by:

- The curvature of the freight equipment;
- The development of lateral instabilities; or
- The response of the vehicle to lateral track irregularities.

In terms of vibration, the longitudinal vibration modes are best interpreted as compression waves within the rails. They considerably affect the fatigue strength in minor axis of the rails, apart from loosening the fastening systems.

It should be noted that the track substructure has a direct influence on the dynamic wheel load, dynamic track stiffness, and track roughness (Esveld, 2001). The substructure components also have a very strong non-linear stress-strain and breakage relationships with non-homogeneous properties (Indraratna and Salim, 2005). Nevertheless, the most important duties of the track substructures are:

- Distributing the dynamic load from the sleepers to the ground;
- Supporting and anchoring the track structure;
- Providing suitable drainage;
- Reducing the severity of frost action on the soil; and
- Providing resilience for the track system.

Hay (1982) found that the majority of the track failures and maintenance cost are ascribed to the track substructures (especially ballast and subgrade). These have raised the concern to observe static and dynamic behaviours of the substructure whereas many theories and methods have been developed (Indraratna and Salim, 2005). The triaxial testing method has been a very successful way of obtaining dynamic characteristics of the ballast and subgrade materials. The test identifies fundamental properties of ballast and subgrade materials such as resilient or elastic behaviour, plastic or permanent strain, breakage, and failure stress level. These properties are imperative for track substructure design and maintenance scheme (Lakenby, 2006).

6.2. In-Situ Dynamic Testing

An in-situ dynamic test was done by Ford (1988b) to determine how systems comprising a concrete sleeper, rail section, and rail pads respond to vibration. In a laboratory, a sleeper was embedded in ballast contained by a wooden box on a concrete floor. The tests were done using an electro-dynamic shaker to excite the top surface of the rail in the frequency of 0-1000 Hz. The frequency of vibration of the rail at the same point and at three points near the rail seat, were measured by the accelerometers.

Sadeghi (1997) observed that by performing modal testing, the natural frequencies obtained from a test on individual concrete sleepers in the laboratory are higher than those obtained from the in-situ track test bed. These results could indicate that ballast and subgrade cause a slight reduction in the natural frequencies of sleepers. It was also found that the components of the substructure have a very strong non-linear stress-strain relationship and non-homogeneous properties.

Plenge and Lammering (2003) developed full scale laboratory experiments in order to measure the dynamic behaviour of a segment ballasted track and its constituents. The effects of voids between the sleepers and underlying ballast were considered, especially in the partially unsupported sleepers. The two kinds of voided sleepers tested were limited to sleepers with a hovering end and with 2 sided hovering ends. The dynamic displacements were measured by the holographic interferometry in addition to the common accelerometers. The results of these experiments agreed with the field data from the previous record. It also showed that deviations from coupling to the subgrade lead to remarkable changes in the dynamic behaviour.

Until now, very few investigations have been conducted to study the dynamic interaction between the ballast and railway track, or its components such as sleepers, adjacent sleepers, and global systems connected by the rail track. Therefore, at the UoW, there are a number of investigations into the dynamic interaction of ballast and sleeper with various types of realistic voids and pockets in a ballasted track system, with varying wet/dry properties and static preloads (Kaewunruen and Remennikov, 2004; Remennikov and Kaewunruen, 2005).

6.3. Field Trials

Apparently the field trials seem to be the most realistic ones because there are no standard methods currently accepted by the industry for determining the dynamic parameters of track components required to successfully model their dynamics. The industry does not have methods for investigating dynamic parameters and responses. Instead, designers usually calculate the dynamic response based on the static loading specified in the Australian Standards and apply factors that incorporate dynamic loading. Nevertheless these experiments may lead to extremely unreliable results due to unpredictable problems in the rail track itself, e.g., the nonlinearity attributed to voided sleepers, rail impurities, corrugation, worn pads, voided sleepers, and so on.

Grassie (1989a) investigated the properties of a number of rail pads with different materials (synthetic and natural rubber, plastics, and composites) and surfaces (plain, grooved and studded surfaces), to find that the relative correlation coefficients of the relationship between data in a laboratory and on the track, were reasonable. The fractional impact attenuation of a particular pad in the field relies considerably on track conditions, and the amplitudes of dynamic and quasi-static loads. It was shown that the impact test is a good, conservative method for determining its performance.

De Man (1996), and De Man and Esveld (2000) carried out the in-field trials to determine dynamic track properties by using excitation hammer testing. In this test the track was simplified as a 2DOF, discretely supported continuous rail system representing two effective masses of rail and sleeper, as well as two dynamic stiffness and two dashpots of rail pad and ballast-formation, respectively. Based on FRF measurements and FFT, the modal parameters of the track were extracted by an automatic curve fitting optimisation procedure. In the test example two resonance frequencies were clearly obtained and the first 'pin-pin' resonance was noticed in the FRFs measured. In 2002 Colla et al. combined several techniques for a site investigation of non-ballasted railway tracks. The radar technique was used for the layer interface bonding tests, ultrasound helped detect the contact conditions between sleepers and subgrade, and an impact echo technique played a vital role in investigating the structural integrity of those tracks.

However Knothe et al. (2003), reported that the values obtained from the field could only be reliable in the frequency range of the second resonance peak. These field measurements could be obtained when:

- The excitation of a train travelling on the track can be measured;
- A shaker can be used to excite the rail; and
- A calibrated impact hammer can be successfully used to hit the rail.

At the UoW two 2DOF models have been developed based on the FFT analysis and mode superposition method to extract the dynamic properties of ballasted track systems. Responses to ambient vibration induced by normal traffic from the Lara site in Victoria are recorded by a CRC-Rail project. The signal analyses are being done to extract the dynamic properties of the track and field tests will be conducted to investigate more precise results (Kaewunruen and Remennikov, 2005).

7. Summary

Rail track is a fundamental part of railway infrastructure and its components are divided into superstructure and substructure. The most observable parts such as the rails, rail pads, concrete sleepers, and fastening systems are referred to as the superstructure while the substructure is associated with a geotechnical system consisting of ballast, sub-ballast and subgrade. The dynamic testing of railway tracks and its components are of interest, particularly the concrete sleepers and rail pads. In this chapter the technical procedures and testing methods for evaluating the components of ballasted track have also been reviewed and summarised. The ballasted track system and the well-known modal testing technique have been described. Previous analytical models and experiments into the vibration of track components have also been included and practical tests methods used in the field and in situ have been highlighted.

Rail pads are usually installed on rail seats to reduce the stress from axle loads and wheel impact from regular and irregular train movements. They are very important because they soften the interaction between the track and the sleepers. Several time and frequency domain modelling of rail pads had been done as well as a number of experiments of their dynamic parameters.

Some investigations dealing with the dynamic properties of concrete sleepers can be found in this paper. Most are based on the experimental modal testing technique. In addition the various boundary conditions of sleepers, such as free-free, perfectly coupled to the subsoil, and voided sleepers, were investigated for their effect on the dynamic behaviour of concrete sleepers. However, some limitations still remain, as reported.

Until now, very few investigations have been conducted into studying the dynamic interaction between the ballast and track or components such as sleepers, adjacent sleepers, and global systems connected by the rail track. In addition, although the field tests seem to be realistic, the experiments may lead to extremely unreliable results due to many unpredictable problems in the track such as nonlinearity that is attributed to voided sleepers, rail impurities, track corrugations, worn pads, and voided sleepers.

Acknowledgement

The authors are grateful to acknowledge the Australian Cooperative Research Centre for Railway Engineering and Technologies for the financial support as part of Project #5/23.

References

Allemang, R. (1993), Modal Analysis – Where do we go from here? *International Journal of Analytical and Experimental Modal Analysis*, **8**(2) April, 79-91.

Allemang, R. and Brown, D. (1986). Multiple – Input experimental modal analysis – A survey, *International Journal of Modal Analysis*, January, 37-43.

Brown, D. (1982). Keynote Speech Modal Analysis – Past, Present and Future, *Proceedings of the 1st International Modal Conference*, Union College, New York.

De Man, A.P. (1996). Determination of dynamic track properties by means of excitation hammer testing, *Railway Engineering International 1996 Edition*, (4), 8-9.

De Man, A.P. and Esveld, C. (2000). Recording, estimating, and managing the dynamic behavior of railway structures, *Proceedings of Symposium Leuven*. The Netherlands.

De Man, A.P. (2002). DYNATRACK: A survey of dynamic railway track properties and their quality, *Ph.D. Thesis*, Faculty of Civil Engineering, Delft University of Technology, The Netherlands.

Esveld, C., Kok, A.W.M. and De Man, A. (1998). Integrated numerical and experimental research of railway track structures, *Proceedings of the 4th International Workshop on*

Design Theories and their Verification of Concrete slabs for Pavements and Railroads, Portugal.

Esveld, C. (2001). *Modern Railway Track*, MRT Press. The Netherlands.

Ewins, D.J. (1995). *Modal Testing: Theory and Practice*, Research Studies Press, Taunton.

Fahey, S. O'F. and Pratt, J. (1998). Frequency domain modal estimation techniques, *Experimental Techniques*, Sep/Oct, 22(5), 33-37.

Fenander, A. (1997). Frequency dependent stiffness and damping of railpads, *Proceedings of Institute of Mechanical Engineering Part F*, **211**, 51-62.

Fenander, A. (1998). A fractional derivative railpad model included in a railway track model, *Journal of Sound and Vibration*, **212**(5), 889-903.

FIP Commission on Prefabrication. (1987). *Concrete Railway Sleepers – FIP State of Art Report*, Thomas Telford Ltd., London, UK.

Ford, R. (1988a). Modal analysis of a concrete railway sleeper, *Research Note AVG/RN881122-1*, School of Mechanical and Industrial Engineering, University of New South Wales, Australia.

Ford, R. (1988b). Vibration responses of systems comprising a concrete sleeper, rail section, and various types of rail pads, *Research Note AVG/RN881122-2*, School of Mechanical and Industrial Engineering, University of New South Wales, Australia.

Grassie, S.L., Gregory, R.W., Harriswon, D., and Johnson, K.L. (1982). The dynamic response of railway track to high frequency vertical excitation, *Proceedings of the Institution of Mechanical Engineers, Part C, Journal of Mechanical Engineering Science*; **24**(2): 77-90.

Grassie, S.L. and Cox, S.J. (1984). The dynamic response of railway track with flexible sleepers to high frequency vertical excitation, *Proceedings of Institute of Mechanical Engineering Part D*, **24**, 77-90.

Grassie, S.L. (1987). Measurement and attenuation of load in concrete sleepers, *Proceedings of Conference on Railway Engineering*, Perth, September 14-16, 125-130.

Grassie, S.L. (1989a). Resilient rail pads: Their dynamic behavior in the laboratory and on track, *Proceedings of Institute of Mechanical Engineering Part F*, **203**, 25-32.

Grassie, S.L. (1989b). Behavior in track of concrete sleepers with resilient rail pads, *Proceedings of Institute of Mechanical Engineering Part F*, **203**, 97-101.

Grassie, S.L. and Kalousek, J. (1994). Rail corrugation: characteristics, causes and treatments, *Proceedings of the Institution of Mechanical Engineers, Part F, Journal of Rail and Rapid Transit*; **207**(F1): 57-68.

Grassie, S.L. (1996). Models of railway track and train-track interaction at high frequencies: Results of benchmark test, *Vehicle System Dynamics*; **25**(Supplement): 243-262.

Gustavson, R. (2000). Static and dynamic finite element analyses of concrete sleepers, *Licentiate of Engineering Thesis*, Department of Structural Engineering, Chalmers University of Technology, Sweden.

Gustavson, R. (2002). Structural behaviour of concrete railway sleepers. *PhD Thesis*, Department of Structural Engineering, Chalmers University of Technology, Sweden.

Hay, W.M. (1982). Railroad Engineering, (2nd Edition), John Wiley & Sons, Inc.

He, H. and Fu, Z. (2001). *Modal Analysis*, Butterworth – Heinemann Publishers, Great Britain.

Indraratna, B. and Salim, W. (2005). *Mechanics of Ballasted Rail Tracks A Geotechnical Perspective*. Taylor & Francis, London.

Kaewunruen, S. and Remennikov, A. (2004). A state-of-the-art review report on vibration testing of ballasted track components, *July-Dec Research Report*, CRC Railway Engineering and Technology, Australia, December, 20p.

Kaewunruen, S. and Remennikov, A.M., (2007). Field trials for dynamic characteristics of railway track and its components using impact excitation technique. *NDT&E International*, **40**(7), 510-519.

Kaewunruen, S. (2007). Experimental and numerical studies for evaluating dynamic behaviour of prestressed concrete sleepers subject to severe impact loading, *PhD Thesis*, School of Civil, Mining, and Environmental Engineering, University of Wollongong, NSW, Australia.

Kaewunruen, S. and Remennikov, A.M. (2008). Nonlinear transient analysis of railway concrete sleepers in track systems. *International Journal of Structural Stability and Dynamics,* in press.

Knothe, K. and Grassie, S.L. (1993). Modelling of railway track and vehicle/track interaction at high frequencies, *Vehicle System Dynamics*, **22**(3-4), 209-262.

Knothe, K., Yu, M. and Ilias, H. (2003). Measurement and Modelling of Resilient Rubber Rail-Pads. In: K. Popp und W. Schiehlen (Ed.), *System Dynamics and Long-term Behaviour of Railway Vehicles, Track and Subgrade*, Springer Verlag, Berlin Heidelberg, Germany, pp. 265-274.

Lackenby, J. (2006). Triaxial behaviour of ballast and the role of confining pressure under cyclic loading, *PhD Thesis*, School of Civil, Mining, and Environmental Engineering, University of Wollongong, NSW, Australia.

Liang, B. and Zhu D. (2001).Dynamic analyisis of the vehicle-subgrade model of a vertical coupled system, *Journal of Sound and Vibration*, (245), 79-92.

Mitchell, L. (1986). Signal processing and the Fast Fourier Transform (FFT) Analyser: A survey, *International Journal of Modal Analysis*, January, 24-36.

Plenge, M. and Lammering, R. (2003). The dynamics of railway track and subgrade with respect to deteriorated sleeper support. In: K. Popp und W. Schiehlen (Ed.), *System Dynamics and Long-term Behaviour of Railway Vehicles, Track and Subgrade,* Springer Verlag, Berlin Heidelberg, Germany, pp. 295-314.

Raymond, G.P. (1978). Design for railroad ballast and subgrade support, *Journal of the Geotechnical Engineering Division ASCE*, **104**(1): 45-60.

Remennikov, A.M. and Kaewunruen, S. (2005). Investigation of vibration characteristics of prestressed concrete sleepers in free-free and in-situ conditions. *Australian Structural Engineering Conference 2005*, Sep 11-14, Newcastle, Australia [CD Rom].

Remennikov, A.M. and Kaewunruen, S. (2007). A review on loading conditions for railway track structures due to train and track vertical interaction. *Structural Control and Health Monitoring*, in press.

Sadeghi, J. (1997). Investigation of characteristics and modeling of railway track system, *PhD Thesis*, School of Civil, Mining, and Environmental Engineering, the University of Wollongong, Australia.

Selig, E.T. and Waters, J.M. (1994). *Track geotechnology and substructure management*, (1st Ed.), Technology Development and Application Committee, on behalf of the Railways of Australia.

Sjoberg, M. (2002). On dynamic properties of rubber isolators, *Ph.D. Thesis*, Department of Vehicle Engineering, Kungl Teknisa Hogskolan Royal Institute of Technology, Stockholm, Sweden.

Smutny, J. (2004). Measurement and analysis of dynamic and acoustic parameters of rail fastening, *NDT&E International*, **37**, 119-129.

Standards Australia (2001). *AS1085.19-2001 Railway track material - Part 19: Resilient fastening assemblies*, Standards Australia.

Standards Australia (2003). *AS1085.14-2003 Railway track material - Part 14: Prestressed concrete sleepers*, Standards Australia

Steffens, D.M. (2005). Identification and development of a model of railway track dynamic behavior, *M.Eng. Thesis*, School of Civil Engineering, Queensland University of Technology, QLD. Australia.

Timoshenko S. (1926). Method of analysis of statistical and dynamical stresses in rail. In: *Proceedings of Second Int. Congress for Applied Mechanics*, Zurich, 407-418.

Thompson, D.J., van Vliet, W.J. and Verheij, J.W. (1998). Developments of the indirect method for measuring the high frequency dynamic stiffness of resilient elements, *Journal of Sound and Vibration*, **213**(1), 169-188.

Van't, Z. (1993). Assessment of dynamic characteristics of rail pads, *Railway Engineering International 1994 Edition*, **23**(4), 15-17.

Verheij, J.W. (1982). Multi-path sound transfer from resiliently mounted shipboard machinery, *Ph.D. Thesis*, TNO Institute of Applied Physics, Delft University of Technology, the Netherlands.

Vincent, G. (2001). Modal analysis and numerical modeling of a concrete railway sleepers, *M.Eng. Thesis*, Department of Structural Engineering, Chalmers University of Technology, Göteborg, Sweden. 128 pp.

Wu, T.X. and Thompson, D.J. (1999). The effects of local preload on the foundation stiffness and vertical vibration of railway track, *Journal of Sound and Vibration*, **219**(5), 881-904.

Wu, T.X. and Thompson, D.J. (2004). The effects of track nonlinearity on wheel/rail impact. *Proceedings of the Institution of Mechanical Engineers Part F: Rail and Rapid Transit*; **218**(1): 1-15.

Zaveri, K. (1985). Modal Analysis of Large Structures – Multiple Exciter Systems, Naerum Offset, Denmark.

In: New Research on Acoustics
Editor: Benjamin N. Weiss, pp. 221-242
ISBN: 978-1-60456-403-7

Chapter 6

ACOUSTIC EVOLUTION OF ANCIENT THEATRES AND EFFECTS OF SCENERY

K. Chourmouziadou and J. Kang
School of Architecture, University of Sheffield, Sheffield S10 2TN, United Kingdom

Abstract

The acoustics of open-air performance spaces, in particular ancient theatres of Classic, Hellenistic and Roman periods, has gained much attention in the past years. However, earlier theatres, situated in the courtyards of the Minoan palaces, in unique shapes, dated around 1500 B.C., have only recently been examined from the acoustic viewpoint. In this chapter, the examination of six identified types of ancient theatre has revealed that theatres evolved architecturally and acoustically through the centuries. Moreover, the excavation of ancient theatre sites in the last century allowed the revival of ancient drama and instituted drama festivals in southern Europe. Scenery is an important component of those drama performances, both from the aesthetic and, as new studies show, the acoustic viewpoint. In this chapter, the effects of temporary scenery design, classified into four generic categories, on the acoustic environment of open-air theatres are investigated, aiming at providing guidelines for architects and scenery designers. In addition to the above two key issues, a general literature review on the acoustics of ancient Greek/Roman theatres is given and relevant research methodology on the subject is discussed.

Introduction

The interest in the acoustic properties of ancient theatres is currently growing. The Minoan, the Pre-Aeschylean, the Classic, the Hellenistic and the Roman types were created during the evolution process that lasted 2,000 years. The identifiable characteristics were either the result of new needs that emerged from the evolution of drama, new construction methods, in terms of materials and position of the theatre in the city or the country, or individual transformation of theatres. In particular, the same theatre could have evolved architecturally when the stage building gained a new shape, new components, or through changes in heights and relative positions between the actors, the chorus and the audience. It is therefore of great interest to examine the effectiveness of the architectural evolution on the acoustics [Chourmouziadou and Kang, 2007a].

The excavation of ancient theatres in the last century allowed the revival of ancient drama, instituted drama festivals and transformed them into social summer events in southern Europe. Scenery is an important component of drama performances, both from the aesthetic and, as new studies show [Chourmouziadou and Kang, 2007b], the acoustic viewpoint. It is therefore vital to systematically consider the effects of various sceneries, especially in terms of aiding design.

In this chapter the above two issues are studied systematically. As a basis of the detailed study, this chapter first reviews the evolution of ancient theatres and general studies on their acoustic characteristics; as well as the methodology of examining acoustics in ancient Greek/Roman theatres.

Ancient Theatres and Their Acoustic Characteristics

Extensive research on ancient theatres in Greece started when the first historical excavations revealed evidence of the spaces used for performances. Previous studies related to ancient performance spaces are found scattered in several fields, including drama, archaeology, architecture, philosophy and acoustics. Changes in performance styles led to innovations that altered the form of the theatre [Baldry, 1971; Chourmouziadou and Kang, 2002; Simpson, 1956]. Ancient Greek and Roman theatres have been discussed and compared in terms of architectural and construction trends [Cailler and Cailler, 1966; Dinsmoor, 1950; Dörpfeld, 1896; George, 1997; Izenour, 1977; Robertson, 1979; Vitruvius, 1st B.C.], although it is still not clear whether the Roman performance spaces were just a transformation or an advanced design of the Greek.

The acoustics in a number of theatres of antiquity has been examined in the past, based on general acoustic principles or on-site measurements [Barron, 1993; Beranek, 1962; Canac, 1967; Cremer, 1975; Egan, 1988; Shankland, 1973], as well as in more recent studies, as will be reviewed below. Although computer simulation has been used for studying the acoustics of ancient theatres [Vassilantonopoulos and Mourjopoulos, 2003; 2004], research regarding the influence of the architectural evolution and material use on the acoustic quality has been limited. Moreover, the relationship between the acoustic knowledge in antiquity [Chadwick, 1981; Guthrie, 1962; Hunt, 1978; O'Meara, 1989] and the design of these theatres has not been systematically investigated.

Architectural Evolution: Theatre Types

The Greeks were renovating their theatres according to the need to accommodate a larger audience, the change in performance style and the use of new materials [Athanasopoulos, 1983; Chourmouziadou, 2002; Dinsmoor, 1950]. Early forms were the Minoan (20th-15th B.C.) and the Pre-Aeschylean (15th-6th B.C.) in rectangular and trapezoid shapes respectively, with limited surviving examples [Athanasopoulos, 1983]. Figure 1 illustrates the plans and sections of these two types of performance space.

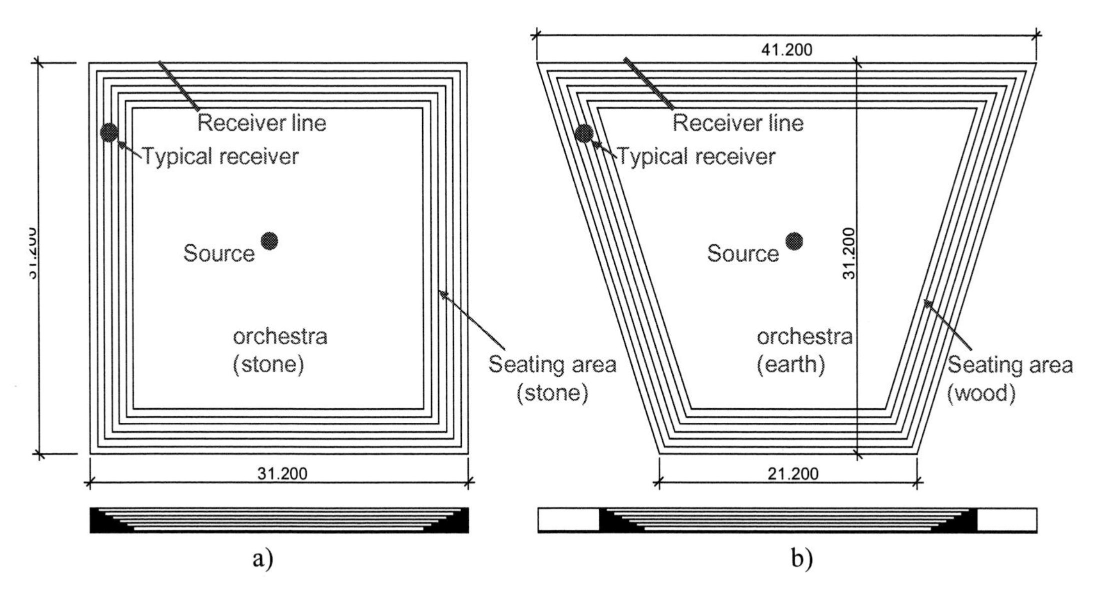

Figure 1. Plan and section of the early forms of performance spaces in Greece, with locations of typical receivers and receiver lines used in the simulation. a) Minoan, and b) Pre-Aeschylean.

The outdoor performance spaces, found through excavations at the palaces of Knossos and Phaestos, can be considered as primitive types of Greek and Roman theatres, as Allen [1963] suggested. The audience area was hardly steep, and there was limited number of seats, which means that it addressed only people from the palace, in a more intimate relationship between them and the performers. The materials were mostly local stone. Contemporary outdoor performance spaces are sometimes built based on rectangular layouts too [Varopoulou, 1991].

The Classic Greek type was created in the late 6th century B.C., comprising two basic elements: the *orchestra* and the *koilon*, namely the hill slope that provided ample amphitheatric space. Stepped tiers were hewn in the shape of concentric circular sections in the hillside around the orchestra to allow the audience a better view of the performers [Simpson, 1956]. They were then replaced by wooden benches. A stage building was constructed later. However, the definite form came with the use of stone and marble in the late 5th century B.C. The seats were laid out concentrically around a circular orchestra in arcs exceeding 180°, often extended around two-thirds of the orchestra circle. Figure 2a illustrates the Classic Greek theatre [Dinsmoor, 1950; Robertson, 1979; Simpson, 1956].

The descendant of the Classic theatre was the Hellenistic. Some characteristics can distinguish it from the previous form, such as the raised stage, and the stage building, or *skene-building*. These innovations were also connected with changes in methods of playwriting in that period. For example, the introduction of the second and the third actor in tragedies moved the centre of the performance from the orchestra to the proscenium and the stage, which were therefore raised [Baldry, 1971]. In some cases the koilon was extended with more seating rows, like in the theatre of Epidaurus, or the orchestra was repositioned, like in the theatre of Dionysus in Athens. Figure 2b illustrates a typical Hellenistic theatre, based on the theatre of Epidaurus, although not including the high podium above the *diazoma* (the intermediate space between the lower and the upper part of the koilon) [Dinsmoor, 1950; Robertson, 1979; Simpson, 1956].

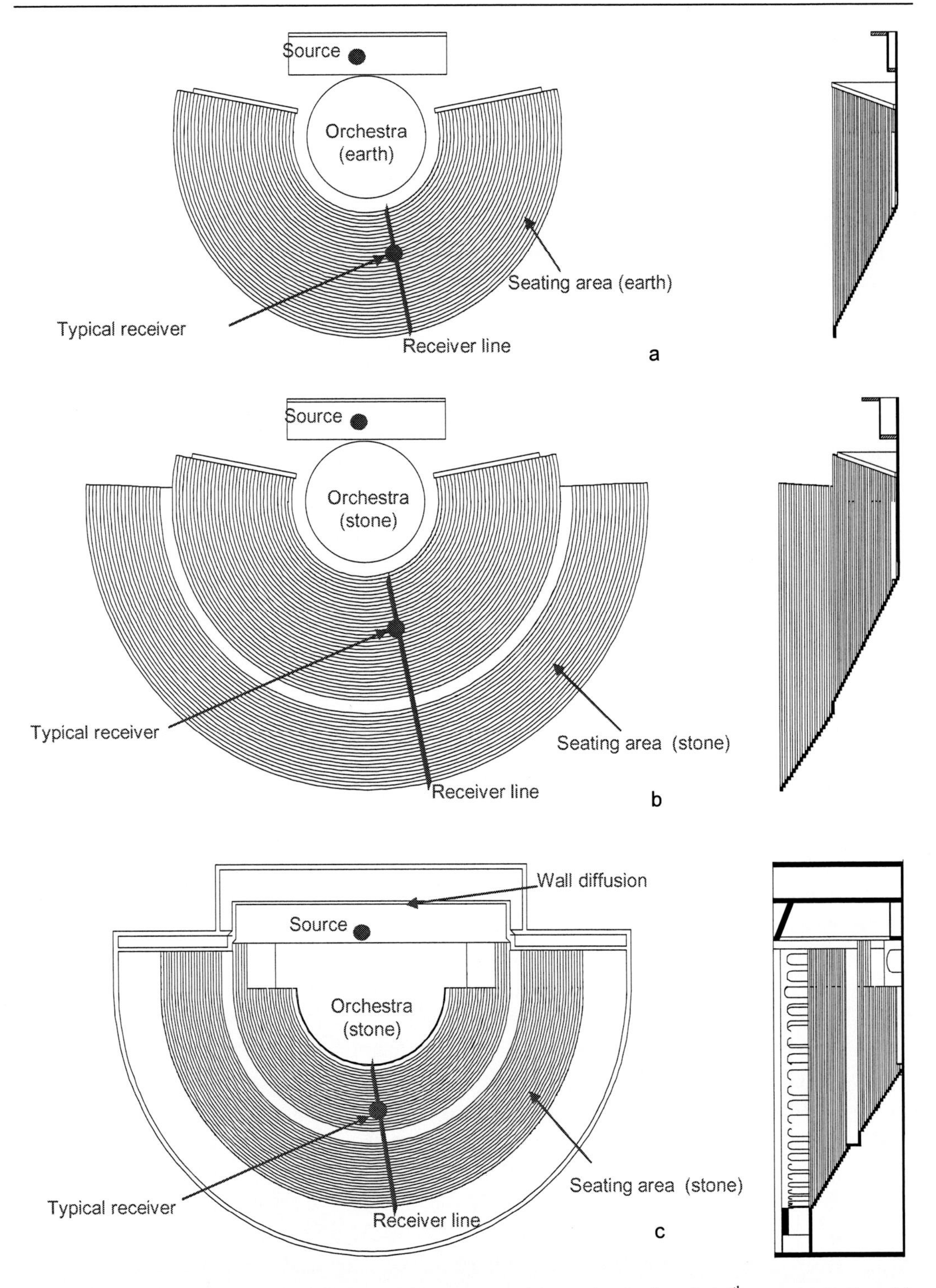

Figure 2. Plan and section of performance space types that evolved after the 6th century B.C., with locations of typical receivers and receiver lines used in the simulation. a) Classic Greek theatre, b) Hellenistic theatre, and c) Roman theatre.

The Roman theatre, like Roman drama, was Greek in origin [Robertson, 1979], although it differed in several aspects from all Greek theatres of Classic or Hellenistic times. The auditorium or *cavea* of the Roman theatre was semicircular and it was united as a single structure with the stage building, as illustrated in Figure 2c [Izenour, 1977; Robertson, 1979]. There were vaulted passages at the point where the orchestra was connected to the cavea, to make the orchestra accessible. The walls of the stage building were at the same height as the cavea and the stage was wide but low in height, projecting much further than the proscenium and affecting the shape of the orchestra, which was reduced to a semicircle.

Early Roman theatres, built from wood in unknown shapes, are not included in this investigation.

Acoustics in Ancient Theatres

In a semi-free field, where there are no reflective boundaries except for the ground, the listening conditions depend largely on the source-receiver distances. With an enclosure the boundaries around the source can reflect the sound to the receivers and increase the evenness of the sound field. In ancient theatres, acoustics cannot be determined solely on the grounds of a semi-free field, given that the acoustic conditions would be comparatively improved due to the hard reflective surfaces. The surfaces of the audience area, usually built from local stone, provided high amplitude reflections, especially enhancing the early sound. Sound could be diffused from the destroyed edges of the seats, resulting in a relatively even distribution [Chourmouziadou and Kang, 2006b]. Under the occupied condition, sound was diffused from the audience's bodies.

When compared to more recent examples of enclosed spaces, like those of shoebox shape, the ancient theatres succeeded in bringing a vast crowd as close to the performers as possible [Pöhlmann, 1995]. Steeply raked seating, like in ancient theatres after the 6th century B.C., increased the angle of incidence, reduced the attenuation caused by the seating audience, and prevented noises from sources situated behind the audience area [Beranek, 1962]. In terms of absorption, the basic difference between ancient theatres and enclosed auditoriums is that the latter have padded absorbing chairs to balance the difference between the occupied and unoccupied condition [Beranek and Hidaka, 1998].

In ancient theatres the audience mainly received the direct sound from both the actor and the chorus, followed by a reflection off the orchestra floor. The more the actor approached the audience, the smaller the part of the audience that received the direct sound, due to its propagation at nearly grazing incidence. Some studies referred to a useful acoustic zone to be used by the actor, the 'Haas zone', a narrow space created by the tangent of the orchestra, the *paraskenia*, the sides of the stage building, and the stage building façade [Canac, 1967; Mparkas, 2004]. Also, the centre of the orchestra was important from the acoustic, the visual and the focal viewpoint. It was a good position *"not to see but to be seen"* [Wiles, 1997], as also recognised by the 5th century (B.C.) playwrights [Rehm, 1992].

It has been suggested by Pöhlmann [1995] that the acoustic conditions, especially at the upper rows of the spectators, were improved by the appearance of the actors on a higher platform, especially since the upper level of the *skene* (stage) served as a sound reflector. Additionally, the use of theatre buildings as a place for assemblies might have contributed to the development of the skene, because an individual speaker stood at the proscenium, so he was clearly visible and audible [Pöhlmann, 1995].

It is noted that in ancient Greek theatres background noise was generally very low, compared to today's conditions, which was important for the unassisted speech to be audible. Noise from birds and animals were the most common, as well as from the audience (up to 30,000 people). Today ancient theatres are mostly situated in cities and/or near busy roads [Chourmouziadou and Kang, 2005]. Air traffic can also be disturbing.

The effects of environmental conditions on outdoor sound propagation are usually more relevant to large distances [Kurze and Beranek, 1971; Kang, 2006], although some previous studies have indicated their influence on theatre acoustics [Goularas, 1995]. Excess attenuation due to ground absorption can be neglected within a distance of 30-70m from the source, which corresponds to typical source-receiver distances in open-air theatres. The trees that surround the audience area in ancient theatres absorb, diffuse and reflect sound, depending on the tree density, the foliage density, the tree type and the height and width of the belt. This can be applied to the theatre of Epidaurus, which is surrounded by a forest of coniferous trees.

Canac [1967] developed a complex trigonometric equation involving the inclination, the axial development of the koilon, the orchestra radius, the width and height of the proscenium and the position of the stage building. Mparkas [1992] and Goularas [1995] analysed the acoustic environment and some theoretical methods used for the evaluation of ancient Greek theatres including Epidaurus, based on previous studies, especially Canac's [1967].

A study on the simulation of ancient theatres was carried out by Chourmouziadou and Kang [2002], which mainly focused on the design and materials of Classic, Hellenistic, Roman as well as Chinese theatres, and the performance evaluation through the method of auralisation. Basic differences between the circular, semicircular and rectangular layouts in terms of distribution and reflection patterns were also examined [Chourmouziadou and Kang, 2003]. Vasilantonopoulos and Mourjopoulos [2003] carried out acoustic simulations for the ancient theatres of Epidaurus, Dodoni and the Roman *odeion* (odium) of Patras, revealing that the excellent acoustics of Epidaurus was possibly related to the increased diffusion, that reverberation time (RT) was below 0.2s and that there were important differences between the three theatre layouts in terms of acoustics. Later, Vasilantonopoulos and Mourjopoulos [2004], Vasilantonopoulos *et al* [2004] and Gade and Angelakis [2006] presented a comparison between acoustic simulation and detailed on-site measurements of the theatre of Epidaurus.

Based on an European Commission funded project on the identification, evaluation and revival of the acoustical heritage of ancient theatres and odeia (ERATO), a comparison between acoustic simulations and measurements of the theatre of Aspendus in Turkey [Gade *et al*, 2004] revealed that the RT of Roman theatres is in the range of 1.4-2.0s. The research team also presented ways of representing the ancient theatres and selecting calculation parameters [Lisa *et al*, 2004], as well as comparing and matching the simulation and measurement results [Gade *et al*, 2005]. Through the latter the importance of diffraction in the acoustics of ancient theatres was indicated. The acoustics of the Roman odeion of Aphrodisias was also studied to examine the differences between the layouts of Roman theatres and odeia [Lisa *et al*, 2005], and to virtually reconstruct the sound [Rindel *et al*, 2006]. Roman odeia in Greece have also been examined by Vasilantonopoulos *et al* [2006].

The present use of ancient theatres and the application of temporary scenery design have also been investigated [Chourmouziadou and Kang, 2004]. By examining the recently excavated theatre of Mieza as well as the theatre of Philippi [Chourmouziadou and Kang, 2005], the effect of scenery, constructed for ancient drama performances, on the acoustic environment of the theatres was identified. Moreover, the increase of reverberation in ancient theatres, due to the installation of temporary scenery, was examined and verified by on-site measurements and subjective evaluation tests [Chourmouziadou and Kang, 2006a]. It has also been indicated that when there is no boundary at the back of the actor, the sound reaching the audience is less intense [Mparkas, 2006]. In reality, restrictions applied by the Boards of Antiquities in Greece do not give the opportunity for the missing parts of the theatre to be restored, at least at a large scale, unless all the materials can be found [Karadedos, 1994], with the exception of organised restoration projects, such as in the case of the palace of Knossos [Bell, 1926].

There have been some discussions about the use of appropriate indices for outdoor performance spaces such as ancient Greek/Roman theatres [Paini *et al*, 2006]. Although outdoor reverberation is perceived in a different manner from that in indoor spaces, RT is still very useful in ancient theatres and can be used for comparative studies between different spaces. While optimum RT values have been studied for many types of space or usage, for outdoor performance spaces the specifications on the optimum values have been rather limited. The central gravity time (TCG) and strength (G) have been used to describe the acoustics of outdoor performance spaces, so are the lateral energy fraction (LEF), clarity (C80), definition (D50) and sound pressure level (SPL) distribution [Chourmouziadou and Kang, 2002; Vasilantonopoulos and Mourjopoulos, 2003].

Acoustic Simulation Methodology

Three software packages, namely CATT, ODEON and Raynoise, have been examined regarding their appropriateness for the acoustic simulation of open-air theatres. Short calculation time, acoustic indices produced, application of diffusion and diffraction were some of the issues considered [Chourmouziadou, 2007]. For the sake of convenience, only Raynoise was used for the simulations in this chapter. Appropriate simulation parameters have been examined, including the number of rays and reflection order. Background noise, source directivity and optimum representation of the theatre model have also been examined. Moreover, absorption coefficient measurements, on-site measurements and subjective evaluation tests have been used for validation.

To validate the influence of source directivity on the acoustics of ancient theatres, a comparison between an omni-directional source and a source with human directivity has been carried out and the results show that the difference in reverberation between the two directivities is generally insignificant. However, regarding the variation in source positions it has been suggested that higher SPL values are created when the actor approaches the audience, as expected, while the opposite phenomenon is observed for reverberation. This is due to the differences in reflection patterns and reflection path distances. For clarity and definition the best results are found when the source is close to the audience and near the centre of the orchestra respectively. Generally speaking, the results seem to correspond to the 'Haas zone', as previously mentioned.

Exact representation of the theatres, especially the audience area, is essential to ensure the accurate simulation of reflection paths. Regarding the representation of circular shapes, the effects differ depending on the size of the enclosure (circular segment) and the proximity to the source/receiver [Chourmouziadou, 2007]. The analysis of the simulation results shows that the representation of the theatre in the unoccupied condition should be based on a minimum of 24 segments. By incorporating high absorption and diffusion in an attempt to represent the state of the theatres in the occupied condition it has been shown that the difference in SPL between 12 and 24 segments is insignificant.

The effects of diffraction on sound distribution in ancient theatres, considering four mechanisms including (a) diffraction to the shadow zone behind a panel, (b) diffusion/diffraction due to limited surface size, (c) diffraction from a panel edge, and (d) diffusion effects due to the acoustic roughness of panel surface, have been investigated in detail. Outdoor performance spaces present few strong reflections, so that even the weak scattered energy would influence decay curves and impulse responses considerably. Diffraction around a barrier is relevant to ancient theatres only when actors are situated behind a scenery panel or object. Small-scale surface roughness becomes negligible with respect to wavelength at low frequencies, while edge diffraction may be significant at higher frequencies, especially for geometries with small surfaces and/or several wedges, like scenery designs that involve several smaller objects/panels.

Based on previous research on diffraction for computer simulations [Christensen and Rindel, 2005a; 2005b], two methods have been developed and tested [Chourmouziadou and Kang, 2006b]. Method 1 involves application of relatively high diffusion coefficients to represent the phenomena of diffraction and Method 2 is a theoretical approach on a combined diffraction/diffusion coefficient for reflective panels. It is shown that RT30 is the index mostly affected, compared to other acoustic indices, with a maximum difference of about 1s compared to a model with flat panel scattering coefficient applied to the surfaces, as suggested by D'Antonio and Cox [Everest, 1994]. Method 2 has been developed to incorporate the 3D characteristics of the reflection paths, by using normalised vectors, and the 'scalar product'. It has been shown that scattering increases as the distance from the reflection surface increases. Since the final values of the coefficients show a pattern relating to the theatre layout, it is possible to divide the theatre into zones and apply the same coefficient to each zone.

Generally speaking, the calculation by using the flat panel diffusion coefficients agrees well with measurements. For other theatre layouts, such as the more enclosed Roman theatres, different diffusion/diffraction coefficients might be more appropriate. Method 2 may be more useful for sceneries and other objects that could determine the acoustic environment in relation to their dimensions and surface roughness.

Acoustic Evolution

The development of drama in Greece, in the forms of tragedy and comedy and new performance styles brought significant changes in theatre design, including size, shape, and site conditions. Material usage and new construction techniques were also important for the theatre's architectural evolution. Moreover, there is evidence that ancient architects studied propagation of sound and could have used it in designing the theatres [Hunt, 1978]. Based on the above review of ancient theatre evolution, a systematic acoustic simulation has been made and correspondingly, the acoustic evolution is summarised below.

For the Minoan and the Pre-Aeschylean theatre types, since there were no enclosures or surrounding walls, sound distribution was mainly determined by the direct sound and reflections from the ground. However, compared with the later theatre types in Greece, the SPLs were relatively high because their size was smaller and the audience was closer to the source, further enhanced by the multiple reflections between parallel seat risers. This is also the case for the simulated reverberation of the unoccupied condition, whereas under the occupied condition, the reverberation time is much shorter and insufficient.

Classic Greek theatres were built on hillsides for performing tragedy during the Dionysian festivals. Their form evolved gradually, with changes in materials and the introduction of the stage building. Their acoustics depended mainly on direct sound. The fan shape of these theatres implied poor visual and acoustic conditions for the audience seated at the sides of the orchestra, but, interestingly, these seats were reserved for latecomers and women, who were not regarded as important guests. A noticeable characteristic of early Classic theatres was the lack of reflections, and consequently the short reverberation, compared to the other types. Nevertheless, since the seating area was steep, sound attenuation over the audience area was less significant. However, there is evidence that the chorus were performing at the orchestra, with decreased angle of incidence of the direct sound as they approached the audience.

Comedy, the new form of drama, brought the Hellenistic renovations, which involved a raised stage, an extended koilon and, in some cases, changes in the seating system. The increase of the stage height resulted in further increased angles of incidence of the direct sound, although the orchestra's reflections reduced their angles to the audience plane. The vertical surfaces of the koilon provided important first reflections, mostly to the lower part of the audience, and the circular shape helped reflections to scatter. The extended part of the diazoma became steeper, so that there was more reflected sound. The use of harder materials, from earth and wood to stone, was useful for providing stronger reflections. Overall, from the early Classic to the Hellenistic theatre, the sound level and reverberation were both increased, and thus the acoustic conditions were improved.

The Roman theatre, in a more enclosed layout than the Hellenistic theatre, presented longer reverberation times for both unoccupied and occupied conditions. The raised stage wall, as mentioned earlier, contributed to the increase of reverberation, due to the multiple reflections between the wall and the audience area [Chourmouziadou, 2002]. Also, the peripheral corridors (colonnade), with low absorption, directed the reflections back to the stage building, and in particular its central axis. Hence, the reflections were repeatedly sent back and forth, resulting in long RT values. The *velarium*, the rope-and-fabric structure covering most of the audience part [Izenour, 1977], was possibly reflective at certain frequencies similar to today's fabric structures, affecting both speech transmission and intelligibility [Mapp, 2000]. For occupied conditions, the acoustic indices in the Roman theatre were rather good, close to those in modern theatres, although there were considerable variations at various receivers. Under unoccupied conditions, however, the reverberation was rather long, around 1.3-1.9s, and this might have led to poor intelligibility depending on the occupancy of the theatre.

Overall, innovations in construction and layout generally resulted in acoustic improvements. Figure 3 shows the changes in reverberation along with theatre evolution, where the average value as well as the variation at various receivers are shown, for both occupied and unoccupied conditions.

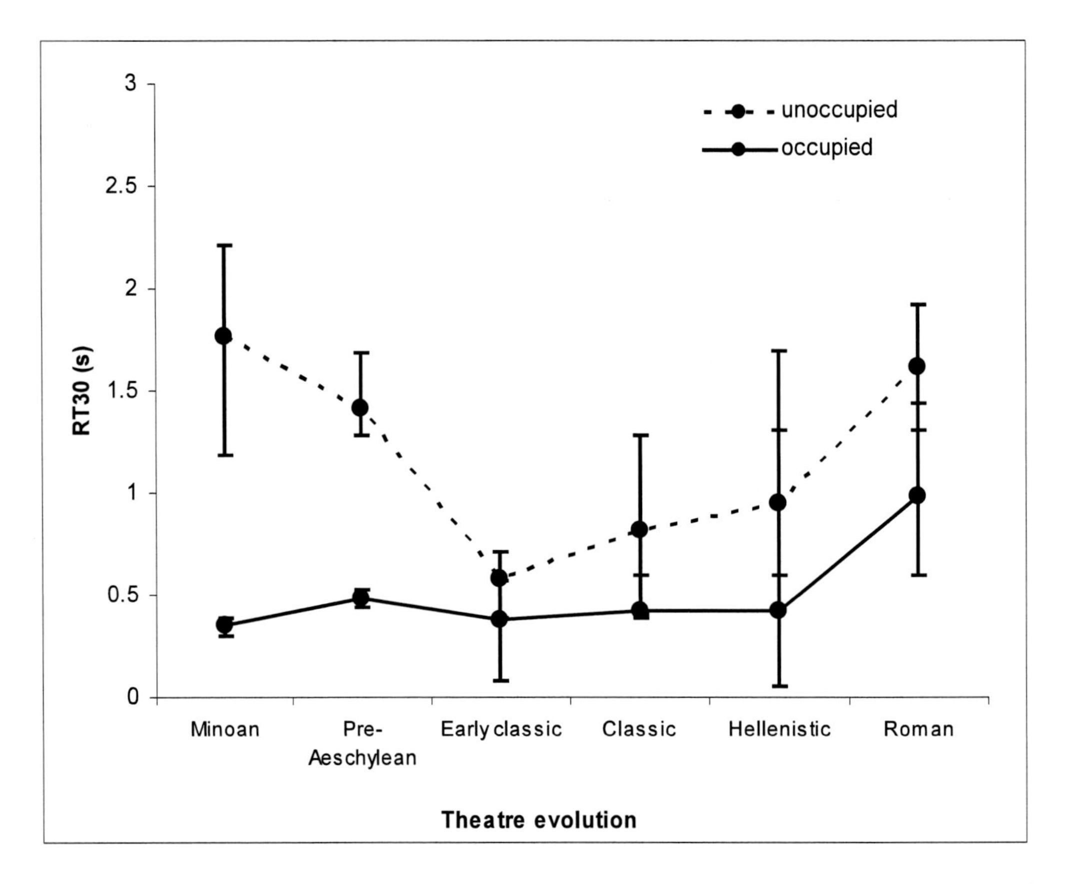

Figure 3. Variation of RT30 (s) at 500Hz with the evolution of ancient Greek and Roman theatres. The error bars indicate maximum and minimum RT30 along the typical receiver line.

The Theatre of Epidaurus

For the theatre of Epidaurus a systematic investigation has been carried out, including acoustic simulation, on-site measurements and subjective evaluation. Refinements in design and construction in relation to early reflections provided by the koilon risers, the orchestra floor and the ancient stage building, as well as multiple reflections between the theatre sides, led to relatively long reverberation time and also good acoustic conditions for theatre and speech. The absorption coefficient of porous stone, a material the theatre of Epidaurus and many other ancient theatres were built from, was measured in an impedance tube as 0.01 at 63, 125 and 250 Hz, 0.052 at 500Hz, 0.071 at 1kHz, 0.87 at 2kHz, 0.085 at 4kHz, and 0.063 at 8kHz.

The reverberation of the theatre was measured when the temporary scenery for the performance of 'Electra' was installed [Chourmouziadou and Kang, 2006]. The results were compared with previous studies [Vasilantonopoulos and Mourjopoulos, 2004], showing that the scenery increased RT30 by about 0.5s compared to the theatre with no scenery.

Consequently, since contemporary performances count on temporary scenery design, which influences the acoustics of open-air theatres, it is important to investigate their effect. Moreover, with appropriately designed scenery, reverberation can be adjusted for other uses, like music.

Simulations were carried out considering the koilon with an inclination of 26°, according to the original drawings by Von Gerkan and Müller-Wiener [1961]. Good agreement was obtained between the simulated and measured results, as shown in Figure 4. A variety of absorption coefficients for hard surfaces was considered, including marble and porous stone, while a range of diffusion coefficients, including the flat panel diffusion and some extreme cases of high and low diffusion were also considered, to explore the appropriate material characteristics to be used in the acoustic simulation of open-air theatres.

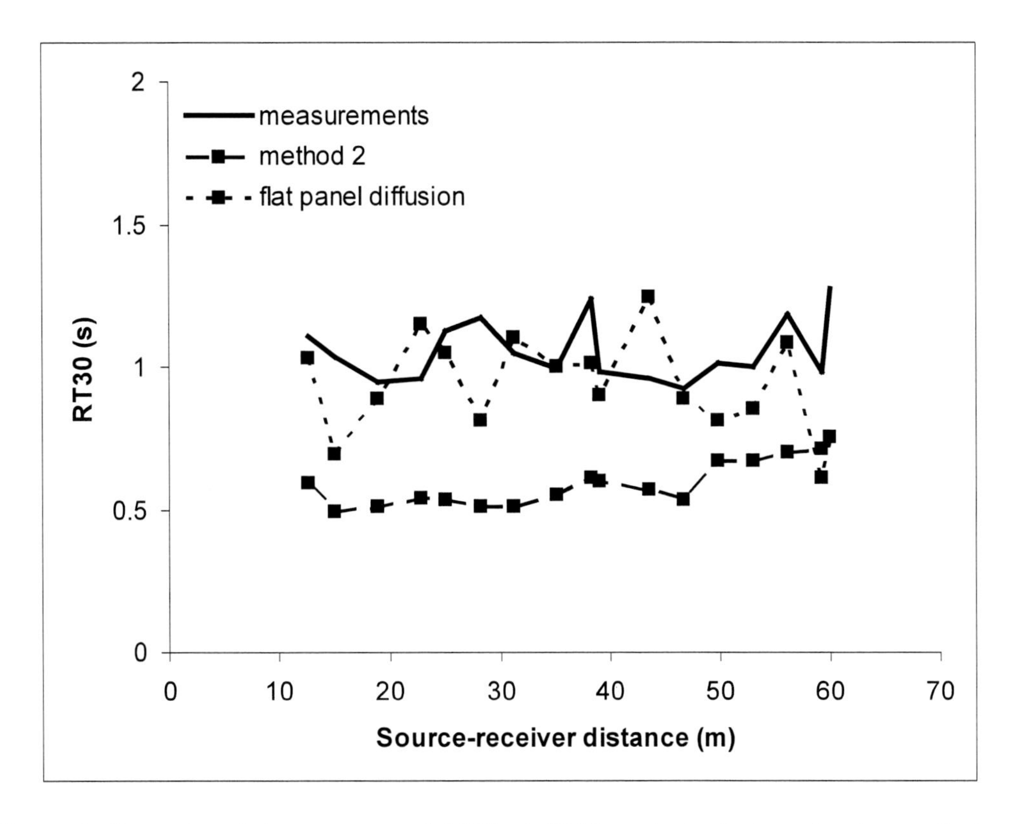

Figure 4. Comparison between measured RT30 at 500Hz and simulation using two methods of considering diffusion for the theatre of Epidaurus.

From the subjective evaluation, carried out during two performances of ancient drama in the theatre of Epidaurus, namely 'Electra' and 'Thesmophoriazousai' [Chourmouziadou and Kang, 2006a], medium sound levels, depending on the actor's position and ability, short but identifiable reverberation, no delayed reflections, noisy background, and 'normal' speech intelligibility were observed. Sudden and intense noises were perceived as more irritating than insects and animals. The uniqueness of the theatre in providing intimacy, regardless of the vast crowd attending, was also indicated [Chourmouziadou, 2007].

The effects of wind and temperature on the acoustics, as previously discussed, have been studied by Cremer [1975] and Goularas [1995] particularly for this theatre. Cremer indicated that air moves from the actor to the audience, while the steep inclination allows for good sound distribution from the actor situated close to the skene-building, with the help of the orchestra's reflection. Goularas investigated the distribution of sound in terms of the temperature increase towards the last rows.

Effects of the Stage Building and Scenery

Whilst considerable research has been carried out on the acoustics of ancient theatres in their original and contemporary conditions, there is very limited consideration on the acoustic impact of temporary scenery design, used for the contemporary performances of ancient drama [Chourmouziadou and Kang, 2004].

Creation of the Stage Building

There was no scenery or skene-building in early Greek theatres. The first attempt to create a specific architectural form of stage was perhaps in the theatre of Poliochni in Lemnos, where a rectangular platform was added to a circular orchestra made of soil. Behind the stage there was a simple structure that was used for changing costumes. In front of the stage some form of painted scenery was installed [Tsouchlou and Baharian, 1985].

It was also suggested that a skene-building was erected in the theatre of Dionysus in Athens, certainly earlier than 458 B.C., due to the need of a palace for the Orestean trilogy of Aeschylus [Allen, 1963]. For the action of one of the tragedies ('Choephori') it was necessary to use several doors, indicating that the skene-building was a substantial structure of considerable size. The addition of such an important feature was significant for drama.

The skene-building was rectangular in plan, with or without projections (at the front, sides or rear). The central part was called *mesoskenion* (μεσοσκήνιον), whose length was usually equal to the orchestra's diameter with the surrounding passageway. In later theatres there were no lateral projections but the length of the skene was increased to one and a half orchestral diameters, while in Roman theatres it was double the diameter of the orchestra [Marquand, 1909].

Greek Theatre Productions in the 19th and 20th Centuries

In Greece mainland, after four centuries of Turkish occupation, theatre started to blossom in the 19th century. In 1860-1870 there were several attempts for the revival of ancient Greek drama. One of the plays performed at the odium of Herodes Atticus in Athens was 'Antigone' by Sophokles for the wedding of George I and Queen Olga. The scenery was archaic but neo-classicist at the same time.

In the 1880s theatrical scripts, adaptations of ancient Greek plays, started evolving. This was accompanied by improvements in stage design. By that time there were evening performances with artificial lighting, with the use of candles and oil lamps. The beginning of the 20th century was very important for the technical characteristics of the theatres: oil lamps were abandoned and electrical lighting was established.

However, the public did not seem to enjoy these performances. Instead they preferred popular theatre, farces and melodrama. In 1927, during the celebrations in Delfoi, 'Prometheus Bound' was performed, receiving negative criticism [Department of Theatrical Studies, University of Athens, 1999]. Figure 5 shows the stage design, created by the sculptor Foskolos. Influenced by Appia, a pioneer in skenograghy with emphasis in sculptural scenery and lighting [Patrikalakis, 1984], the scenery was considered melodramatic and unsuccessful both aesthetically and functionally. In terms of acoustics, however, the irregular shape and relatively hard materials for the background allowed sound diffusion for most of the theatre area, which was helpful for creating a relatively even sound field.

In the same year 'Ekabi' was performed in the stadium in Athens. The scenery, initially designed by two painters, Kastanaki and Spahi, was changed many times. The material was rather hard, which could be useful for sound reflection.

Figure 5. Prometheus Bound in Delfoi, directed by Eva Sikelianou in 1927. Adopted from [Department of Theatrical Studies, University of Athens, 1999].

Figure 6. Kontoleon's scenery design for the performances of 'Prometheus Bound' and 'Iketides' in Delfoi in 1930.

In the 1930's the 2nd Celebrations of Delfoi were organised, and 'Prometheus Bound' and 'Iketides' were performed. Kontoleon designed the scenery, as shown in Figure 6, which was simple and 'classically modern', functionally and aesthetically effective, by combining the abstract ideas of Appia with the landscape of Delfoi [Department of Theatrical Studies, University of Athens, 1999].

The performance of 'Electra' by Sophokles in 1938 was the first in the theatre of Epidaurus after its excavation. Klonis, the scenery designer, decided to minimise the scenery as much as possible. He used the remains of the ancient stage and the *thymele*, the altar at the centre of the orchestra, adding only some necessary objects for the performance, like vases. As expected, this scenery would not contribute to the acoustics compared to the contemporary state of the theatre.

By the beginning of the 2nd World War performances were unachievable for curfew and economic reasons and people of theatre were arrested and sometimes murdered. The 1940s began with the performances of 'Antigone' and 'Oedipus Tyrant' in the Odeum of Herodus Atticus with similar concepts in stage design – a multilayered stage for the appearance of the actors. Nevertheless, it was the high proscenium and the general layout of the specific performance space, with a high stage wall, that allowed this decision.

Since the 1950s a wide variety of ancient plays have been performed in ancient sites. The most important change though was the institution of the Festival of Epidaurus in 1955 by the National Theatre. This allowed Klonis to use an almost permanent stage design that represented the front of an ancient palace or temple, with a central and two side entrances [Kontogiorgi, 2000]. In front of it extra architectural parts were added or taken away, like columns, openings, stairs, according to the needs of each performance. Figure 7 presents sketches of the sceneries used for the plays of 'Ekabi' and 'Oresteia' in the ancient theatre of Epidaurus in 1955, performed in front of the wall of Troy and an ancient Greek city, respectively [Kontogiorgi, 2000].

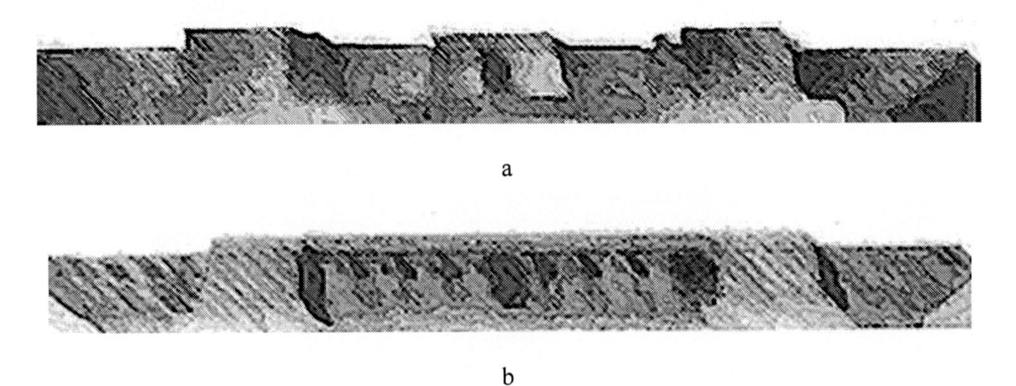

Figure 7. 1955 was the first year of the Festival of Epidaurus. a) Model of 'Ekabi', and b) Model of 'Oresteia'. Adopted from [Kontogiorgi, 2000].

Skenography is usually based on art movements. The turn of skenography into the painted background was sometimes characterised as 'abstract and meaningless' for the non-realistic appearance. Cubism created new problems in space representation with its crooked perspectives, while constructivism used new stage mechanisms and established the idea that the actors should move freely, without any limitations [Chourmouziadou and Sakantamis, 2006]. Nowadays skenography is the visual language of the performance and the audience's first impression. The contribution of skenography to the theatre is the aesthetic aspect of the action and the meaning of the play.

Classification of Scenery Design for Acoustic Examination

A variety of sceneries that have been designed for performing ancient tragedies and comedies in the last 40 years are presented in Figure 8. There are few characteristics that can classify the sceneries, including background, foreground, object, size and orchestra. In general, the sceneries must be adaptable to many theatres. The architecturally apparent and aesthetically impressive scenery can, through its acoustic attributes, become a determining factor for the sensory success of the performance. The sceneries can be classified into four generic categories:

1. Visually created by a background wall, with a variety of materials and heights (Bacchae, 1993; Medea, 1997; Prometheus Bound, 1983).
2. Represented by one large-size object, placed on the orchestra, in several shapes and volumes. This can partly create a background without interfering with the surroundings (Oedipus Tyrant, 1985; Eleni, 2006).
3. Composed by many small objects. These can be scattered in several places on the orchestra, or lined up to create the image of a wall. The variables are width and density, which in their combination may either provide a transparent effect, allowing the actors to move in-between the objects, or create the sense of an impenetrable boundary [Chourmouziadou and Sakantamis, 2006]. Several combinations in width and distribution can be used (Trojan Women, 1985; Iketides, 1964; Ecclesiazousai, 1993).
4. Concentrated on the orchestra's floor. There are two variables, the material and, in the case of a raised orchestra level, the height from the original orchestra's floor (Ploutos, 1985; Rhesus, 1981; Iphigeneia en Taurois, 2006).

The four generic categories of scenery design have been examined to identify their effect on the acoustic environment of ancient theatres [Chourmouziadou and Kang, 2007b]. Several variables were investigated from the viewpoint of acoustics in ancient theatres of Classic and Hellenistic types.

Figure 8. Classification of scenery design for acoustic examination. Sources of the scenery photos: [Kontogiorgi, 2000; Patsas,1995; Photopoulos, 1987].

Acoustic Effects

For Category 1, namely the scenery in the form of a background wall, it has been shown that soft materials used in the construction of the theatres lead to low values in SPL curves, although smooth decays are mainly found with hard materials and higher inclinations. Wide theatre layouts, in terms of the fan shape, produce comparatively long reverberation. The effect of the materials used for the scenery wall is generally trivial on the SPL and more significant on reverberation. Higher variations due to different background scenery wall characteristics are found in theatres built from hard materials. The height of the wall is particularly important for RT30, especially for the rear seats of the koilon, because it can contribute to a better sound distribution through early reflections. Figure 9 illustrates the RT30 for a variety of wall materials and heights. It can be seen that, compared to the theatre conditions without any scenery, the effect of the wall is high, even on SPL. The scenery is valuable for the occupied condition since it increases reverberation. On the other hand, the audience absorption is of less importance when there is no scenery. Linear walls are more effective, compared with other shapes, due to the sound distribution patterns.

A series of simulations has also shown that it is possible to maintain the area of the wall that 'functions' acoustically and, in the meantime, create a scenery that will take any form required.

In Category 2 four geometric objects, namely a cube, a sphere, a cylinder and a pyramid, have been applied to the theatre. The sphere and the cylinder increase the SPL due to the distribution of early reflections throughout the koilon. The cylinder provides the maximum number of reflections, using most of its surface area, and consequently produces long RT30 values. The reverberation with the cube and the sphere is shorter, while with the pyramid it is similar to the theatre with no scenery, due to the high number of lost reflections. In general, objects that provide diffusion distribute sound more evenly to the audience area, increase SPL and decrease RT30, and consequently bring higher STI and C80. Figure 10a shows the effects of different scenery shapes and surface conditions on the SPL.

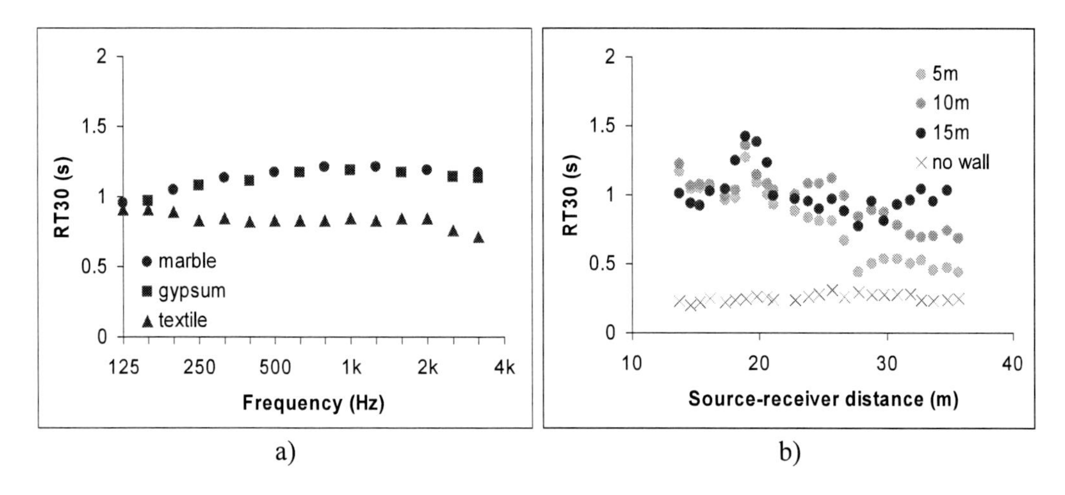

Figure 9. RT30 (s) results for Category 1 with two variables. a) Background wall materials, and b) Background wall heights.

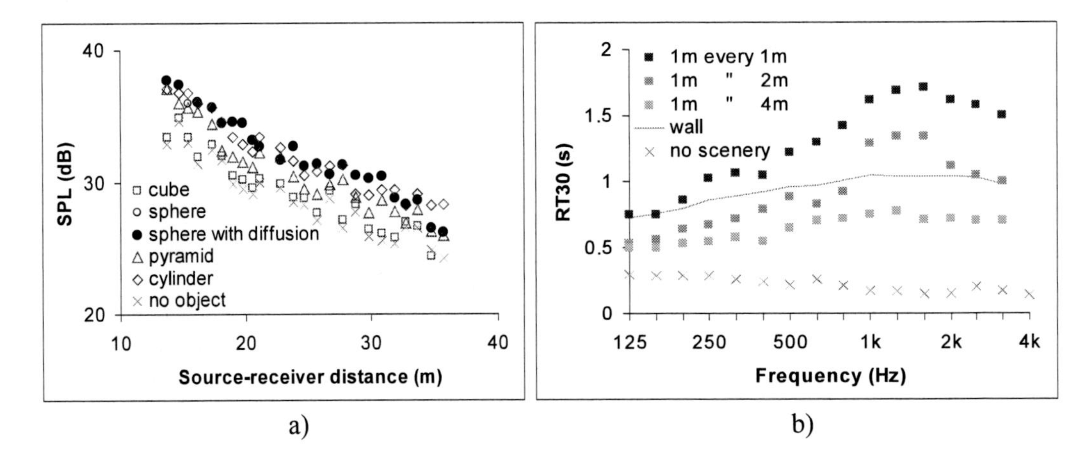

Figure 10. SPL (dB) and RT30 (s) results for Categories 2 and 3. a) Orchestra objects, and b) Panel arrangements.

In Category 3 small objects in the form of panels have been installed in the orchestra, in different densities, sizes and distributions. It is indicated that dense objects contribute to relatively long reverberation, as shown in Figure 10b, while the effect of edge diffraction is related to the number of the panels. Uneven SPL and RT30 distribution, when the objects are not dense, can be avoided by providing surface diffusion. Variations in acoustic indices are generally less when a receiver is closer to the source.

For Category 4 the parameters investigated are the orchestra material and the heights of the orchestra level. The SPL and the RT30 are not particularly influenced by material characteristics, since the reflections from the orchestra's floor are less in number than the ones from the koilon. Corresponding to the variations in absorption coefficients, the variations in SPL and RT30 are more distinct at high frequencies. On the contrary, the height of the orchestra's floor presents much more possibilities to a designer, since it can influence the acoustic indices considerably. Similar to the raised stage of the late Classic and the Hellenistic theatres, the increase in height can affect both SPL and RT substantially.

Because of the complexity scenery designs usually present, certain simplifications have been explored, by representing the scenery with a flat wall, divided into segments that stand for the solids and voids the original scenery creates. By applying the absorption coefficients of the scenery panels and of air to the corresponding wall areas, acceptable accuracy of acoustic results has been achieved.

Conclusions

The ancient performance spaces combine two contradictive functions: large occupancy and optimum visual and acoustic comfort. It has been shown that basic principles of acoustic design of outdoor performance spaces were used for the planning and construction of ancient theatres, including reducing background noise; elevating the direct sound and enhancing it with early reflections; and developing the space around the performers using geometric principles regarding the orchestra's radius, the inclination of the koilon and the dimensions of the stage building. It is interesting to note that the computer simulation suggests that the acoustics had been improved in the evolution of ancient Greek and Roman theatres. The SPL increased gradually along with theatre evolution, and there was a general increase in reverberation.

It has been indicated that, through appropriately placed reflectors or stage enclosures incorporated in the scenery application, the acoustic quality can be enhanced. Through the investigation of four generic categories of scenery design it has been revealed that the most effective scenery in terms of acoustics is a background wall placed on the orchestra, for an abstract representation of the ancient stage building. On the contrary, simple geometric shapes and soft orchestra materials are less effective.

The systematic examination of the computer simulation methodology for ancient Greek and Roman theatres has shown that with careful considerations of theatre representation and boundary absorption/diffusion conditions, good agreement can be obtained between simulation and measurement.

References

Allen, J. T. (1963) *Stage Antiquities of the Greeks and Romans and Their Influence.* New York: Cooper Square Publishers Inc.

Athanasopoulos, C. G. (1983) *Contemporary Theatre, Evolution and Design.* New York; Chichester: John Wiley & Sons.

Baldry, H. C. (1971) *The Greek Tragic Theatre.* London: Chatto & Windus.

Barron, M. (1993) *Auditorium Acoustics and Architectural Design.* London: E & FN Spon.

Bell, E. (1926) *Prehellenic Architecture in the Aegean.* London: G. Bell and Sons Ltd.

Beranek, L. L. (1962) *Music Acoustics & Architecture.* New York; London: John Wiley & Sons.

Beranek, L. L., & Hidaka, T. (1998). Sound absorption in concert halls by seats, occupied and unoccupied, and by the hall's interior surfaces. *Journal of the Acoustical Society of America,* **104** (6), 3169-3177.

Cailler, P., & Cailler, D. (1966) *Les Theatres Greco-Romains de Grece. Style,* no.1.

Canac, F. (1967) *L'acoustique des theatres antiques: Ses enseignements.* Paris: Èditions du Centre National de la Recherche Scientifique.

Chadwick, H. (1981) *Boethius: The Consolations of Music, Logic, Theology, and Philosophy.* Oxford: Clarendon Press.

Chourmouziadou, K. (2002) *Acoustic Simulation of Ancient Performance Spaces.* MPhil dissertation, University of Sheffield, UK.

Chourmouziadou, K. (2007) *Ancient and Contemporary Use of Open-air Theatres: Evolution and Acoustic Effect of Scenery Design.* PhD dissertation, University of Sheffield, UK.

Chourmouziadou, K., & Kang, J. (2002) Acoustic simulation of ancient performance spaces. In: *Proceedings of the International Conference on Auditorium Acoustics: Historical and Contemporary Design and Performance,* London, UK, 24 (4).

Chourmouziadou, K., & Kang, J. (2003) Effect of surface characteristics on the acoustic environment of ancient outdoor performance spaces. In: *Proceedings of the Research Symposium on Acoustic Characteristics of Surfaces: Measurement, Prediction and Applications,* Salford, UK, 25 (5).

Chourmouziadou, K., & Kang, J. (2004) Acoustic evolution of ancient theatres and the effects of scenery: A case study of the theatre of Mieza. In: *Proceedings of Acoustics 2004 Conference,* Thessaloniki, Greece, 351-358.

Chourmouziadou, K., & Kang, J. (2005) Soundscape of contemporary use of ancient Greek theatres with scenery design. In: *Proceedings of the 12th International Congress on Sound and Vibration,* Lisbon, Portugal.

Chourmouziadou, K., & Kang, J. (2006a) The contribution of ephemeral sceneries to the increase of reverberation of ancient theatres: In: *Proceedings of Acoustics 2006 Conference,* Heraklion, Greece, 183-190.

Chourmouziadou, K., & Kang, J. (2006b) Simulation of surface diffusion and diffraction in open-air theatres. In: *Proceedings of the 6th International Conference on Auditorium Acoustics,* Copenhagen, Denmark, **28** (2), 97-108.

Chourmouziadou, K., & Kang, J. (2007a) *Acoustic evolution of ancient Greek and Roman theatres. Applied Acoustics*, in press, also available in: <http://www.sciencedirect.com>.

Chourmouziadou, K., & Kang J. (2007b) Is scenery important for ancient drama? In: *Proceedings of the 19th International Congress on Acoustics* (ICA), Madrid, Spain.

Chourmouziadou, K., & Sakantamis, K. (2006) Non-apparent – Transparent – Apparent presence of scenery in the acoustics of ancient theatres. In: *Proceedings of the 2nd International Conference on Transparency and Architecture: Challenging the Limits,* Thessaloniki, Greece, 337-342.

Christensen, C. L., & Rindel, J. H. (2005a) Predicting acoustics in class rooms: Introducing a new method for scattering and diffraction in ODEON. In: *Proceedings of Internoise* 2005, Rio de Janeiro, Brasil.

Christensen, C. L., & Rindel, J. H. (2005b) A new scattering method that combines roughness and diffraction effects. In: *Proceedings of Forum Acusticum*, Budapest, Hungary.

Cremer, L. (1975). The different distributions of the audience. *Applied Acoustics,* **8**(3), 173-191.

Department of Theatrical Studies, University of Athens (1999). *Greek Skenographers, Costume Designers and Ancient Drama.* Athens: Ministry of Culture (in Greek).

Dinsmoor, B. W. (1950) T*he Architecture of Ancient Greece: An Account of Its Hhistory Development.* 3rd edition. London: B.T. Batsford Ltd.

Dörpfeld, W., & Reisch, E. (1896) *Das griechische Theater.* Athens: Barth und Von Hirst.

Egan, D. M. (1988) *Architectural Acoustics.* New York; London: McGraw-Hill.

Everest, F. A. (1994) *The Master Handbook of Acoustics.* London: TAB Books.

Gade, A. C., Lisa, M., Lynge, C., & Rindel, J. H. (2004) Roman theatre acoustics: Comparison of acoustic measurement and simulation from the Aspendos theatre, Turkey. In: *Proceedings of the 18th International Congress on Acoustics (ICA),* Kyoto, Japan, 2953-2956.

Gade, A. C., Lynge, C., Lisa, M., & Rindel, J. H. (2005) Matching simulations with measured acoustic data from Roman theatres using the ODEON programme. In: *Proceedings of Forum Acusticum,* Budapest, Hungary.

Gade, A. C., & Angelakis, K. (2006) Acoustics of ancient Greek and Roman theatres in use today. In: *Proceedings of the 4th Joint Meeting of ASA and ASJ,* Honolulu, Hawaii.

George, M. (1997) *The Acoustics of Greek and Roman Theatres.* MA dissertation, University of Sheffield, UK.

Goularas, D. (1995) *Acoustics of Ancient Theatres*. Diploma dissertation, Aristotle University of Thessaloniki, Greece (in Greek).

Guthrie, W. K. C. (1962) *A History of Greek Philosophy. Vol. 1: The Earlier Pre-Socratics and the Pythagoreans*. Cambridge: Cambridge University Press.

Hunt, F. V. (1978) *Origins in Acoustics: The Science of Sound from Antiquity to the Age of Newton*. London: Yale University Press Ltd.

Izenour, G. C. (1977) *Theatre Design*. New York; London: McGraw- Hill.

Kang, J. (2006) *Urban Sound Environment*. London: Taylor & Francis incorporating Spon.

Karadedos, G. (1994) Evidence on the scene-building of the Hellenistic theatre of Dion. In: *Archaeological Project in Macedonia and Thrace,* 1993, Thessaloniki, Greece. Thessaloniki Ministry of Culture, Ministry of Macedonia and Thrace and Aristotle University of Thessaloniki, 5, 157-170 (in Greek).

Kontogiorgi, A. (2000) *The Skenography of the Greek Theatre 1930-1960*. Thessaloniki: University Studio Press (in Greek).

Kurze, U., & Beranek, L. L. (1971) Sound propagation outdoors. In: Beranek, L. L., *Noise and Vibration Control*. New York: McGraw-Hill Book Company.

Lisa, M., Rindel, J. H., & Christensen, C. L. (2004) Predicting the acoustics of ancient open-air theatres: The importance of calculation methods and geometrical details. In: *Proceedings of the Joint Baltic-Nordic Acoustics Meeting*, Mariehamn, Åland.

Lisa, M., Rindel, J. H., Christensen, C. L., & Gade, A. C. (2005) How did the ancient Roman theatres sound? In: *Proceedings of Forum Acusticum,* Budapest, Hungary. 2179-2184.

Mapp, P. A. (2000). The acoustics of the outdoor Shakespeare theatre – Rutland, UK. *Journal of the Acoustical Society of America*, **108** (5), 2566.

Marquand, A. (1909) *Greek Architecture*. New York: The Macmillan Company.

Mparkas, N. K. (1992) *Ancient Greek Theatre: Design and Function*. PhD dissertation, Dimokriteio University of Thrace, Greece (in Greek).

Mparkas, N. K. (2004) Acoustic comfort in the contemporary use of ancient theatres: Sound protection and scenography applications. In: *Proceedings of Mild Interventions for the Protection of Historic Constructions,* Thessaloniki, Greece. (in Greek).

Mparkas, N. M. (2006) The acoustic function of the stage roof of the ancient Greek theatre: Were the gods sound intense in ancient drama? In: *Proceedings of Acoustics 2006 Conference,* Heraklion, Greece, 191-198 (in Greek).

O'Meara, D. J. (1989) *Pythagoras Revived: Mathematics and Philosophy in Late Antiquity*. Oxford: Clarendon Press.

Paini, D., Gade, A. C., & Rindel, J. H. (2006) Is reverberation time adequate for testing the acoustical quality for unroofed auditoriums? In: *Proceedings of the 6th International Conference on Auditorium Acoustics,* Copenhagen, Denmark. 66-73.

Patsas, G. (1995) *Costumes – Stage Design*. Athens: Ergo Publications (in Greek).

Patrikalakis, P. (1984) *History of Skenography: 19th-20th Century*. Athens: Aigokeros (in Greek).

Photopoulos, D. (1987) *Skenography in the Greek Theatre*. Athens: Emporiki Bank of Greece Publications (in Greek).

Pöhlmann, E. (1995) *Studien zur Buhnendichtung und zum theaterbau der antike*. Frankfurt: Peter Lang.

Rehm, R. (2002) *The Play of Space: Spatial Transformation in Greek Tragedy*. Princeton: Princeton University Press.

Rindel, J. H. Gade, A. C., & Lisa, M. (2006) The virtual reconstruction of the ancient Roman concert hall in Aphrodisias, Turkey. In: *Proceedings of the 6th International Conference on Auditorium Acoustics, Copenhagen, Denmark,* **28** (2), 316-323.

Robertson, D. S. (1979) *Greek and Roman Architecture.* 2nd edition. Cambridge: Cambridge University Press.

Shankland, R. S. (1973). Acoustics of Greek theatres. *Physics Today,* **26**, 30-35.

Simpson, F. M. (1956) *Ancient and Classical Architecture: Vol.1. Simpson's History of Architectural Development.* London: Logmans, Green and Co.

Tsouchlou, D., & Baharian, A. (1985) *The Skenography in Neo-Hellenic Theatre.* Athens: Apopsis (in Greek).

Varopoulou, E. (1991) In: *Theatres architectural creation in the created: Special edition for Manos Perrakis' participation in the international exposition Prague Quadrennial of Theatre Design and Architecture 1991.* Athens: M. Perrakis (in Greek).

Vasilantonopoulos, S. I., & Mourjopoulos, J. N. (2003). A study of ancient Greek and Roman theatre acoustics. *Acta Acustica united with Acustica,* **89**, 123-136.

Vasilantonopoulos, S. I., Zakynthinos, T., Chatziantoniou, P., Tatlas, N., Skarlatos, D., & Mourjopoulos, I. (2004) Measurements and analysis of the acoustics of the ancient theater of Epidauros. In: *Proceedings of Acoustics 2004 Conference,* Thessaloniki, Greece, 359-366 (in Greek).

Vasilantonopoulos, S. I., & Mourjopoulos, J. N. (2004) Acoustic simulation and analysis of the open ancient theatres. In: *Proceedings of Acoustics 2004 Conference,* Thessaloniki, Greece, 367-373 (in Greek).

Vasilantonopoulos, S. I., Mourjopoulos, I., Chatziantoniou, P., & Zarouchas, T. (2006) Acoustics of the Roman Odeion of Patras. In: *Proceedings of Acoustics 2006 Conference,* Heraklion, Greece, 199-206 (in Greek).

Vitruvius (1st century B.C.) *The Ten Books on Architecture.* Translated from the Latin by M. H. Morgan (1914), London; Cambridge, Mass: Harvard University Press.

Von Gerkan, A., & Müller-Wiener, W. (1961) *Das theater von Epidauros.* Stuttgart: W. Kohlhammer Verlag.

Wiles, D. (1997) *Tragedy in Athens: Performance and Theatrical Meaning.* Cambridge: Cambridge University Press.

In: New Research on Acoustics
Editor: Benjamin N. Weiss, pp. 243-283
ISBN: 978-1-60456-403-7

Chapter 7

HYBRID PASSIVE-ACTIVE ABSORPTION OF WIDEBAND NOISE

Pedro Cobo and María Cuesta
Instituto de Acústica, CSIC. Serrano 144, 28006 Madrid, Spain

Abstract

Porous and fibrous materials provide sound absorption within a frequency band which lower limit depends mainly on its thickness. In general, low frequency absorbers require such a large thickness than they are not installed in practice, except perhaps for large anechoic chambers. Active systems, on the other hand, allow to control the input impedance of multilayer absorbers, this affording absorption in the low frequency range. Combining appropriately the properties of the passive material with those of the active system, it is possible to design efficient absorbers for broadband noise, including low frequencies, with a reduced thickness. These systems are named hybrid passive-active absorbers. This chapter describes different implementations of such hybrid absorbers using porous and microperforated panels (MPP) as the passive material. Both *pressure-release* and *impedance-matching* are analyzed as the active control condition. Experimental results measured both in 1D (impedance tube) and 2D (anechoic chamber), which validate the predictions of the theoretical model, are presented.

I. Introduction

Passive control has been traditionally used to reduce the noise that inhabitants are exposed to daily. The weight and/or size requirements for passive control at low frequencies are often prohibitive. The demand for noise reduction, especially at low frequencies, motivates the use of active control (Nelson and Elliot, 1992; Hansen and Snyder, 1997; Cobo, 1997; Elliott, 2001). Active control is a complementary, rather than alternative, technique to passive control of noise. Passive control provides noise reduction at medium and high frequencies with moderate performance-to-cost ratio, whereas active control is capable of canceling low frequency noise. Therefore, a broadband noise problem requires a hybrid passive-active noise control solution.

Noise in rooms is an example requiring a broadband solution. Passive absorption can be achieved using a two-layer system consisting of a fibrous layer in front of an air layer, backed by a rigid wall. Depending on the thickness of the porous and air layers and the constitutive properties of the porous layer, such a two-layer system provides high absorption at medium and high frequencies. The maxima of the absorption curve are at those frequencies for which the air layer thickness is an odd-integer multiple of the quarter-wavelength (*λ/4 resonance absorber)*. The absorption curve can be moved to lower frequencies by increasing the thickness of the air layer. Thus, although fully passive low frequency absorbers can be designed, the resulting liner is too bulky. A solution to this problem is to design the two-layer passive system to yield high absorption above medium frequencies and complement the low frequency range with an active system. Paul Lueg already proposed to construct walls as vibrated membranes to silence rooms in 1933.

An active controller drives an actuator with a signal synthesized from the noise signal according to some performance objective and control strategy. Previous authors have discussed active absorption with different control strategies. Guicking and Lorenz (1984) described the active equivalent of the λ/4 resonance absorber by replacing the rigid wall with a rigid piston driven to achieve zero sound pressure at a microphone on the rear face of the porous layer. Using an analogue feedback controller, they accounted absorption coefficients of 0.6-0.7 in the frequency band 200-500 Hz. Guicking *et al.* (1985) applied adjustable electronic to afford active control of the acoustic impedance at the end of a standing wave tube. This controller afforded active absorption below 1000 Hz. Furstoss *et al.* (1997) demonstrated that minimizing the sound pressure at the entrance of the porous layer (*pressure-release condition*) gave absorption coefficients above 0.9 at frequencies between 200 and 900 Hz.

Beyene and Burdisso (1997) demonstrated that active absorption can be obtained also by canceling the reflected plane wave in the air layer (*impedance-matching* condition). They compared experimentally the two control conditions on a two layer hybrid passive-active absorbing system (pressure-release and impedance-matching). Note that to implement the impedance-matching condition requires two microphones in the air gap, together with a wave deconvolution circuit, to separate the plane waves propagating down- and up-stream in the air layer. The experimental results of Beyene and Burdisso, further corroborated by Smith *et al.* (1999), showed that the impedance-matching condition yielded higher hybrid passive-active absorption than the pressure-release condition of broadband noise. Nevertheless, the authors recognized that a confirmation of this experimental result would require an analytical model including the physical properties of the absorption material. Cobo *et al.* (2003) carried out such a theoretical model which allows comparing both conditions. They concluded that the performance of the hybrid passive-active absorber depends on ratio of the flow resistance to the air acoustic impedance. Thus, the pressure-release condition provides higher values of absorption than the impedance-matching condition when the flow resistance of the porous layer matches the acoustic impedance of air. On the other hand, the impedance-matching condition yields more absorption when the flow resistance of the porous layer is smaller than about 0.7 times the acoustic resistance of air. Using two-layer system consisting on a 3 cm thick porous layer and a 7 cm thick air cavity, they obtained an average absorption coefficient (with the pressure release condition) of 0.97 in the frequency band from 190 to 1700 Hz.

The total thickness of the hybrid system can be further decreased by using microperforated panels (MPP) as the absorbent material (Cobo *et al.*, 2004; Cobo and Fernández, 2006). A MPP consists of a thin sheet panel perforated with many submillimeter holes distributed along its surface (Maa, 1987; 1998). Such MPPs can be designed to have the high acoustic resistance and low acoustic mass reactance required for broadband sound absorption without additional porous material. Thus, this absorber can be made lightweight and inexpensive, producing also less health-related concerns than fibrous materials (Kang and Fuchs, 1999).

The panel thickness should be close to the diameter of the holes to obtain a MPP with proper absorption properties (Maa, 1987). Therefore, such rather thin (submillimetric) panels could be susceptible to suffer from structural managing problems. Pfretzschner *et al.* (2006) proposed an alternative strategy to develop more structurally manageable MPPs. They named this alternative a Microperforated Insertion Unit (MIU). A MIU can be obtained by combining two perforated panels with appropriate constitutive parameters. Each one operating individually provides rather bad absorption performance. One of them, the supporting plate, is too thick and has perforations too large. The other, the micrometric mesh, is too thin and has too high perforation ratio. However, both can be combined to produce absorption in two or three octaves. Since the resulting MIU has the thickness of the supporting plate, it lacks of the mechanical constraints of the equivalent MPP. Cobo *et al.* (2004) designed a MIU with a supporting panel 1 mm thick, with perforation ratio of 10 %, which provided an average absorption coefficient of 0.82 in the frequency bandwidth 100 to 1600 Hz. Cobo and Fernandez (2006) increased the average absorption coefficient to 0.94 designing a MIU with a supporting panel of thickness 1 mm and perforation ratio 23 %.

Most of the published papers on hybrid passive-active absorption use a loudspeaker as secondary source. The hybrid passive/active absorber can be made even thinner by replacing the loudspeaker by a plate actuator (Cuesta *et al.*, 2006; Sellen *et al.*, 2006). Experimental results demonstrated that is possible to obtain hybrid passive-active absorption of broadband noise using a thin plate moved by two piezoceramics as the actuator. For instance, a 4.6 cm thick hybrid absorber, made up with an MIU as the passive system and a plate actuator as secondary source, provided an average absorption of 81 % in the frequency range from 100 Hz to 1600 Hz (Cuesta *et al.*, 2006).

The main goal of this chapter is to analyze the fundamentals of the design of hybrid passive-active absorbers, using both porous materials and MPPs as the passive absorbent. A theoretical model is presented which allows comparing the performance of such systems as a function of the constitutive/geometrical properties of the system and the control conditions. The model provides the plane-wave absorption coefficient at both normal and oblique incidence. Then, a discussion of the comparison of the measurement methods of the absorption coefficient in an impedance tube and in an anechoic chamber is provided. Finally, experimental results which validate the model both in an impedance tube (normal incidence) and in an anechoic room (oblique incidence) are given.

II. Plane Wave Model at Normal Incidence

Let's consider the impedance tube sketched in Figure 1. A loudspeaker at the left side of the tube (not shown in the Figure) generates plane waves which propagate downstream, A_i, and upstream, B_i, in the i layer. The passive absorber, at the opposite side of the tube, consists on a porous layer, Figure 1a, or an MPP, Figure 1b, in front of an air layer of thickness D. The porous layer of thickness d is characterized by an acoustic impedance, Z_a, and propagation constant, Γ_a. The MPP is defined by its acoustic impedance, Z_m. Let Z_0 and k_0 be the acoustic impedance and the wavenumber, respectively, of the air filling the tube. In the passive case, the right termination of the tube is a rigid wall. When this termination is replaced by a rigid piston with velocity V_p, the system becomes active. In the active case, this piston moves according the control condition which will be analyzed in the following. The input impedances to the air layer and the absorber are Z_2 and Z_1, respectively.

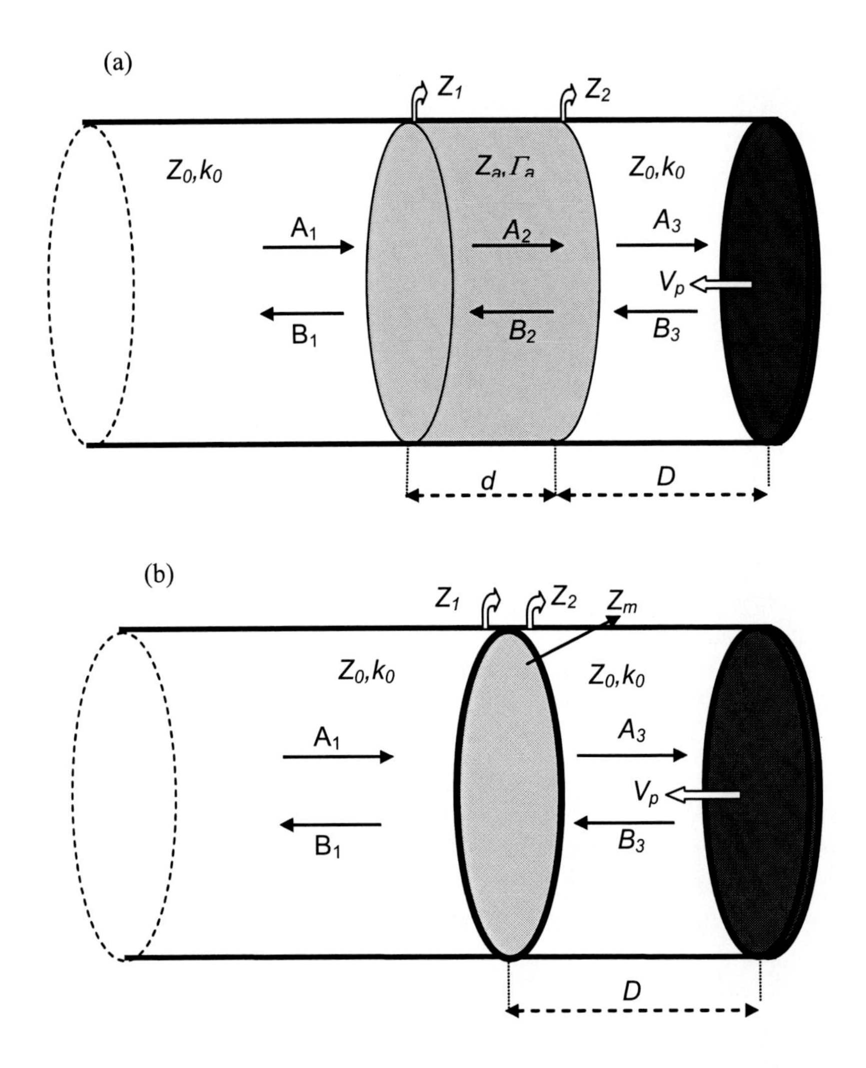

Figure 1. Plane waves in an impedance tube with a hybrid passive-active absorbing termination using a) a porous layer or b) a MPP as the passive material.

The absorption coefficient of the two-layer system of Figure 1 is (Cobo *et al.*, 2003, 2004)

$$\alpha = 1 - |R|^2, \tag{1}$$

where

$$R = \frac{Z_1 - Z_0}{Z_1 + Z_0}, \tag{2}$$

is the reflection coefficient,

$$Z_1 = \begin{cases} Z_a \dfrac{Z_2 \cosh(\Gamma_a d) + Z_a \sinh(\Gamma_a d)}{Z_a \cosh(\Gamma_a d) + Z_2 \sinh(\Gamma_a d)} & \text{porous layer} \\ Z_m + Z_2 & MPP \end{cases}, \tag{3}$$

is the acoustic impedance at the entrance of the absorber (Figure 1), and

$$Z_2 = Z_0 \frac{Z_0 V_p + 2B_2 e^{jk_0 d} \cos(k_0 D)}{Z_0 V_p + 2jB_2 e^{jk_0 d} \sin(k_0 D)}, \tag{4}$$

is the acoustic impedance at the entrance of the air layer (Figure 1).

The acoustic characterization of the passive element is required to complete the propagation model. In the porous layer case, an empirical model relating the acoustic impedance, Z_a, and the propagation constant, Γ_a, with the absorbing variable, $E=\rho_0 f/\sigma$, where σ is the flow resistivity of the porous material, ρ_0 is the air density, and f is the frequency, could be used. Although most of the work on porous and fibrous material assume the empirical model of Delany and Bazley (1970) its validity region, $0.01<E<1$, is not enough for active control, since the interest is focused at low frequencies where $E<0.01$. Therefore, the semi-empirical equations of Allard and Champoux (1992), will be assumed here, which are valid for any $E<1$, and provide similar results to those of Delany and Bazley in the E region where both models overlap. The equations of Allard and Champoux (1992) are

$$\Gamma_a = j2\pi f \sqrt{\rho(f)/K(f)} \tag{5a}$$

$$Z_a = \sqrt{\rho(f)K(f)}, \tag{5b}$$

where

$$\rho(f) = 1.2 + \left[-0.0364E^{-2} - j0.1144E^{-1}\right]^{1/2} \tag{6a}$$

is the complex density, and

$$K(f) = 101320 \frac{j29.64 + \left[2.82E^{-2} + j24.9E^{-1}\right]^{1/2}}{j21.17 + \left[2.82E^{-2} + j24.9E^{-1}\right]^{1/2}}, \tag{6b}$$

is the complex bulk modulus of the porous medium.

In the MPP case, Maa (1987, 1998) proposed the next equation for the acoustic impedance

$$Z_m = j\frac{\omega\rho_0 t}{p}\left[1-\frac{2\ J_1(x\sqrt{-j})}{x\sqrt{-j}\ J_0(x\sqrt{-j})}\right]^{-1}+\frac{\sqrt{2}\eta x}{pd}+j\frac{0.85\omega\rho d}{p}, \tag{7}$$

where η is the air viscosity coefficient, ($1.789\ 10^{-5}$ kg/m·s), ω is the angular frequency, t is the panel thickness, d is the hole diameter, p is the perforation ratio of the panel surface, and J_0 and J_1 are the Bessel functions of first kind and orders zero and one, respectively. The perforation constant, x, is defined as

$$x = d/\sqrt{\frac{4\eta}{\rho_0\omega}}, \tag{8}$$

and represents the ratio of the hole diameter to the viscous boundary layer thickness of the air inside the holes. For a large range of x values, the MPP has an acoustic resistance close to that of the air, and a mass reactance slightly smaller. An efficient absorber is obtained when its mass reactance is much smaller than the air acoustic resistance. This is provided by the air cavity behind the MPP, which greatly reduces the mass reactance of the system in the frequency range of interest.

The plane wave model described by Eqs. (1)-(8) allows to analyze both the passive and active performance of the two-layer absorber. The control condition can be implemented in the input acoustic impedance of the air cavity, Eq. (4), (Cobo *et al.*, 2003, 2004):

- Passive control, V_p=0. In this case $Z_2 = -jZ_0 \cot(k_0 D)$.
- Active control by pressure-release condition. In this case the acoustic pressure at the entrance of the air cavity is zero, and $Z_2 = 0$.
- Active control by impedance-matching condition. In this case the reflected component in the air cavity (B_2 in Figure 1a) is zero. Therefore, from Eq. (4), $Z_2 = Z_0$.

The above described model, Eqs. (1)-(8), allows to obtain the reflection coefficient of either a passive or active system as a function of geometrical and constitutive parameters of the passive system (thickness and flow resistivity in the case of a porous layer; hole diameter, panel thickness and perforation ratio in the case of a MPP; air cavity thickness), the frequency and the control condition. Mechel (1988) demonstrated the convenience of grouping these parameters in nondimensional variables which facilitate the graphic representation of the absorption coefficient (design charts). In the porous case, Mechel proposed the nondimensional variables $F=f\,d/c_0$ and $R=\sigma\,d/Z_0$. F represents the ratio of the porous layer thickness to the wavelength. R is the ratio of the porous layer flow resistance to the air acoustic impedance.

Figure 2 shows the design chart of a two-layer absorber with $D=2.5\ d$, for the passive, active by pressure-release, and active by impedance-matching, control conditions. The passive system provides maximum absorption for $R\approx1$ and $F\approx0.1$. The active control of the input impedance of the two-layer system increases meaningfully the absorption at low frequencies with both active control conditions. However, notice the distinct behavior of both active control conditions with the normalized flow resistance. With the pressure-release condition (Figure 2b), maximum absorption is obtained for $R\approx 1$, like the passive condition. With the impedance-matching condition (Figure 2c), on the other hand, the lower the flow resistance, the higher absorption is obtained. This behavior with the impedance-matching condition could be advanced, since in the limit of zero flow resistance, the porous layer disappears, and cancel the reflected wave, B_3, is equivalent to cancel the input reflection, this providing maximum absorption.

Figure 3 shows the contours of the differences between the active absorption provided by both control conditions, $\alpha_{\text{pressure-release}}-\alpha_{\text{impedance-matching}}$. The thick line denotes the zero contour. As it can be seen in this Figure, the pressure-release condition affords more absorption than the impedance-matching condition at low frequencies for $R>0.63$.

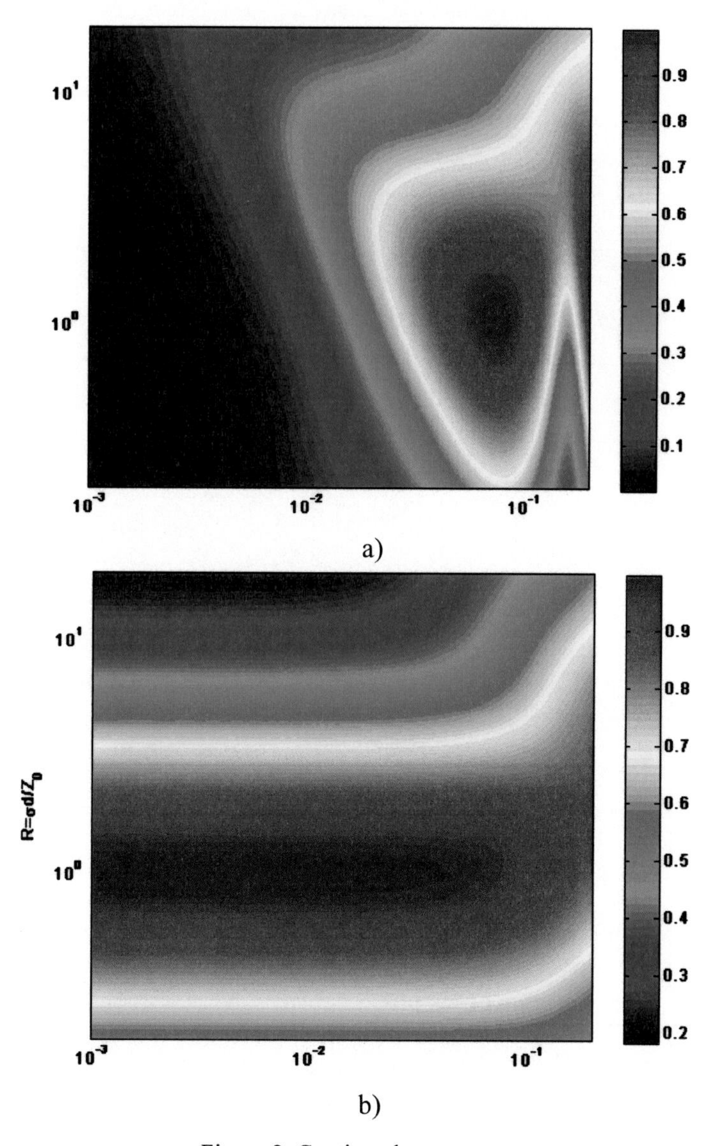

a)

b)

Figure 2. Continued on next page.

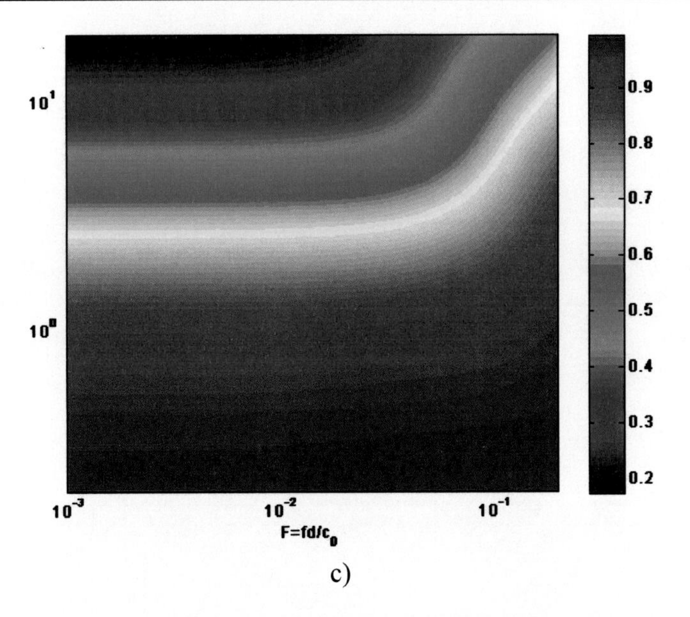

c)

Figure 2. Design charts of a two-layer porous absorber with D=2.5 d, with a) passive, b) active by pressure-release, and c) active by impedance-matching, control conditions.

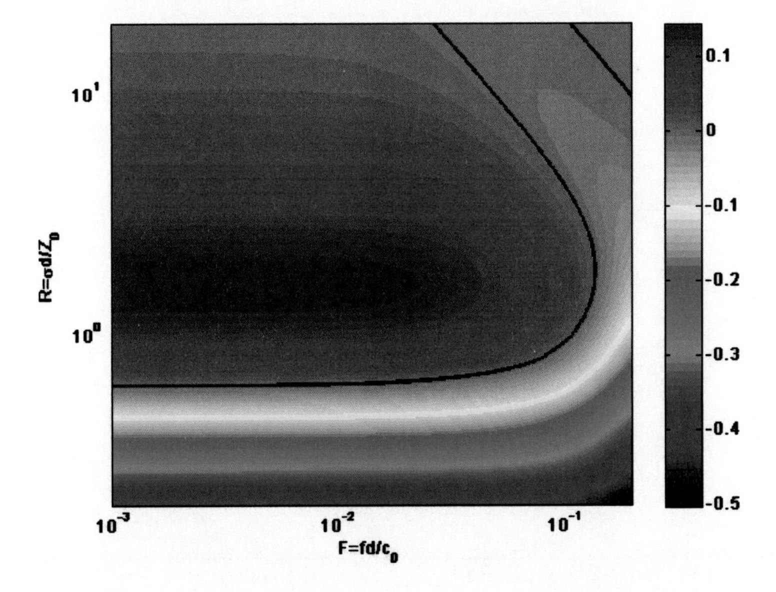

Figure 3. Contours of the differences between the design charts of the active absorber of Figure 2 for the pressure-release and impedance-matching control conditions. The thick line denotes the zero contour.

Classical absorption curves as a function of frequency can be obtained from the design charts by specifying the parameters of the passive system. As an example, Figure 4 shows the absorption curves of a two-layer system with σ=13500 N·s/m^4, d=3 cm and D=7 cm, under passive, active by pressure-release and active by impedance-matching control conditions.

The passive condition affords maximum absorption at about 835 Hz. Below 677 Hz, and above 1130 Hz, the active by pressure-release control condition gives more absorption than the passive condition. Also, below 1200 Hz, the pressure-release condition yields higher absorption than the impedance-matching condition.

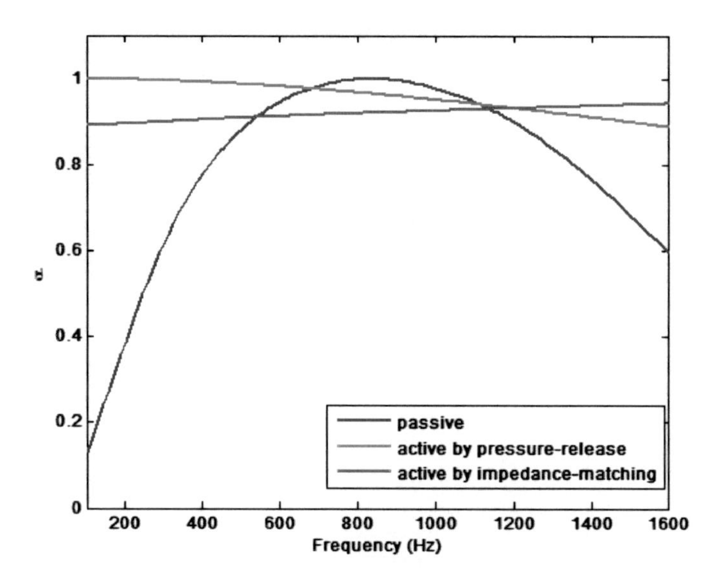

Figure 4. Absorption curves of a two-layer porous system with σ=13 500 N s/m^4, d=3 cm and D=7 cm.

Since the performance of practical active control systems is limited to low frequencies (Nelson and Elliott, 1992; Hansen and Snyder, 1997) the absorption excess of the active system over the passive one at high frequencies should not be taken into account. Therefore, if the active system is restricted to work below the frequency where the active and passive curves intersect each other (677 Hz, in the example of Figure 4), the overall passive-active system will afford high absorption in a wide band. This system is said to provide hybrid passive-active absorption.

In the case where the passive system is a MPP, one of the non-dimensional variables can be the perforation constant, x. Cobo *et al.* (2004) used the following as the second non-dimensional variable

$$\varsigma = \frac{\eta}{Z_0 d}. \tag{9}$$

The normalized acoustic impedance of the MPP as a function of these two nondimensional variables is

$$z_m = \frac{Z_m}{Z_0} = \frac{\varsigma}{p}\left\{\sqrt{2}x + j4x^2\left[0.85 + \left[1 - \frac{2\ J_1(x\sqrt{-j})}{x\sqrt{-j}\ J_0(x\sqrt{-j})}\right]^{-1}\right]\right\}, \tag{10}$$

and the air cavity impedance is

$$z_2 = \frac{Z_2}{Z_0} = -j\cot\left(\frac{\omega\rho D}{Z_0}\right) = -j\cot\left(\frac{4D\varsigma\, x^2}{d}\right). \tag{11}$$

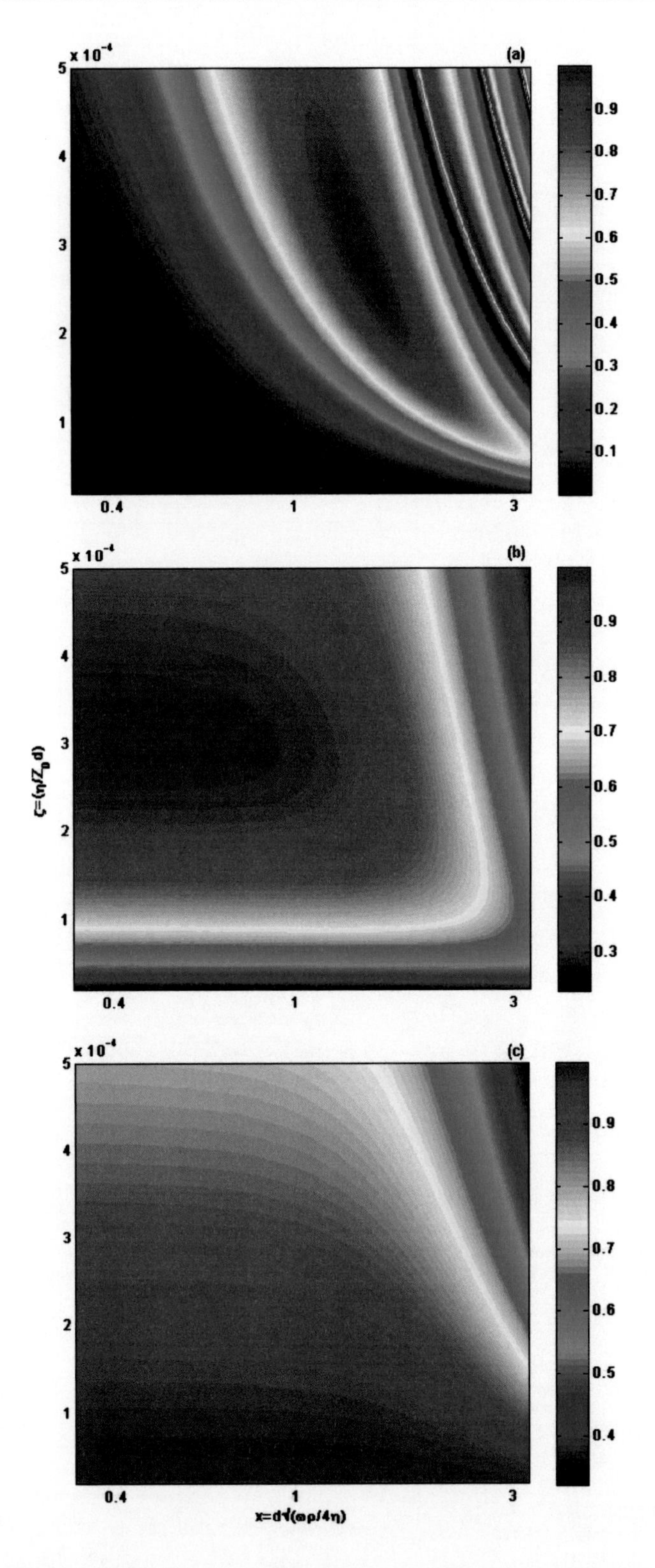

Figure 5. Design charts of a MPP absorber with p=1 %, d=0.1 mm and D=5 cm with a) passive, b) active by pressure-release, and c) active by impedance-matching control conditions.

Figure 5 shows the design charts of a MPP absorber with passive, active by pressure-release, and active by impedance-matching, control conditions for p=0.1 %, t=d=0.1 mm and D=5 cm.

The following can be concluded from Figure 5:

- The active system tends to move the absorbing band towards the low frequencies (lower perforation constant).
- The performance of the active system depends critically on the control condition. To illustrate this, Figure 6 shows the contours of the differences between the active absorption provided by both control conditions, $\alpha_{\text{pressure-release}}-\alpha_{\text{impedance-matching}}$. The pressure-release condition affords more absorption than the impedance-matching condition at low frequencies for $\varsigma > 0.00019$.

A classical frequency curve can be obtained by cutting the design chart of Figure 5 for a line parallel to the x axis, and converting it into a frequency axis. Figure 7 shows such a classical frequency curve for d=t=0.1 mm, p=1 % and D=5 cm. This MPP provides an absorption curve very similar to that of a porous layer, with a reduced thickness and without any material susceptible of health-related concern. The passive and active by pressure-release curves intersect at f=987 Hz. Thus, if the active control filter is low-passed at this frequency, the whole passive-active system would afford hybrid absorption close to 100 % in the frequency band from 100 to 1600 Hz.

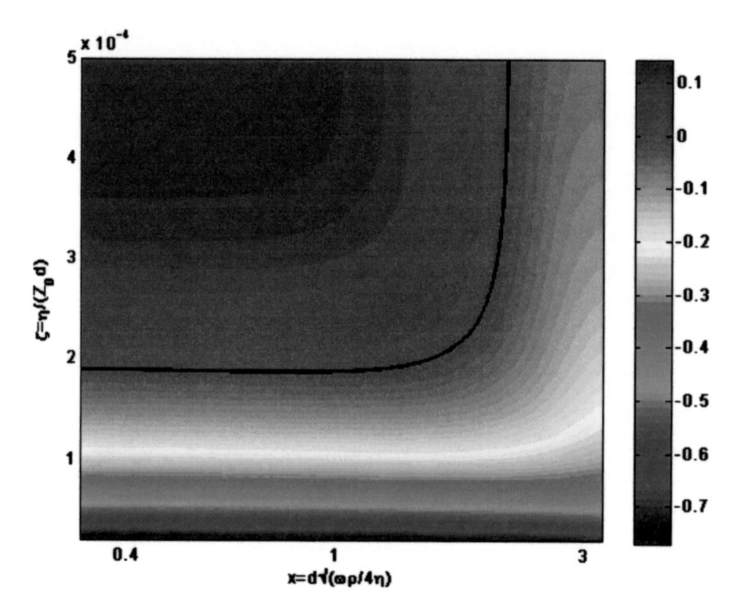

Figure 6. Contours of the differences between the design charts of the active absorber of Figure 5 for the pressure-release and impedance-matching control conditions. The thick line denotes the zero contour.

As mentioned in the Introduction, a MPP with thickness 0.1 mm is not rigid enough to be used as a practical absorber. Pfretzschner *et al.* (2006) proposed the design of a MPP with less structural concerns by assembling two MPPs, Figure 8. One of them, the supporting panel, has hole diameter, d_1, thickness, t_1, and perforation ratio, p_1. The other, the micrometric mesh, has hole diameter, d_2, thickness, t_2, and perforation ratio, p_2. When the micrometric mesh is glued in the rear face of the supporting panel (when $D_1 \approx 0$, or $D_2 \approx D$ in Figure 8), a new MPP is obtained, named a MIU by Pfretzschner *et al.* (2006). Its acoustic impedance is

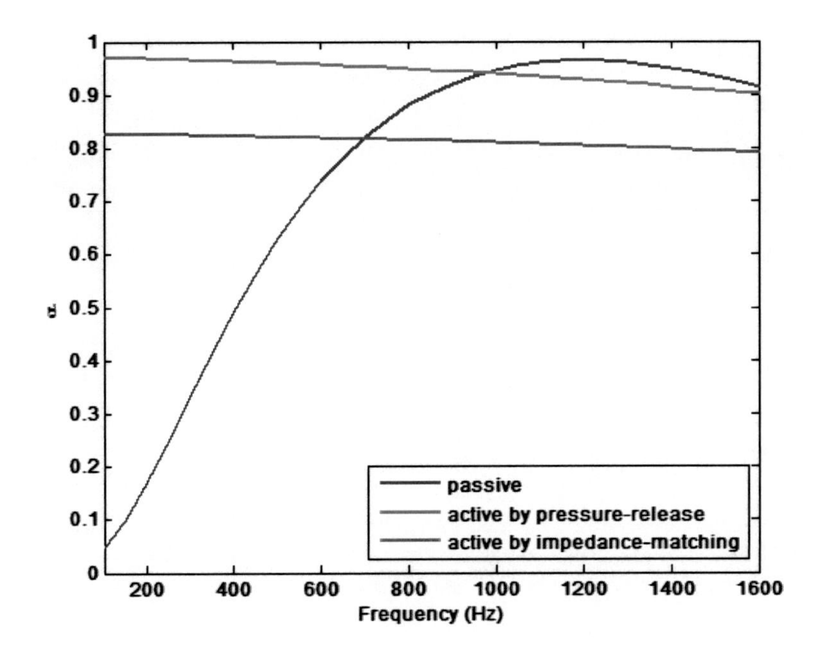

Figure 7. Absorption curves of a MPP with p=1 %, d=t=0.1 mm, and D=5 cm.

$$
\begin{aligned}
Z_m &= j\frac{\omega\rho t_1}{p_1}\left[1-\frac{2\ J_1(x_1\sqrt{-j})}{x_1\sqrt{-j}\ J_0(x_1\sqrt{-j})}\right]^{-1}+\frac{\sqrt{2}\eta x_1}{p_1 d_1}+j\frac{0.85\omega\rho d_1}{p_1} \\
&+j\frac{\omega\rho t_2}{p_2}\left[1-\frac{2\ J_1(x_2\sqrt{-j})}{x_2\sqrt{-j}\ J_0(x_2\sqrt{-j})}\right]^{-1}+\frac{\sqrt{2}\eta x_2}{p_2 d_2}+j\frac{0.85\omega\rho d_2}{p_2}
\end{aligned}, \tag{12}
$$

where

$$
x_1 = \frac{d_1}{\sqrt{\frac{4\eta}{\rho\omega}}}, \text{ and } \ x2 = \frac{d_2}{\sqrt{\frac{4\eta}{\rho\omega}}}. \tag{13}
$$

Therefore, the design of a MIU depends on the seven parameters $(d_1,t_1,p_1,d_2,t_2,p_2,D)$. In general, commercial micrometric meshes are used with fixed (d_2,t_2,p_2), so that the design depends only on the four parameters (d_1,t_1,p_1,D). Figure 9 shows the passive, active by pressure-release and active by impedance-matching absorption coefficients of a MIU with $(d_1,t_1,p_1,d_2,t_2,p_2,D)=(6\ mm,1\ mm,23\ \%,35\ \mu m,39\ \mu m,14\ \%,5\ cm)$. The MIU consists of a commercial micrometric mesh of $(d_2,t_2,p_2)=(35\ \mu m,39\ \mu m,14\%)$ glued to a supporting panel of $(d_1,t_1,p_1)=(6\ mm,1\ mm,23\ \%)$ in front of an air cavity 5 cm deep. Note as a practicable MIU, Figure 9, is able to afford an absorption curve similar to that of a MPP, Figure 7.

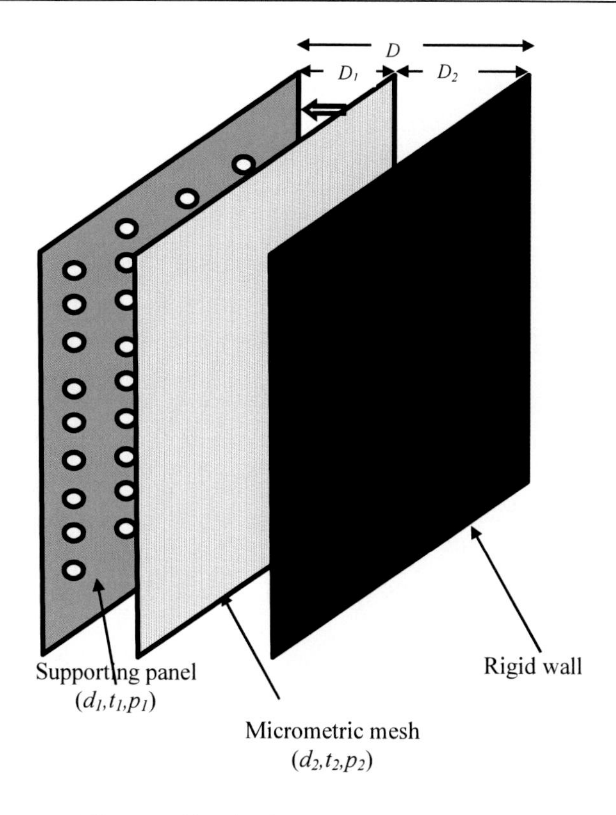

Figure 8. Sketch of the assembly of a MIU.

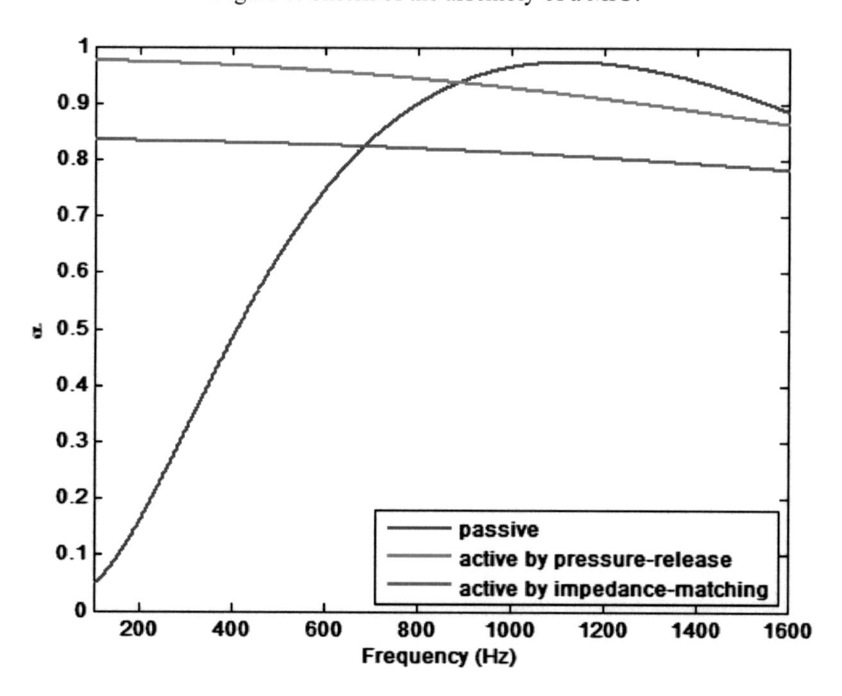

Figure 9. Absorption curves of a MIU with $(d_1,t_1,p_1,d_2,t_2,p_2,D)$=(6 mm,1 mm,23 %,35 μm,39 μm,14 %,5 cm).

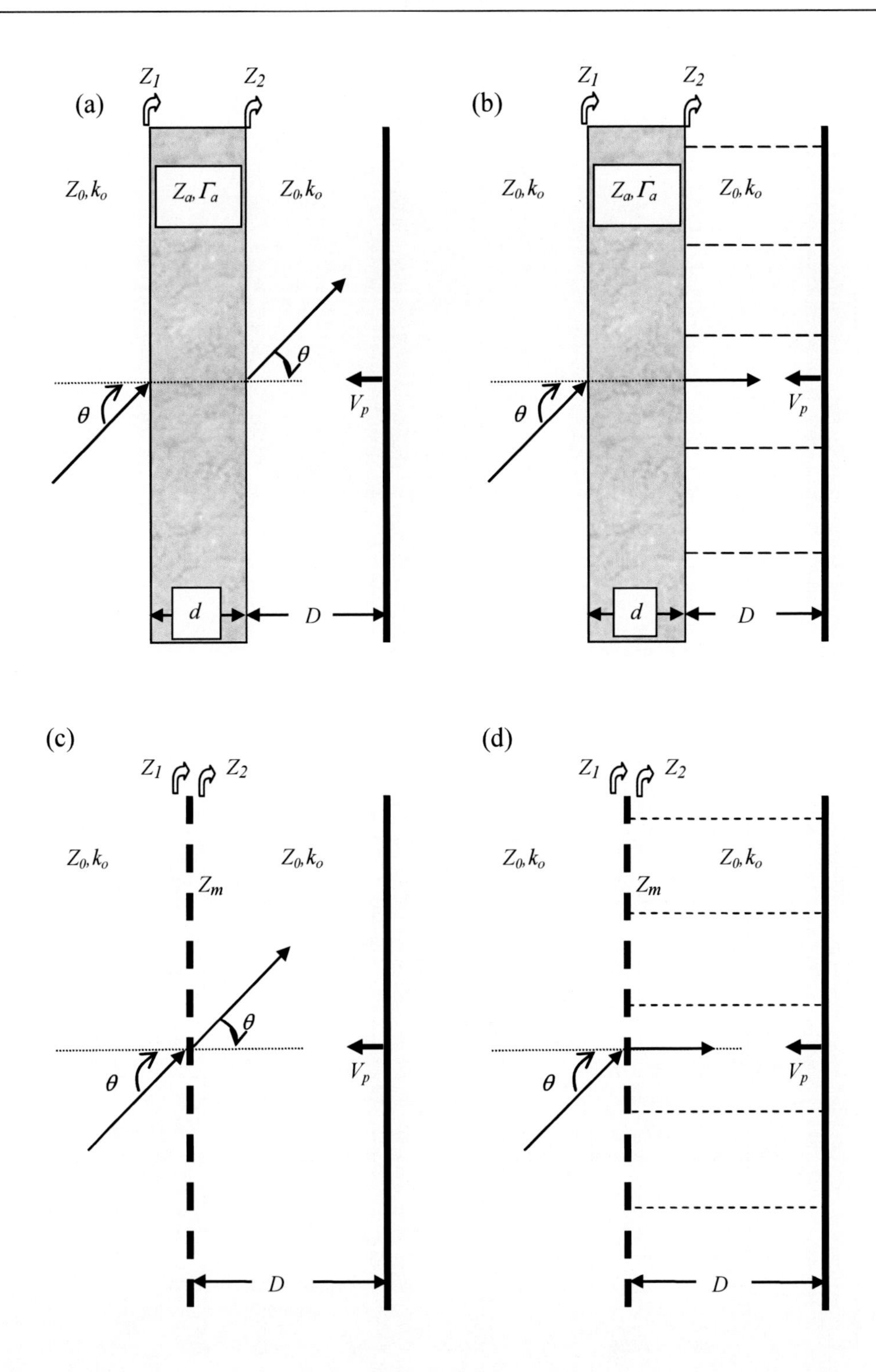

Figure 10. Plane waves at a (a,c) bulk-reacting and (b,d) locally-reacting two-layer porous (a,b) and MPP (c,d) system.

III. Plane Wave Model at Oblique Incidence

Figure 10 shows a sketch of either a two-layer porous or a MPP absorber, both bulk-reacting and locally-reacting, at oblique incidence. Plane waves propagate normally to the surface at locally-reacting absorbers. At bulk-reacting absorbers, waves propagate obliquely across the layers following the Snell law. The absorption coefficient at oblique incidence is

$$\alpha(\theta) = 1 - |R(\theta)|^2, \tag{14}$$

where θ is the incidence angle,

$$R(\theta) = \frac{Z_1 \cos\theta - Z_0}{Z_1 \cos\theta + Z_0}, \tag{15}$$

is the reflection coefficient at oblique incidence,
and,

$$Z_1 = \begin{cases} \dfrac{Z_a\Gamma_a}{q} \dfrac{Z_2 \cosh(qd) + \dfrac{Z_a\Gamma_a}{q}\sinh(qd)}{\dfrac{Z_a\Gamma_a}{q}\cosh(qd) + Z_2 \sinh(qd)} & \text{porous layer} \\ Z_m + Z_2 & MPP \end{cases}, \tag{16}$$

is the acoustic impedance at the entrance of the absorber. Z_a and Γ_a are the acoustic impedance and the propagation constant, respectively, of the porous layer (defined by Eqs. (5)-(6)), Z_m is the acoustic impedance of the MIU (defined by Eqs. (12)-13)),

$$q = \Gamma_a \sqrt{\frac{\Gamma_a^2 + k^2 \sin^2\theta}{\Gamma_a^2}}, \tag{17}$$

and

$$Z_2 = \begin{cases} \dfrac{Z_0}{\cos\theta} \dfrac{\dfrac{Z_0}{\cos\theta} V_p + 2B_2 e^{jk_0 d\cos\theta} \cos(k_0 D\cos\theta)}{\dfrac{Z_0}{\cos\theta} V_p + 2jB_2 e^{jk_0 d\cos\theta} \sin(k_0 D\cos\theta)} & bulk-reacting \\ Z_0 \dfrac{Z_0 V_p + 2B_2 e^{jk_0 d} \cos(k_0 D)}{Z_0 V_p + 2jB_2 e^{jk_0 d} \sin(k_0 D)} & locally-reacting \end{cases}, \tag{18}$$

is the acoustic impedance at the entrance of the air layer.

Cobo et al. (2006a) showed that usual absorbers perform as bulk-reacting. Locally-reacting absorbers can be obtained by partitioning the system normally to their surface (honeycomb absorbers). Therefore, in the following, the analyzed systems are assumed to be bulk-reacting absorbers. Equations (14)-(18) together with Eqs. (5)-(6) and Eqs. (12)-(13) allow to model a hybrid passive-active porous or MPP absorber at oblique incidence. As an example, Figure 11 shows the design charts of a porous absorber with d=3 cm, D=5 cm, σ=14000 N·s/m^4, as a function of F and θ, under passive, active by pressure-release and active by impedance-matching, control conditions. The behavior of the two-layer porous absorber with the incidence angle is rather complex. For the passive case, the maximum absorption is obtained at normal incidence (Figure 11a). For the active by pressure-release case (Figure 11b), the maximum average absorption coefficient is obtained for $\theta \approx 30°$.

For the active by impedance-matching case (Figure 11c), the maximum average absorption coefficient is obtained for $\theta \approx 50°$. Both active control conditions provide absorption close to zero for incidence angle close to 90°.

Figure 12 shows the design charts of a MIU absorber with $(d_1,t_1,p_1,d_2,t_2,p_2,D)$=(6 mm,1 mm,23 %,35 mm,39 mm,14 %,5 cm), as a function of f and θ, under passive, active by pressure-release and active by impedance-matching, control conditions. Like with the porous absorber, the behavior of the MPP absorber with the incidence angle is rather complex. For the passive case, the maximum absorption is obtained for $\theta \approx 40°$ (Figure 12a). For the active by pressure-release case (Figure 12b), the maximum average absorption coefficient is obtained for $\theta \approx 45°$. For the active by impedance-matching case (Figure 12c), the maximum average absorption coefficient is obtained for $\theta \approx 67°$. The absorption of both active control conditions tends to zero when the incidence angle tends to 90°.

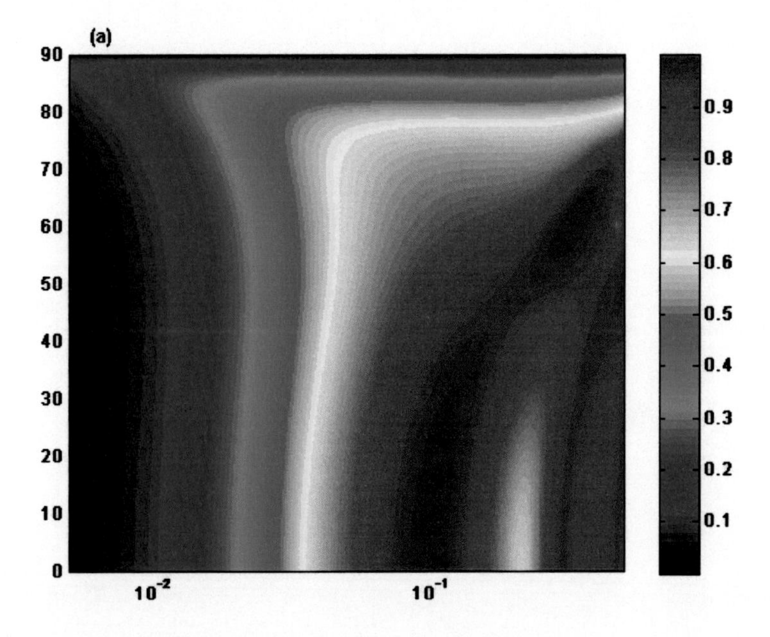

Figure 11. Continued on next page.

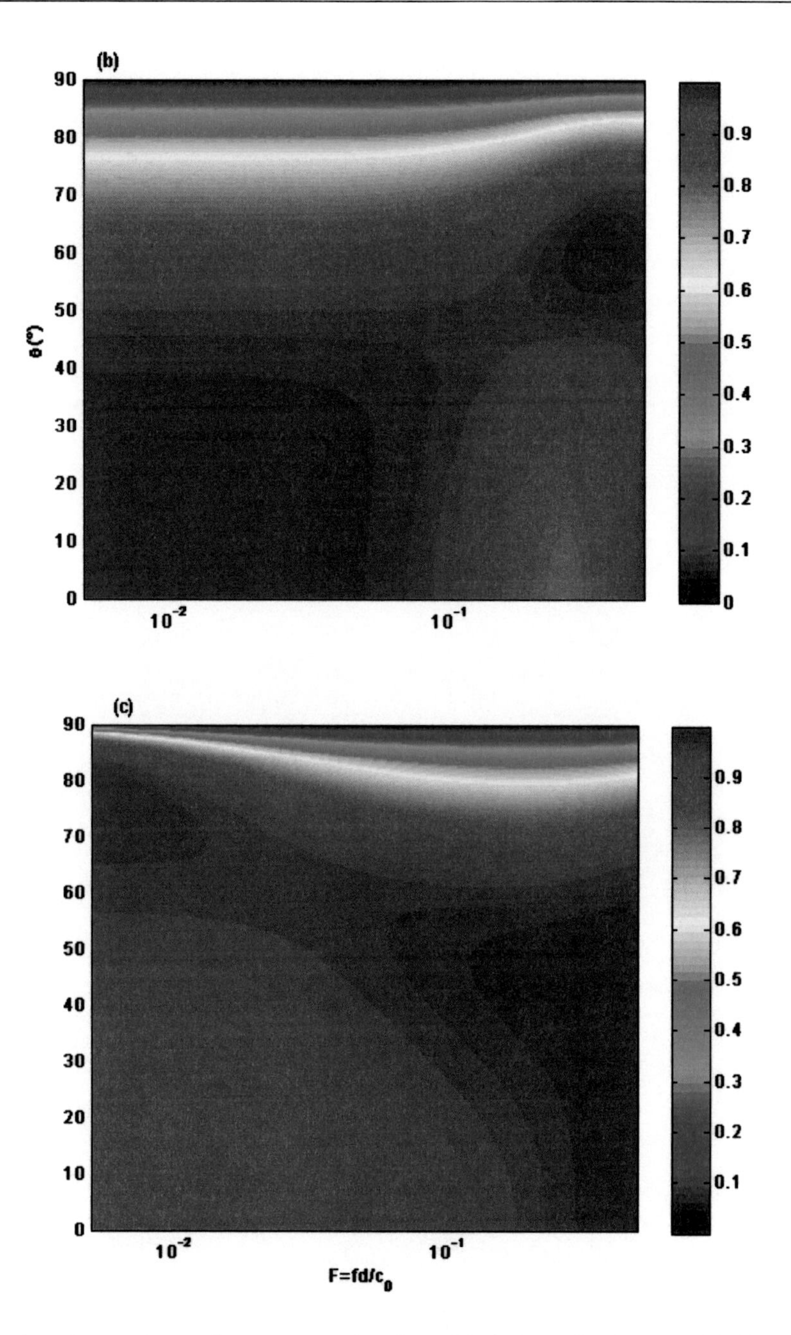

Figure 11. Design charts of a two-layer porous absorber with d=3 cm, D=5 cm, σ=14000 N·s/m^4, for a) passive, b) active by pressure-release, and c) active by impedance-matching, control conditions.

The absorption coefficient at any incidence angle, Eq. (14), can be integrated to compute the statistical absorption coefficient, α_{st},

$$\alpha_{st} = \int_0^{\pi/2} \alpha(\theta) \sin 2\theta d\theta \,. \tag{19}$$

Equation (19) represents the ratio of the energy absorbed by the sample and that incident of a pure diffuse field. Figure 13 shows the design charts for the statistical absorption of a porous absorber with d=3 cm and D=5 cm, under passive, active by pressure-release and active by impedance-matching, control conditions. The effect of the active control is similar to that at normal incidence. Namely, the active control increases the low frequency part of the absorption chart (lower F), the active control by the pressure-release condition affords higher absorption for the flow resistance close to the air impedance ($R\approx1.58$), and the active control by the impedance-matching condition performs the better for the lower values of R.

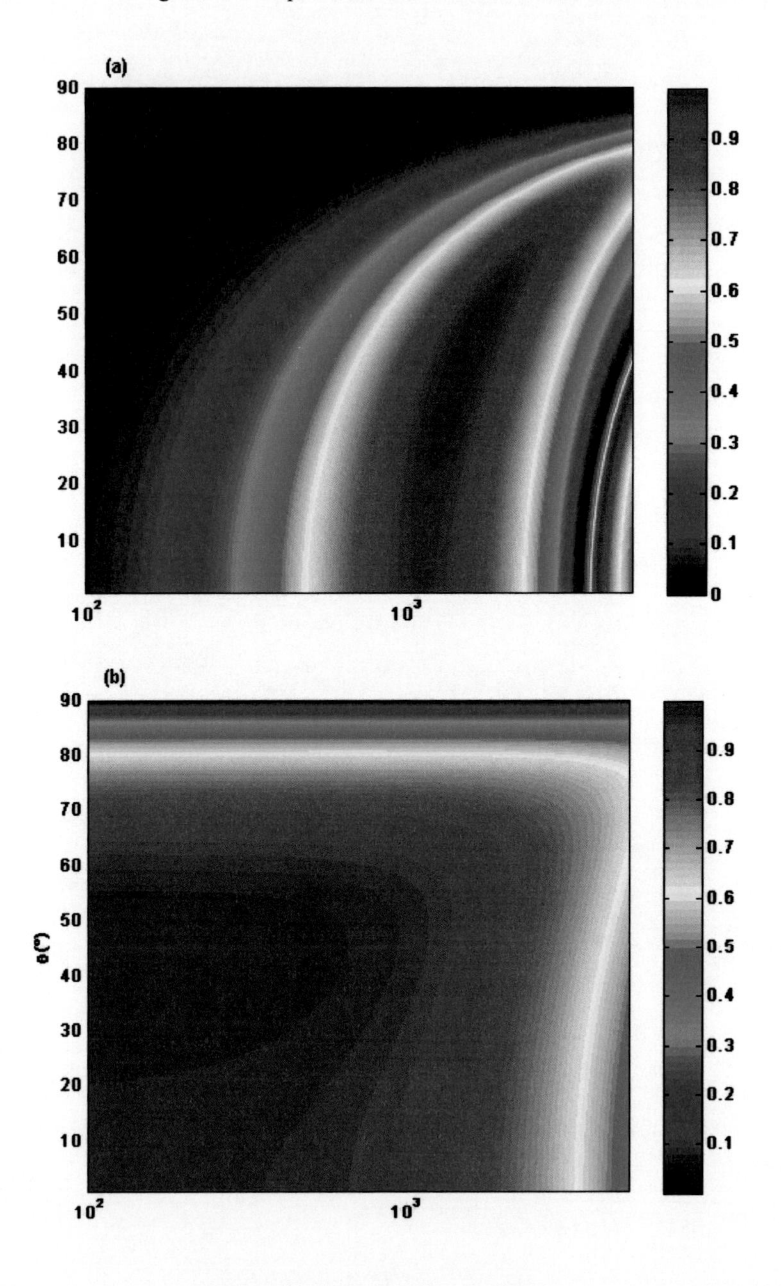

Figure 12. Continued on next page.

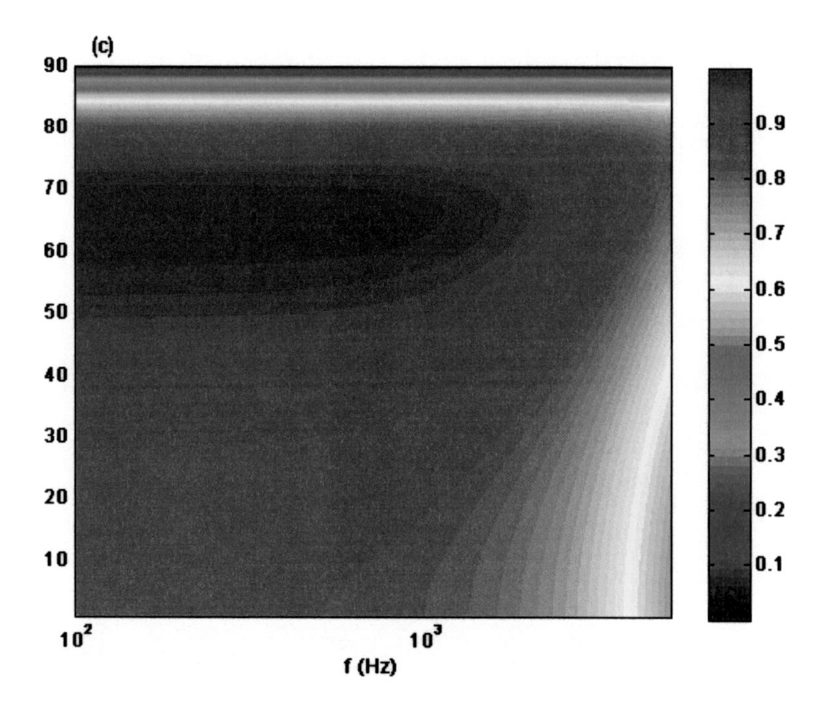

Figure 12. Design charts of a MIU absorber with (d_1,t_1,p_1,d_2,t_2,p_2,D)=(6 mm,1 mm,23 %,35 mm,39 mm,14 %,5 cm), for a) passive, b) active by pressure-release, and c) active by impedance-matching, control conditions.

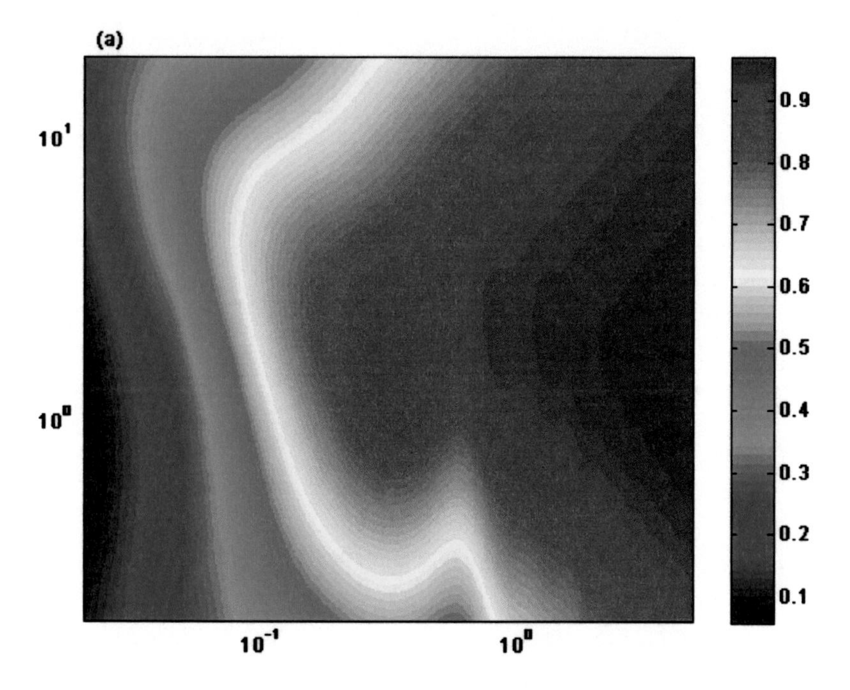

Figure 13. Continued on next page.

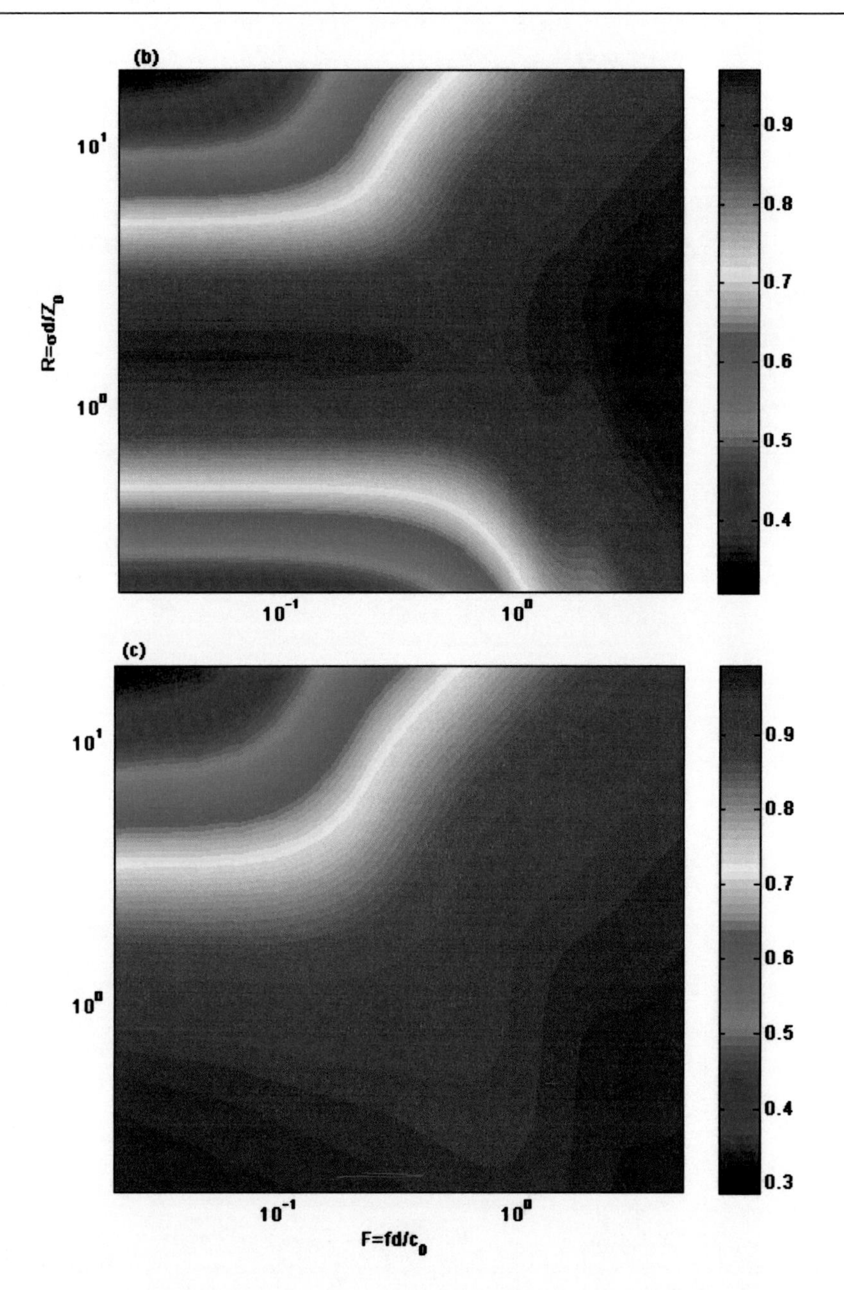

Figure 13. Design charts for the statistical absorption of a two-layer porous absorber with d=3 cm and D=5 cm, for a) passive, b) active by pressure-release, and c) active by impedance-matching, control conditions.

Figure 14 shows the design charts for the statistical absorption of a MIU absorber with (t_1,d_2,t_2,p_2,D)=(1 mm,35 μm,39 μm,14 %,5 cm), under passive, active by pressure-release and active by impedance-matching, control conditions, as a function of the perforation constant, x_1, and perforation ratio, p_1, of the supporting panel. Again, the active control increases the low frequency part of the passive chart (lower x_1).

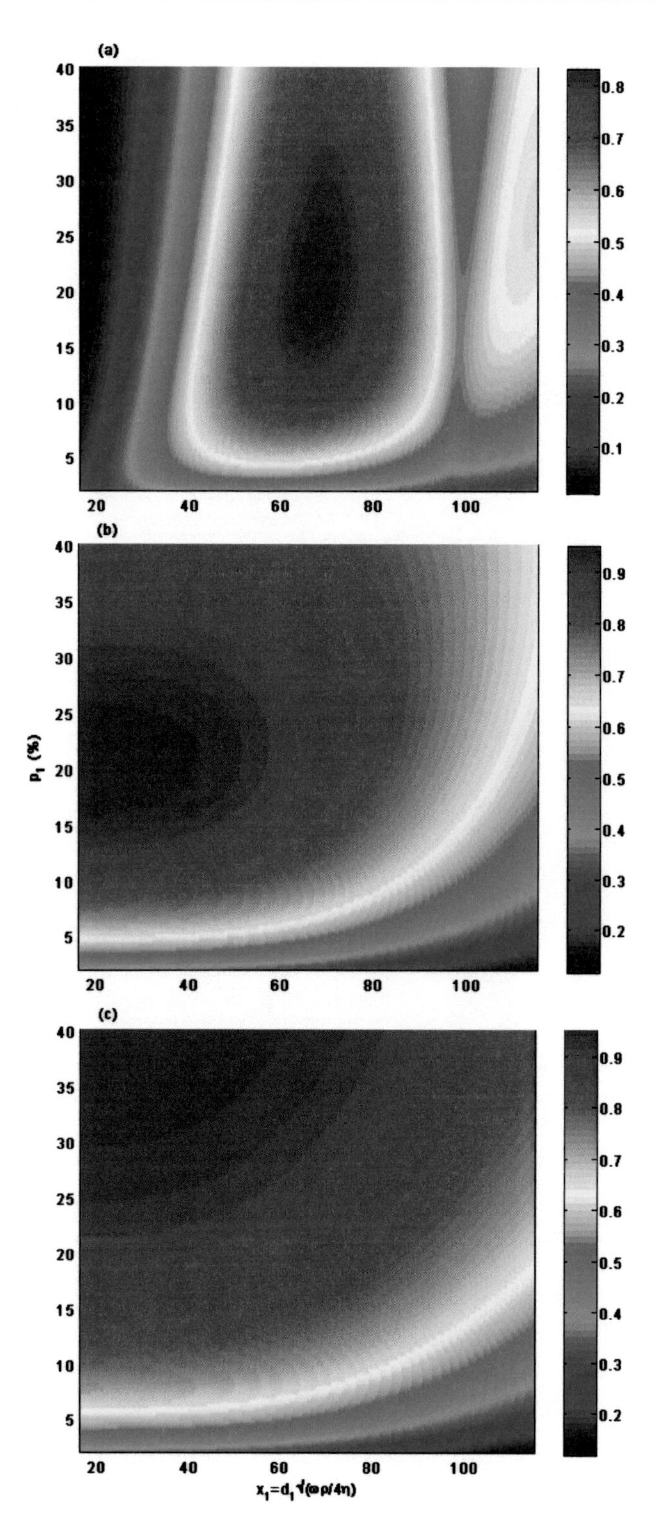

Figure 14. Statistical absorption coefficients of a MIU absorber with (t_1,d_2,t_2,p_2,D)=(1 mm,35 μm,39 μm,14 %,5 cm), for a) passive, b) active by pressure-release, and c) active by impedance-matching, control conditions.

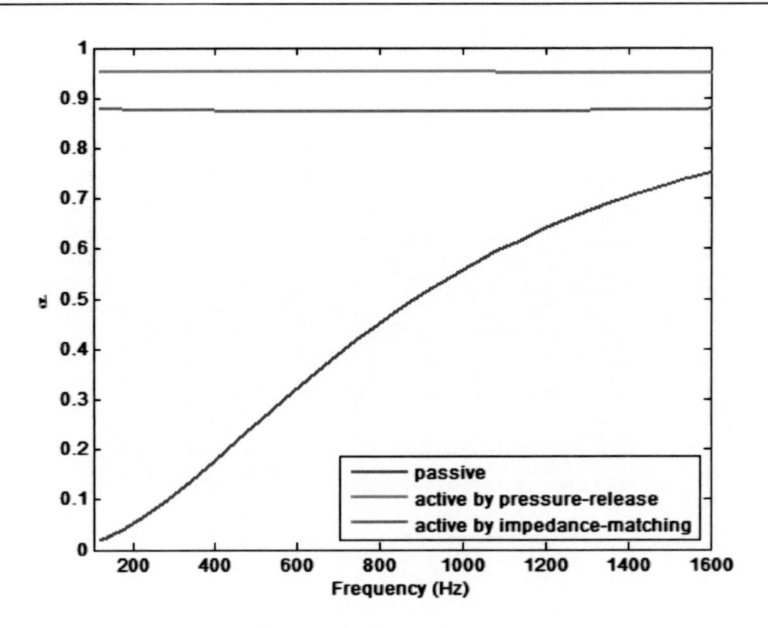

Figure 15. Statistical absorption curves of a two-layer porous absorber with σ=22 100 N s/m^4, d=3 cm and D=5 cm.

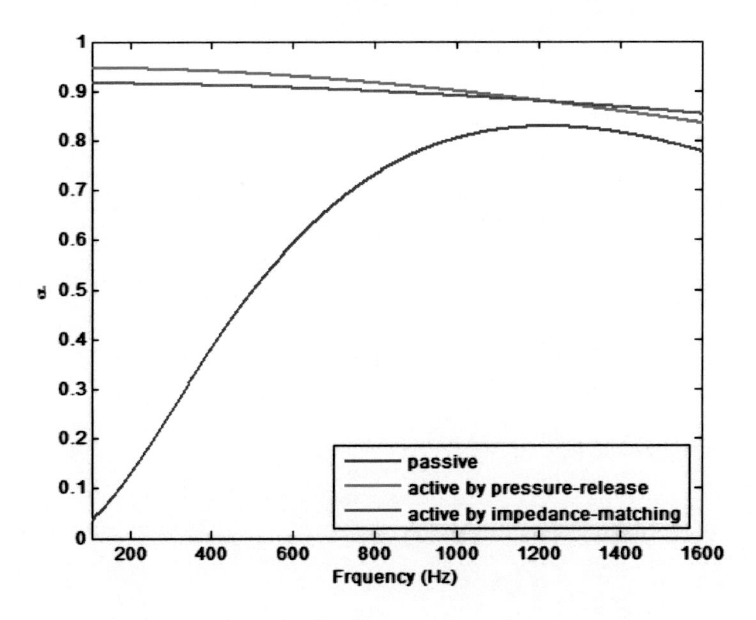

Figure 16. Statistical absorption curves of a MIU absorber with (d_1,t_1,p_1,d_2,t_2,p_2,D)=(6 mm,1 mm,23 %,35 μm,39 μm,14 %,5 cm).

Classical curves of the statistical absorption coefficient as a function of frequency can be obtained by cutting the design charts by a line parallel to the *R* axis, in the case of a porous absorber, or the x_I axis, in the case of a MIU absorber.

Figures 15 and 16 show such classical curves for the combination of parameters stated in their captions.

Statistical absorption curves show distinctive features when compared with their corresponding normal incidence ones. Namely:

- The passive statistical absorption curve is reduced in amplitude and its first maximum is moved to higher frequencies.
- Both control conditions, pressure-release and impedance-matching, provide high absorption in the low frequency range. Besides, the differences between both conditions tend to decrease.

Therefore, the pressure-release condition is able to afford slightly higher absorption than the impedance-matching condition with less electronic and computational burden (the impedance-matching requires two microphones and a deconvolution circuit). Thus, only the pressure-release condition will be implemented in the experimental rig of the next Section.

IV. Measurement Methods

The plane-wave normal-incidence sound absorption coefficient of acoustic materials is usually measured in an impedance tube using either the standing wave ratio or the transfer function method (ISO 10534-1 and -2). These methods use continuous signals, assume a sample of infinite size, and proceed in the frequency domain. When measuring the absorption coefficient at oblique incidence or in situ, a finite size sample must be used, and diffractions on the edges of the sample must be taken into account. Thus, a time domain reflection method, based on the measurement of a loudspeaker-microphone impulse response, in free-field and in front of the sample, is better employed. Since in this method the measurement microphone is close to the absorbing sample, it is important than the loudspeaker-microphone has an impulse response as shorter as possible. Pulse synthesis by inverse filtering techniques is able to provide such short impulse responses. More details of these methods and techniques are given in the following.

IV.A. The Transfer Function Method

The transfer function method follows the international standard ISO 10534-2. The sample to be measured is fitted in one end of the tube, within a test sample holder, Figure 17. A loudspeaker in the opposite end generates a broadband signal (for instance a white noise). Two microphones measure the acoustic pressure at two locations along the tube. Let s and x be the separation between microphones and the distance between the microphone 2 and the sample, respectively (Figure 17).

Let $H_{12}(f)$ be the transfer function between both microphones. To remove the errors from the distinct sensitivities of the two microphones

$$H_{12} = \sqrt{H_{12}^{I} \cdot H_{12}^{II}} = \left|H_{12}\right| \cdot e^{j \cdot \phi}, \tag{20}$$

where H_{12}^{I} and H_{12}^{II} are two measurements with the microphones 1 and 2 interchanged. From this measurement of the transfer function, the reflection coefficient is

$$R = \frac{H_{12} - H_i}{H_r - H_{12}} \cdot e^{2 \cdot j \cdot k_0 \cdot (x+s)}, \tag{21}$$

where

$$\begin{aligned} H_i &= e^{-j \cdot k_0 \cdot s} \\ H_r &= e^{j \cdot k_0 \cdot s} \end{aligned}, \tag{22}$$

the acoustic impedance and the absorption coefficient, Eq. (1), can be calculated (Chung and Blaser, 1980). According to the standard ISO 10534-2, the frequency range covered by this method is

$$\frac{0.05c}{s} < f < \frac{0.45c}{s}. \tag{23}$$

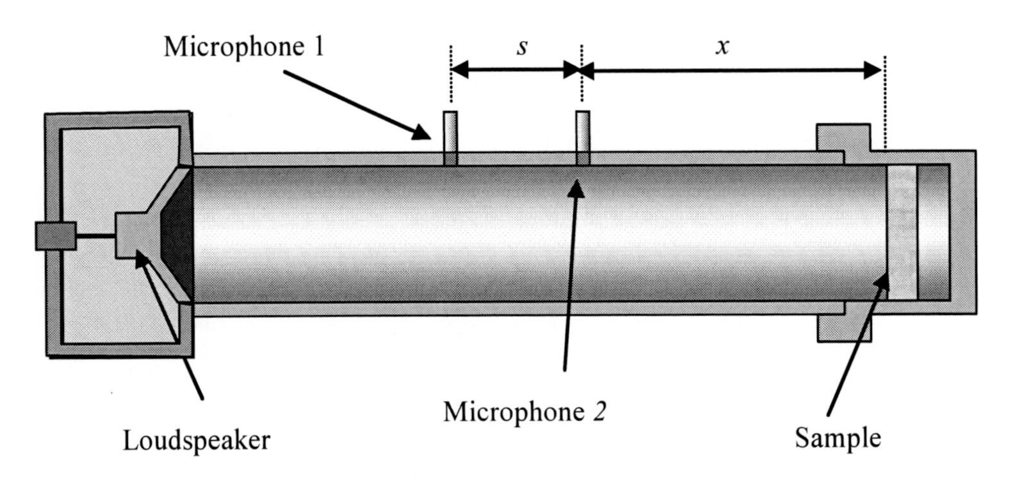

Figure 17. Impedance tube for the measurement of sound absorption.

IV.B. The *in situ* Reflection Method

The *in situ* reflection method follows the European standard CEN/TS 1793-5 to measure the intrinsic characteristics of antinoise devices. It is also implemented in the international standard ISO 13472-1 to determine the absorption properties of road surfaces in situ. It is based in the measurement of two impulse responses of a loudspeaker-microphone system, Figure 18. The microphone is kept at a fixed distance from the loudspeaker by thin rigid rods. First, the direct impulse response is measured with the loudspeaker-microphone system turned 180° with respect to the sample (direct signal). Then, the loudspeaker-microphone system is located at the desired angle in front of the sample, and the impulse response is measured again (reflection trace). Since the direct event is located exactly in the same place in both measurements, subtracting the direct signal from the reflection trace will remove the direct event from it (reflection signal).

Besides the direct and sample reflected events, both measurements will also contain edge diffracted and other reflected events. Time windowing allows to extract the desired direct and sample reflected events. The absorption coefficient is then

$$\alpha(f) = 1 - \left| \frac{\Im\{s_r(t) \cdot y(t) \cdot w_r(t)\}}{\Im\{s_d(t) \cdot d(t) \cdot w_d(t)\}} \right|^2 , \tag{24}$$

where $w_r(t)$ and $w_d(t)$ are the time windows applied to the direct and reflected signals, respectively, $s_r(t)$ and $s_d(t)$ are functions to correct for spreading effects, and $\Im$ denotes Fourier transform. The most simple spreading correction functions are $s_r=d_{lm}$ and $s_d(t)=d_{ms}$, the loudspeaker-microphone and microphone-sample distances, respectively (Figure 18).

Cobo (2007) shows that the truncation effect of time windowing is to remove low frequencies from the measured traces. Thus, the lower reliable frequency of the absorption curve provided for this method is approximately the first notch of the magnitude spectrum of the applied time window.

IV.C. Pulse Synthesis by Inverse Filtering

Since the microphone is close up the sample in the reflection method, the direct, sample reflected and edge diffracted events will overlap likely in the measured traces. This complicates the positioning of the time windows. To avoid this, the loudspeaker-microphone impulse response can be shortened by inverse filtering (Cobo *et al.*, 2006b; 2007). The loudspeaker-microphone system can be modeled as a linear filter. Let $h(t)$ be the impulse response of the loudspeaker-microphone system, and $H(f)$ its Fourier transform. If the loudspeaker is driven by a conventional voltage, $x_c(t)$, and the microphone receives a conventional waveform, $y_c(t)$, then

$$Y_c(f) = H(f) X_c(f), \tag{25}$$

where $X_c(f)$ and $Y_c(f)$ are the Fourier transforms of the input and output signals, respectively. If now the shape of the radiated spectrum received by the microphone is prescribed, $Y_s(f)$, the loudspeaker must be driven with a spectrum, $X_s(f)$, such that

$$X_s(f) = \frac{Y_s(f)}{H(f)}. \tag{26}$$

A positive constant, p, can be added to the denominator of Eq. (26) to avoid instabilities (regularization). Thus

$$X_s(f) = Y_s(f)\frac{H^*(f)}{|H(f)|^2 + p^2}. \tag{27}$$

The loudspeaker should be driven then with the electrical signal

$$x_s(t) = \Im^{-1}\left\{Y_s(f)\frac{H^*(f)}{|H(f)|^2 + p^2}\right\}, \tag{28}$$

where $\Im^{-1}$ denotes inverse Fourier transform.

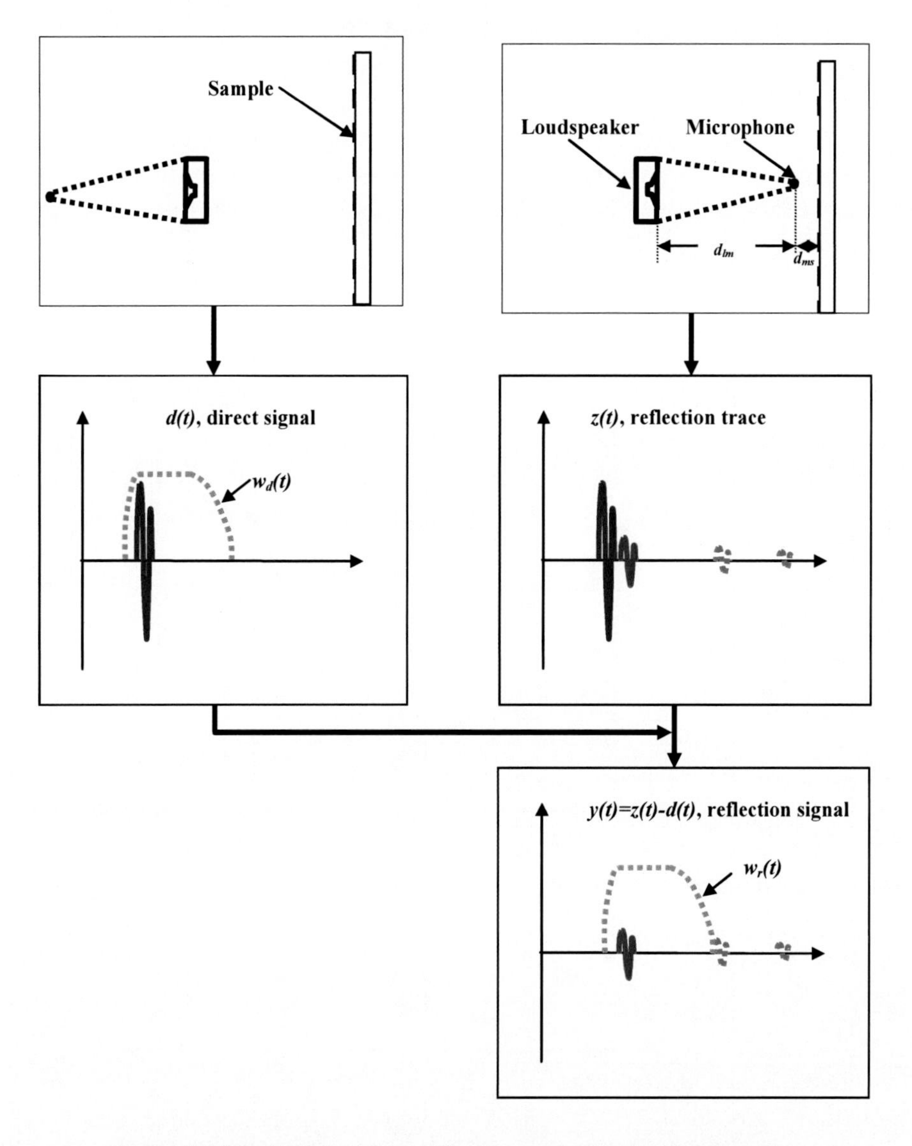

Figure 18. Sketch of the in situ reflection method in the time domain.

Therefore, this method needs to measure the loudspeaker-microphone transfer function and to specify the desired spectrum. For instance, cosine-magnitude pulses have a magnitude spectrum given by

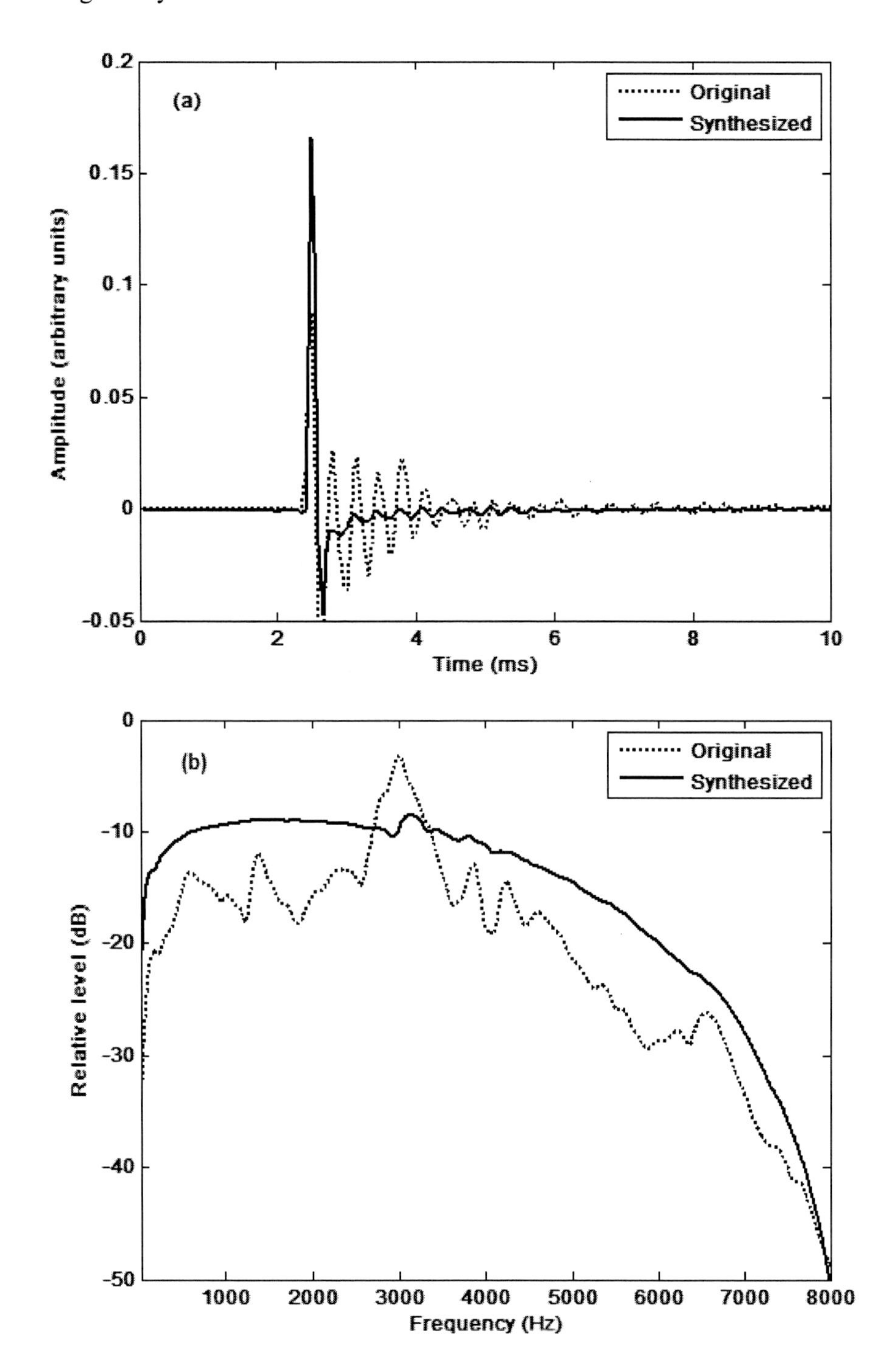

Figure 19. Original (dotted line) and synthesized by inverse filtering (solid line) pulses (a) and spectra (b) of a loudspeaker.

$$|Y_s(f)| = \begin{cases} A\cos^g\left(\dfrac{\pi(f-f_0)}{B}\right) & f_1 \le f \le f_2 \\ 0 & f < f_1,\ f > f_2 \end{cases}, \tag{29a}$$

with

$$\begin{aligned} f_0 &= \frac{f_2+f_1}{2} &&, \text{the central frequency} \\ B &= (f_2 - f_1) &&, \text{the bandwidth} \end{aligned} . \tag{29b}$$

The parameter g in Eq. (29a) determines the trade-off between the main lobe and the lateral oscillations of the pulse (Cobo *et al.*, 2006b). Zero-phase cosine-magnitude pulses have a phase spectrum $\Psi_{Y_S}(f) = 0$. Minimum-phase cosine-magnitude pulses have a phase spectrum

$$\Psi_{Y_s}(f) = \frac{-1}{\pi}\int_{-\infty}^{\infty}\frac{|Y_s(\xi)|}{f-\xi}d\xi . \tag{30}$$

Whilst cosine-magnitude shaping provides symmetrical magnitude spectral, most of the loudspeakers have a non symmetrical magnitude spectrum. Therefore, it could be preferable to shape the magnitude spectrum of the loudspeaker with two half-cosine functions, one for the roll-on and other for the roll-off parts of the spectrum, matched to some intermediate frequency. Figure 19 shows an example. The original impulse response of the loudspeaker has a large oscillating tail. When the loudspeaker is driven with an inverse filter with two half-cosine-magnitude minimum-phase spectrum, the length of the radiated impulse response is drastically decreased (Figure 19a), and its magnitude spectrum has been smoothed (Figure 19b).

V. Experimental Results

V.A. Hybrid Passive-active Impedance Tube

Hybrid passive-active absorption can be measured by modifying properly a passive impedance tube. Figure 20 shows the hybrid passive-active absorption tube at the *Instituto de Acústica* (Madrid). The system consists of an aluminum tube of length 1 m and inner diameter 10 cm. The primary loudspeaker at the right side generates a white noise in the frequency range of interest. In the opposite side is the absorber to be measured. In the passive case, it is a test sample holder (either a porous layer or a MPP in front of an air layer). In the active case, there are, besides the passive sample, a control source, an error microphone at the entrance of the air layer, and an active controller. The control source can be either a loudspeaker (Figure 20) or a plate actuator (Figure 21).

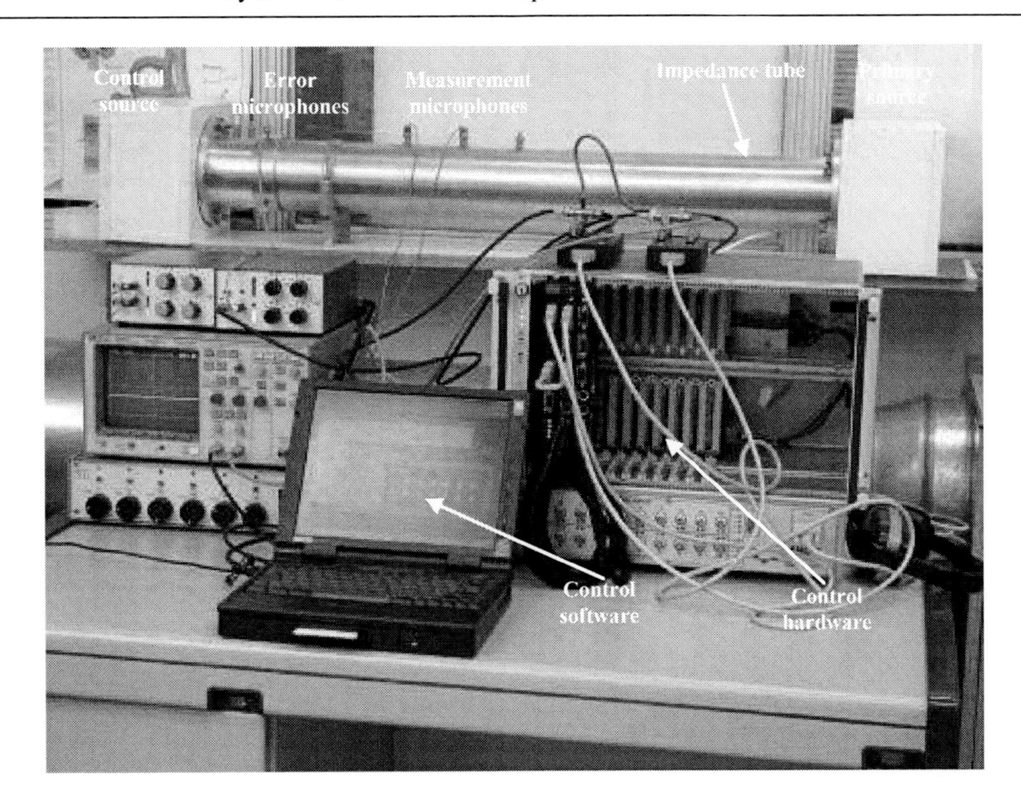

Figure 20. Hybrid passive-active impedance tube.

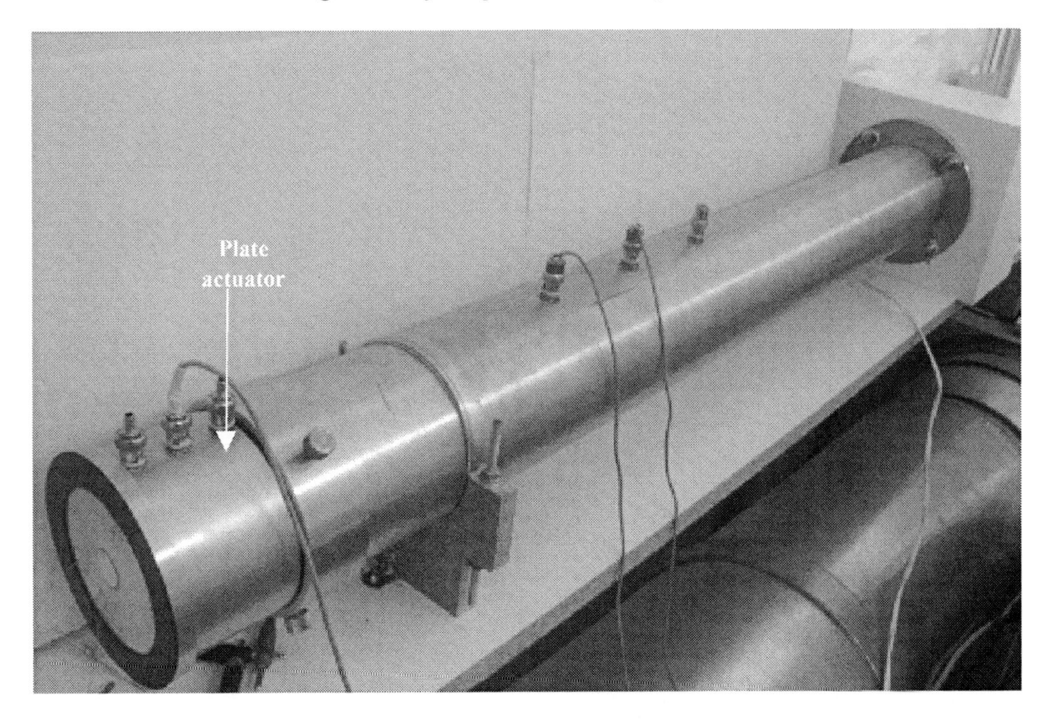

Figure 21. The hybrid passive-active impedance tube with a plate actuator as the control source.

The signal picked up by the error microphone is the error signal for the active controller, based on a DSP TI TMS320 C40 by Texas Instruments©. This controller, which implements the adaptive algorithms FX-LMS and FU-LMS (Elliott, 2001), allows to configure the number of taps for the identification and control adaptive filters, and the number of taps and cut-off frequency for the low-pass filter. The controller uses the same signal driving the primary loudspeaker as the reference signal. Thus, when the active controller is working, the adaptive filter generates a control signal which drives the control source to cancel the acoustic pressure at the error microphone (pressure-release condition).

The passive, purely active, or hybrid passive-active absorption coefficient is measured using the transfer function method above described. Three measurement microphones are used in the centre of the impedance tube (Figures 20 and 21). The separation between each two microphones is 9 cm. Therefore, two frequency ranges can be covered using either two consecutive or two extreme microphones. These frequency ranges are (95,857) Hz for s=9 cm, and (190, 1715) Hz for s=18 cm.

V.B. Results in the Impedance Tube

Figure 22 shows the passive, active, and hybrid passive-active absorption coefficient of a two-layer porous system made of a 4 cm thick melamine foam layer, with flow resistivity σ=12 000 N s/m^4, and a 4 cm thick air layer. Notice that the flow resistance of the porous layer is 480 N s/m^3, close to acoustic impedance of air, 410 N s/m^3. The passive system affords the maximum absorption at about 900 Hz. The active controller provides absorption close to 100 % at about 200 Hz. The active absorption curve decreases slowly and intersects the passive one at about 680 Hz. Above this frequency, the active system yields less absorption than the passive one. When the active controller is low-pass filtered at this frequency, the hybrid passive-active system gives absorption close to 100 % in the whole frequency band. The absorption coefficient averaged in the frequency band (200, 600) Hz is 0.97 with a two-layer porous system with full thickness 8 cm.

Figure 23 shows the passive, active and hybrid passive-active absorption curves of a MIU with $(d_1,t_1,p_1,d_2,t_2,p_2,D)$=(6 mm,1 mm,23 %,35 μm,39 μm,14 %,5cm). The passive system affords the maximum absorption at about 1200 Hz. The active controller provides absorption close to 100 % at about 200 Hz. The active absorption intersects the passive one at about 850 Hz. When the active controller is low-pass filtered at this frequency, the hybrid passive-active system gives absorption close to 100 % in the whole frequency band. The absorption coefficient averaged in the frequency band (200, 600) Hz is 0.94 with a MIU with full thickness 5.1 cm (Cobo and Fernández, 2006).

V.C. Results in the Anechoic Chamber

Performance of hybrid passive-active absorbers has also been demonstrated in free field conditions (Cobo *et al.*, 2006b; Cobo and Cuesta, 2007). Two prototypes of hybrid absorbers using MIUs as passive layers have been designed and manufactured to be tested in anechoic room.

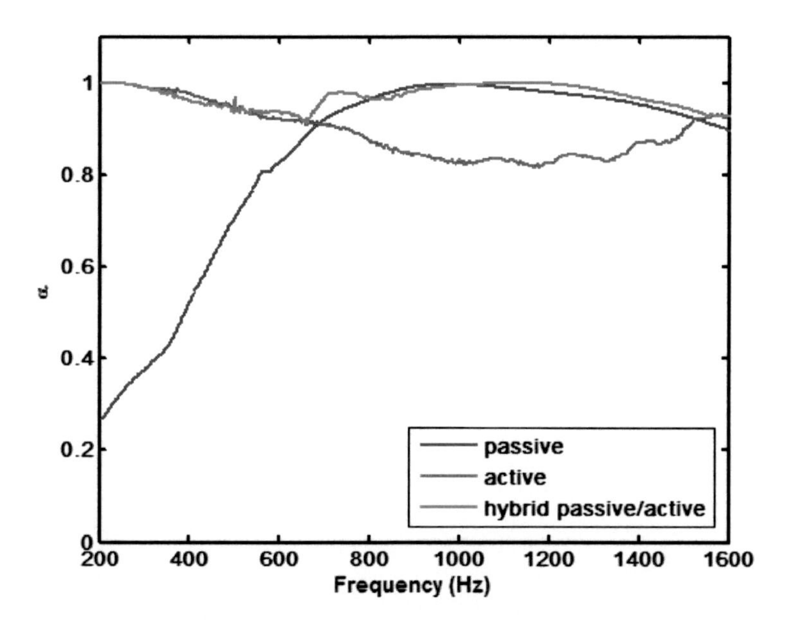

Figure 22. Passive, active, and hybrid passive-active absorption coefficients of a two-layer porous system with σ=12 000 N s/m^4, d=4 cm, and D=4 cm.

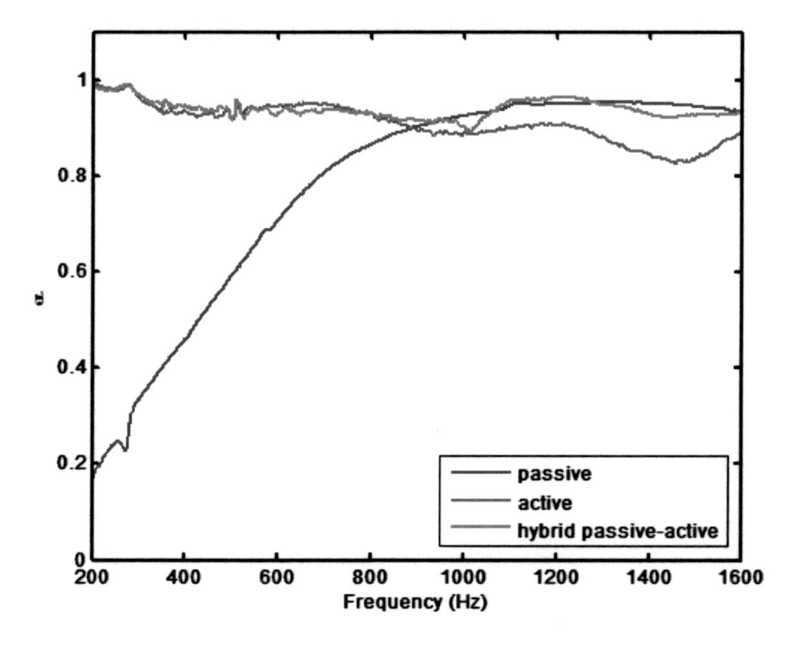

Figure 23. Passive, active and hybrid passive-active absorption coefficients of a MIU with (d_1,t_1,p_1,d_2,t_2,p_2,D)=(6 mm,1 mm,23 %,35 μm,39 μm,14 %,5 cm).

Each one consists of a (2.44 m x 2.44 m x 1 mm) MIU, backed by a 5 cm deep air cavity, and a rigid (passive condition) or active termination. The MIUs have been designed to provide high absorption around 1 kHz according to the model described in Section 2. Each one is constructed with a 1 mm thick perforated plate with 2.44 m of square section, and a micrometric mesh covering its holes (Pfretzschner *et al.*, 2006). The main difference between these MIUs is the perforation rate of the carrying plate, being 10 % (Figure 24 (up)) or 23 % (Figure 24 (down)). Theoretically, the 23% perforated MIU provides higher absorption above 800 Hz.

Figure 24. Two MIUs: (up) with $(d_1,t_1,p_1,d_2,t_2,p_2,D)$=(6 mm,1 mm,10 %,35 μm,39 μm,14 %,5 cm) and (down) with $(d_1,t_1,p_1,d_2,t_2,p_2,D)$=(6 mm,1 mm, 23 %,35 μm,39 μm,14 %,5 cm).

In practice, both the passive (rigid termination) and active (pressure release) conditions in the air cavity, have been implemented using a 1.5 cm thick wood panel (Figure 25 (up)) and an active wall (Figure 25 (down)), respectively. For practical reasons, the active wall is only applied to a reduced area at the back of the prototype. In fact, a (62 cm x 54 cm) central section of the wooden panel acting as rigid ending in the passive prototype, has been removed and replaced by an active cell of the same dimension (Cobo and Cuesta, 2007). It contains 4 loudspeakers (secondary sources), inserted symmetrically; and 4 microphones, numbered from M1 to M4, placed just behind the material, each one in front of the corresponding loudspeaker. When this active cell is completely sealed to the remaining wooden panel (note that Figure 25 (down) presents the inserting procedure and thus it is not completely attached), the air gap between the MIU and the active cell continues to be 5 cm deep. In order to create local units, the air cavity contained in the area cell is separated into 4 equal cavities using a wooden cross, each one containing a loudspeaker and a microphone. Each loudspeaker is driven to reduce the sound pressure field in the related microphone, according to the FX-LMS algorithm implemented in the controller described in Figure 20. Therefore, this 4I4O (4 Input/4 Output) multichannel active cell is configured as a 4 local 1I1O units, this allowing to decrease the computational burden of this active cell.

Figure 25. Views of the control conditions at the back of the two-layer absorber: passive (up) and active (down).

Both the passive and active absorption coefficient of such prototypes are measured according the procedure described in Sections IV.B and IV.C, using the loudspeaker-microphone device located in front of the absorber (Figure 24). The microphone is kept at a fixed distance from the loudspeaker by thin rigid rods. The distance microphone-sample also remains constant. First, the direct impulse response of the loudspeaker-microphone is measured, turning the device 180° from the absorber. An MLS signal is used for this purpose. Then, the absorber response is recorded by addressing the same device to the sample. This procedure permits to measure up to 25° of incidence from the central axis of the panel. The absorption coefficient can be extracted by windowing appropriately the direct and reflected events from the respective traces.

The hybrid passive-active performance of such two MIU prototypes is described as follows:

- First, the passive absorption of each prototype using the wooden panel at the back is measured at normal and oblique incidence. Results are presented in Section V.C.a.
- Then, the active wall is set up, as previously described, and optimized to reduce the primary field (MLS signal) at each error microphone. This requires configuring the parameters of the controller. As reference signal, the feedforward strategy implements the same MLS signal driving the loudspeaker. Once this optimization is attained, the active absorption coefficient is measured following the same methodology of the passive case. Finally, in order to evaluate the effect of the active wall in the passive configuration, the absorption coefficient is measured with the active control switched off. Results of these measurements are presented in Section V.C.b.

V.C.a. Passive Absorption at Oblique Incidence

Figures 26 and 27 show both the experimental and theoretical passive absorption coefficients of the 10% perforated MIU (Figure 24 up), at normal and 21° of incidence, respectively. For these measurements, the distances loudspeaker-microphone (d_{lm}) and microphone-sample (d_{ms}) are 87 cm and 5 cm, respectively. This geometry limits the minimum reliable frequency to 217 Hz (See Section V.B). This passive absorber provides an absorption coefficient between 0.2 and 0.85, from 200 Hz to 1 kHz. In both cases, good agreement between prediction and measurement is seen. Some results at oblique incidence for the 23% perforated MIU are represented in Figures 28 and 29. In order to avoid scattering effects of the plate around the microphone, it is moved 15 cm away from the MIU. This augments slightly the lowest reliable frequency of the measurement (253 Hz). As seen, this new MIU provides higher performance above 800 Hz, attaining maximal absorption at around 1 kHz. The comparison between experimental and theoretical curves in this prototype presents some discrepancy, especially at lower frequencies. This should be associated to some changes carried out in the wooden panel in this configuration. In any case, experimental results confirm the excellent performance of such an absorber.

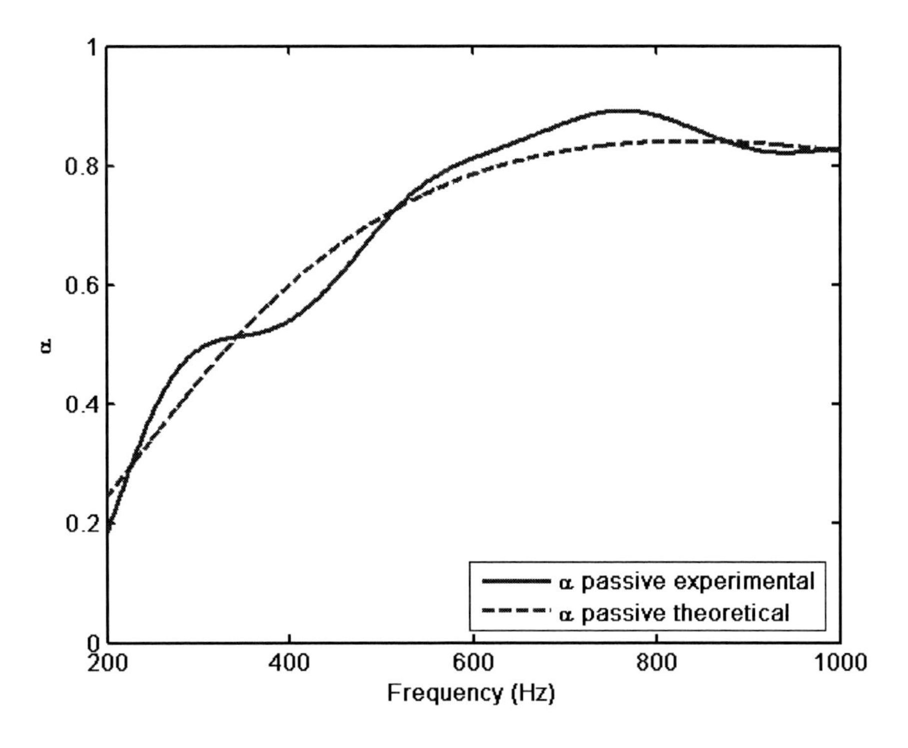

Figure 26. Experimental and theoretical passive absorption coefficients of the 10% perforated MIU (see Figure 24), at normal incidence.

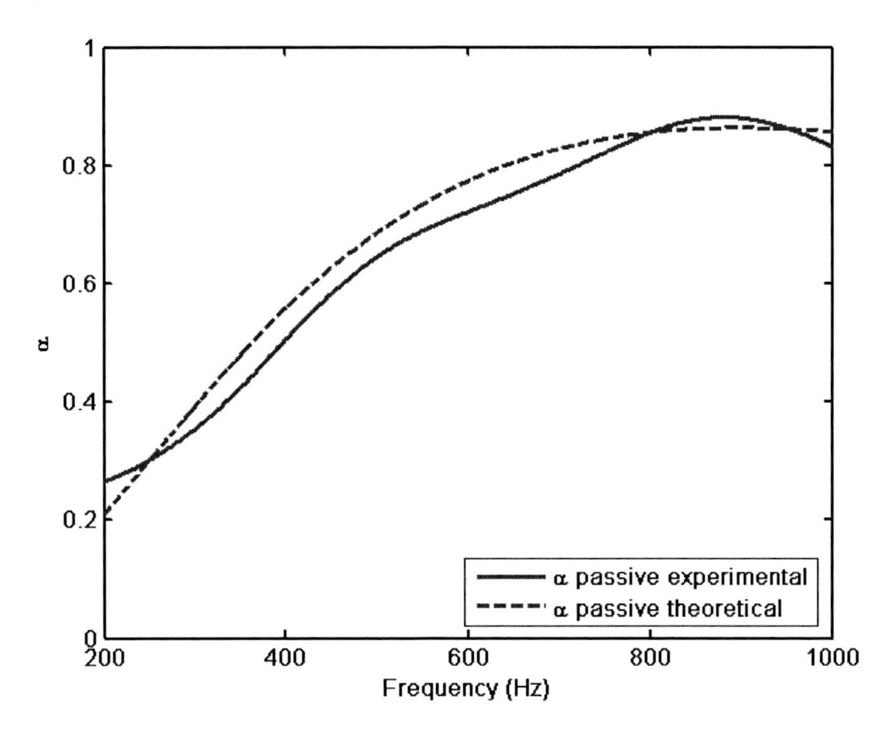

Figure 27. Idem for incidence at 21°.

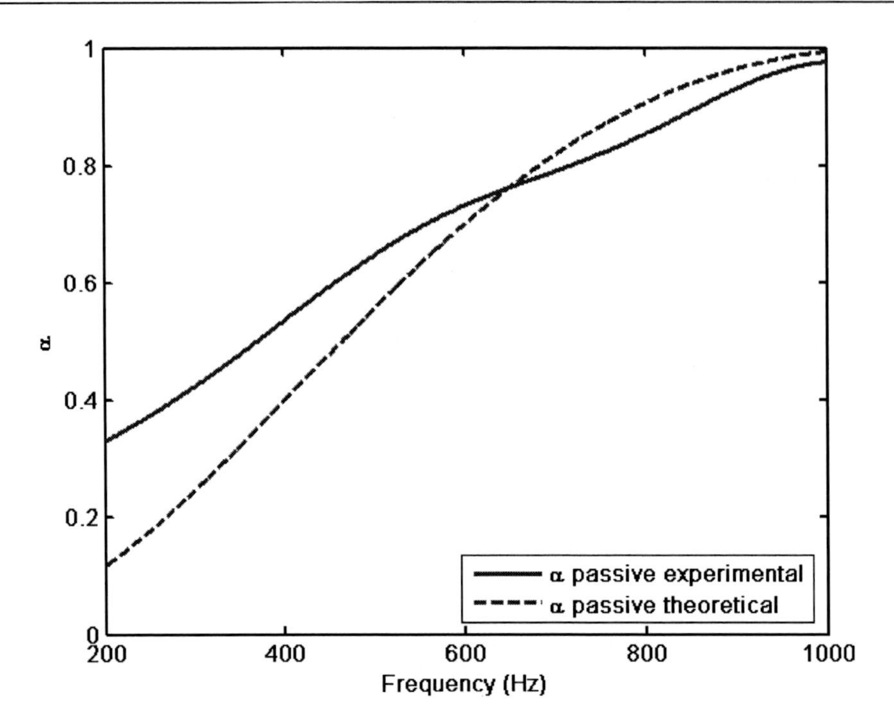

Figure 28. Experimental and theoretical passive absorption coefficients of the 23% perforated MIU (see Figure 24), for incidence at 13°.

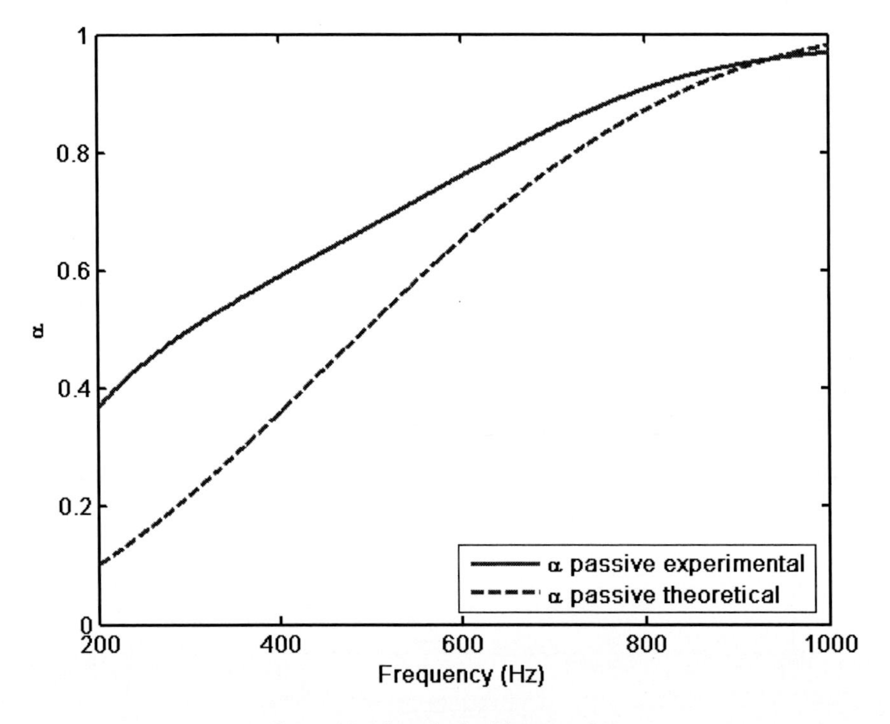

Figure 29. Idem for incidence at 21°.

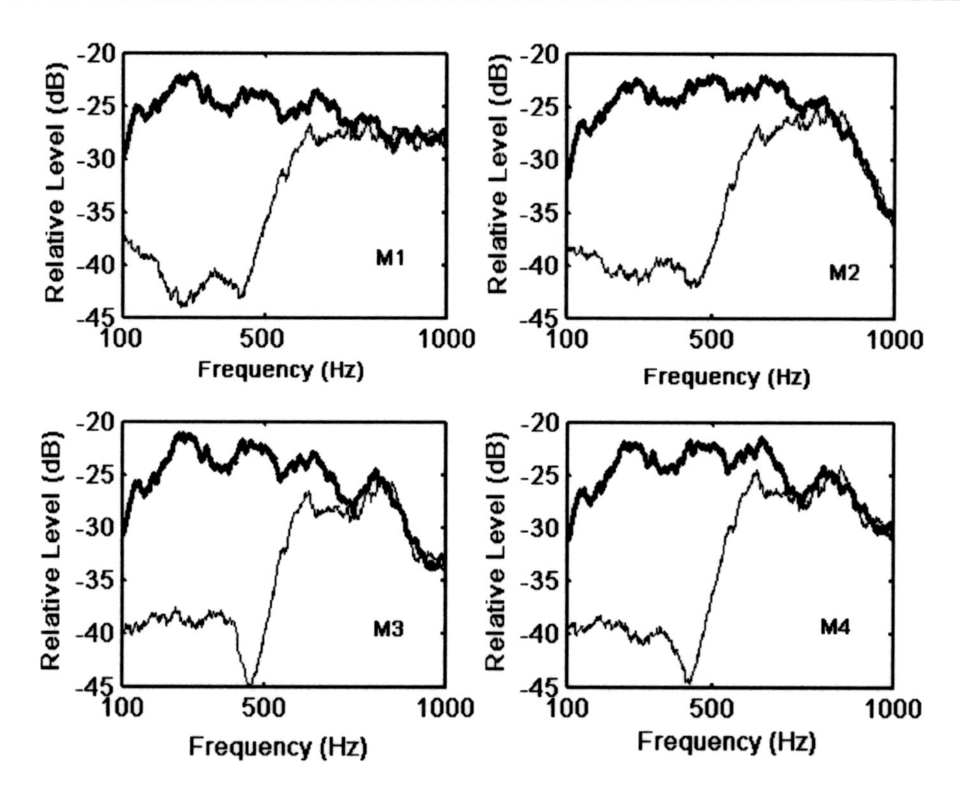

Figure 30. Error signal spectra without (solid line) and with (thin line) active control at the rear face of the 10% perforated MIU, at normal incidence.

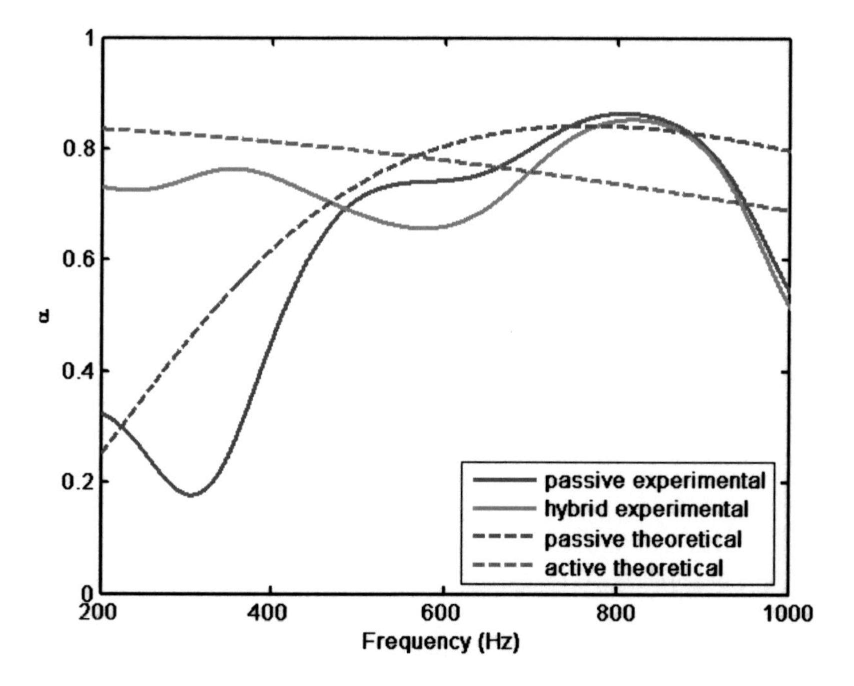

Figure 31. Passive (experimental and theoretical), active (theoretical), and hybrid passive-active (experimental) absorption coefficients of the 10% perforated MIU, at normal incidence.

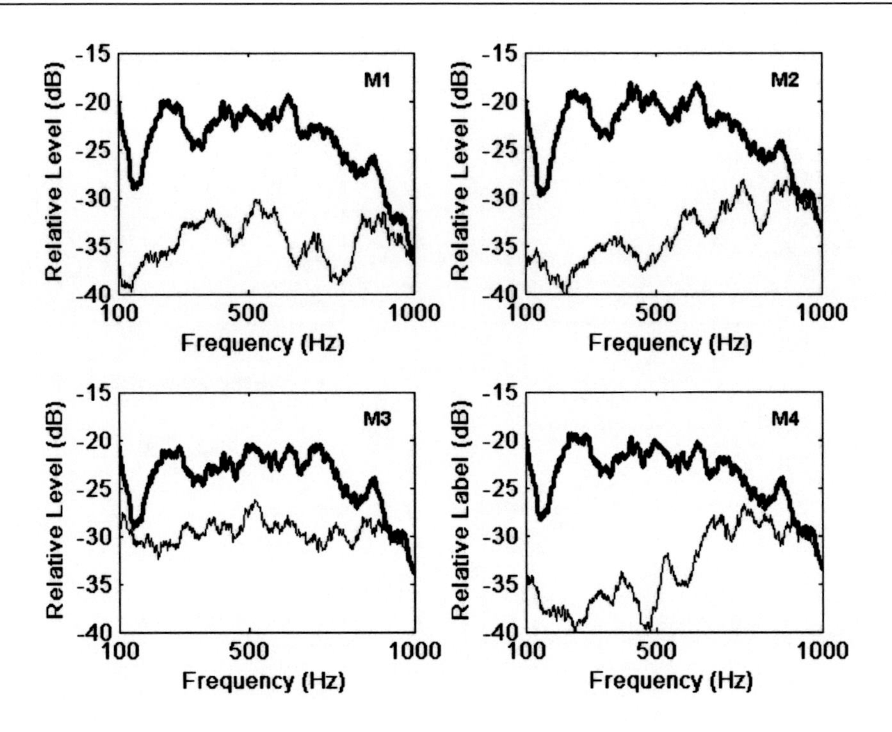

Figure 32. Error signal spectra without (solid line) and with (thin line) active control at the rear face of the 23% perforated MIU, for incidence at 13°.

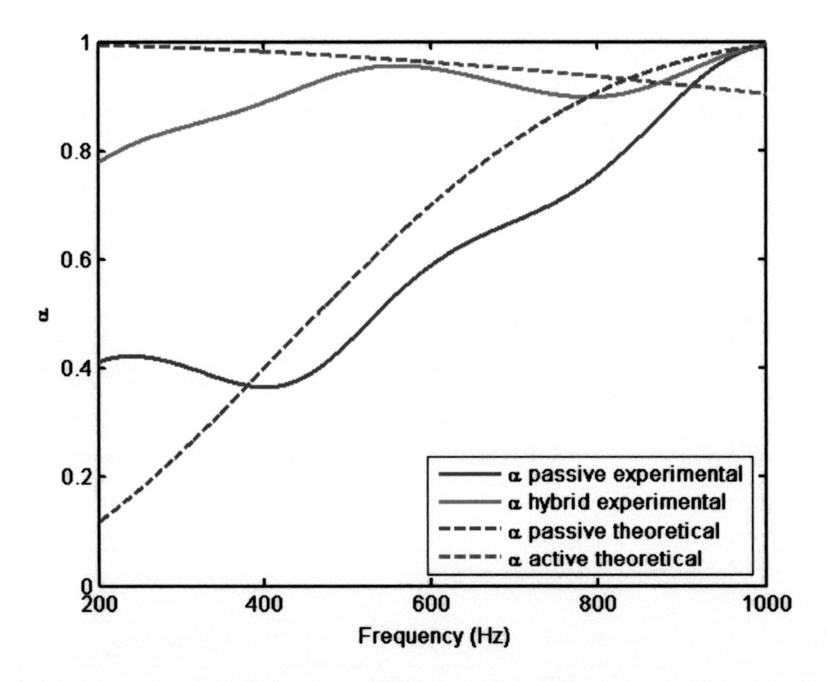

Figure 33. Passive (experimental and theoretical), active (theoretical), and hybrid passive-active (experimental) absorption coefficients of the 23% perforated MIU, for incidence at 13°.

V.C.b. Hybrid Passive-active Absorption at Oblique Incidence

Figures 30 and 31 describe the active performance of the 10 % perforated MIU at normal incidence (Cobo and Cuesta, 2007). Figure 30 shows the error signal spectra of each microphone inside the active wall, with the active control off (solid line) and on (thin line). The error signals are low-pass filtered at 550 Hz, which is the cut-off frequency between the passive and active absorption coefficient curves of the prototype. This permits to design the hybrid absorber, functioning as a passive system above the cut-off frequency, where this configuration is optimal; and as an active absorber below this frequency, where active control is more suitable. Results of Figure 30 are obtained with 70 taps in the control and error path filters and a convergence parameter of 0.1. The relative level of the primary MLS field picked up the microphones is reduced between 15 and 20 dB below 500 Hz. Figure 31 compares the hybrid absorption of the prototype, measured with this active wall optimized, to the passive one. Experimentally, the cut-off frequency of such a prototype appears around 470 Hz. As expected, the active wall improves the passive absorption below this frequency. The averaged absorption coefficient from 217 Hz to 1 kHz is 0.6 and 0.75, for the passive and hybrid absorbers, respectively. Therefore, these results confirm the benefit of designing hybrid strategies.

Figures 32 and 33 show the active results of the 23 % perforated MIU, for an MLS field incident at 13°. In this case, the cut-off frequency between the passive and active absorption coefficient curves of the prototype is theoretically set-up around 800 Hz. The error signals are therefore low-pass filtered at this frequency (Figure 32). The controller is configured with similar parameters. The relative level of the primary MLS field picked up the microphones is reduced between 10 and 20 dB below 800 Hz. The M3 microphone seems to perform somewhat worse, probably due to be more distant than the others from the measuring microphone. Figure 33 shows the passive and hybrid absorption coefficient curves. As expected theoretically, this MIU prototype provides higher performance in both passive and active configurations than the previous one. Again, the active wall improves the passive absorption below the cut-off frequency of the prototype. In average, around 90% of hybrid passive-active absorption is achieved between 253 Hz and 1 kHz. These results show the feasibility of designing hybrid absorbers for free field conditions. More experiments should be conducted to improve the absorption coefficient at the lowest frequency range, where a lack of absorption is observed in relation to predictions. More active cells or different positions could be tried for this purpose.

VI. Conclusion

This chapter provides theoretical models and experimental results of hybrid passive-active absorbers. A traditional absorber consists of a porous material backed by an air cavity and a rigid termination. According to the acoustic properties of the material as well as the size of each layer, the absorption spectrum of such an absorber can be tuned in a specific medium to high frequency bandwidth. To extend absorption to the lower frequencies, the thickness of the material as well as that of the air cavity should be drastically augmented, rendering the prototype unpractical. However, this constraint can be avoided constructing a hybrid absorber which combines passive and active methods in order to cancel a broadband noise including low frequencies. In such a prototype, the active system allows to control the input impedance of the multilayer absorber, this affording absorption at the lower frequencies. Two active strategies can be implemented for this purpose: the *pressure-release condition* at the rear face of the passive layer or the *impedance-matching* condition requiring the cancellation of the reflected wave inside the air cavity. Practically, the active system consists of a secondary source (loudspeaker or actuator) acting as a moving wall at the back of the prototype, one *(pressure-release)* or two (*impedance-matching)* microphones inside the cavity, and an adaptive controller. Both porous or microperforated (MPP) material can be implemented as passive layers. MPPs have the practical advantage to provide similar absorption as porous absorber with thinner and cleaner prototypes. This chapter describes the methodology to design such hybrid passive-active absorbers using both porous and MPPs, and presents some practical designs tested at normal and oblique conditions. Good agreement between the experimental and predicted absorption coefficient curves is achieved, confirming the feasibility of designing hybrid passive-active absorbers with the proposed methodology.

Acknowledgement

This work has been funded by the Spanish Ministry of Education and Science, through Grants No. DPI2001-1613-C02-01 and DPI2005-05505-C02-01.

References

Allard, J.F.; Champoux, Y. New empirical equations for sound propagation in rigid frame fibrous materials *J. Acoust. Soc. Am.* 1992, 91, 3346-3353.

Beyene, S.; Burdisso, R.A. A new hybrid passive/active noise absorption system *J. Acoust. Soc. Am.* 1997, 101, 1512-1515.

Cobo, P. *Control Activo del Ruido. Principios y Aplicaciones*; CSIC, Colección Textos Universitarios Nº 26, Madrid, 1977.

Cobo, P.; Fernández, A.; Doutres, O. Low frequency absorption using a two-layer system with active control of input impedance *J. Acoust. Soc. Am.* 2003, 114, 3211-3216.

Cobo, P.; Pfretzschner, J.; Cuesta, M.; Anthony, D.K. Hybrid passive-active absorption using microperforated panels *J. Acoust. Soc. Am.* 2004, 116, 2118-2125.

Cobo, P.; de la Colina, C.; Pfretzschner, J.; Cuesta, M.; Fernández, A. Characterization of a Microperforated Insertion Unit (MIU) in a reverberation room *Acust. Acta Acust.* 2006a, 92 (Supp. 1), S108.

Cobo, P.; Cuesta, M.; Pfretzschner, J.; Fernández; Siguero, M. Measuring the absorption coefficient of panels in free field by using inverse filtered MLS signals Noise Control *Eng. J.* 2006b, 54, 414-419.

Cobo, P.; Fernández, A. Hybrid passive-active absorption of broadband noise using MPPs *Noise&Vibration Worldwide* 2006, 37, 19-23.

Cobo, P., Fernández, A., Cuesta. Measuring short impulses responses with inverse filtered *Maximum-Length Sequences App. Acoust.* 2007, 68, 820-830.

Cobo, P. A model comparison of the absorption coefficient of a Microperforated Insertion Unit in the frequency and time domains *App. Acoust.* 2007, in press.

Cobo, P.; Cuesta, M. Hybrid passive-active absorption of a microperforated panel in free field conditions *J. Acoust. Soc. Am.* 2007, 121, EL251-EL255.

Chung, J.Y.; Blaser, D.A. Transfer function method of measuring in-duct acoustic properties *J. Acoust. Soc. Am.* 1980, 68, 907-921.

Cuesta, M; Cobo, P; Fernández, A; Pfretzschner, J. Using a thing actuator as secondary source for hybrid passive-active absorption in an impedance tube *App. Acoust.* 67, 12-27.

Delany, M.E.; Bazley, E.N. Acoustical properties of fibrous absorbent materials *App. Acoust.* 1970, 3, 105-116.

Elliott, S.J. *Signal Processing for Active Control;* Academic Press, London, 2001.

Furstoss, M.; Thenail, D.; Galland, M.A. Surface impedance control for sound absorption: direct and hybrid passive/active strategies *J. Sound Vib.* 1997, 203, 219-236.

Guicking, D.; Lorenz, E. An active sound absorber with porous plate *J. Vib. Acoust. Stress&Reliab. Des.* 1984, 106, 389-392.

Guicking, D.; Karcher, K.; Rollwage, M. Coherent active methods for application in room acoustics *J. Acoust. Soc. Am.* 1985, 78, 1426-1434.

Hansen, C.H.; Snyder, S.D. *Active Control of Noise and Vibration;* E&FN Spon, London, 1997.

Kang, J.; Fuchs, H.V. Predicting the absorption of open weave textiles and micro-perforated membranes backed by an air space *J. Sound Vib.* 1999, 220, 905-920.

Maa, D.Y. Microperforated-panel wideband absorbers *Noise Control Eng. J.* 1987, 29, 77-84.

Maa, D.Y. Potential of microperforated panel absorber *J. Acoust. Soc. Am.* 1998, 104, 2861-2866.

Mechel, F.P. Design charts for sound absorber layers *J. Acoust. Soc. Am.* 1988, 83, 1002-1013.

Nelson, P.A.; Elliott, S.J. *Active Control of Sound;* Academic Press, London, 1992.

Pfretzschner, J.; Cobo, P.; Simón, F.; Cuesta, M.; Fernández, A. Microperforated Insertion Units: a new step in the design of microperforated panels *App. Acoust.* 2006, 67, 62-73.

Sellen, N.; Cuesta, M.; Galland, M.A. Noise reduction in a flow duct: Implementation of a hybrid passive/active solution *J. Sound Vib.* 2006, 297, 492-511.

Smith, J.P.; Johnson, B.D.; Burdisso, R.A. A broadband passive-active sound absorption system *J. Acoust. Soc. Am.* 1999, 106, 2646-2652.

In: New Research on Acoustics
Editor: Benjamin N. Weiss, pp. 285-298

ISBN: 978-1-60456-403-7

Chapter 8

PENGUINS AND OTARIIDS AS MODELS FOR THE STUDY OF INDIVIDUAL VOCAL RECOGNITION IN THE NOISY ENVIRONMENT OF A COLONY

Thierry Aubin, Isabelle Charrier, Hélène Courvoisier and Fanny Rybak
Université Paris-Sud, F-91405 Orsay, France

Abstract

In animals, recognition between individuals is essential to the settlement of sexual and social relationships. Due to their physical properties and their potentiality to encode any kind of information, sounds are an effective mean to reliably transmit the identity of the emitter. Nevertheless, in colonial animals, the vocal signal produced by an adult seeking its young or its partner among thousands of individuals is transmitted in a particularly noisy context generated by the colony. Such a background noise drastically reduces the signal-to-noise ratio and masks the signal by a noise with similar spectral and temporal characteristics. In this review, we report how this extreme acoustic environment constrains the transfer of information by sounds. To illustrate the problem of precise acoustic identification in the noise, we have chosen two representative biological models: the penguins and the otariids. We examine solutions found at the level of the emitter to improve the efficiency of communication and we report how the receiver can optimize the collected information. On the basis of the results obtained in numerous field studies, we show that penguins and otariids use a particularly efficient "anti-confusion" and "anti-noise" acoustic coding system, allowing a quick and accurate identification and localization of individuals on the move in a noisy crowd. The study of these biological models allows us to highlight the basic rules that govern the identification of a precise acoustic message in the noise. The differences in the coding-decoding strategies are also discussed with respect to the social structure and the environment of the different studied species.

Introduction

In natural environments, animals are often exposed to a stream of noise from which they have to extract pertinent sounds. Ambient noise arises from environmental factors such as wind or water (rain, sea, torrent) or biotic sources (Morton 1975, Ryan & Brenowitz 1985, Slabberkoorn 2004, Brumm & Slabbekoorn 2005). The background noise affects the signal-to-noise ratio and may impair the transmission of acoustic information from the emitter to the receiver (Holland et al. 1998).

Some of the most challenging conditions for communication by sounds are found in coastal colonies of animals. Seabirds and seals forage at sea, but breed on land in colonies numbering several hundreds or thousands of individuals. After foraging for several hours, days or weeks, adults come back to the crowded colony to find their mates and/or their young. The arriving parent has thus to recognize its family and to be recognized by it. The ability to recognize without ambiguity the mate or the young is particularly important in colonies, where individuals are densely packed, making the possibility of confusion great (Hutchison et al. 1968).

Penguins and otariids find their kin mate among thousands of individuals using an acoustic signal (Jouventin 1982, Insley et al. 2003). Calls involved in individual recognition in both penguins (i.e., "display call") and otariids ("pup or female attraction call") present a similar structure (see figure 1): a complex signal, composed of a fundamental frequency and its relative harmonics .The frequency band ranges from 250 Hz to 6000 Hz, with a maximum energy concentrated over the first harmonics (i.e., 300-2000 Hz). All calls present both amplitude and frequency modulations, with strong gaps of amplitude and slow frequency modulations with noisy parts occurring in some species (in penguins: Aubin & Jouventin 2002a; in otariids: Trillmich 1981, Charrier et al.. 2001, 2002, 2003a,b; Phillips & Stirling 2000; Page et al.. 2002).The display call in penguins is composed of a succession of different syllables, whereas the attraction call in otariids is a single unit that is repeated several times during mother-pup reunion. These calls are produced at a high amplitude level, averaging 90 dB SPL and 80 dB SPL measured at 1 meter in penguins (Aubin & Jouventin 1998; Jouventin & Aubin 2002) and in otariids (Charrier et al. 2002, Insley 2001) respectively.

The signals are transmitted in a context involving a huge noise generated by the colony and a screening effect of the bodies of animals. Both these factors drastically reduce the signal-to-noise ratio. Furthermore the signals are masked by a background noise with similar acoustic characteristics. In addition a rendezvous site in the colony does not exist in all species of penguins and in the otariids, and the absence of visual landmarks enhances the difficulty to locate the right individual on the move in a noisy crowd.

From their ability to recognise a particular call in this constraining environment, it is assumed that penguins and otariids use peculiar strategies of communication. To fully understand how these animals solve this problem of vocal recognition, we have done behavioural observations, acoustic analysis, propagation and playback experiments in the field with different species of penguins and otariids during ten years. We will discuss the differences in the acoustic strategies of communication with respect to the territorial habits and the environmental constraints of the species.

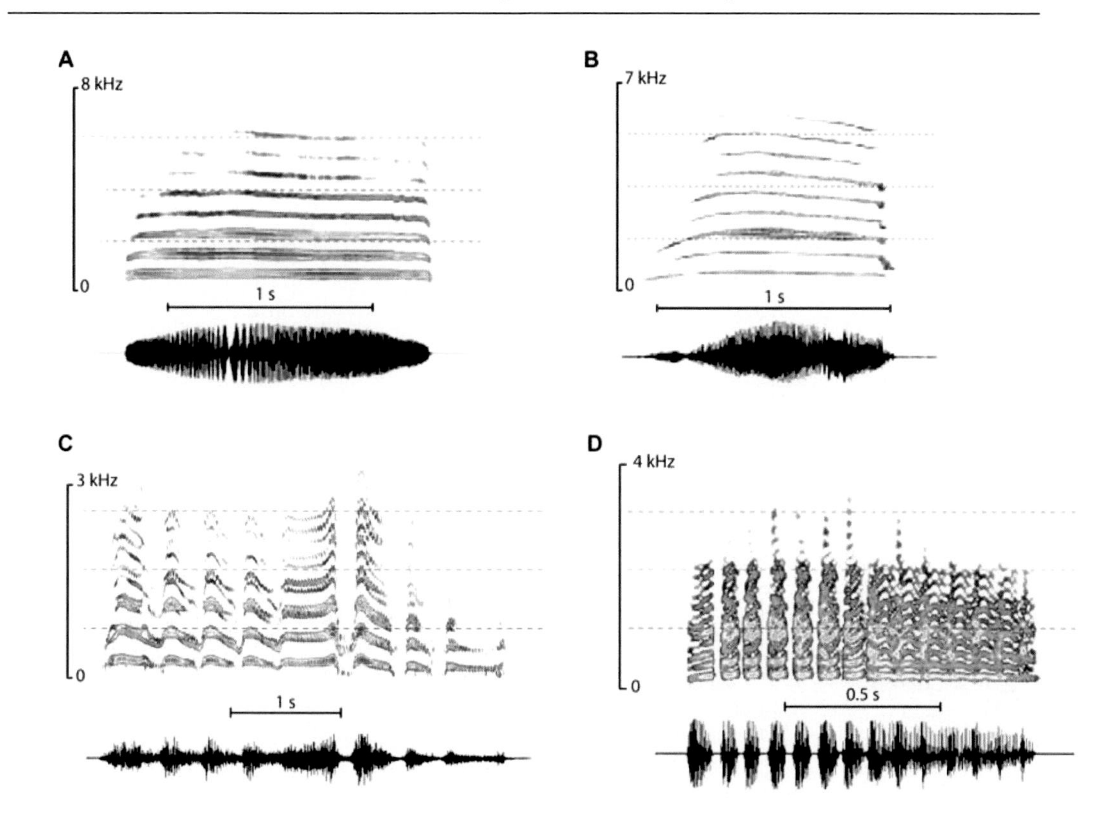

Figure 1. Sonographic and oscillographic representations of vocal signals involved in individual recognition. Pup attraction calls of A) an Antarctic fur seal *Arctocephalus gazella* female and B) an Subantarctic fur seal *Arctocephalus tropicalis* female. Display calls of C) a King penguin *Aptenodytes patagonicus* adult and D) a Gentoo penguin *Pygoscelis papua* adult.

The Problem: Environmental Constraints on Acoustic Recognition

Locate Each other in the Crowd

Almost all species of seabirds have a nest. The penguin family is thus unusual in having 15 nesting and 2 non-testing species, the king penguin, *Aptenodytes patagonicus*, and the emperor penguin, *A. forsteri.* These two large species brood the egg and then the small chick on their feet until chicks are large enough to move on their own. As a consequence, these two "non-nesting" species have to find their partner or their chick without the help of topographical cues. Although penguin species breed in dense colonies numbering hundreds, thousands or even in some colonies, up to two millions pairs, an adult coming from the sea limits the search of its family members to some meeting places: either the nest or previous feeding site for non-nesting penguins. According to the pioneering experiments of Jouventin (1982), penguins are not able to recognize each other by smell or by sight. Visual signals seem to be used only for species recognition and mate choice (Dresp et al. 2005) . Vocal signals are thus the only way for partners to identify each other and their offspring with certainty (Speirs & Davis 1991, Jouventin 1982).

Table 1. Responses of penguins and otariids to experimental signals corresponding to modified natural calls modified (see text)

	Penguins					Otariids		
Experimental signals	**EP**	**KP**	**MP**	**AP**	**GP**	**SFSf**	**SFS p**	**AFS p**
No AM	-	+	-	-	-	+	+	-
No FM	+/-	-	+/-	+/-	+/-	nt	nt	-
FM reversed	+/-	-	+	+	+	-	+/-	-
Low Pass	+	+	+	+	-	+	+	+
High Pass	+/-	-	+	-	-	+/-	+	+
2 H (Fo+H1)	nt	nt	nt	nt	nt	+	+/-	Nt
1H (Fo only)	nt	+/-	nt	-	-	+/-	-	Nt
Energy distribution						+/-	-	-
1 voice suppressed	-	-	nt	nt	nt			
Shift ± 25	nt	nt	+	+/-	+	nt	nt	Nt
Shift ± 50	+	+	+	-	-	nt	+	Nt
Shift ±75	+	+/-	+	-	-	nt	nt	p: + ; n: +
Shift ± 100	-	-	+/-	nt	nt	nt	p: + ; n: +/-	Nt
Shift ± 150						nt	nt	p: + ; n: -
Shift ± 200						nt	p: + ; n: +/-	Nt
Shift ± 300						nt	p: - ; n: -	p: - ; n: -
Half call (1st part)	+/-	+	+	+	nt	nt	+	Nt

Table 1. Continued

	Penguins					Otariids		
1st quarter of call	nt	nt	nt	nt	nt	nt	+	Nt
Last quarter of call	nt	nt	nt	nt	nt	nt	-	Nt
One syllable	-	+	+/-	+	+			
Half syllable	nt	+	-	-	-			

Behavioural responses: +: strong response, +/-: moderate response, -: no response, nt: not tested, p: positive shift, n: negative shift.

EP: Emperor penguin (Robisson et al. 1993, Aubin et al. 2000); KP: King penguin (Jouventin et al. 1999, Lengagne et al. 2000, 2001); MP: Macaroni penguin (Searby, Jouventin & Aubin 2004); AP: Adélie penguin (Jouventin & Aubin 2002); GP: Gentoo penguin (Jouventin and Aubin 2002); SFSf: Subantarctic fur seal female (Charrier et al. 2002); SFSp: Subantarctic fur seal pup (Charrier et al. 2003b); AFSp: Antarctic fur seal pup (Aubin & Charrier unpublished data).

Even if some otariids species show certain site fidelity (i.e., area around where the parturition or the last suckling occurred), there is no precise rendezvous site as a "nest" as shown in some penguin species. Moreover pups are highly mobile in the colony, and they can be anywhere when their mother comes back from the sea. Nevertheless, several observations on fur seal or sea lion species have shown that females tend to come back to the birth place area or the last suckling spot (Peterson & Bartholomew 1967, Marlow 1975, Baker et al. 1985, Wolf and Trillmich 2007). Both protagonists use vocalisations to reunite as penguins do and can also use olfactory signals but only at short range (Miller 1991, Dehnhardt 2002). Olfactory cues are not reliable at long distances since they are confounded by the direction and the speed of the wind. They seem nevertheless to be essential for the pup acceptance by the female. Indeed, observations on mother-pup reunion have shown that a female always smells her pup before accepting to suckle it (Stirling 1971, Riedman 1990, Renouf 1991, Dehnhardt 2002).

Hearing Each other in the Noise

In colonies with a high density of individuals, the noise is high, almost continuous and periods of relative silence are rare, short and unpredictable. Thus, for a King penguins' colony numbering 40000 pairs, it has been shown that the average value of the ambient noise measured at a distance of 2 m from the edge of the colony is around 74 dB_{SPL} (Aubin & Jouventin 1998). For comparison, it has been measured 70 dB_{SPL} in a colony of Emperor penguins (Robisson 1991), 57 dB_{SPL} in a colony of Adélie penguins *Pygoscelis adeliae* (Jouventin & Aubin 2002) and between 69 and 78 dB in a colony of northern fur seal *Callorhinus ursinus* (Kajimura & Sinclair 1992). This high and continuous noise comes first from the calling behaviour of individuals, and second from non-biologically significant sounds like wind and flipper flaps (Aubin & Jouventin 1998, Charrier et al. 2003b). The only noises that have the same spectral and temporal characteristics as that of an adult call, and which would thus theoretically lead to a masking effect, are the calls produced by other adults. The acoustic properties of the masking signals and those of the signal to be detected being similar, the extraction of information can be impeded. The recognition process of the signal is even made more difficult by propagation problems due to the distance between individuals and to the screening effect of bodies of animals, which together impose a particularly difficult problem of acoustic communication. According to propagation tests realised in the field, the signal emitted showed strong degradation with increasing broadcast distance. As the signal propagates, it is degraded mainly by blurring of amplitude and frequency parameters induced by selective frequency-filtering, reverberation and atmospheric turbulence (Willey & Richards 1982, Michelsen & Larsen 1983). In addition, the screening effect of the crossed bodies, often neglected or underestimated in studies dealing with colonial animals, produces an excess attenuation enhancing degradation of the signal transmitted (Aubin & Jouventin 1998, Lengagne et al. 1999a, 1999b). Thus, correlations between a signal recorded at 1 m (reference signal) and propagated signals decrease as distance and number of bodies obstacles increase (see the results for two penguin species and one seal species in Table 1). For amplitude parameters, correlations are weak after 14 m of propagation in penguin colonies and after 32m in seal colonies. Frequency parameters and particularly those related to slow frequency modulations (see correlation values of propagated calls, Figure 2) resist a little bit better to degradation. Nevertheless, according to these propagation experiments, communication involving individual recognition in colonies seems possible only at a short or moderate range. These propagations problems increase thus the difficulty of the meeting between individuals on the move in a crowd.

The Solution: The Acoustic Communication Strategy

The Vocal Stereotypy

The analysis of temporal and frequency features in penguins and otariids have revealed that some particular parameters are highly individualised, showing a great inter-individual variation and a weak intra-individual variation (Figure 3). Coefficient of variation and PCA analyses reveal that individual signatures can be found in the temporal and frequency domains as well for the display calls of all the species of penguins studied (Robisson et al. 1989, Brémond et al. 1990, Jouventin and Aubin 2000, Aubin & Jouventin 2002a) as for the contact calls of all the species of seals studied (for review see Insley et al. 2003, Charrier & Harcourt 2006). Individual stereotypy is a prerequisite for individual recognition. However, such stereotypy does not attest that individual recognition really occurs, as it has been shown in some species of seals (Grey seal *Halychoerus grypus*, McCulloch et al. 1999; Hawaiian monk seal *Monachus schauinslandi*, Job et al. 1995). Thus, playback experiments are essential to assess if vocal signatures are used by the animals for the identification process

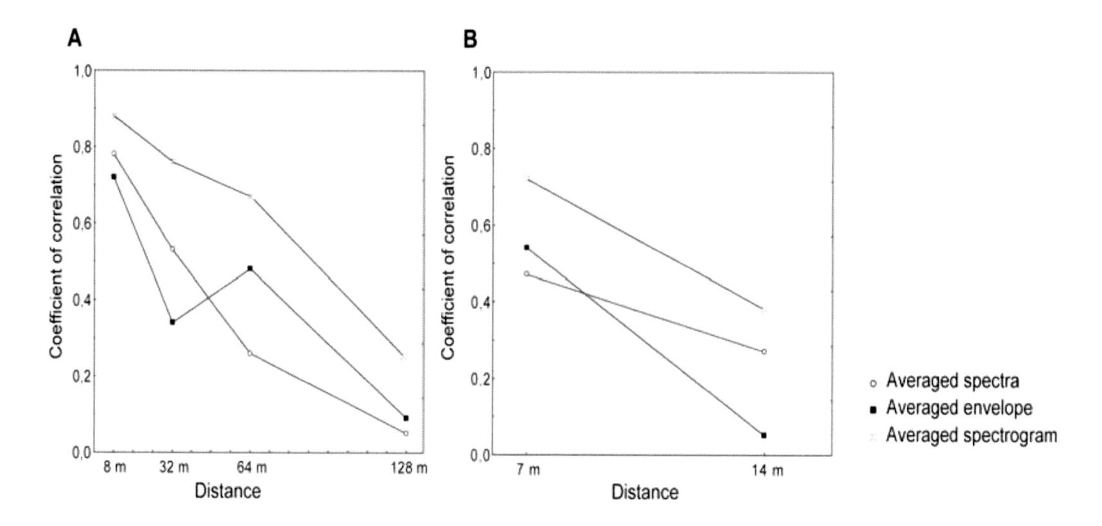

Figure 2. Correlations between reference (i.e., calls recorded at 1m) and propagated signals for averaged envelopes, spectra and spectrograms in A) Antarctic fur seal and B) King penguin (Pearson product-moment correlations, $p<0.05$ in all cases).

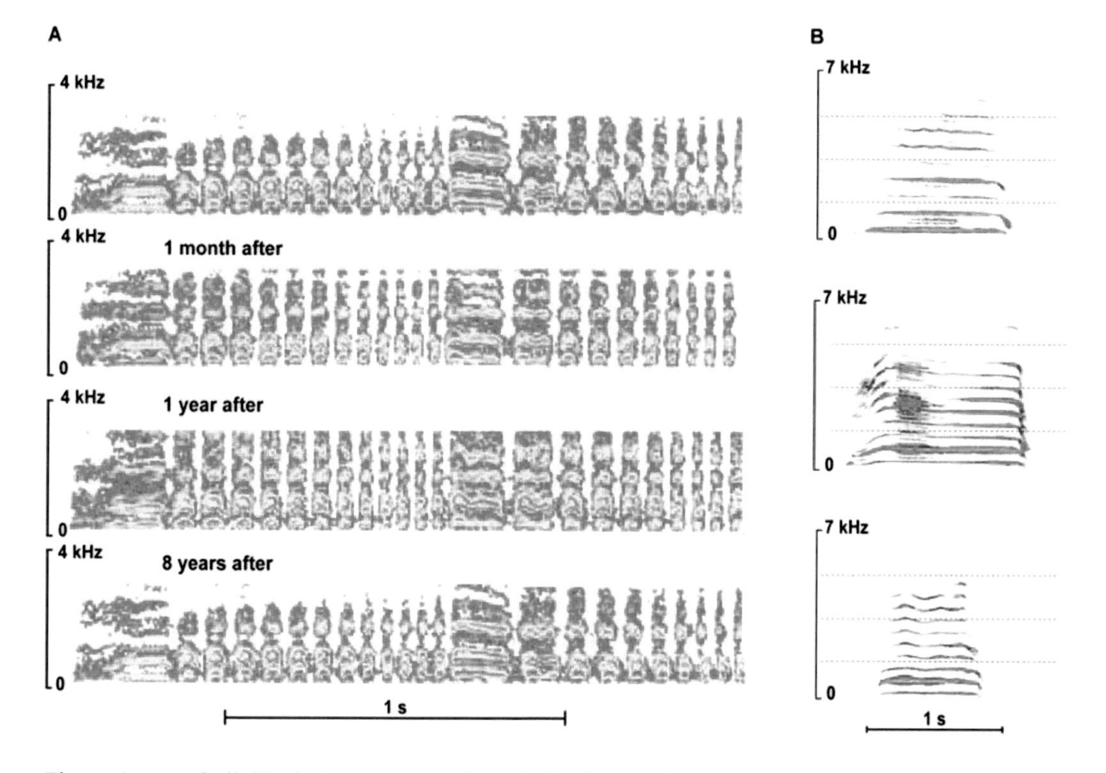

Figure 3. Intra-individual stereotypy and inter-individual variations in signals used by penguins and otariids. A) Sonograms of Emperor penguin *Aptenodytes forsteri* female calls recorded at 4 different ages. B) Sonograms of female attraction calls of three different Subantarctic fur seal pups.

The Coding Process

From their ability to discriminate a target call in the constraining environment of a colony, it is assumed that penguins and otariids use peculiar strategies of coding/decoding. Based on the analyses of calls, some particular acoustic parameters have been proposed to be involved in the recognition process (see above). To assess the effective use of these parameters, one needs to perform playback experiments using signals in which one of these parameters has been modified or removed (e.g., signals without amplitude or frequency modulation, high-pass or low-pass filtered, frequency shifted, – see Table 1). By comparing the behavioural responses obtained with these experimental signals to those obtained with natural signals (i.e. unmodified calls), it is possible to conclude on which acoustic parameters the individual recognition process relies.

In penguins, the individual recognition is observed between mates and between parents and chicks. Experimental studies were realised in the field on five species of penguins: two species without a nest, the Emperor Penguin (EP) (Robisson et al. 1993, Aubin et al. 2000) and the King Penguin (KP) (Jouventin et al. 1999, Lengagne et al. 2000, 2001) and three species with a nest, the Macaroni Penguin (MP) *Eudyptes chrysolophus* (Searby et al. 2004), the Adélie penguin (AP) (Jouventin & Aubin 2002) and the Gentoo Penguin (GP) *Pygoscelis papua* (Jouventin & Aubin 2002).

If we summarize the results obtained, two acoustic code categories emerge in the penguin family, depending of the breeding ecology of birds. The first one concerns the nesting penguins (MP, AP, GP). They identify the mate or the parent by analyzing in the frequency domain mainly the spectral profile and the pitch of the call (timber analysis). The second one concerns the two non-nesting penguins (EP, KP). Both these species have developed a complex "twin-pass system" allowing a reliable coding of individual identity. They use a vocal signature in the time domain based on amplitude/time (AM) analysis for the EP and on frequency/time (FM) analysis for the KP. This temporal analysis is complemented by another sophisticated system: the "two-voice" system. It is well known that the sound-producing organ of birds located at the junction of the two primary bronchi, the syrinx, is potentially a double sound source (Greenewalt 1968, Goller & Larsen 1997). In both EP and KP, the two acoustic sources of the syrinx produce two simultaneous fundamental frequencies and their respective harmonics series (Brémond et al. 1990, Aubin 2004). Playback experiments with natural and modified penguin calls have shown that the two-voices are of a primary importance for the transmission of individual identity information in both species of non-nesting penguins (Aubin et al. 2000, Lengagne et al. 2001). This two-voice system associated with the temporal pattern of syllables creates a huge variety of vocal signatures. This is necessary to distinguish between several thousand birds breeding without nests, i.e. no topographical cues to aid identification (like the presence of a fixed nest). At the opposite, with nesting penguins, the identification by one dimensional parameter as the timber does not offer a great variety of vocal signatures. Possibilities of confusion should exist but, since the nest is used as a meeting place, the probability that a bird emits the right call at the wrong place is weak.

Individual vocal signature has been investigated in only two otariids species: the Subantarctic fur seal *Arctocephalus tropicalis* (Charrier et al. 2002, 2003b) and the Antarctic fur seal *Arctocephalus gazella* (Aubin & Charrier, unpublished data). In the temporal domain, parameters such as amplitude and frequency modulations have been found to be essential in mother-pup recognition. If both Subantarctic and Antarctic fur seals are unable to recognize their mother's or pup's voice when the frequency modulation has been reversed or removed (see table 1), only Antarctic fur seal seems to pay attention to the amplitude modulation pattern for individual recognition. In the frequency domain, both species show a high tolerance for frequency shift, meaning that they do not analyse the accurate frequency values of the different harmonics. Calls in which some harmonics are filtered or in which the distribution of energy between harmonics is randomly changed do not elicit recognition (see table 1). So, energy distribution appears as another important parameter in the identification process.

As shown in penguins without nest, otariids use a complex, multi-parametric signature. With multiple cues conveying a message, there is more possibility that the signal-to-noise ratio for one of them is sufficient to reach the receiver.

These studies have pointed out that vocal signatures depend on the breeding ecology of the animals: the timber analysis of the nesting penguins suits for the easier problem of the individual recognition at a precise location whereas the complex acoustic code of the non-nesting penguins and otariids is an adaptation to breeding conditions that do not allow the use of topographical landmarks for individual identification.

The Serial Redundancy of Information

As predicted by information theory (Shannon & Weaver 1949), animals must repeat the same information to ensure an efficient communication in a noisy channel. In all penguins and colonial seals investigated up to now, redundancy is performed by repeating the same information, at the level of the syllable for penguins or at the level of the call for otariids (in penguins: Aubin & Jouventin 2002b; in otariids: Charrier et al. 2002, 2003b). For instance, the organisation of the signal in a succession of more or less similar syllables is highly redundant in the display calls of penguins. This is particularly obvious for the Emperor, the King, the Adélie and the Gentoo penguins (Aubin et al. 2000, Aubin & Jouventin 2002b, Jouventin & Aubin 2002). In the King penguin, chicks need only 230 ms, that corresponds to half a syllable, to vocally identify their parents without any background noise (Jouventin et al. 1999) although the call duration of the adults lasts from 3 to 6 seconds on average (see table 1). In the Subantarctic fur seal, only a short part of the call is sufficient to elicit recognition (first 25% of the call, see Table 1; Charrier et al. 2003b) and more precisely, only the ascending frequency modulation present at the beginning of the call. It has been experimentally demonstrated that redundancy increases the probability of signal detection in a high background noise. For instance, the King penguins display call corresponds usually to a succession of 4-8 syllables. The broadcast of only one syllable of parents against a normal background noise of a colony elicits few responses from the corresponding chicks, whereas 75% of the tested chicks respond to 2 successive syllables and, with 4 syllables broadcasted, all tested chicks called in reply (Aubin & Jouventin 2002b). Moreover, redundancy can be an active process showing great plasticity. Thus, in King penguin, the duration of the calls changes as wind speed and consequently background noise increases. Lengagne et al. (1999b) showed that the call duration (and thus the number of syllables in the call) is multiplied by 2 when the wind speed changes from less than 8 $m.s^{-1}$ to 11 $m.s^{-1}$. Moreover, the number of emitted calls is also greater in windy conditions than in calm days. This is also true for

Otariids (Charrier, personal observation). This redundancy process allows the individuals to increase the probability of communicating during a short time-window of silence.

The Signal Extraction

On the receiver side, there are some physiological mechanisms that diminish the impact of noise (Klump 1996). The auditory system can be viewed as a selective band-pass filter allowing animals to tune into the frequency bandwidth used by conspecifics and/or as a temporal filter picking-up in the time domain small parts of information to reconstitute the whole message. Penguins and otariids are surprisingly efficient in extracting a target signal in jamming situations even when frequencies of signal and noise overlap. In field experiments with penguins, the parental call was artificially combined with five extraneous adult calls with different emergence levels between the signal to detect and the masker. The chick was able to detect the presence of the parental call at a distance of 7m even when amplitude level of five mixed extraneous call was 6 dB greater than the signal of interest for the King penguin (Aubin &Jouventin 1998) or at the same level for the Emperor and the Adélie penguins (Aubin & Jouventin 2002a, Jouventin & Aubin 2002). This capacity to extract the focus signal from signals of other conspecifics is termed "cocktail-party" effect in speech intelligibility tests.

The Signal Localization

Locating someone in a crowd is not easy without precise knowledge of their approximate location. In an experiment with King penguin (Aubin & Jouventin 2002b), it has been shown that some particular acoustic cues of the call, although not absolutely necessary for vocal recognition, are implicated in the localization of the parents by the chick. These cues are the harmonic structure, the amplitude modulations and the repetition of syllables. Thus, the number of broadcast calls necessary for the chick to localize the source is significantly more important 1) with a signal with AM removed than with a signal with the natural AM and 2) with a signal with the fundamental frequency kept only than with a signal showing the complete harmonic series. These acoustic features, while not directly implicated in the individual recognition process, help the chick to better localise the parental call. These results are in agreement with those obtained in psychoacoustic experiments: animals can locate wide-spectrum signals better than any pure tone, and can locate sounds with sharp amplitude changes better than sounds weakly modulated in amplitude (Wiley & Richards 1982, Dooling 1982). Both penguins and seals exhibit signals that have sharp amplitude changes and wide frequency bandwidth, which both make the call locatable.

Conclusion

In penguins and otariids colonies, the ability to acoustically recognize individuals is crucial for relocating mates or relatives among a multitude of similar-looking individuals. Although acoustic communication is fully hampered by the huge ambient noise of the colony, penguins and otariids succeed in using efficient acoustic strategies: a precise vocal signature, a well-matched code for identification of the signal in the noise, a redundant and locatable acoustic structure of the signal. These findings clearly demonstrate the capacity of colonial animals for detection and recognition of acoustic signals despite severe temporal and spectral jamming, even when the use of topographical cues is strongly limited.

Acknowledgements

Logistic supports were provided by the CNRS and the Institut Paul-Emile Victor. Many thanks to Jean-Claude Brémond, Steve Dobson, Christophe Hildebrant, Pierre Jouventin, Jacques Lauga, Thierry Lengagne, Nicolas Mathevon, Patrice Robisson, Sébastien Santmann and Amanda Searby for their participation to the studies. Special thanks to the members of the 50th and the 51st scientific missions on Amsterdam Island for their help in the field.

References

Aubin, T. (2004). Penguins and their noisy world. In J.M.E. Vielliard, M. L. Silva ML da & R. A. Suthers RA (Eds.), *Advances in Bioacoustics* **76**(2) (pp. 279-283). Rio de Janeiro: Anais da Academia Brasileira de Ciências.

Aubin, T., & Jouventin P. (1998). Cocktail-party effect in King penguin colonies. *Proceedings of the Royal Society of London B: Biological Sciences,* **265**, 1665-1673.

Aubin, T., & Jouventin, P. (2002a). How to identify vocally a kin in a crowd? The penguin model. *Advances in the Study of Behavior* **31**: 243-277.

Aubin, T., & Jouventin, P. (2002b). Localisation of an acoustic signal in a noisy environment: the display call of the King penguin. *Journal of Experimental Biology,* **205**, 3793-3798.

Aubin, T., Jouventin, P., & Hildebrand, C. (2000). Penguins use the two-voice theory to recognise each other. *Proceedings of the Royal Society of London B: Biological Sciences,* **267**, 1081-1087.

Baker, J.D., Antonelis, G.A., Fowler, C.W & York, A.E. (1995). Natal site fidelity in northern fur seals, Callorhinus ursinus. *Animal Behaviour,* **50**, 237–247

Brémond, J-C., Aubin, T., Mbu Nyamsi, R., & Robisson, P. (1990). The song of the Emperor penguin: research of parameters likely to be used for individual recognition. *Comptes Rendus de l'Académie des Sciences,* **311**, 31-35.

Brumm, H. & Slabbekoorn, H. (2005). Acoustic communication in noise. *Advances in the Study of Behavior,* **35**, 251-209.

Charrier, I. & Harcourt, R. G. (2006). Individual Vocal Identity in Mother and Pup Australian sea lion Neophoca cinerea. *Journal of Mammalogy,* **87**, 929-938.

Charrier, I., Mathevon, N. & Jouventin, P. (2001). Mother's voice recognition by seal pups. *Nature,* **412**, 873.

Charrier, I., Mathevon, N. & Jouventin, P. (2002). How does a fur seal mother recognize the voice of her pup ? An experimental study of Arctocephalus tropicalis. *Journal of Experimental Biology,* **205**, 603-612.

Charrier, I., Mathevon, N. & Jouventin, P. (2003a). Individuality in the voice of fur seal females: an analysis study of the Pup Attraction Call in Arctocephalus tropicalis. *Marine Mammal Science,* **19**, 161-172.

Charrier, I., Mathevon, N. & Jouventin, P. (2003b). Vocal signature recognition of mothers by fur seal pups. *Animal Behaviour,* **65**, 543-550.

Dehnhardt, G. (2002). Sensory Systems. In R. Hoezel (Ed.), *Marine Mammal Biology, An Evolutionary Approach* (pp. 116-141). Oxford: Blackwell.

Dooling, R.J. (1982). Auditory perception in birds. In D.E. Kroodsma & E.H. Miller (Eds.), *Acoustic communication in birds (*vol.2, pp. 94-130). Ithaca, NY: Academic Press.

Dresp, B., Jouventin, P., & Langley, K. (2005). Ultraviolet reflecting photonic microstructures in the King Penguin beak. *Biology Letters,* **1**, 310-313.

Goller, F., & Larsen, O.N. (1997). A new mechanism of sound generation in songbirds. *Proceedings of the National Academy of Sciences of the USA*, **94**, 14787-14791.

Greenewalt, C. H. (1968). *Bird song: acoustics and physiology*. Washington, TN: Smithonian Institute Press.

Holland, J., Dabelsteen, T., Pedersen, S.B., & Larsen, O.N. (1998). Degradation of song in the wren *Troglodytes troglodytes*: Implications for information transfer and ranging. *Journal of the Acoustical Society of America,* **103**, 2154-2166.

Hutchison, R. E., Stevenson, J. G., & Thorpe, W. (1968). The basis for individual recognition by voice in the Sandwich tern (*Sterna sandvicensis*). *Behaviour,* **32,** 150-157.

Insley, S. J. (2001). Mother-offspring vocal recognition in northern fur seals is mutual but asymmetrical. *Animal Behaviour,* **61**, 129-137.

Insley, S. J., Phillips, A. V. & Charrier, I. (2003). A review of social recognition in pinnipeds. *Aquatic Mammals,* **29**, 181-201.

Job, D. A., Boness, D. J. & Francis, J. M. (1995). Individual variation in nursing vocalisations of Hawaian monk seal pups, Monachus schauinslandi (Phocidae, Pinnipedia), and lack if maternal recognition. *Canadian Journal of Zoolgy,* **73**, 975-983.

Jouventin, P. (1982). *Visual and vocal signals in Penguins, their evolution and adaptive characters.* Berlin and Hamburg: Paul Parey.

Jouventin, P., & Aubin, T. (2000). Acoustic convergence in the calls of two nocturnal burrowing seabirds. Experiments with a penguin and a shearwater. *Ibis,* **142**, 645-656.

Jouventin, P., & Aubin, T. (2002). Acoustic systems are adapted to breeding ecologies : individual recognition in nesting penguins. *Animal Behaviour,* **64**, 747-757.

Jouventin, P., Aubin, T., & Lengagne, T. (1999). Finding a parent in a King penguin colony : the acoustic system of individual recognition. *Animal Behaviour,* **57**, 1175-1183.

Kajimura, H., & Sinclair, E. 1992. *Fur Seal Investigations, 1990.* US Dep. Commer. NOAA Technical. Memorandum. NMFS-AFSC-2, 192 p.

Klump, G.M. (1996). Bird communication in the noisy world. In D.E. Kroodsma & E.H. Miller (Eds.), *Ecology and Evolution of Acoustic Communication in Birds (*pp. 321-338). Ithaca, NY: Cornell University Press.

Lengagne, T., Lauga, J., & Aubin, T. (2001). Intra-syllabic acoustic signatures used by the King penguin in parent-chick recognition: an experimental approach. *Journal of Experimental Biology,* **204**, 663-672.

Lengagne, T., Aubin, T., Jouventin, P., & Lauga, L. (1999a). Acoustic communication in a King penguin colony: importance of bird location within the colony and of the body position of the listerner. *Polar Biology,* **21**, 262-268.

Lengagne, T., Aubin, T., Lauga, L., & Jouventin, P. (1999b) How do King penguins apply the Mathematical Theory of Information to communicate in windy conditions? *Proceedings of the Royal Society of London B: Biological Sciences,* **266**, 1623-1628.

Lengagne, T., Jouventin, P., & Aubin, T. (1999c). Finding one's mate in a King penguin colony: efficiency of acoustic communication. *Behaviour,* **136**, 833-846.

Lengagne, T., Aubin, T., Jouventin, P., & Lauga, L. (2000). Perceptual salience of individually distinctive features in the calls of adult King penguins. *Journal of the Acoustical Society of America,* **107**, 508-516.

Marlow, B. (1975). The comparative behaviour of the Australasian sea lions Neophoca cinerea and Phocartos hookeri (Pinnepedia: Otariidae). *Mammalia,* **39**, 159-230.

McCulloch, S., Pomeroy, P. P. & Slater, P. J. B. (1999). Individually distinctive pup vocalizations fail to prevent allo-suckling in grey seals. *Canadian Journal of Zoology.* **77**, 716-723.

Michelsen, A. & Larsen, O. N. (1983). Strategies for acoustic communication in complex environments. In F. Huber & H. Markl (Eds.) *Neuroethology and behavioural physiology. Roots and growing points* (pp. 321-331). Berlin: Springer.

Miller, E. H. (1991). Communication in pinnipeds, with special reference to non-acoustic signalling. In D. Renouf (Ed.), *Behaviour of Pinnipeds* (pp. 128-235). London: Chapman and Hall.

Morton, E.S. (1975). Ecological sources of selection on avian sounds. *American Naturalist,* **109**, 17-34.

Page, B., Goldsworthy, S. D. & Hindell, M. A. (2002). Individual vocal traits of mother and pup fur seals. *Bioacoustics,* **13**, 121-143.

Peterson, R. S., & Bartholomew, G. A. (1967). The natural history and behavior of the California sea lion. *American Society of Mammalogy.* Spec. Pub. No. 1.

Phillips, A. V., & Stirling, I. (2000). Vocal individuality in mother and pup South American fur seals, Arctocephalus australis. *Marine Mammal Science,* **16**, 592-616.

Renouf, D. (1991). *Behaviour of Pinnipeds.* London: Chapman and Hall.

Riedman, M. (1990). *The Pinnipeds: Seals, Sea Lions and Walruses.* Berkeley, CA: University of California Press.

Robisson, P. (1991). Broadcast distance of the mutual display call in the Emperor penguin. *Behaviour,* **119**, 302-316.

Robisson, P., Aubin, T., & Brémond, J-C. (1989). Individual recognition by voice in the Emperor penguin: respective parts of the temporal pattern and the sound structure of the courtship song. *Comptes Rendus de l'Académie des Sciences,* **309**, 383-388.

Robisson, P., Aubin, T., & Brémond, J-C. (1993). Individuality in the voice of the Emperor penguin *Aptenodytes forsteri*: adaptation to a noisy environment. *Ethology,* **94**, 279-290.

Ryan, M.J., & Brenowitz, E.A. (1985). The role of body size, phylogeny, and ambient noise in the evolution of bird song. *American Naturalist,* **126**, 87-100.

Searby, A., Jouventin, P., & Aubin, T. (2004). Acoustic recognition in macaroni penguins: an original signature system. *Animal Behaviour,* **67**, 615-625.

Shannon, C.E., & Weaver, W. (1949). *The mathematical theory of information.* Urbana, IL: University of Illinois Press.

Slabbekoorn, H. (2004). Singing in the wild: the ecology of bird song. In P. Marler & H. Slabbekoorn (Eds.) *Nature's music, the science of birdsong* (pp. 178-205). San Diego, London: Elsevier Academic Press.

Speirs, A. H., & Davis, L. S. (1991). Discrimination by the Adélie penguins, *Pygoscelis adeliae*, between the loud mutual calls of mates, neighbours and strangers. *Animal Behaviour,* **41**, 937–944.

Stirling, I. (1971). Studies on the behaviour of the south australian fur seal, Arctocephalus forsteri. 2- Adult females and pups. *Australian Journal of Zoology,* **19**, 267-273.

Trillmich, F. (1981). Mutual mother-pup recognition in Galapagos fur seals and sea lions: cues used and functional significance. *Behaviour,* **78**, 21-42.

Wiley, R. H., & Richards, D. G. (1982). Adaptations for acoustic communication in birds: transmission and signal detection. In D. E. Kroodsma & E. H. Miller (Eds.) *Acoustic communication in birds* (vol. 1, pp. 131-181). Ithaca, NY: Academic Press.

Wolf J. B. W & Trillmich, F. (2007). Fine-scale site fidelity in a breeding colony of the Galápagos sea lion (Zalophus californianus wollebaeki): a prerequisite for social networking? *Oecologia,* **152**, 553-567.

In: New Research on Acoustics
Editor: Benjamin N. Weiss, pp. 299-316
ISBN 978-1-60456-403-7

Chapter 9

ENGINEERING SOLUTIONS USING ACOUSTIC SPECTRAL FINITE ELEMENT METHODS

Andrew T. Peplow[*]
Marcus Wallenberg Laboratory for Sound and Vibration Research,
Dept of Aeronautics & Vehicle Engineering,
KTH, S-100 44, Stockholm, Sweden

Abstract

The spectral finite element method is an advanced implementation of the finite element method in which the solution over each element is expressed in terms of a priori unknown values at carefully selected spectral nodes. These methods are naturally chosen to solve problems in regular rectangular, cylindrical or spherical regions. However in a general irregular region it would be unwise to turn away from the finite element method since models defined in such regions are extremely difficult to implement and solve with a spectral method. Hence for a complex waveguide the method uses the efficiency and accuracy of the spectral method and is combined with the flexibility of finite elements to produce a high–performance engineering tool. Contemporary examples from engineering including fluid-filled pipes, tyre acoustics, silencers and waveguides. Some of these will be reviewed, presented and analysed. From simple examples to complex mixed materials configurations the study will highlight the strengths of the method with respect to standard methods.

1. Introduction

The advantage of the spectral finite element method is that stable solution algorithms and high accuracy can be achieved with a low number of elements under a broad range of conditions. Spectral element techniques are high order methods which allow for either obtaining very accurate results or reducing the number of degrees of freedom for fixed standard precision. The work described in this chapter is concerned with the development of the waveguide spectral finite element method (SFEM). Its beginnings in structures as a combination of the dynamic stiffness method and the finite element method based on a variational

[*]E-mail address: atpeplow@kth.se

formulation for a non-conservative motion. A dynamic stiffness approach for frame structures has been developed originally by Richard and Leung (25). However, a major advance, was made by Gavric (17), where the cross-sectional motion of a given waveguide was approximated by standard finite element polynomials. Wave propagation along the waveguide could then be studied by finding eigenvalues, corresponding to propagating wavenumbers, from a system of differential equations. This innovation inspired the study for beam and plate structures where Finnveden used the spectral finite element method in (10) and (12). Orrenius and Finnveden (21) and more recently Nilsson and Finnveden (20) used this approach for rib–stiffened plate structures used in train wagons. For two-dimensional modelling, one-dimensional finite element shape polynomial functions describe the motion's z–dependence where, without loss of generality, it is assumed that the waveguide is aligned with the x–axis. It follows that nodal displacements, vertical and longitudinal, are functions of the x variable and may be found by the elastic waveguide boundary spectral finite element method. The underlying is through the solution of a matrix polynomial eigenvalue problem. This novel approach has been used to describe the dynamic motion of sandwich composite structures Bonfiglio *et al.* (6) and ground vibration in layered geomechanic media Peplow and Finnveden (22) modelled masses lying within bedrock layers of infinite extent.

Handling complex geometries by spectral finite element methods are now an established alternative to finite difference and finite element methods to solve elliptic Partial Differential Equations (PDE). Spectral methods are naturally chosen to solve problems in regular rectangular, cylindrical or spherical regions. However in a general irregular region it would be unwise to turn away from the finite element method since models defined in such regions are extremely difficult to implement and solve with a spectral method. The examples here show how the spectral method uses boundary data to devise trial functions so that, in essence, no discretization of the interior domain is necessary. Thus the method maybe viewed as a boundary element method as the method is "meshless" as possible but unlike most boundary element methods it is not derived from Greens theorem. Hence for a complex waveguide the method uses the boundary data to devise an efficient and accurate spectral method and is combined with the flexibility of finite elements to produce a high–performance engineering tool. In complex acoustic systems the underlying polynomial eigenvalue problem for acoustic problems are straightforward, however, see section 2. and 5.1. for large–scale three–dimensional problems. Waveguides with uniform cross sectional properties will be described here using the spectral finite element method. For such an analysis, a natural starting point is the variational formulation. In this context, this type of finite element analysis first appeared in Finnveden (13) and (11) for fluid–filled flexible pipes , and was used to compare with experimental techniques by Finnveden and Pinnington (15). The effect of including distributed loading by turbulent boundary layers on pipes can be found in (3). Peplow and Finnveden developed the method to study sound transmission in various acoustic waveguide configurations in (23). Recently the SFEM has been used for the study of and comparison with measurements, by Birgersson *et al.* to predict the response of a structure by turbulence excitation (14). (5), and (4). Applications of the method can be found in the theses by Birgersson (2) and Nilsson (19) on applications in fluid–structure interaction and Fraggstedt (16) on modelling wave propagation in car tyres. For three-dimensional problems, two-dimensional polynomial shape functions describe the motion's y– and z– dependence. Kirby used collocation to determine transmission loss for

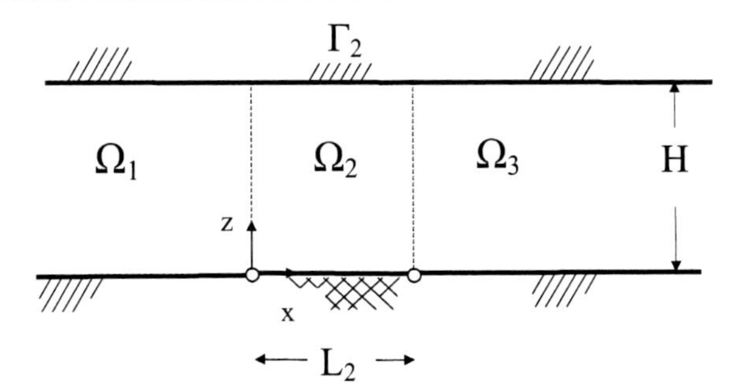

Figure 1. Two dimensional acousti ducts with expansion chamber.

elliptic dissipative silencers with and without mean flow (18) and (8). Although a specific three–dimensional finite element was not constructed the efficiency of the method is clearly described.

In general the SFEM technique is an important development in engineering analysis. The flexibility of the waveguide boundary spectral finite element and the versatility of the modelling applications give new solutions to engineering dynamics and acoustic problems.

2. Weak Formulation for the Spectral Finite Element

A weak formulation is based on introducing the weight function $\chi(\vec{x})$ and testing it with the Helmholtz operator such that integrating by parts gives

$$\begin{aligned} L_\Omega &= \int_\Gamma \chi(\vec{x})\, a v_f(\vec{x}) d\Gamma(\vec{x}) - \int_\Omega \left[\vec{\nabla}\chi(\vec{x}) \cdot \vec{\nabla} p(\vec{x}) - k^2 \chi(\vec{x}) p(\vec{x})\right] d\Omega(\vec{x}) \\ &= k^2 \int_\Omega \chi(\vec{x}) p(\vec{x}) d\Omega(\vec{x}) - \int_\Omega \chi_z(\vec{x}) p_z(\vec{x}) d\Omega(\vec{x}) \\ &+ \int_\Gamma \chi(\vec{x})\, a v_f(\vec{x}) d\Gamma(\vec{x}) - \int_\Omega \chi_x(\vec{x}) p_x(\vec{x}) d\Omega(\vec{x}) = 0. \end{aligned} \quad (1)$$

Often, equation (1) represents the starting point for conventional finite element discretizations, e.g. Galerkin method. The second part (lower row) consists of a domain integral and a boundary integral.

The variational statement Eq. (1) is used to obtain *wave influence* basis functions $[W(x)]$. An approximate solution to the original problem is also found by selecting a solution $p_j(x,z)$ from a discrete set of trial functions determined by a finite element discretization of a region with a uniform cross–section Ω_j.

2.1. Approximation Functions

We approximate the sound pressure $p(\vec{x})$ as

$$p(\vec{x}) = \sum_{i=1}^{N} \phi_i(\vec{x})\, p_i = \phi^T(\vec{x}) p \quad (2)$$

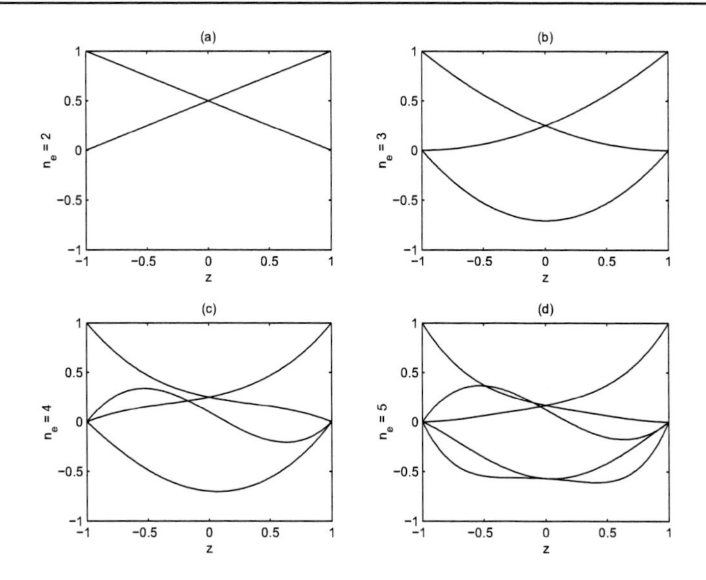

Figure 2. Linear to quartic functions, $\phi(s)$, $N = n_e$. Polynomials in (4) is illustrated in (b).

where p_i represents the discrete sound pressure at point $\vec{x}_i$ and ϕ_i is the $i-$th basis function for our approximation. First we start the analysis with a two–dimensional formulation and then followed by an three–dimensional example.

Figure 1 shows a typical two–dimensional waveguide section where $\Omega = \Omega_1 \cup \ldots \cup \Omega_5$. In the super–spectral method this section will define one *waveguide finite element*. Within one *super–spectral finite element*, $\Omega_2 := D \times C$, $D := [0, L_2]$, $C := [0, H]$ for example, the approximate solution $p(.)$ may be represented by an expression of the form

$$p(\vec{x}) = \sum_{i=1}^{N} p_i \phi_i(z) W_i(x) \tag{3}$$

using piecewise quadratic polynomial shape functions. Shape functions, $\phi(s)$, are defined sub–locally over the cross–sections. Elemental shape functions $W(x)$ are defined later in this section. The complete set of quadratic polynomials defined over $[-1, 1]$ is $\phi(s) = [\phi^1(s)\, \phi^2(s)\, \phi^3(s)]$ where

$$\phi^1(s) = \frac{1}{4}\left(1 - 2s + s^2\right), \quad \phi^2(s) = \frac{1}{4}\left(1 + 2s + s^2\right), \quad \phi^3(s) = \frac{-1}{\sqrt{2}}\left(1 - s^2\right). \tag{4}$$

Consider the functional L_Ω, (1), for a single arbitrary region Ω_2 say. Substitution of expression (3) into the resulting form, where subscript $_x$ denotes x–derivative, yields the approximation

$$\begin{aligned} L_{\Omega_j} &= k^2 \int_{\Omega_i} W^T K_1 W \, dx - \int_{\Omega_i} W^T K_2 W \, dx \\ &+ ik \int_{\Omega_i} W^T K_3 W\} \, dx - \int_{\Omega_i} W_x^T K_4 W_x \, dx, \end{aligned} \tag{5}$$

where the matrices are defined by

$$K_1 = \int_C \phi^T(z)\phi(z)\,dz, \tag{6}$$

$$K_2 = \int_C \frac{d\phi}{dz}^T \frac{d\phi}{dz}\,dz, \tag{7}$$

$$K_3 = \left[\frac{1}{\zeta}\right]_0^H, \tag{8}$$

$$K_4 = \int_C \phi^T(z)\phi(z)\,dz. \tag{9}$$

Explicitly the *mass* and *stiffness* matrices in (6) and (7), defined for the interval $C = [0,H]$, are given by

$$K_1 = H\int_{-1}^{+1} \phi^T(s)\phi(s)\,ds, = \frac{H}{15}\begin{bmatrix} 6 & 1 & -6\sqrt{2} \\ 1 & 6 & -6\sqrt{2} \\ -6\sqrt{2} & -6\sqrt{2} & 8 \end{bmatrix}, \tag{10}$$

$$K_2 = H\int_{-1}^{+1} \frac{d\phi}{ds}^T \frac{d\phi}{ds}\,ds = \frac{H}{3}\begin{bmatrix} 2 & -1 & \sqrt{2} \\ -1 & 2 & \sqrt{2} \\ \sqrt{2} & \sqrt{2} & 4 \end{bmatrix}. \tag{11}$$

For each geometrical sector Ω in Fig.1 the finite element cross–sectional matrices $K_1,\ldots,K_4$ are frequency independent, real valued and are fairly small in size (for a two–dimensional problem at least) and hence may be stored. In the case here these are clearly 3×3 real–valued symmetric matrices. For a full problem a dynamic stiffness matrix requires assemblage. To do this wave influence functions W for each sector Ω are required.

3. Wave Influence Functions & the Dynamic Stiffness Matrix

Construction of the wave influence functions follows by consideration of the ordinary differential equations which correspond to Eq. (5) found by taking an appropriate first variation and *ignoring* any boundary conditions:

$$K_4\frac{d^2}{dx^2}W(x) + k^2K_1W(x) - K_2W(x) + ikK_3W(x) = 0. \tag{12}$$

Crucial to the fundamental principle of determining waveguide boundary spectral finite elements is that the system of equations Eq. (12) are **not** defined over a specific region. It could be argued that Eq. (12) is defined over a region of infinite length. The differential equations have constant coefficients in the form of symmetric positive and semi–definite $(N\times N)$ real–valued matrices $K_1,\ldots,K_4$. Hence, the solutions of the linear homogeneous system may be written as :

$$W_m(x) = \Phi_m \; e^{i\lambda_{m-1}x}, \quad m = 1,\ldots,N \tag{13}$$

where Φ_m is a vector representing the cross–sectional mode shapes. Under this assumption Eq. (13) reduces to a linear eigenvalue problem, K:

$$K(\lambda)\Phi = \{k^2K_1 - K_2 + ikK_3 - \lambda^2K_4\}\Phi = 0 \tag{14}$$

of order N for the parameters λ^2. The solutions of Eq. (14) yield values for λ that occur in pairs, $\lambda^{\pm} = \pm\lambda$, indicating that pairs of eigenmodes result with the same phase speed propagating in the positive and negative axial directions. The dimension of the eigenvalue problem is $(N \times N)$ and a finite $2N$ number of propagating wavenumbers are obtained.

The resolution of the matrix eigenvalue problem Eq. (14) itself may be achieved by a number of standard computational routines. In the present analysis a *QZ* algorithm was used as implemented in Matlab 7.0.2 and requires $46N^3$ operations to determine all eigenvalues and right eigenvectors. For *large* problems this numerical analysis procedure can dominate the total computation time for a single problem. This produces a complete set of cross–sectional mode shapes and corresponding eigenvalues λ_m^2, $m = 0, \ldots, N-1$.

Now the finite element trial functions have been constructed the dynamic stiffness for a general spectral waveguide element Ω is described. The local dynamic stiffness matrix defined over a region Ω shown in Fig. 1 will now be described. A simple translation to $T := \{x : -D \leq x \leq +D\}$, where $2D$ is the length of the sector, is a key element in defining the wave influence trial functions. By consideration of the eigensolutions in Eq. (13) it is clear that each wave influence function may be written as

$$W_{jk}(x) = \sum_{l=1}^{2N} \overline{\Phi}_{jl} E_{ll}(x) A_{lk}\, p_k, \quad j = 1, \ldots, N, \quad k = 1, \ldots, 2N, \tag{15}$$

where entries $\overline{\Phi}_{jl}$ and E_{ll}, a diagonal matrix, take the values of the eigenmodes and wavefunctions respectively. Coefficients A_{lk} are determined by appropriate scaling of the set of wave influence functions, and p_k are the unknown coefficients. The local dynamic stiffness matrix for a certain element may be written in matrix form as:

$$L = B^T\left(k^2K_1 - K_2\right)B - \lceil\Lambda\rfloor B^T K_4 B \lceil\Lambda\rfloor + ikB^T K_3 B, \tag{16}$$

where the matrix B, of order $2N \times 2N$, combines wave influence function matrices

$$B = \overline{\Phi} \ \lceil E \rfloor \ A. \tag{17}$$

The *ansatz* dynamic stiffness matrix above may seem a little contrived compared to direct FEM but it can be constructed very simply from Eq. (16) using matrix algebra operations. Within the computation of local dynamic stiffness matrix entries use is made of an important matrix generating function for the combination of matrix diagonal exponential terms which has an analytic form.

The global dynamic stiffness matrix is generated by calculating local dynamic stiffness matrices for each region and enforcing continuity of pressure (and velocity) across neighbouring interfaces. For example, consider a uniform waveguide geometry consisting of three spectral elements, as in Fig. 1, with N degrees of freedom across each element cross–section. A $N \times N$ matrix eigenvalue problem is solved and a $2N \times 2N$ local dynamic stiffness matrix generated for each element. Enforcing continuity across neighbours results in a $4N \times 4N$ global dynamic stiffness matrix. Acoustic sources may be modelled as volume point sources but in the following examples normal accelerations are applied on the left–hand boundary. The total number of operations for solving this problem for an arbitrary length waveguide are around $57N^3$ for all the Gaussian elimination operations and $138N^3$ operations for all eigenvalue and eigenfunction computations. Note that any code written can be made efficient by re–using elements and finite element matrices. The estimates above represent over–estimates of real computations.

Construction of dynamic stiffness matrices for layered media follows exactly as with construction of stiffness matrices for a single layer waveguide as above. Hence it is possible to solve multi–layered waveguide problems using super–spectral elements bearing in mind the large generalised matrix eigenvalue problems to be solved. This is performed in Example 2 for multi–layered dissipative elements.

4. Examples for Two–Dimensional Spectral Finite Elements

One area in acoustical engineering that is well–suited to spectral waveguide finite element methods is the design of silencer systems for noise control. There is much work that has been done for smaller systems such as those used in automobiles and small engines, however, the design of much larger systems (such as the parallel baffle type used for gas turbines and other large industrial machines) is still largely guesswork and empirical extensions of previous results. Due to the large size, difficulties in testing and high costs of these silencer systems, the ability to accurately predict the performance before construction and commissioning would be very beneficial. To properly predict the performance of a silencer system, many factors need to be involved in the calculation. Geometrical concerns, absorptive material characteristics, flow effects (turbulence), break out noise, self-generated noise, and source impedance all need to be included in the design calculations of insertion loss (IL). It is very important to note that the method derivations, and their use with the numerical methods are based on plane wave propagation sound sources (i.e. the entire face of the inlet section moving in unison) and an anechoic termination, *i.e.* $\zeta = \rho c$ at the left and right hand end sections of regions Ω_1 and Ω_4 respectively. Anechoic termination at the outlet and the inlet pipes are assumed.

In both two–dimensional examples cubic polynomials corresponding to $N = 4$ for SFEM and cubic polynomials defined over 9–noded triangular elements for FEMLAB 3.1 were used. The total number of degrees of freedom (DOF) for the spectral waveguide finite element totalled 68 and using FEMLAB 3.1 (9) 2210 DOF over 472 triangles were required. The length of the inlet and outlet pipes for the SFEM and FEMLAB geomtries were 1.0 m. Note that the pipe lengths for SFEM could have been considerable longer but for the purposes of sensible computations it was necessary to keep the DOF for FEMLAB 3.1 to a minimum. The characteristic impedance $\zeta = \rho c$ was applied to both inlet and outlet pipes outer boundaries. The height of the outer pipes was fixed at $H = 0.05$ m and chamber extended by an extra $H_1 = 0.2$ m depth. In keeping with Belawchuk (1) the length of the chamber was fixed at $L_3 = 1.2$ m. To compare efforts of the two methods the CPU timing for 55 frequencies was 3.9 s for SFEM and 26.3 s CPU expenditure for FEMLAB (9). It seems that the profit speedup for SFEM is rather conservative. However, if the chamber was

$L_3 = 2.4$ m long the CPU time would remain 3.9 s for the waveguide finite element method, but FEMLAB 3.1 would now require 849 elements with 3928 DOF and using UMFPACK CPU expenditure would increase to 46.5 s. Both a substantial increase in both storage and computation time compared to the spectral element method.

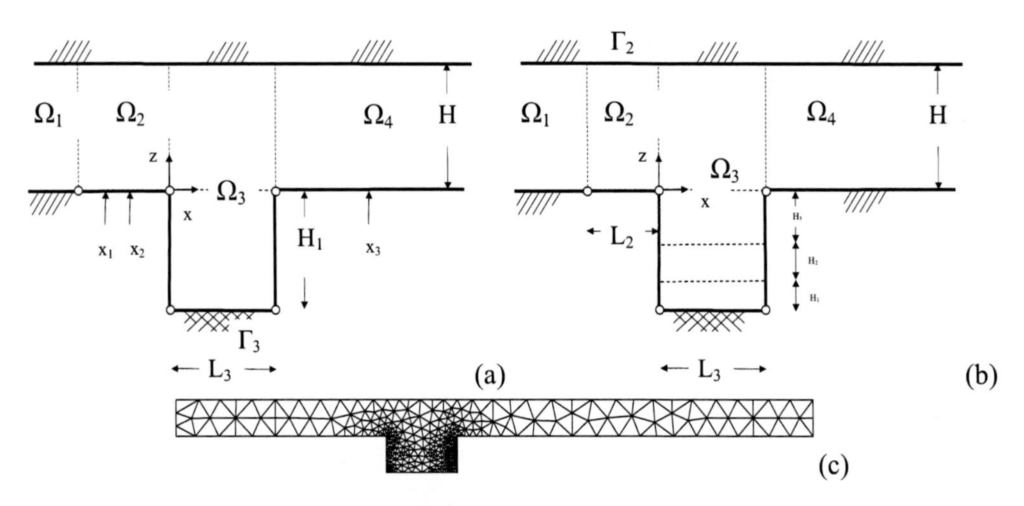

Figure 3. (a) Geometry used for expansion chamber examples. Inlet and outlet pipes are regions $\Omega_{1,2}$ and Ω_4 respectively and Ω_3 is the expansion chamber. Absorbing material may be placed along the boundary Γ_3; (b) Geometry used for expansion chamber example including dissipative lining in chamber. Lining material material may be included in the regions of height $H_1 = 0.05$ m and $H_2 = 0.1$ m. Region of height $H_3 = 0.05$m is air ; (c) typical standard FE mesh for silencer problem with 586 triangles and 354 nodes. Note that (a) and (b) are *meshes* for SFEM.

4.1. Theory for Transmission Loss

The definition of transmission loss is the ratio of the incident sound power to the transmitted sound power. As long as the inlet and outlet regions of the silencer are of the same cross section, and the properties of the fluid (density, temperature) do not change, then the TL can be expressed as:

$$TL = 20\log 10 \left| \frac{P_i}{P_t} \right| \tag{18}$$

where P_i is the rms pressure of the incident wave without silencer in place, P_t is the rms pressure of the transmitted wave with silencer in place. This can be simplified to the following equation: $TL = SPL_i - SPL_t$ where it is understood that SPL_i is obtained without the silencer in place, and SPL_t is obtained with the silencer in place, on the exhaust side of the silencer. Fig. 3(a) illustrates the geometry used to calculate SPL_i in region Ω_2 and SPL_t in Ω_4. The SPL_i is calculated with the straight pipe (no expansion chamber) and the SPL_t is calculated with the expansion chamber (no straight pipe). The inlet and outlet sections have the characteristic impedance ($\zeta = \rho c$) boundary condition applied. This models a completely anechoic source and termination. Also the inlet section is given a unit acceleration amplitude to model a sound source.

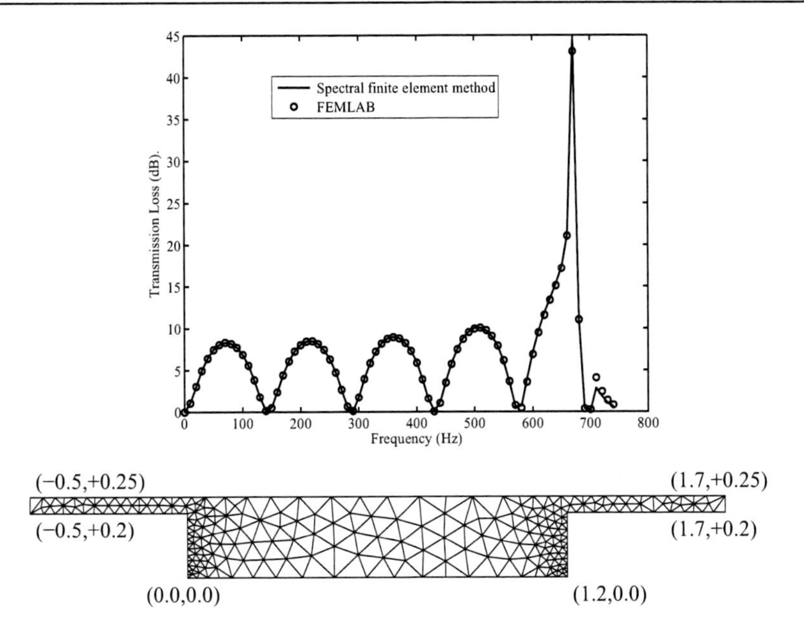

Figure 4. Transmission loss results for expansion chamber shown in Fig. 3(a), height and length of chamber $H_3 = 0.2$, $L_3 = 1.2$m respectively and width of pipes $H = 0.05$m. Mesh for FEMLAB results, # of triangles = 402, # nodes = 260.

4.2. Example 1. Absorbing Boundary Material Lining Silencer Chamber

Starting from the one-dimensional wave equation, the so called 3-point method can be derived as Bilawchuk (1):

$$p_i = \frac{p_1 - p_2 e^{ikx_{12}}}{1 - e^{2ikx_{12}}} \tag{19}$$

where referring to Fig. 3: p_i is the incoming contribution of rms sound pressure wave; p_1 is the rms sound pressure at location x_1; p_2 =rms sound pressure at location x_2; ; $x_{12} = x_2 - x_1$ (microphone spacing). Now that the incoming rms pressure values are known, the exiting rms pressure can be obtained and the TL can be calculated simply as follows:

$$TL = 20\log 10 \left| \frac{p_i}{p_3} \right| \tag{20}$$

where p_3 is the rms sound pressure at point x_3. The rms pressure this point can obtained directly since the termination at the exit is given the characteristic impedance ($Z = \rho c$). The FEM and/or BEM calculations can then be started. In the post-processing stage, the pressures at points x_1, x_2, and x_3 can be calculated and, knowing the distances $x_1 = -0.12$ m and $x_2 = -0.1, x_3 = 1.3$ m and the wave number, k, the transmission loss can be determined. Fig. 4 shows results from SFEM and FEMLAB. Accuracy of the waveguide finite element method is clearly evident. Fig. 5 shows the change in transmission loss where the chamber lining, Γ_3, is lined with absorbing material. Flow resistivity values taken were from Delany and Bazley formula (7).

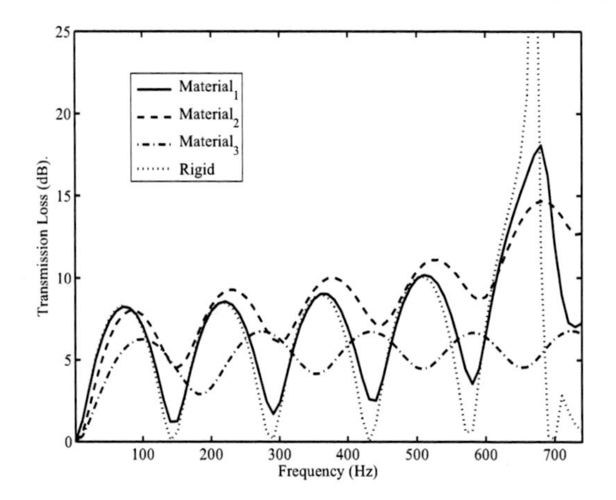

Figure 5. Transmission loss for rigid expansion chamber and chamber with three types of absorbing lining along Γ_3. Flow resistivity values, from (7), for materials are $\sigma_{1,2,3} = 400, 50, \& 1 \times 10^3$ Nsm^{-4}.

4.3. Example 2. Dissipative Fibrous Material Chambers

The acoustic performance of a dissipative expansion chamber lined with two layers of fibrous material with different resistances is investigated as a two–dimensional version of Selamet *et al.* (26). A two–dimensional numerical approach is used to determine the transmission loss of this dissipative silencer. The flow resistivity of the fibre in the dissipative chamber greatly influences the acoustic performance. The model used describes complex valued characteristic impedance and wavenumber leading to complex values of wavespeed and density to input into SFEM scheme:

$$\hat{Z} = \rho c\left(1+0.0855(f/R)^{-0.754}\right) - 0.0765(f/R)^{-0.732} i, \tag{21}$$

$$\hat{k} = \omega\left(1+0.1472(f/R)^{-0.577}\right) - 0.1734(f/R)^{-0.595} i \tag{22}$$

Generally, the increasing resistance of fibre in the dissipative chamber improves the sound attenuation in the mid to high frequency range, while deteriorating to a degree at low frequencies.

Thus, to improve the sound attenuation performance at all frequencies, it is a paradox to design a dissipative expansion chamber filled completely with a unique fibre. The present study considers a layered dissipative silencer to investigate the potential trade-offs. Thus a single–pass expansion chamber lined with two fibre layers of different fibre resistance is examined primarily by the SFEM approach, in Fig.6. Dissipative material be included in the regions of height $H_1 = 0.05$ m and $H_2 = 0.1$ m, see Fig.3(b). Respectively Dissipative material$_1$ comprised fibrous materials R_1 and R_1, Material$_2$ comprised fibrous materials R_2 and R_1, Material$_3$ comprised materials R_3 and R_1, and Material$_4$ comprised materials R_4 and R_1.

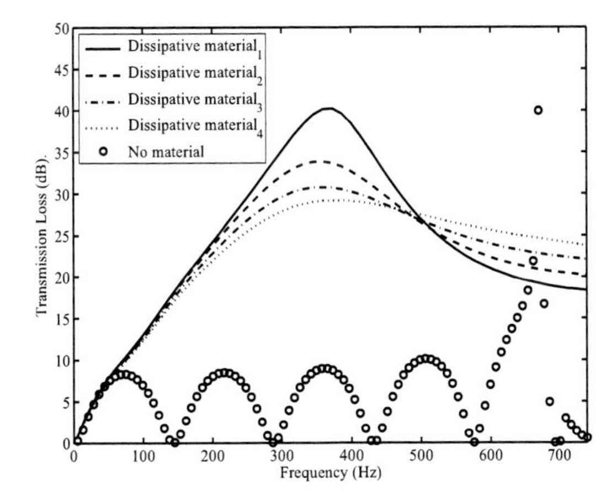

Figure 6. Transmission loss for rigid expansion chamber and chamber with four types of dissipative lining material configurations in Fig. 3(b). Resistivity values used $R_1 = 5000$ rayls/m, $R_2 = 10,000$ rayls/m, $R_3 = 17,000$ rayls/m, and $R_4 = 25,000$ rayls/m. See text for nomenclature.

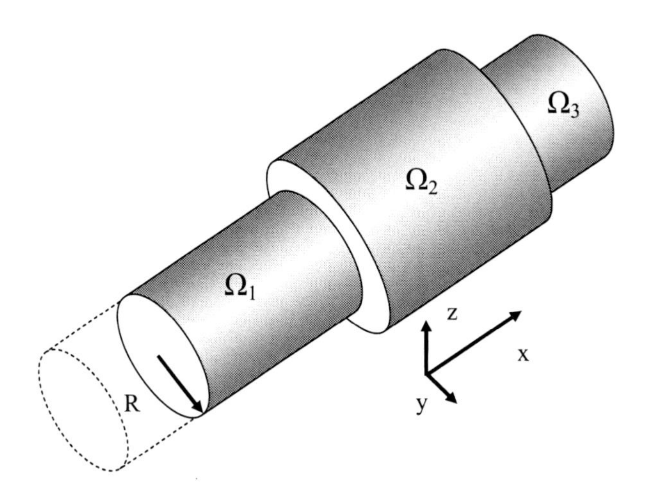

Figure 7. Geometry for the three–dimensional waveguide problem. The system is finite in length with circular inlet and outlet pipes with radius R. The geometry of the silencer is arbitrary, but has an axially uniform cross–section.

5. Results from a Three Dimensional Analysis

A numerical technique has been developed for the analysis of rigid and absorbent lined silencers of arbitrary, but axially uniform, cross–section. The analysis begins by employing the spectral finite element method to extract the eigenvalues and associated eigenvectors for a silencer chamber. It is demonstrated also that the technique presented offers a considerable reduction in the computational expenditure when compared to a three-dimensional finite element analysis. The method for determining transmission loss predictions from section 4.2. may be used to compare with experimental measurements taken for automotive dissipative silencers with elliptical cross sections.

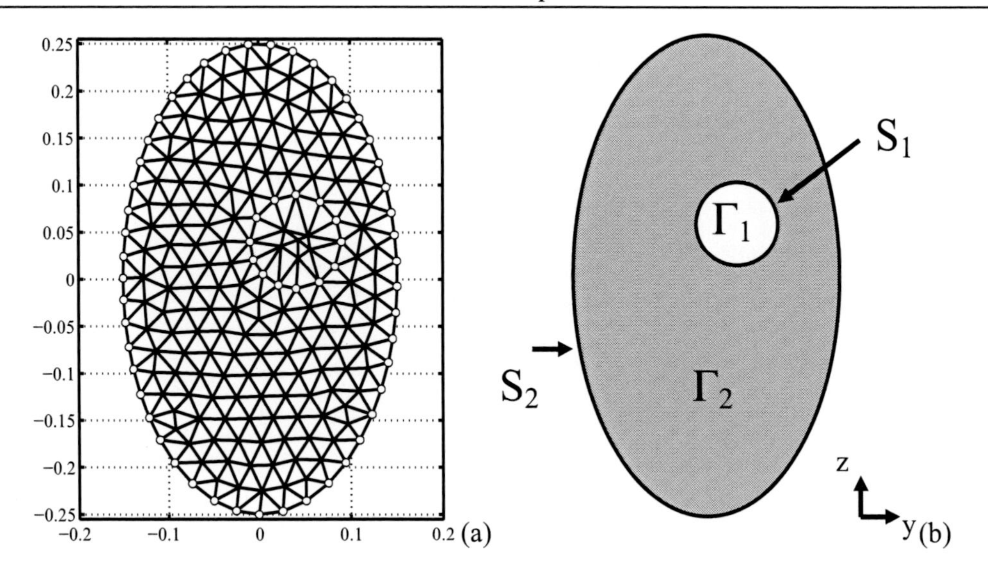

Figure 8. Typical mesh for expansion chambers and cross–sections Γ_2, Γ_1 clearly shown in the right–hand figure.

The dissipative silencer consists of a concentric tube of arbitrary cross section is surrounded by absorbent lined material, on S_2 see Fig. 8(b). The silencer chamber, which has a length L, is assumed to be uniform along its length, the outer walls of which are assumed to be absorbent except for the final example. The inlet and outlet pipes regions Ω_1 and Ω_3 are identical, each having a circular cross section radius R with rigid walls.

5.1. Dispersion Relations for Three-Dimensional Examples

If the outcome of an analysis is a dispersion relation between frequency and propagating wavenumber it is prudent to use a sparse eigensolver, such as **eigs** in MATLAB 7.0.2. The finite element mesh for the silencer chamber cross–section consisted of three-noded triangular elements and was generated using *distmesh* by Persson and Strang (24), see Fig.8(a). The elliptical cross–section had major-axis radius 0.25 m and minor–axis radius 0.15 m. For the silencer, 470 elements equating to 310 nodes (DOFs) were used to mesh the chamber. Finite element meshes, not shown here, for a square cross–section of width 0.5 m and circular cross–section, radius 0.25 m, were also constructed.

The finite element mesh (including inlet and outlet circular section) with 470 triangles and 310 nodes is shown in Fig. 8. The acoustic pressure in the three–dimensional problem is approximated by piecewise linear triangular elements in the cross–section and wave influence functions in the axial direction, similar to the two–dimensional case, Eq. (3)

$$p(\vec{x}) = \sum_{J=1}^{N} p_J \phi_J(y,z) W_J(x). \tag{23}$$

To arrive at the eigenvalue problem, in order to derive the wave trial functions, the following

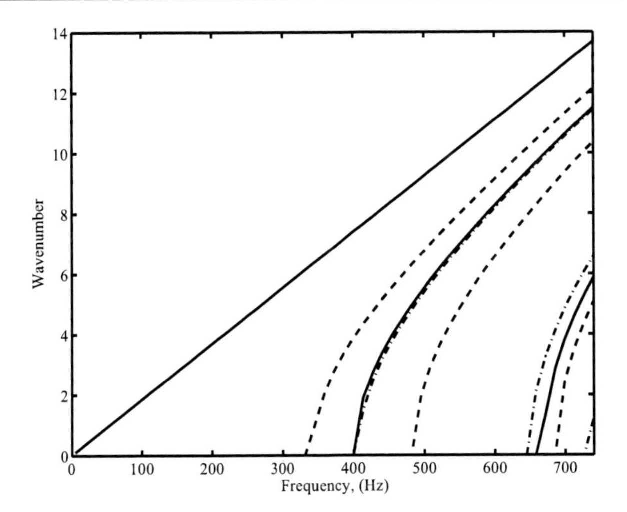

Figure 9. Dispersion curves for various silencer chambers. Square cross–section of width 0.5 m (dashed line), circular cross–section radius 0.25 m (solid line), and elliptical cross–section major-axis radius 0.25 m minor–axis radius 0.15 m. Mesh (including inlet and outlet circular section) used with 470 triangles and 310 nodes shown in Fig. 8.

matrix entries are assembled across the elliptic and cylindrical cross–sections, see Eq. (6)

$$K_{1(IJ)} = \int_{\Gamma_{1,2,3}} \phi_I^T(y,z)\phi_J(y,z)\,dy\,dz, \tag{24}$$

$$K_{2(IJ)} = \int_{\Gamma_{1,2,3}} \nabla\phi_I^T(y,z)\cdot\nabla\phi_J(y,z)\,dy\,dz, \tag{25}$$

$$K_{3(IJ)} = \int_{S_{1,2,3}} \frac{1}{\zeta(y,z)}\phi_I^T(y,z)\phi_J(y,z)\,dy\,dz, \tag{26}$$

$$K_{4(IJ)} = \int_{\Gamma_{1,2,3}} \phi_I^T(y,z)\phi_J(y,z)\,dy\,dz \tag{27}$$

where $\zeta(y,z)$ is the function defining specific surface impedance on the walls of the pipes and the chamber, that is the boundaries of $S_{1,2,3}$. The corresponding eigenvalue problem for the problem becomes

$$K(\lambda)\Phi = \left\{k^2K_1 - K_2 + ikK_3 - \lambda^2K_4\right\}\Phi = 0 \tag{28}$$

Energy transmission through a waveguide system is possible when the propagating wavenumber λ, in Eq. (29) has zero imaginary component. For a given frequency or wavenumber k it is possible to solve the eigenvalue problem below seeking real–valued wavenumbers and their corresponding mode shapes

$$W_m(x) = \Phi_m \quad e^{i\lambda_{m-1}x} \quad , \quad m = 1,\ldots,N \tag{29}$$

giving a dispersion relation between frequency $\omega = 2\pi f$ and wavenumber λ. The eigenvalue problem corresponding to the elliptic expansion chamber equated to solving a 310×310 *sparse* eigenvalue problem. For 55 frequencies the sparse eigensolver, for the first seven

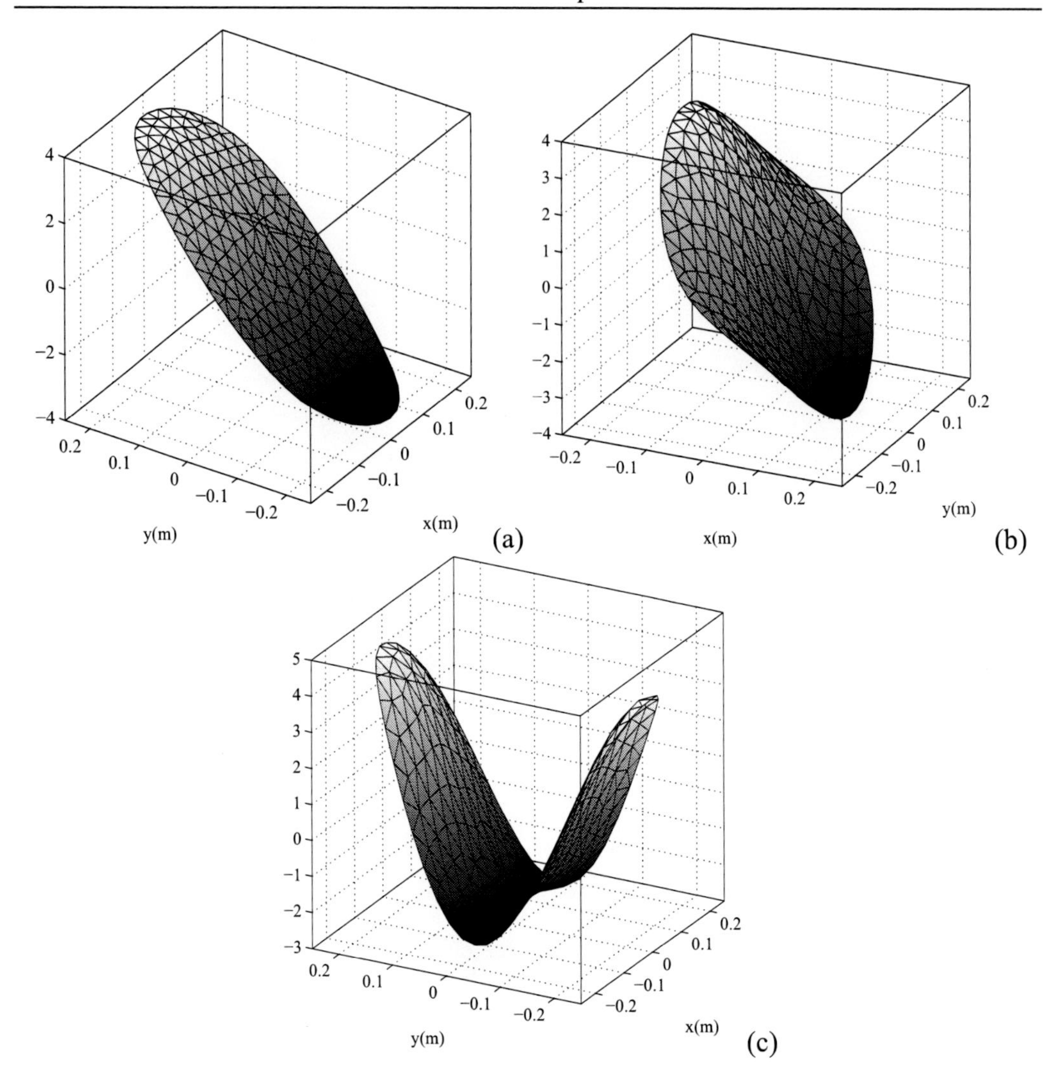

Figure 10. Propagating modes for elliptic cross–section expansion chamber, at excitation frequency $f = 740$ Hz for geometry in Fig. 8

eigenvalues λ^2 with smallest imaginary part, CPU time costing around 15 s for the ellipse. For the circular and square cross–sections the CPU time increased to around 20 s due to the increased number of DOFs. Figure 9 shows dispersion relations for the three square, circular and elliptical configurations. The curves for the square and circular cross–sections could be easily compared with known solutions. Note that four propagating modes, $\boldsymbol{\Phi}$, exist for the elliptic geometry at 740 Hz. Omitting the plane–wave mode, corresponding to $\lambda_0 = 0$, these are shown in Fig. 10 from lowest wavenumber to highest wavenumber respectively.

5.2. Solutions for Elliptic Cylinder Silencer Problem

This section illustrates computations for a finite length combination of pipes and chamber assuming unit normal acceleration at the end of the inlet pipe. The mesh used, Fig. 9,

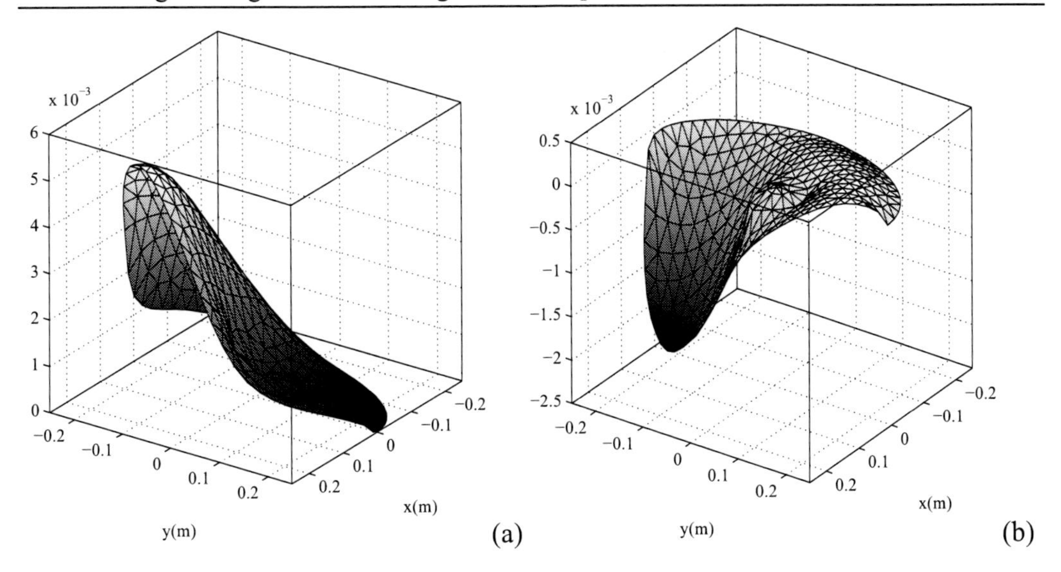

Figure 11. Acoustic pressure at two cross–sections of lined expansion chamber (length $L = 0.4$ m) due to unit normal acceleration at inlet pipe (length $L = 1.0$ m) at excitation frequency 740 Hz. Left hand plot shows acoustic pressure at entry of chamber and right hand plot acoustic pressure at exit of chamber. Outlet pipe length of 4.0 m with rigid termination, geometry of pipes shown in Fig. 7 and Fig. 8.

shows the finite elements for the chamber and the cross–section mesh for the rigid inlet and outlet pipes, located just above the centre. The computation of acoustic pressure for the silencer problem is dominated by the assemblage of the wave influence functions Eq. (28) and the corresponding dynamic stiffness matrix for the silencer chamber Eq. (16). All the frequency independent matrices $K_1,\ldots,K_4$ are stored as sparse matrices. However, all the eigenvectors and eigenvalues are required for the elliptic chamber problem of size $N_2 = 310$, requiring $46N_2^3$ operations. The dynamic stiffness matrix for the chamber costs a little over $5N_2^3$ operations, due to re–use of *LU* decomposition, from Eq. (**??**). For the inlet and outlet pipes the numbers of degrees of freedom are somewhat lower, $N_{1,3} = 20$. Computing the wave influence functions and the dynamic stiffness matrices for the inlet and outlet pipes (of lengths 1.0 m and 4.0 m) are negligible in comparison to the chamber.

If one assumes the full dynamic stiffness matrix for this problem the total numbers of DOFs amount to $N_{tot} = 660$. For a unit normal acceleration at the left–hand end of Ω_1 and a given excitation frequency the CPU expenditure time for solving the system of equation is 8.1 s, the total time for finding wave functions for chamber takes 30.3 s, and the total CPU time for solving the problem in MATLAB 7.0.2 on Pentium M machine took 54.8 seconds. Although the number of operations are proportional to $50N_2^3$ as discussed in the section on two–dimensional analysis the CPU expenditure is generally higher than for an equivalent analytic matching procedure. However, CPU expenditure compares favourably with alternative fully three-dimensional treatments.

Figures 11–12 show solutions at cross–sections of the silencer assembly. Specifically, solutions are shown at the beginning of the chamber (a) and at the midway point (b) for elliptic chambers of lengths $L = 0.8, 1.0$ and 1.2 m respectively with absorbent liner material

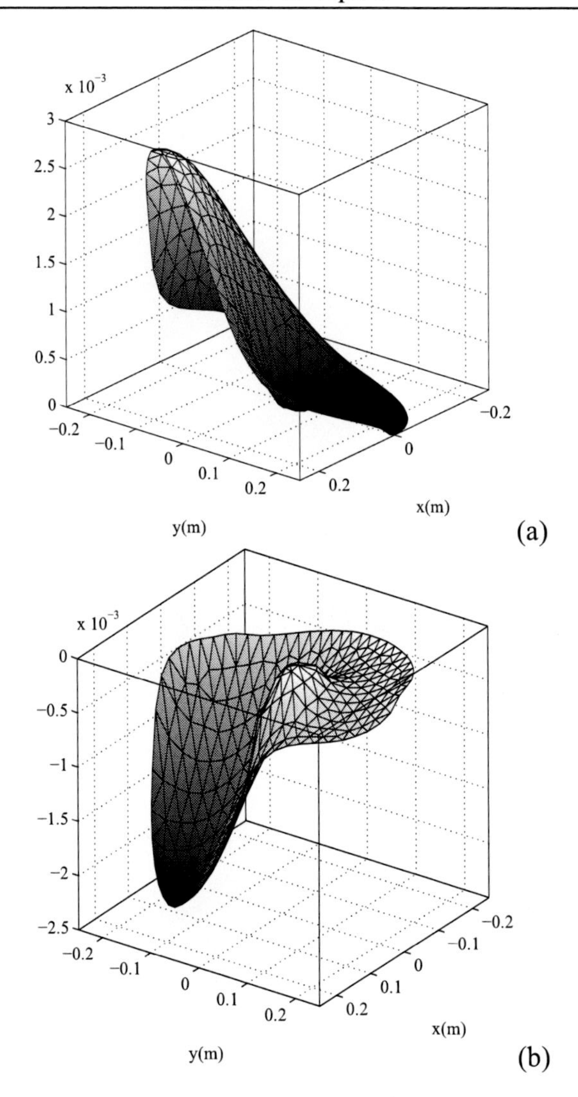

Figure 12. Acoustic pressure at two cross–sections of lined expansion chamber (length $L =$ 0.8 m) due to unit normal acceleration at inlet pipe (length 1.0 m) at excitation frequency 740 Hz. Left hand plot shows acoustic pressure at entry of chamber and right hand plot acoustic pressure at exit of chamber. Outlet pipe length of 4.0 m with rigid termination, geometry of pipes shown in Fig. 7 and Fig. 8.

(flow resistivity $\sigma = 400 \times 10^3$ Nsm^{-4}) covering the entire outer surface. Note that the size of the computational problem is not changed for each expansion chamber length.

6. Conclusion

A new spectral method in the form of a finite element scheme has been used to treat the problem of sound transmission in non–uniform waveguides or ducts. A unique feature of the super–spectral finite element approach is the use of basis functions generated from linear eigenvalue calculations. The basis functions, themselves solutions to the homogeneous reduced wave equations, may be defined over regions of arbitrary length with sound absorbing sides.

The waveguide geometries that have been given the most attention in the present study are that of a rectangular duct partially lined on one side or a geometrical non–uniform rectangular duct. These configurations are not easily amenable to analytical treatment and the use of a numerical method is appropriate here. The finite element formulation in this investigation is, however, sufficiently general that it can be extended to any non–uniform rectangular waveguide with varying linings and fluid densities, and may be coupled with a standard finite element code. The conclusions and remarks may be listed as:

1. Acoustic problems defined such that a propagating in one direction is amenable to SFEM in two or three dimensions.
2. Muffler problems with varying density or complex characteristic impedances may be solved extremely with SFEM.
3. Where a mesh generator is available three–dimensional duct problems are easily within reach using SFEM. Computational expenditure is reduced such that problems may be solved on Desktop PCs.

This new finite element technique extends the family of computational methods for predicting acoustic wave transmission and may possess wider applications in general wave propagation analysis for solids, fluid–structure interaction and fluid flow analysis.

References

[1] Bilawchuk S, Fyfe KR, *Appl. Acoust* **64**, 903 (2003).

[2] Birgersson F, *Trita-AVE, ISSN 1651-7660* PhD thesis, KTH, Stockholm (2004).

[3] Birgersson F, Ferguson NS, Finnveden S, *J Sound Vib.* **259**, 873 (2003).

[4] Birgersson F, Finnveden S, *J Sound Vib.* **287**, 315 (2005).

[5] Birgersson F, Finnveden S, Nilsson C–M, *J Sound Vib.* **287**, 297 (2005).

[6] Bonfiglio P, Pompoli F, Peplow AT, Nilsson AC, *J Sound Vib.* **303**, 780 (2007).

[7] Delany ME, Bazley EN, *Appl. Acoust* **3**, 105 (1976).

[8] Denia FD, Selamet A, Fuenmayor FJ, Kirby R, *J Sound Vib* **302**, 1000 (2007).

[9] FEMLAB 3.1 (2004) Users manual.

[10] Finnveden S, *Acta Acust* **2**, 461 (1994).

[11] Finnveden S, *J. Sound Vib* **208**, 685 (1997).

[12] Finnveden S, *Acust. Acta Acust* **82**, 479 (1996).

[13] Finnveden S, *J. Sound Vib* **199**, 125 (1997).

[14] Finnveden S, Birgersson F, Ross U, Kremer T, *J. Fluid Struct* **20**, 1127 (2005).

[15] Finnveden S, Pinnington RJ, *J Sound Vib* **229**, 147 (2000).

[16] Fraggstedt M, *Trita-AVE, ISSN 1651-7660* Lic. thesis, KTH, Stockholm (2006).

[17] Gavric L, *J. Sound Vib* **173**, 113 (1994).

[18] Kirby R, *J. Acoust. Soc. Am* **114**, 200 (2003).

[19] Nilsson C–M, *Trita-AVE, ISSN 1651-7660* PhD thesis, KTH, Stockholm (2004).

[20] Nilsson C–M, Finnveden S, *J Sound Vib* **305**, 641 (2007).

[21] Orrenius U, Finnveden S, *J Sound Vib* **198**, 203 (1996).

[22] Peplow AT, Finnveden S, *Int. J. Num. Anal. Meths Geom* DOI: 10.1002/nag.643, (2008).

[23] Peplow AT, Finnveden S, *J. Acoust. Soc. Am* **116**, 1389 (2004).

[24] Persson PO, Strang G, *SIAM Review* **46**, 329 (2004).

[25] Richard TH, Leung AYT, *J Sound Vib* **55**, 363 (1979).

[26] Selamet A, Xu MB, Lee IJ, Huff NT, *Int. Veh. Noise Vib* **1**, 341 (2005).

In: New Research on Acoustics
Editor: Benjamin N. Weiss, pp. 317-341
ISBN: 978-1-60456-403-7

Chapter 10

RESEARCH ON THE CRACKED PLATE'S STRUCTURE-BORNE INTENSITY FIELDS USING SOLID FINITE ELEMENTS

Xiang Zhu, TianYun Li, Yao Zhao and Jing Xi Liu
Department of Naval Architecture & Ocean Engineering
Huazhong University of Science and Technology, Wuhan 430074, P.R.China

Abstract

Structure-borne intensity fields indicate the magnitude and direction of structure-borne sound in vibraioning structures. The structural intensity fields can be used to identify the energy sources, sinks and indicate the distribution of the energy in structures. It can guide the application of vibration and noise control treatments. In this research, the structrual intensiy concept is utilized to investigate the cracked plate's vibratino characteristics. Generally speaking, a existing crack may change the dynamic characteristics of a structure, therefore the existance of the crack in a structure will change vibrational wave in the structure and substantially affect the power flow or structural intensity characteristics. As a result, the investigation of the structural intensity in cracked structures will have the potential to crack detection. In this chapter, the structure-borne intensity fields of a simply-supported thin aluminium plate with a surface crack are investigated by using solid finite elements. The structural intensity conectp is introduced at first and the formulas of basic structural elements (beam, shell and solid) are given in detail. The structural intensity streamline is introduced to visualize the structural intensity fields. To describe the internal element stress fields more accurate, the isoparametric solid elements are used to model the plate. The intact plate's structural intensity patterns obatined by shell element and solid element are respectively computed, which validates the accuracy of the solid element calculation for structural intensity. The crack is modelled by quarter point crack tip element. Based on solid finite element, the cracked plate's displacement vector field, structural intensity vector field and structural intensity streamline field are obtained under a point excitation harmonic force applied at the centre of the plate. The calcuations show that the structural intensity field is dependent on the vibraion mode, and the vibrating source can be successfully indicated. The cracked plate's intensity vector and streamline patterns show that the existance of crack changes the structural intensity in the plate. At the location of the crack, the intensity vector and streamline have abruptly changes in magnitude and direction. Cases of different crack location are considered to investigate the relationship bewteen crack's location with the structural intensity patterns, which indicates that the structural intensity pattern can successfully identify the location of the crack in the plate.

Introduction

Crack is a kind of damages that often occurs on members of structures due to different causes, which presents a serious threat to the performance of structures. Cracks must be detected in the early state. In the last two decades, a lot of research efforts have been devoted to develop effective approaches to early detection and localization of the cracks in the structures. A large number of studies have been carried out on conventional (magnetic particle induction, ultrasonic, etc.) and modern approaches to non-destructive testing. The conventional methods have been well developed, implemented in widely marketed equipment, and accepted by industry and regulatory agencies as practically applicable nondestructive evaluation (NDE) methods. The modern NDE methods are still under development, implemented in a limited manner in some equipment and not fully accepted by the industry and regulatory agencies as practicably applicable NDE methods. One of these modern methods is the vibration-based detection methodology.

It is well known that a crack may change the dynamic characteristics of a structure. In other words, the crack results in the changes of frequencies and mode shapes for the vibration. In the past thirty years, the detection of a crack based on the structural vibration has been widely studied. Doebling et al. [1] presented a review of the state of the art of vibration - based damage detection methods. This survey reviewed the numerous technical literatures available on detection, sizing and location of structural damage via vibration-based testing. It categorized the various methods available for crack detection according to the measured data and analysis techniques.

In recent literatures, most of the researches for crack detection were concentrated on the simple beam structures and plate structures. In the beam or plate structures, many methods are used to model the crack, such as the short beam model, finite element method and rotational spring model, et al. Gounaris and Dimarogonas [2] developed a finite element for a cracked prismatic beam for structural analysis based on the compliance matrix for the crack. Chaudhari and Maiti [3] modelled the crack section by a rotational spring to analyze the transverse vibrations of a geometrically segmented slender beam and identify the crack in the beam. The linear spring model was also used to determine the local flexibility to research the cracked plate structures by Khadem and Rezaee[4], Douka et al.[5], Krawczuk et al.[6]. From kinds of literatures it can be found that the local linear spring model is by far the most commonly used model in dynamic analysis of cracked beams and plate.

In recent years, the structure-borne sound analysis and control of flexible structures such as cabins of marine-structures and aeronautical crafts are becoming an important topic. The usage of vibrational power flow in the problem of this type is very valuable. Generally, the vibration of a structure can be regarded as a typical example of structural wave propagation. The presence of the crack in the structure will in some way change the motion of the wave. Consequently, the changes of the wave will in turn influence the power flow characteristics in the structure. Based on this, the research on the power flow characteristics of the cracked structures will be of great value for crack detection. Li et al. [7, 8] first researched the power flow of the cracked periodic beam structures and cracked infinite beam structures. The relations of the vibrational power flow, the position and the characteristic size of the crack were obtained to detect the crack. Li et al. [9] investigated the power flow characteristics of the circular plate structure with peripheral surface crack. Zhu et al.[10] studied the structural power flow characteristics of cracked Timoshenko beams, and contours of input power flow with different frequencies are constructed to identify the location and depth of the crack. The method is promising for the detection and location of structural damage.

Structural intensity is the power flow per cross-sectional area. The structural intensity field indicates the magnitude and direction of vibrational energy flows at any point of a structure. The structural intensity was introduced by Noiseux[11], Pavic[12], Verheij[13] to solve structural borne sound problems. Computation of the structural intensity using the finite element method was developed by Hambric[14]. Not only flexural but also torsional and axial power flows were taken into account in calculating the structural intensity of a cantilever plate with stiffeners.

Pavic and Gavric [15] evaluated the structural intensity fields of a simply supported plate by using the finite element method. Normal mode summations and swept static solutions were employed for computing the structural intensity fields and identifying the source and the sinks of energy. The first effort to use solid finite elements to compute structural power flow was performed by Hambric and Szwerc [16] in a T-beam model. Li and Lai [17] calculated the surface mobility for a thin plate by using structural intensity approach. The structural intensity fields of plates with viscous dampers and structural damping were computed using finite element analysis. Xu et al. [18] calculated the structural intensities in isotropic and orthotropic composite laminated plates with and without a hole using the finite element method. Xu et al. [19] studied the energy transmission within rotating hard disk systems through the structural intensity vectors direction and vectors magnitude. Khun et al.[20] used the finite element method to predict the structural intensity of a plate with multiple discrete and distributed spring-dashpot systems. Xu et al.[21] also calculated the structural intensity of a rectangular plate with stiffeners attached using the finite element method. Liu et al.[22] used the structural intensity techniques to study the transient dynamic characteristics of plates under low-velocity impacts. Recently, Lee et al.[23] utilized the structural intensity method to explore the positioning of dampers in vibrating thin plates to divert the vibration energy flow away from crack tips. A higher order mesh refinement is used at the crack tip region. This approach is proposed as a temporary measure to prevent the further propagation of the crack before repair of the crack can be done. More Recently, the thermally induced vibration and its control for thin isotropic and laminated composite plates was studies by Tran et al.[24] using structural intensity pattern.

The pioneering work on structural intensity measurement was carried out by Noiseux[11], most earlier works on structural intensity measurements, e.g. the two-transducer method were undertaken using the contact method. Accelerometers are usually used to obtain velocity or acceleration of vibrating bodies. The finite difference scheme is usually used to obtain spatial derivatives of measuring quantities. In the early 1980s, some alternative method of intensity measurement were proposed, for example, non-contact ways of detecting surface motions suitable for intensity applications. found either by near-field acoustic holography (NAH) or by an optical method (laser). Non-contact measurements of vibration intensity using the laser vibrometer are widely researched. The development of the optical laser Doppler vibrometer offered a good alternative to ordinary transducers in intensity measurements. Structural intensity can be measured by an automated laser vibrometer in plates by Pascal et al.[25] for bending waves. Freschi et al.[26] analyzed the total structural intensity in beams using a homodyne laser doppler vibrometer. A z-shape beam was used in order to analyze the propagation of all types of waves in measuring structural intensity. Arruda and Mas [27] presented an experimental method especially adapted for the computation of structural power flow using spatially dense vibration data measured with scanning laser Doppler vibrometers. This non-contact method is free from surface loading problems, and it helps to measure both in-plane and out-of-plane quantities, useful for structural intensity. This method can be used to obtain structural intensity for simple and complex structures, so it is very attractive in the field of source identification and path prediction to viration control.

This chapter researches the cracked plate's structure-borne intensity fields using solid finite elements. The structural intensity concept is introduced and the basic structural elements (beam, shell and solid)'s formulas which can be used in the finite element analysis are given in detail at first. The structural intensity streamline is introduced to visualize the structural intensity fields. A simply-supported thin aluminium plate with a surface crack is studies. To describe the internal element stress fields more accurate, the isoparametric solid elements are used to model the plate. The intact plate's structural intensity patterns obatined by shell element and solid element are respectively computed at first to validate the accuracy of the solid element calculation of structural intensity. Then the plate with a part-through surface crack is considered. The crack is modelled by quarter point crack tip element. Based on solid finite element, the cracked plate's intensity vector plat and streamline plots are given under a point excitation harmonic force applied at the centre of the plate. Cases of plates with different crack location and driving frequencies are considered to investigate the relationship bewteen crack's infromtion with the structural intensity patterns.

2. Formulations of Structural Intensity

2.1. Concept of Structural Intensity

Analogous to acoustic intensity in a fluid medium, the instantaneous structural intensity is a time-dependent vector quantity equal to the vibrational power flow per unit area of a dynamically loaded structure, that is,

$$i_n(t) = -\sigma_{nl}(t) v_l(t) \quad l = 1, 2, 3 \quad n = 1, 2, 3 \tag{1}$$

where $\sigma_{nl}(t)$ and $v_l(t)$ are the stress and velocity in the direction n at time t, respectively.

The temporal average of the kth instantaneous intensity component I_n represents the net power flow through the structure,

$$I_n = \langle i_n(t) \rangle = \frac{1}{T}\int_0^T i_n(\tau)\,\mathrm{d}\tau \tag{2}$$

where $\langle \cdots \rangle$ denotes time average.

For a steady state vibration, the n direction of active structural intensity inside an elastic medium in a frequency domain is given by [28]

$$I_n = -\frac{1}{2}\mathrm{Re}(\sigma_{nl} v_l^*) \tag{3}$$

where σ_{nl} is the complex amplitude of stress, v_l^* is the complex conjugate of the velocity, and Re denotes a real part of a complex number.

2.2. Formulation of Structural Intensity

Since the structural intensity concept was developed mostly as a measurement technique, the variables that used in the formulations of structural intensity are mainly velocities and accelerators, which are easily measured in practice. The expressions of internal forces are in the form of velocities and accelerators and the spatial derivatives are usually obtained by using finite difference concept. However, for the purpose of finite element calculation of structural intensity, the variables used in the formulations of structural intensity are always internal force, internal moments, angular displacement and translational displacement. For elements of beams, plates and shells, the distributions of stresses and displacements in the direction perpendicular to the surface are known. Instead of displacements and stresses at each point of the structure, unknown variables becomes stress resultants (i.e. moments and forces) and generalized displacements (translational and angular displacement) of the structural mid-surface of the shell or the beam's centerline[15].

2.2.1. Structural Intensity of Beam Element

Integration of the structural intensity over the cross-section of a beam gives the total net power flow in the beam. Since a beam is a one-dimensional element, the power flows are along the centerline of the beam. In a single case, the velocities can be replaced by displacements by using the commonly adopted complex algebra. The formulation of structural intensity in beams can be expressed as follows [15]:

$$I_x = -(\omega/2)\mathrm{Im}(Nu^* + Q_y v^* + Q_z w^* + M_x\theta_x^* + M_y\theta_y^* + M_z\theta_z^*) \tag{1}$$

where N is complex axial force, Q_y and Q_z are complex transverse shear force, M_x is complex twisting moment, M_y and M_z are complex bending moments, u, v and w are complex translational displacements, θ_x, θ_y and θ_z are complex rotational displacement about x, y and z directions, Im denotes the imaginary part.

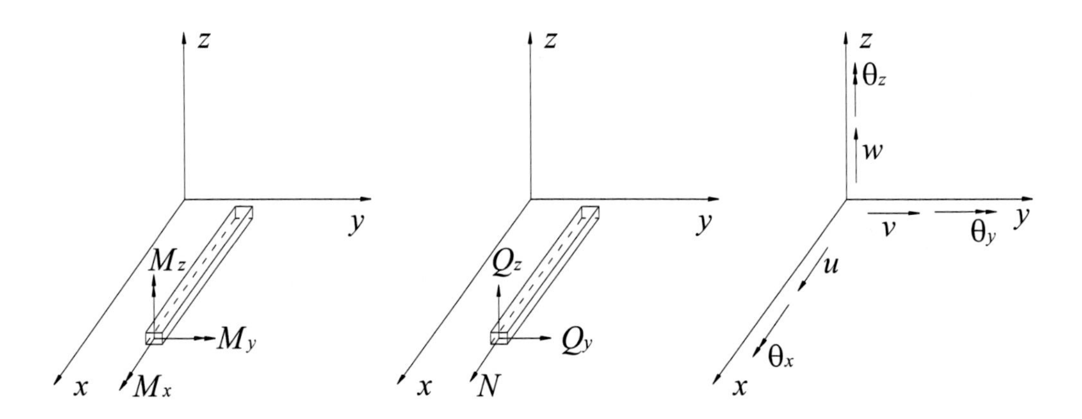

Figure 1. Beam element with positively defined internal forces and displacements.

From the above analysis, structural intensities are dependent on displacements (velocities) and stresses (internal forces). Most often, in finite element programs the displacements (velocities) are calculated at the nodes, and the forces in the beam are calculated on the element level. Computation of structural intensity requires all the response quantities at the same points, so this difference causes a problem. As a result, it is required to convert the element-associated stresses into node-associated values, or convert node-associated velocities into element associated values. In the research, the average of the nodal response at one element give the average quality for an element, then the structural intensities at each element can be calculated according to the above formulas.

2.2.2. Structural Intensity of Plate and Shell Elements

An analogous approach can be used for plate and shell elements. For the flexural vibration of a thin plate, the power flow in the x direction can be expressed in terms of the out of plane displacement w as

$$P_x = D\left[\frac{\partial(\nabla^2 w)}{\partial x}\dot{w} - \left(\frac{\partial^2 w}{\partial x^2} + \mu\frac{\partial^2 w}{\partial y^2}\right)\frac{\partial \dot{w}}{\partial x} - (1-\mu)\frac{\partial^2 w}{\partial x \partial y}\frac{\partial \dot{w}}{\partial y}\right] \tag{1}$$

where ∇^2 is the Laplace operator, $D = Eh^3/12(1-\mu^2)$ is the flexural stiffness, E elastic modulus, μ poisson ratio and h thickness of the plate.

The intensity for the y-direction can be obtained by interchanging the subscript x and y.

Since stresses and displacements are usually determined as stress results and movements of the mid-surface and the integration is carried out over the thickness, the structural intensity in the plates and shells can be expressed in the form of power flow per unit width. Besides flexural deformations of the plate, the membrane effect is also considered in the formulation of structural intensity for shell elements. The two-dimensional components of the structural intensity in the local x and y directions for a vibrating plate can be expressed as[15]

$$I_x = -(\omega/2)\,\mathrm{Im}(N_x u^* + N_{xy} v^* + Q_x w^* + M_x \theta_y^* - M_{xy} \theta_x^*) \tag{2}$$

$$I_y = -(\omega/2)\,\mathrm{Im}(N_y v^* + N_{yx} u^* + Q_y w^* - M_y \theta_x^* + M_{yx} \theta_y^*) \tag{3}$$

where N_x, N_y, N_{xy} and N_{yx} are complex membrane forces per unit width of plate, $N_{xy} = N_{yx}$; M_x, M_y are complex bending moments per unit width of plate; M_{xy} and M_{yx} are complex twisting moments per unit width of plate, $M_{xy} = M_{yx}$; Q_x and Q_y are complex transverse shear forces per unit width of plate; u, v and w are complex conjugate of translational displacements in *x*, *y* and *z* directions; θ_x and θ_y are complex conjugate of rotational displacement about x and y directions.

All forces, both flexural and membrane are per unit width. Therefore the unit for power flow in shell element are power per unit length. Analogous to the case for beam elements, displacements (velocities) at the nodes must be transformed to the element coordinate systems to be used in the structural intensity calculations. Here the 8-noded quadrilateral isoparametric shell element is used to model the plate structure, therefore in a element there are eight nodes. When the eight nodal displacements in a shell element are obtained from finite element analysis, the averaged element displacement in a direction can be expressed as

$$u_j = \frac{1}{8}\sum_{i=1}^{8} u_i \tag{4}$$

where u_i is the generalized displacement of node i at element j.

2.2.3. Structural Intensity of Solid Element

For solid elements, structural intensity indicates the vibrational power flow in a given infinitesimal volume. According to the describe of Pavic[12], the active structural intensity normal to the element face is

$$I_n = -\omega/2\,\mathrm{Re}(\sigma_n u_n^* + \tau_{n1} u_1^* + \tau_{n2} u_2^*) \tag{1}$$

where σ_n is the stress normal to the element face, τ_{n1} and τ_{n2} are the shear stresses on the element face in directions 1 and 2, and u_n, u_1 and u_2 are the complex displacement in the normal, 1 and 2 directions. The minus sign is due to stress orientation conventions. The term $\sigma_n u_n^*$ is the intensity carried by the normal stress and the $\tau_{n1} u_1^* + \tau_{n2} u_2^*$ term is the intensity carried by shear stresses. When I_n is positive, energy is flowing in the position n direction.

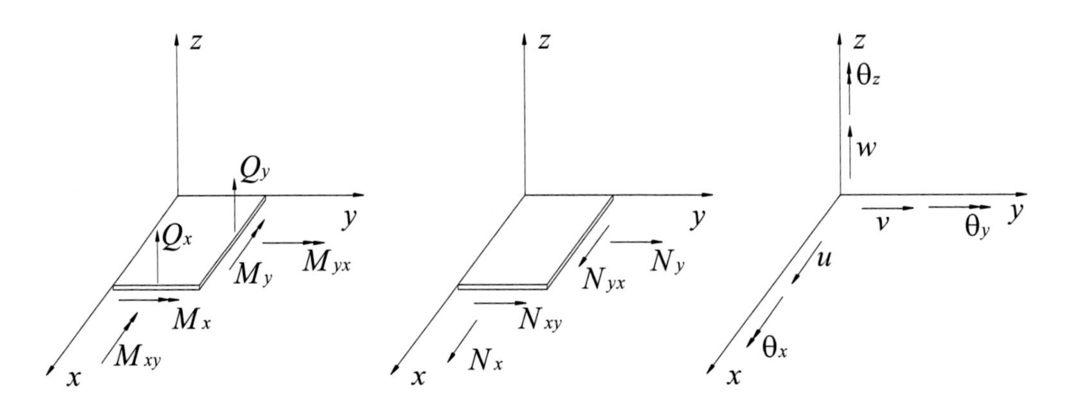

Figure 2. Forces (moments) and displacements for shell element.

The active structural intensity (see Figure 3) in the x, y and z directions can then be expressed as

$$I_x = -\omega / 2\text{Re}(\sigma_x u_x^* + \tau_{xy} u_y^* + \tau_{xz} u_z^*) \tag{2}$$

$$I_y = -\omega / 2\text{Re}(\sigma_y u_y^* + \tau_{yx} u_x^* + \tau_{yz} u_z^*) \tag{3}$$

$$I_z = -\omega / 2\text{Re}(\sigma_z u_z^* + \tau_{zx} u_x^* + \tau_{zy} u_y^*) \tag{4}$$

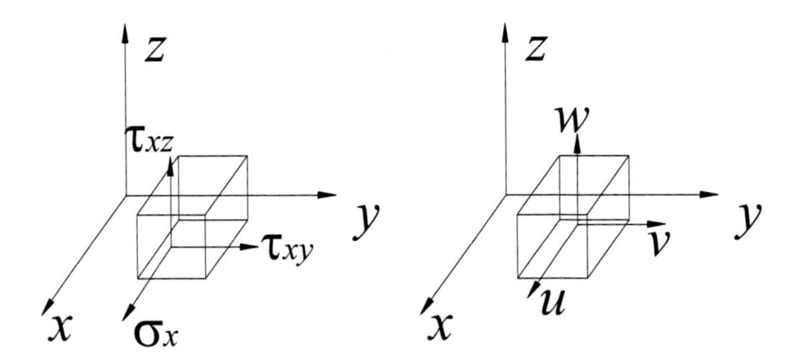

Figure 3. Solid element with internal forces and displacements.

In this study, all the solid models used isoparametric brick elements, which is actually defined by 20 nodes having three degrees of freedom per node. Linear element models are not considered due to their poor representation of internal element stress fields. As mentioned above, the element displacement in a direction can be obtained from the average of the nodal displacements at each nodes of this element,

$$u_j = \frac{1}{20}\sum_{i=1}^{20} u_i \tag{5}$$

where u_i is the generalized displacement of node i at element j. When the degradation happened in the element, the total number of nodes in a element would not be twenty, the element displacement still can be derived from the average of N nodes in element j.

2.3. Visualization of Structural Intensity Field

When the internal forces and displacements in the structures are obtained from finite element analysis, the structural intensity at the position of every element can be then calculated according to the above formulas. Since the structural intensity is a vector indicating the magnitude and direction of power flow inside a dynamically loaded structure, the structural intensity vector map can be plotted to indicate the energy distribution in the structures.

The concept of field lines has been used in the studies of various field theories to improve their understanding. A line in a structural intensity vector field is drawn so that its direction at any point is the same as the direction of the field at that point. This concept is very useful for visualizing the structural intensity fields determined by numerical methods at discrete points in space.

The streamline technique displays the flow as lines everywhere parallel to the velocity field. The structural intensity streamline can be mathematically expressed as

$$\mathbf{dr} \times I(\mathbf{r},t) = 0 \tag{1}$$

where $\mathbf{r}$ is the energy flow particle position. For the steady state power flows, the cross product can be written as

$$\begin{vmatrix} i & j & k \\ I_x & I_y & I_z \\ \mathrm{d}x & \mathrm{d}y & \mathrm{d}z \end{vmatrix} = 0 \tag{2}$$

Thus, the differential equation describing structural intensity stream line is

$$\frac{\mathrm{d}x}{I_x} + \frac{\mathrm{d}y}{I_y} + \frac{\mathrm{d}z}{I_z} = 0 \tag{3}$$

For the two-dimensional plate and shell structures, the differential equation for a streamline is

$$\frac{\mathrm{d}x}{I_x} + \frac{\mathrm{d}y}{I_y} = 0 \tag{4}$$

The streamline representation of the structural intensity can be taken into account according the structural intensity vector. A picture of the streamlines may help us to understand and obtain more information on vibrational energy flow than the previous vector representation.

3. Modeling of a Cracked Plate

Modelling of a crack using efficient element is an important part of the analysis of cracked structure. Many researchers have using different kinds of elements in finite element analysis to model the crack.

A general approach is to increase the mesh density at the crack region. Malone et al. [29] used triangular linear elements to predict the vertical displacement and stress intensity factor of an infinite plate. The sizes of elements were increased for regions approaching the crack tip. Vafai and Estenkanchi [30] carried out a parametric study using four-noded shell elements on cracked plates and shells. Stress and displacement fields were predicted by a high order of mesh refinement at the crack tip with no singular or special elements. Lee et. al [23] researched the structural intensity of a vibrating rectangular plate with a crack using the finite element method. A higher order mesh refinement is used at the crack tip region.

Three-dimensional quadratic solid isoparametric elements, such as tetrahedral and hexahedral families of elements are often used to model the crack. This element gives more accurate results of stress and displacement fields near the crack region. On the other hand, the singular elements or special elements at the crack tip are widely used in the finite element modeling, such as the quarter point crack tip element, which was first introduced by Henshell and Shaw [31]. This elements is actually an 8-noded quadrilateral isoparametric shell element. In order to obtain $1/\sqrt{r}$ singularity, the mid-nodes of these elements are moved to the quarter point near the crack tip. The crack opening displacement mathematically is a function of the square root of the distance to the tip and this element successfully simulates this behaviour. Therefore this method is preferred in the finite element modeling because of its accuracy and simplicity. In this study, the commercial finite element software ANSYS is used to calculate the complex dynamic response of the cracked structure. The quarter point crack tip element is used to model the surface crack in the vibrating plate structures. The isoparametric solid element SOLID 95 is used to model the plate structure.

4. Numerical Results and Discussions

4.1. Validation of Solid Model

The use of commercial FEM code to computing the power flow in a truss and a beam-stiffened cantilever plate was first introduced by Hambric [14] using NASTRAN. The numerical computation of structural intensity in plate structures with a damper was first presented by Gavric and Pavic [15] They used the normal mode summations with the sweeping procedure to calculate the complex dynamic response of the structure. In the modal approach, the number of modes used in the analysis has to be appropriately chosen. Generally, a large number of modes has to be used to obtain a good approximation of the structural intensity distribution which would increase both the computer storage and computing time substantially.

Unlike the mode superposition, full matrices of the system are used for solution in the full method without simplifications, thus giving more accurate results. Li and Lai [17] computed the structural intensity from the field variable using the commercial finite element software ANSYS. The full method for harmonic response solution was chosen for the calculation of the structural intensity and surface mobility of a finite plate. In this study, the full method for harmonic response solution is chosen for the calculation of the structural intensity to predict the structural intensity in cracked plate structures.

In most literature about structural intensity calculation in plates and shells, the element used in the finite element analysis is shell element. Hambric and Szwere [16] first predicted structural intensity fields in T-beam using solid finite elements to consider the situations that wave numbers are significant in all directions of a structure. Both the beam and solid element models were shown to be good predictors of the structural-borne power flow field in the T-beam at low wave numbers. However, the validation of the plate and shell's intensity fields computed by using solid element are not verified. Here the analysis of the structural intensity for a flat plate is first conducted to validate the solid finite element method used for the computation of the structural intensity.

The computation of the structural intensity is carried out on a simply-supported thin aluminium plate which is 0.7m long and 0.5m wide with a thickness of 10mm. The material properties of the plate are as follows: the Young's modulus $E = 7\times10^{10}$Pa, Poisson's ratio $\mu = 0.3$, mass density $\rho = 2100\text{kg/m}^3$. The plate is simply supported along the two short edges and free along the two long edges. The constant structural damping η in the plate is assumed to be 0.005. A point excitation harmonic force with an amplitude of 1 N is applied at the centre of the plate.

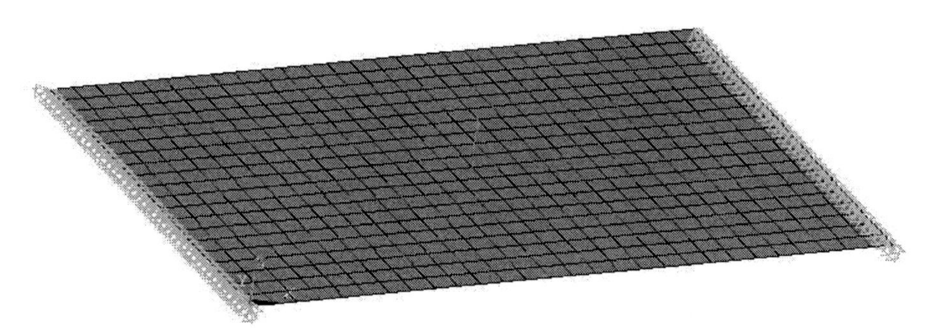

Figure 4. Shell model of a plate.

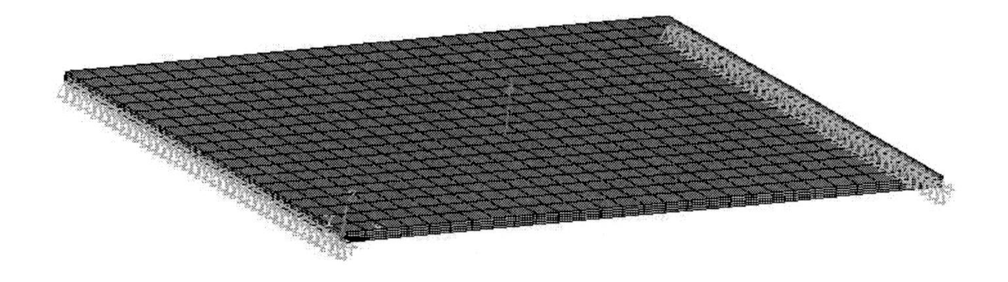

Figure 5. Solid model of a plate.

In Figure 4 the plate is modelled using 560 eight-noded isoparametric shell elements SHELL93 with 1776 nodes. The solid model of the plate is modelled by SOLID95 isoparametric element, as shown in Figure 5. In literature [16], Hambric and Szwerc pointed out that the accurate modeling of the total cross-sectional power carried by flexural waves requires at least two, and preferably three or four solid elements through the thickness of the cross section carrying the waves. If the structural intensity field is dominated by longitudinal waves, a single element should be sufficient to compute the total power flow through a cross section. In the study, bending moments also contribute significantly to normal stresses, and bending wave stresses and velocities vary through the cross section of the plate, therefore at least two elements are necessary to represent the normal intensity field due to bending and shear waves. Here the plate is divided into four elements through the thickness of the cross section.

The first 10th natural frequencies of the plate computed by shell element and solid element are given in Table 1. It can be seen that the natural frequencies from the two model show a good agreement.

The structural intensity vectors distributions for 60 and 480 Hz computed using shell element are shown in Figure 6(a) and (b) respectively. Because of the different driving frequencies, the patterns of the structural intensity vectors of two frequencies are different. But the two figures can both clearly reveal the source of the power flow being located at the centre of the plate.

In Figure7 (a) and (b), the surface structural intensity vectors for 60 and 480 Hz are computed from the solid finite element model. The figures also reveal the location of source. From the comparison between Figure 6 and Figure 7, it can be seen that the two results shows a good agreement. The example indicate that the plate structure's structural intensity can be successfully obtained from the calculation of solid element model.

Table 1. Natural frequencies of the plate

Mode No.	Frequency(Hz) shell model	Frequency(Hz) solid model
1	54.26	54.27
2	114.76	115.09
3	219.56	219.68
4	298.51	299.04
5	332.75	333.19
6	496.00	496.48
7	540.92	541.99
8	580.57	581.44
9	760.15	760.45
10	845.06	846.49

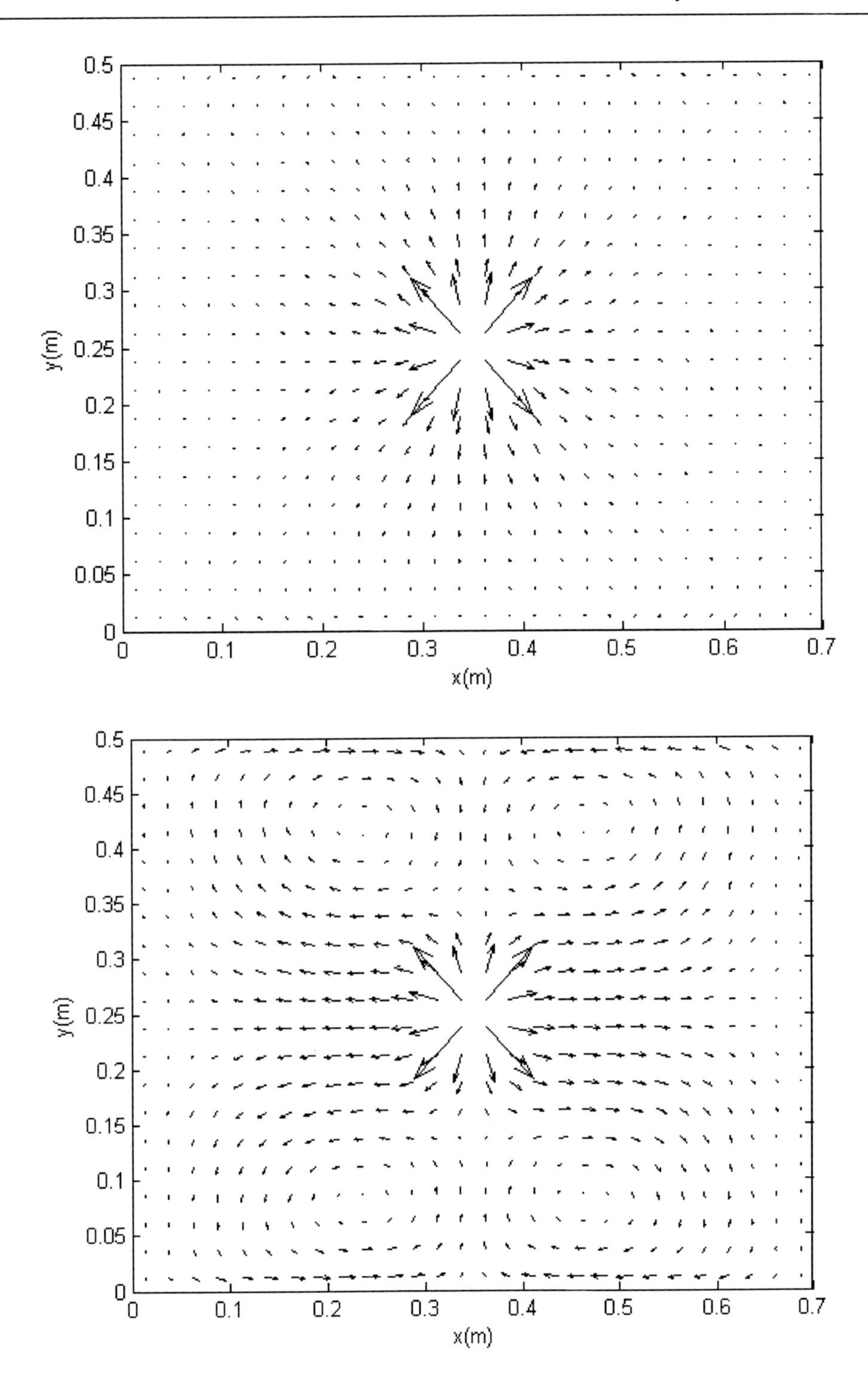

Figure 6. structural intensity vectors plot computed by using shell element (a) f=60Hz, (b) f=480Hz.

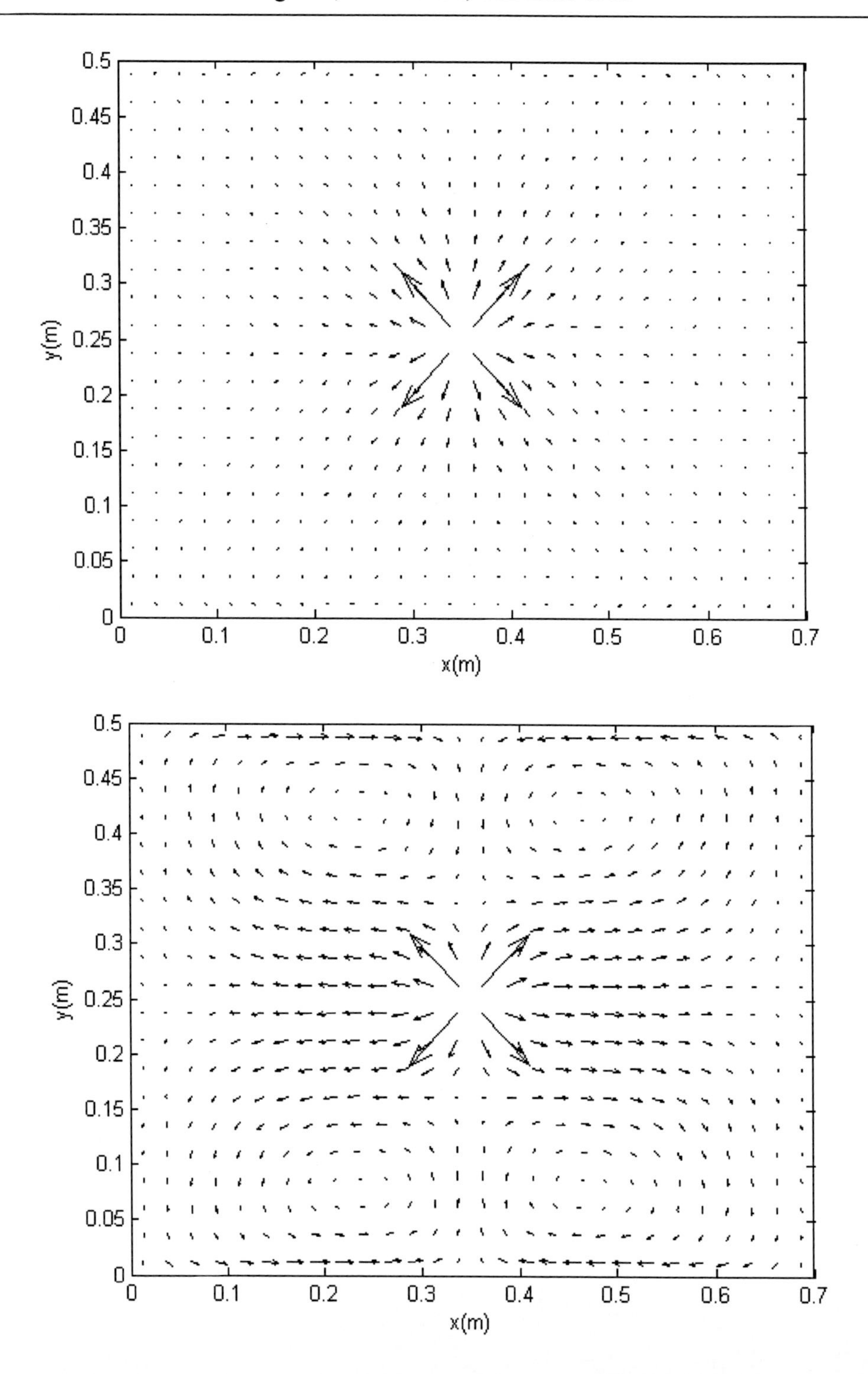

Figure 7. structural intensity vectors plot computed by using solid element (a) f=60Hz, (b) f=480Hz.

4.1. Cracked Plate's Structural Intensity Using Solid Element

In this study, it is assumed that the crack is a surface crack and the shape of the crack is "V" type, and the crack is running perpendicular to the long edge through one short edge to the other, as can be seen from Figure 8. The crack's relative depth a/h is uniform along the width direction for the sake of briefness, here $a/h = 1/2$, h is the thickness of the plate. The distance between the crack and one shorter edge of the plate is c. In the first case, $c/a = 1/8$ is assumed. The crack is modelled by quarter point crack tip element, which can be seen from Figure 8. The plate is modelled using 3660 isoparametric solid elements with 17648 nodes, in which the number of the singular crack tip elements is 160. Though the thickness of the plate's cross section the plate is divided into four elements.

Figure 9 (a) and (b) give the normal velocity distributions pattern of cracked plate with the 60 Hz and 480Hz driving frequency respectively. From the figures, it can be found that the response distribution of the plate is related to the vibration mode, as 60Hz is close the first flexural natural frequency and 480Hz is close the third mode frequency. However, from the velocity distributions pattern it is hard to realize the source of the energy and the path of the energy propagation.

Figure 10 (a) and (b) give the structural intensity vector plot and streamline plot respectively when the driving frequency is 60Hz. From Figure 10(a) and Figure 10(b) the driving source can be obviously found at the center of the plate, furthermore the structural intensity streamline in Figure 10(b) can show the power flow path more clearly. It can be also found from the figure that near the location of $x = 0.0875$ m, the structural intensity vector's direction change abruptly and the structural intensity streamline are no longer smooth and continuous. The location of the discontinuity in the figures is just the location of the crack. Therefore from the structural intensity vector and streamline plot the crack's position can be readily located, and the streamline plot can show the location of the source and crack's position more straightforwardly.

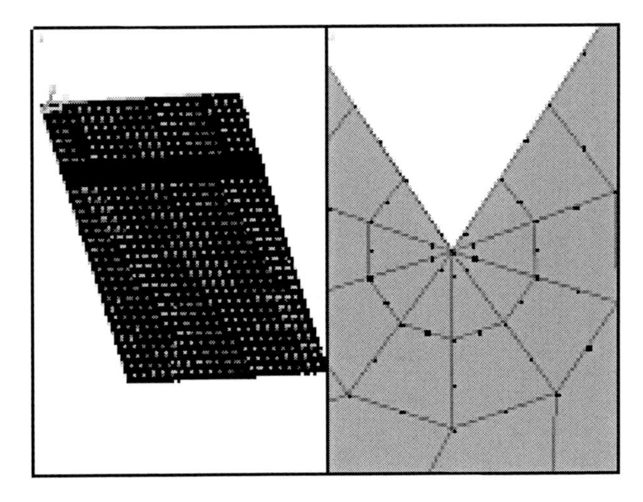

Figure 8. Solid element model of cracked plate and local crack model.

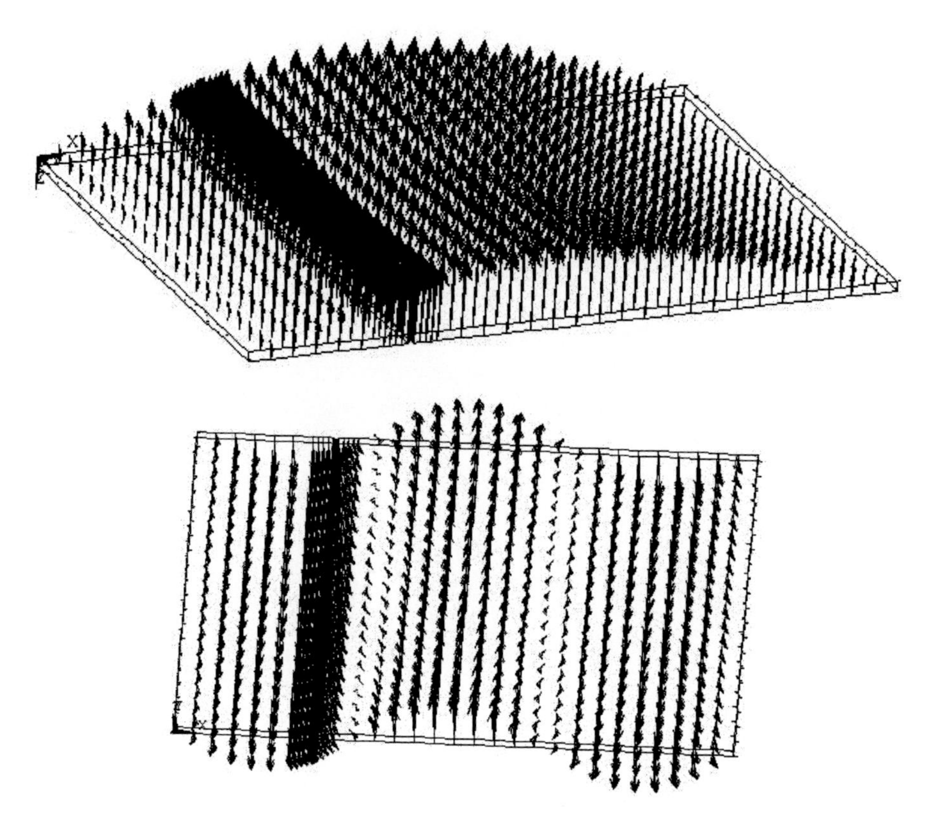

Figure 9. Normal velocity distributions pattern of cracked plate (a) f=60Hz, (b) f=480Hz.

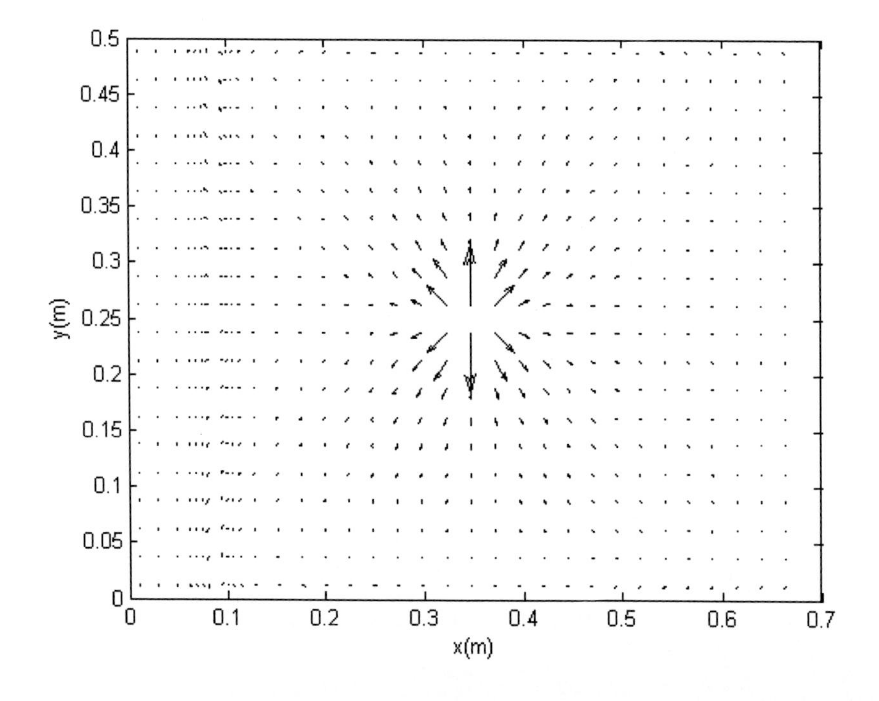

Figure 10. Continued on next page.

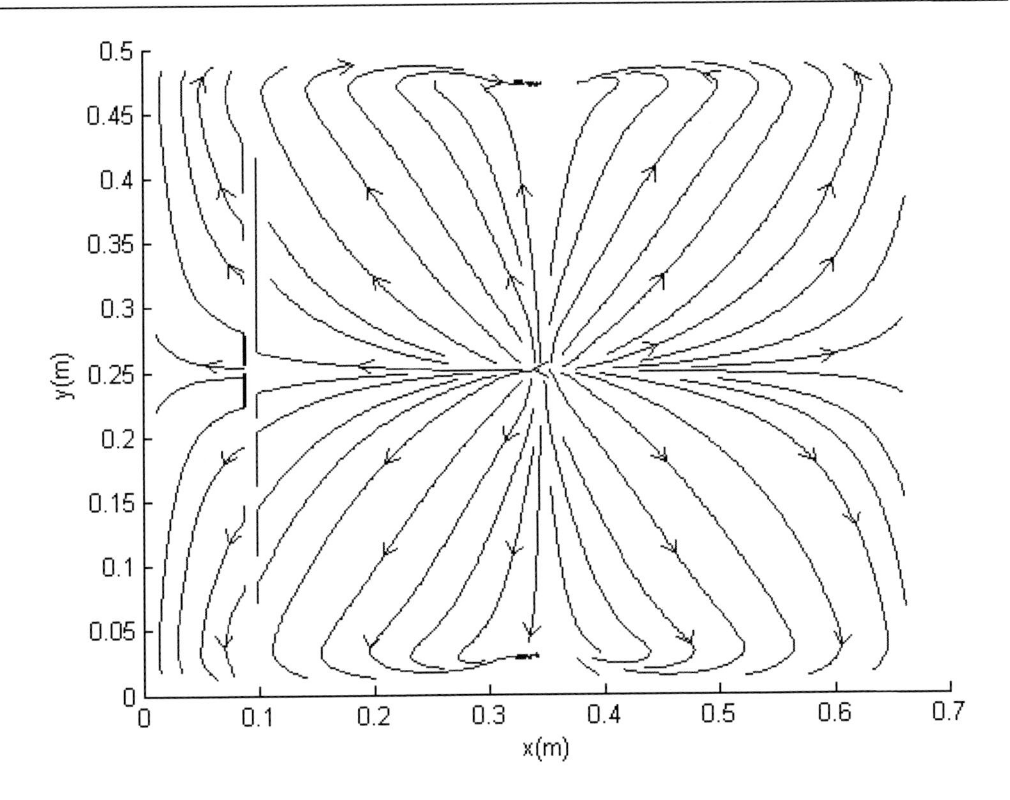

Figure 10. (a) Structural intensity vector plot, (b) Structural intensity streamline plot (a/h=1/2, c/a=1/8, f=60Hz).

When the driving frequency is 480Hz, the structural intensity maps are shown in Figure 11. By compared with Figure 10, the structural intensity vector plot and streamline plot show a complex pattern with four recirculation regions, and the source can still be identified to be located at the centre of the plate. It can be also found at the location of the crack, the structural intensity vector and the streamline change abruptly because of the exist of the crack. The location of the discontinuity in the figures indicates the position of the crack.

In Figure 12 and 13, the crack's location is $c/a = 1/4$ and the driving frequency is 60Hz and 480Hz respectively. The crack's location is $c/a = 3/8$ in Figure 14 and 15. The crack's depth is always kept constant. From these figures, it can seen that the cracked plate's structrul intensiy is highly related to the crack's location, the structural intensiy vector and streamline can easliy indicate the energy souce and the location of the crack where the map is abruptly changed. The results also show that the power flow path in plate structures are frequency dependent. As different modes, the nature of structural intensity field changes. To better understand structural intensity, the structural intensity vector plots and streamline plots over frequency can be used.

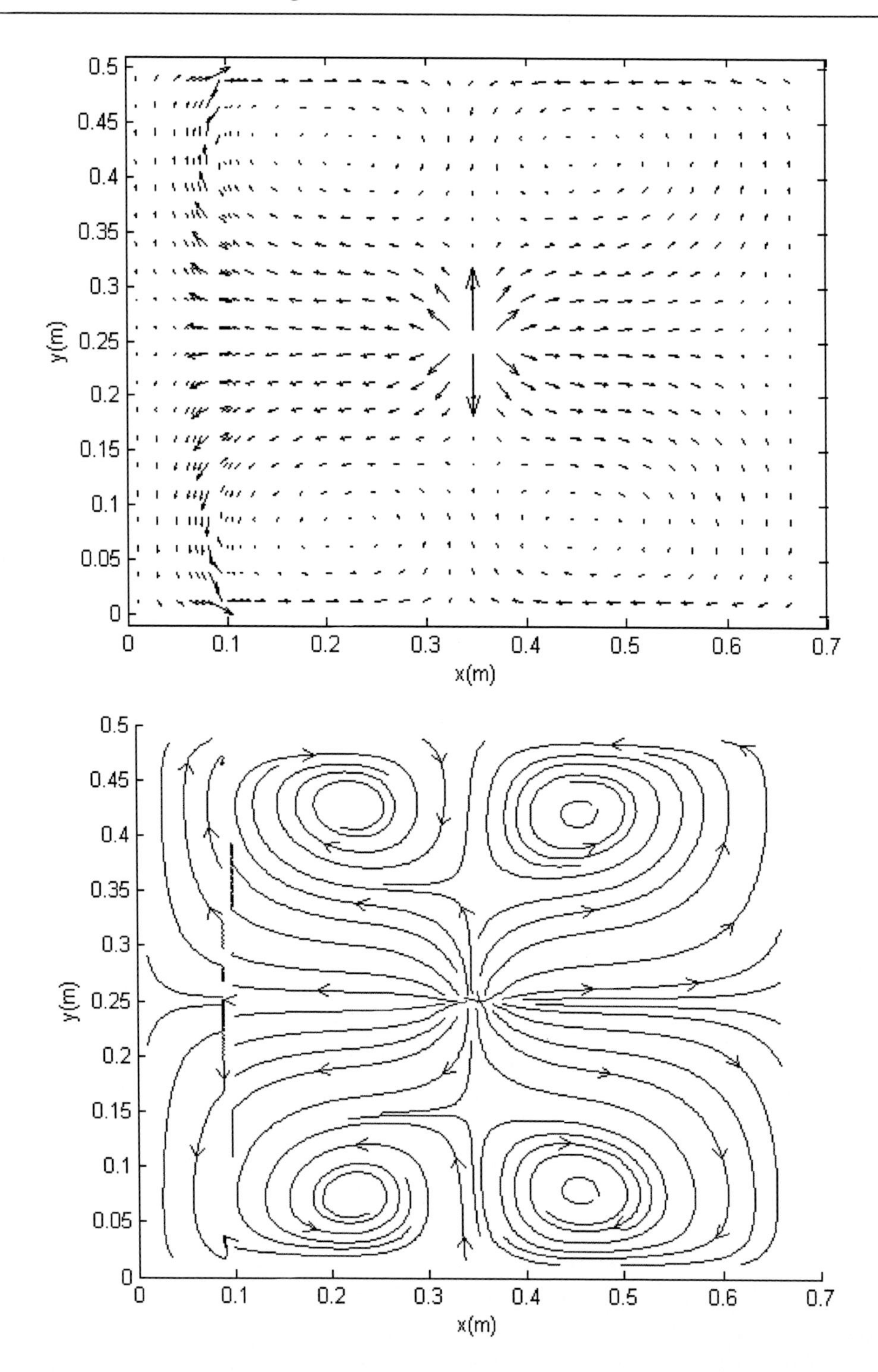

Figure 11. (a) Structural intensity vector plot, (b) Structural intensity streamline plot (a/h=1/2, c/a=1/8, f=480Hz).

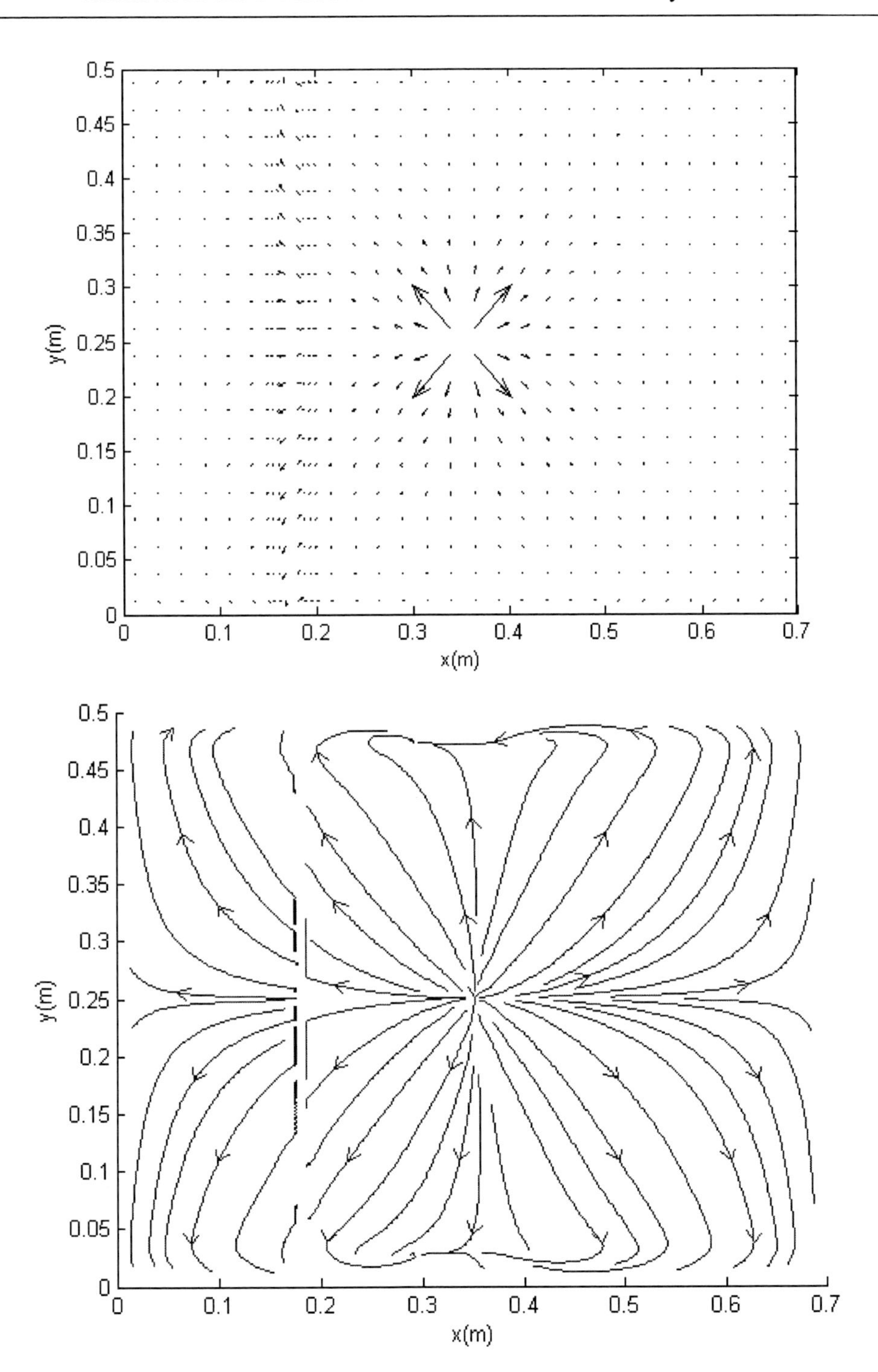

Figure 12. (a) Structural intensity vector plot, (b) Structural intensity streamline plot (a/h=1/2, c/a=1/4, f=60Hz).

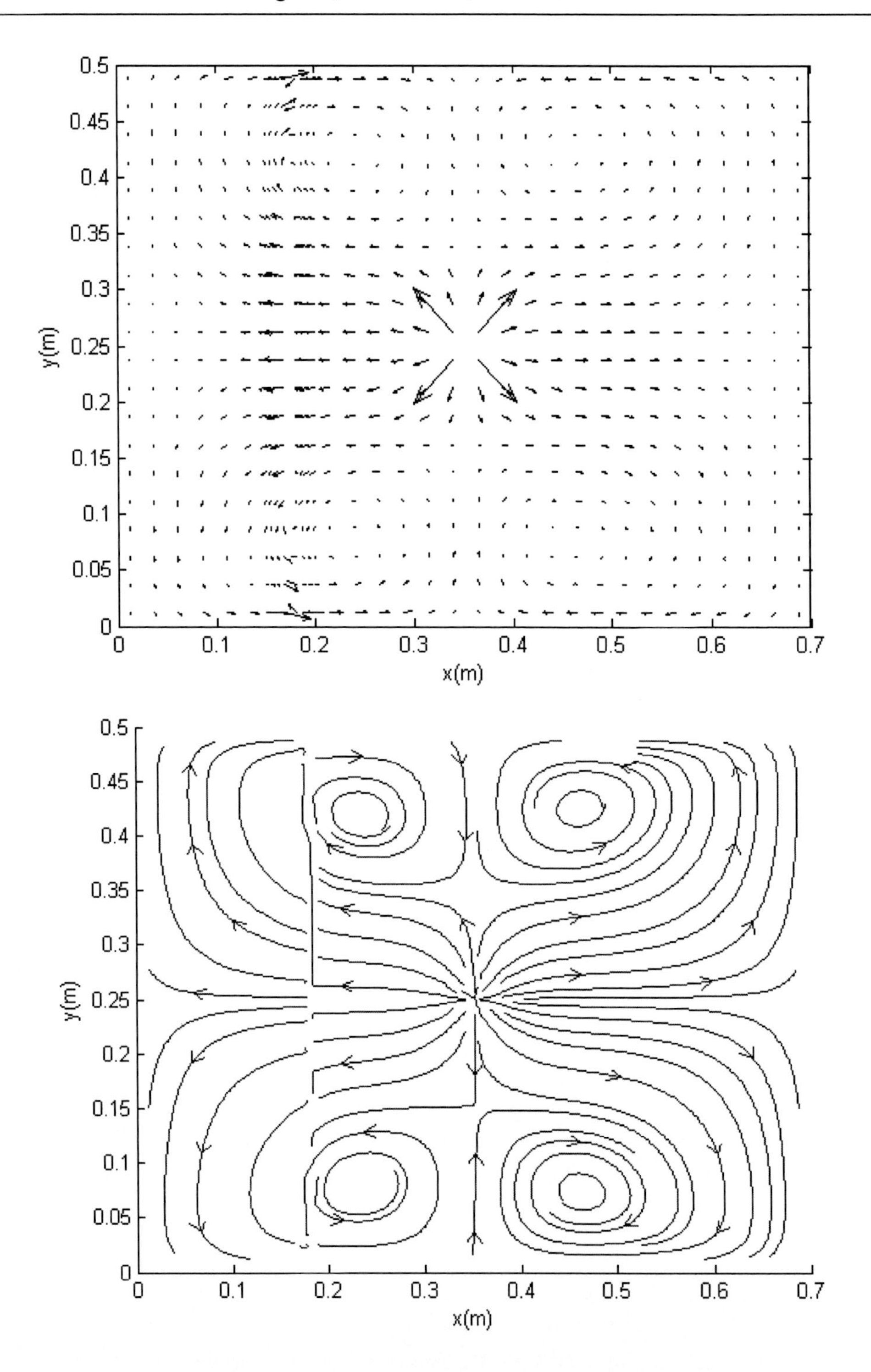

Figure 13. (a) Structural intensity vector plot, (b) Structural intensity streamline plot (a/h=1/2, c/a=1/4, f=480Hz).

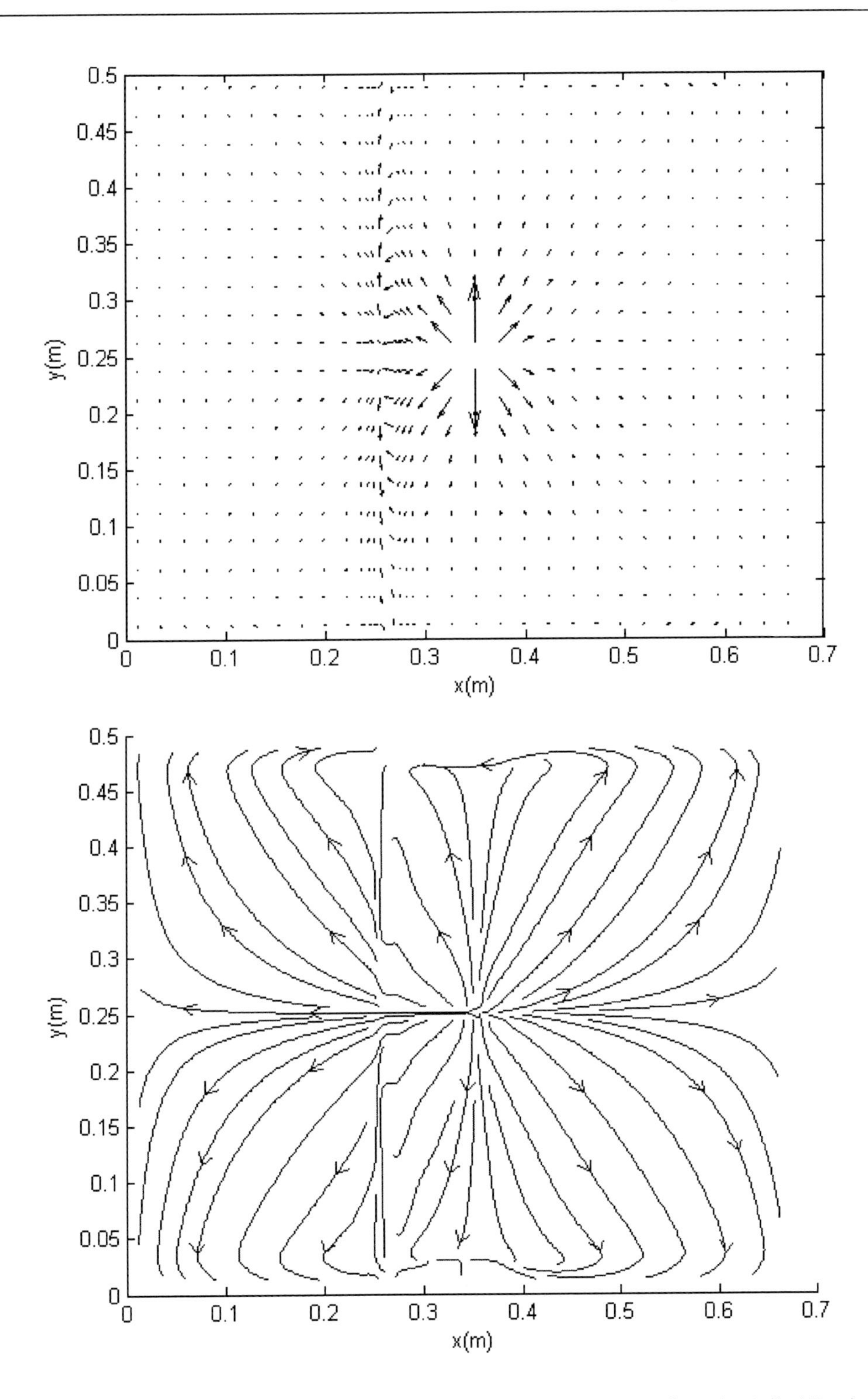

Figure 14. (a) Structural intensity vector plot, (b) Structural intensity streamline plot (a/h=1/2, c/a=3/8, f=60Hz).

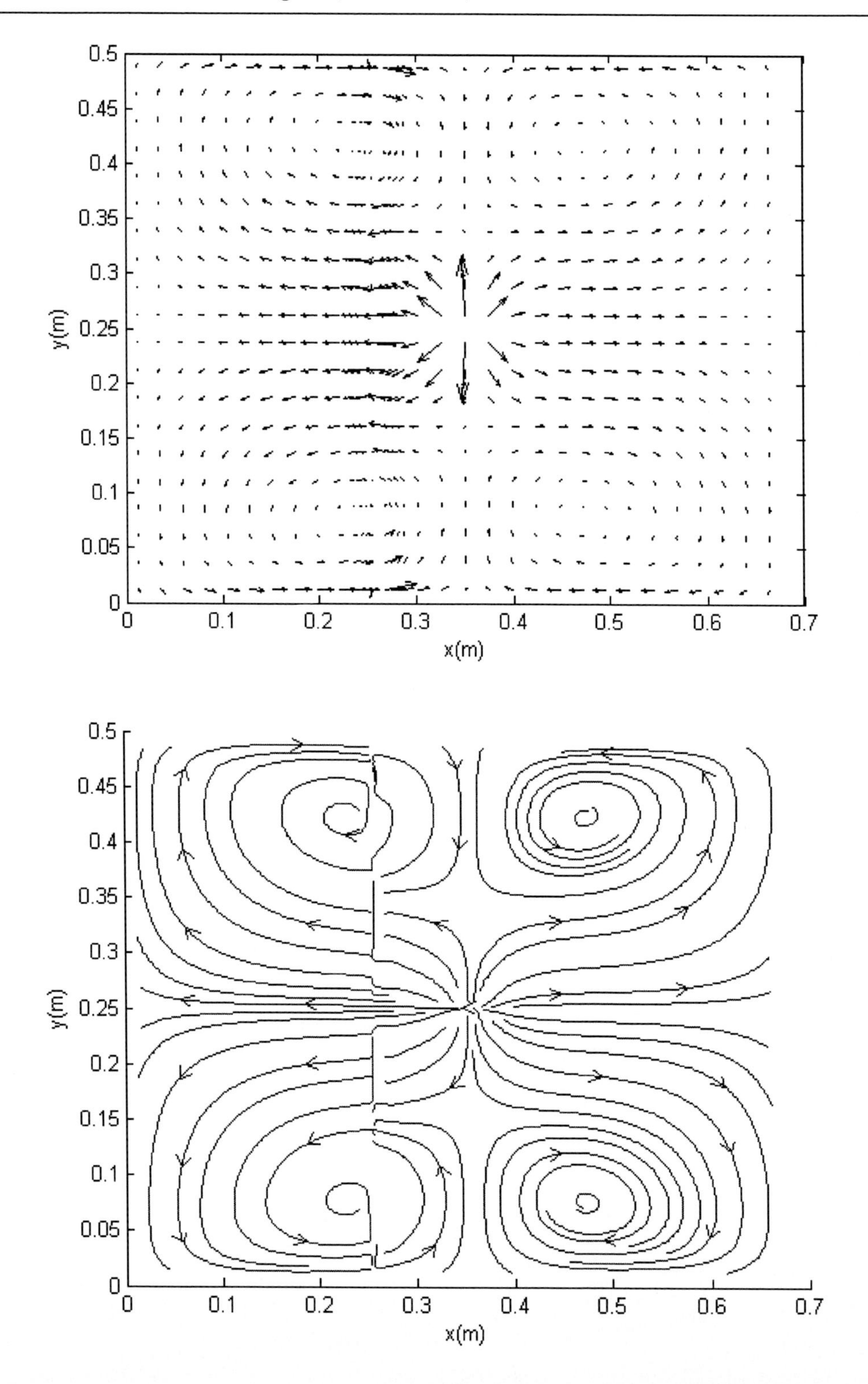

Figure 15. (a) Structural intensity vector plot, (b) Structural intensity streamline plot (a/h=1/2, c/a=3/8, f=480Hz).

Conclusion

In this chapter, the structure-borne intensity fields of a simply-supported thin aluminium plate with a surface crack are investigated by using solid finite elements. To describe the internal element stress fields more accurate, the isoparametric solid elements are used to model the plate. The intact plate's structural intensity patterns obatined by shell element and solid element are respectively computed, which validates the accuracy of the solid element calculation of structural intensity. The crack is modelled by quarter point crack tip element. Based on solid finite element, the cracked plate's displacement vector field, structural intensity vector field and structural intensity streamline field are obtained under a point excitation harmonic force applied at the centre of the plate.

The calcuations show that the velocity fields is related to the vibraion modes but can not indicate the source and crack's location. The cracked plate's intensity vector and streamline patterns show that the existance of crack changes the structural intensity in the plate. At the location of the crack, the intensity vector have abruptly changes in magnitude and direction and the structural intensity streamline are no longer smooth and continuous. Moreover, the structural intensity streamline can show the power flow path and crack location more clearly. Cases of different crack location indicates that the structural intensity pattern is highly related to the crack's location and it can clearly identify the location of the crack in the plate. The research implies that the investigation of the structural intensity in cracked structures have the potential to crack detection. For the measurement of structural intensity field, the laser Doppler vibrometer can offer a good alternative to ordinary transducers in intensity measurements, and it helps to measure both in-plane and out-of-plane quantities, useful for structural intensity in simple and complex structures.

Acknowledgments

The authors are pleased to acknowledge the support of the National Natural Science Foundation of P. R. China (Contract No. 50375059).

References

[1] Doebling, S. W.; Farrar, C. R.; Prime, M. B. Summary review of vibration-based damage identification methods. *Shock and Vibration Digest.* 1998, 30(2), 91-105.

[2] Gounaris, G.; Dimarogonas, A. Finite element of a cracked prismatic beam for structural analysis. *Computers and Structures.* 1988, 28(3), 309-313.

[3] Chaudhari, T. D.; Maiti, S. K. Study of vibration of geometrically segmented beams with and without crack. *International Journal of Solids and Structures.* 2000, 37(5), 761-779.

[4] Khadem, S. E.; Rezaee, M. An analytical approach for obtaining the location and depth of an all-over part-through crack on externally in-plane loaded rectangular plate using vibration analysis. *Journal of Sound and Vibration.* 2000, 230(2), 291-308.

[5] Douka, E.; Loutridis, S.; Trochidis, A. Crack identification in plates using wavelet analysis. *Journal of Sound and Vibration.* 2004, 270(1-2), 279-295.

[6] Krawczuk, M.; Palacz, M.; Ostachowicz, W. Wave propagation in plate structures for crack detection. *Finite Elements in Analysis and Design.* 2004, 40(9-10), 991-1004.

[7] Li, T. Y.; Zhang, W. H.; Liu, T. G. Vibrational power flow analysis of damaged beam structures. *Journal of Sound and Vibration.* 2001, 242(1), 59-68.

[8] Li, T. Y.; Zhang, T.; Liu, J. X.;etc. Vibrational wave analysis of infinite damaged beams using structure-borne power flow. *Applied Acoustics.* 2004, 65(1), 91-100.

[9] Li, T. Y.; Liu, J. X.; Zhang, T. Vibrational power flow characteristics of circular plate structures with peripheral surface crack. *Journal of Sound and Vibration.* 2004, 276 (3-5), 1081-1091.

[10] Zhu, X.; Li, T. Y.; Zhao, Y.;etc. Structural power flow analysis of Timoshenko beam with an open crack. *Journal of Sound and Vibration.* 2006, 297(1-2), 215-226.

[11] Noiseux, D. U. Measurement of power flow in uniform beams and plates. *Journal of the Acoustical Society of America.* 1970, 47(1), 238-247.

[12] Pavic, G. Measurement of structure borne wave intensity 1. Formulation of the methods. *Journal of Sound and Vibration.* 1976, 49(2), 221-230.

[13] Verheij, J. W. Cross-spectral density method for measuring structure borne power flow on beams and pipes. *Journal of Sound and Vibration.* 1980, 70(1), 133-139.

[14] Hambric, S. A. Power flow and mechanical intensity calculations in structural finite element analysis. *Journal of Vibration, Acoustics, Stress, and Reliability in Design.* 1990, 112(4), 542-549.

[15] Gavric, L.; Pavic, G. Finite element method for computation of structural intensity by the normal mode approach. *Journal of Sound and Vibration.* 1993, 164(1), 29-43.

[16] Hambric, S. A.; Szwerc, R. P. Predictions of structural intensity fields using solid finite elements. *Noise Control Engineering Journal.* 1999, 47(6), 209-217.

[17] Li, Y. J.; Lai, J. C. S. Prediction of surface mobility of a finite plate with uniform force excitation by structural intensity. *Applied Acoustics.* 2000, 60(3), 371-383.

[18] Xu, X. D.; Lee, H. P.; Lu, C. The structural intensities of composite plates with a hole. *Composite Structures.* 2004, 65(3-4), 493-498.

[19] Xu, X. D.; Lee, H. P.; Lu, C. Numerical study on energy transmission for rotating hard disk systems by structural intensity technique. *International Journal of Mechanical Sciences.* 2004, 46(4), 639-652.

[20] Khun, M. S.; Lee, H. P.; Lim, S. P. Structural intensity in plates with multiple discrete and distributed spring-dashpot systems. *Journal of Sound and Vibration.* 2004, 276(3-5), 627-648.

[21] Xu, X. D.; Lee, H. P.; Lu, C. Power flow paths in stiffened plates. *Journal of Sound and Vibration.* 2005, 282(3-5), 1264.

[22] Liu, Z. S.; Lee, H. P.; Lu, C. Structural intensity study of plates under low-velocity impact. *International Journal of Impact Engineering.* 2005, 31(8), 957-975.

[23] Lee, H. P.; Lim, S. P.; Khun, M. S. Diversion of energy flow near crack tips of a vibrating plate using the structural intensity technique. *Journal of Sound and Vibration.* 2006, 296(3), 602-622.

[24] Tran, T. Q. N.; Lee, H. P.; Lim, S. P. Structural intensity analysis of thin laminated composite plates subjected to thermally induced vibration. *Composite Structures.* 2007, 78(1), 70-83.

[25] Pascal, J. C.; Loyau, T.; Carniel, X. Complete determination of structural intensity in plates using laser vibrometers. *Journal of Sound and Vibration.* 1993, 161(3), 527-531.

[26] Freschi, A. A.; Pereira, A. K. A.; Ahmida, K. M.;etc. Analyzing the total structural intensity in beams using a homodyne laser doppler vibrometer. *Shock and Vibration.* 2000, 7(5), 299-308.

[27] Arruda, J. R. F.; Mas, P. Localizing energy sources and sinks in plates using power flow maps computed from laser vibrometer measurements. *Shock and Vibration.* 1998, 5(4), 235-253.

[28] Pavic, G. Vibrational energy flow in elastic circular cylindrical shells. *Journal of Sound and Vibration*. 1990, 142(2), 293-310.

[29] Malone, J. G.; Hodge, P. G.; Plunkett, R. Finite element mesh for a complete solution of a problem with a singularity. *Computers and Structures.* 1986, 24(4), 613-623.

[30] Vafai, A.; Estekanchi, H. E. A parametric finite element study of cracked plates and shells. *Thin-Walled Structures.* 1999, 33(3), 211-229.

[31] Henshell, R. D.; Shaw, K. G. Crack tip finite elements are unnecessary. International *Journal for Numerical Methods in Engineering.* 1975, 9(3), 495-507.

In: New Research on Acoustics
Editor: Benjamin N. Weiss, pp. 343-357
ISBN: 978-1-60456-403-7

Chapter 11

ULTRASOUND SAFETY INDICES IN CLINICAL OBSTETRICS: A CRITICAL REVIEW OF THEIR VALUES AND END-USERS FAMILIARITY

Eyal Sheiner[1,2,*], Ilana Shoham-Vardi[3], and Jacques S. Abramowicz[1]

[1]Department of Obstetrics and Gynecology, Rush University Medical Center, Chicago, IL
[2]Department of Obstetrics and Gynecology, Soroka University Medical Center, Beer-Sheva, Israel, Faculty of Health Sciences, Ben-Gurion University of the Negev, Israel
[3]Epidemiology and Health Services Evaluation Department, Faculty of Health Sciences, Ben-Gurion University of the Negev, Israel

Abstract

Background
As a form of energy, diagnostic ultrasound (DUS) has the potential to have effects on living tissues, e.g. bioeffects. The two most likely mechanisms for bioeffects are heating and cavitation. The thermal index (TI) expresses the potential for rise in temperature at the ultrasound's focal point. Since an output of TI over 1.5 is considered hazard, the question is what the settings in which such hazardous exposure occurs are. The mechanical index (MI) indicates the potential for the ultrasound to induce inertial cavitation in tissues. Nevertheless, cavitation has not been documented in mammalian fetuses, since there is not an air-water interface, which is needed for the cavitation mechanism.

Objective
This review presents data regarding ultrasound end-users familiarity with safety issues, acoustic output and safety of obstetrics ultrasound.

Conclusions
There are scarce data on instruments acoustic output (nor patient acoustic exposure) for routine clinical ultrasound examinations. Ultrasound end-users are poorly informed regarding

* Corresponding author: **Eyal Sheiner, MD,** Departments of Obstetrics and Gynecology, Soroka University Medical Center, Beer-Sheva 84103, Israel, Tel: (972-54) 8045074; E-Mail: **sheiner@bgu.ac.il**

safety issues during pregnancy. While first trimester ultrasound is associated with negligible rise in the thermal index, increased TI levels are reached while performing obstetrical Doppler studies. In particular, TI levels may reach 1.5 and above. Acoustic exposure levels during 3D/4D ultrasound examination, as expressed by TI are comparable to the two-dimensional B-mode ultrasound. However, it is very difficult to evaluate the additional scanning time needed to choose an adequate scanning plane and to acquire a diagnostic 3D volume.

Keywords: Ultrasound; pregnancy; safety; acoustic output; thermal index; mechanical index

Introduction

It has been more than four decades since ultrasound (US) has been used to provide fetal imaging [1]. Over the years, it has become widespread, in the labor rooms, private offices, emergency departments and even, recently, in the shopping mall [1,2]. Indeed, most pregnant women have, at least one US scan during pregnancy and almost 40% of total US scans performed are for obstetric use [1-3]. Proven clinical benefits for first trimester ultrasound include accurate dating, assessment of fetal viability in cases of threatened abortions, assessment of multiple gestations and early identifications of congenital malformations [3]. Moreover, most new technologies in prenatal diagnosis such as the nuchal translucency screening are ultrasound based [3]. They are recommended as part of a comprehensive prenatal screening and counseling programs by experienced operators [3].

Acoustic Output

Diagnostic ultrasound (DUS) is considered safe for the fetus [1,2]. However, data regarding effects on living tissues, e.g. bioeffects, are inconclusive [4-7], although no objective findings justify withholding scanning for clinical indications [8]. As a form of energy, ultrasound has the potential to have bioeffects, the two most likely mechanisms for these being heating and cavitation [9,10].

The cavitation mechanism involves the presence of gaseous bubble in an air-water interface [1]. Cavitation is the oscillation of gas bubbles caused by ultrasound waves due to alternating positive and negative pressures. Inertial (previously known as transient) cavitation is the growth of bubbles, which undergo large variations in their size. The variation in size can be such that the bubble can "collapse" in a violent fashion, resulting in a tremendous rise in temperature (thousands of degree Kelvin), albeit for a fraction of time in a small area. Furthermore, since the internal mass of the bubble is so small, the actual temperature rise is negligible (adiabatic reaction). However, the result may be high-pressure shock waves and/or free radicals that may damage surrounding tissues. *The mechanical index (MI)* indicates the potential for the ultrasound to induce inertial cavitation in tissues [11-14]. Cavitation has not been documented in mammalian fetuses, since there is no air-water interface, which is needed for the cavitation mechanism [14].

Hyperthermia is a proven teratogen in experimental animals, and although controversial, is considered teratogenic in human fetuses [15,16]. Human body normal core temperature is generally accepted to be 37° Celsius with a diurnal variation of ± 0.5-1 °Celsius, although 36.8 ±0.4°Celsius (95% confidence interval) may be closer to the actual mean for large populations. Temperature in the human fetus is higher than maternal core-body temperature by 0.3-0.5°Celsius during the entire gestation but in the third trimester (near-term) temperature of the fetus is higher by 0.5 °Celsius than that of its mother. As the waveform travels through tissue, it loses amplitude by absorption and scatter. With absorption, energy is converted into heat. The latter can raise the temperature of the tissue being scanned [17]. Several studies have suggested a general threshold of temperature elevation of 1.5-2 ° C before any evidence of developmental effect occurs. An increase of 2.5° C and above is possible with one hour of exposure to ultrasound [1]. Sensitivity of the fetus to external insults changes noticeably during different stages of pregnancy, the highest being during organogenesis in the first trimester [15]. The World Federation for Ultrasound in Medicine and Biology has summarized that exposure that produces a maximum temperature elevation of no more than 1.5° C above normal physiological levels may be used without reservation on thermal ground [18]. *The thermal indices (TI)*_express the potential for rise in temperature along the ultrasound's beam. There are 3 thermal indices: TIS, thermal index for soft tissues, when the ultrasound beam does not impinge on bone and is appropriate mostly for the first trimester; TIB, thermal index for bones, when the beam impinges on bone at or near its focus and should be displayed in the second and third trimesters; and TIC, thermal index for cranial bone, when the transducer is very close to the bone, such as when scanning in the adult [10,12,13].

Prior to 1976, there were no limits to the permissible acoustic output from ultrasonic diagnostic equipment. In 1976, the FDA began regulating the output levels of machines to be no more than 94 mW/cm^2, spatial-peak temporal-average intensity (I_{SPTA}) for fetal use. In 1992, under pressure from manufacturers and end-users, the FDA changed this limit to 720mW/cm^2 but mandated that machines capable of producing higher outputs display to the diagnostician some indication of the likelihood of ultrasound-induced bioeffects. This Output Display Standard (ODS) was implemented in 1992-1993 [13,19, 20]. It consists of on-screen voluntary labeling by manufacturers. The ODS comprises the mechanical and the thermal indices, because of the 2 main possible bioeffects of ultrasound in tissues.

While the rise in temperature during first trimester ultrasound (performed at time of organogenesis) is assumed to be low, data actually documenting temperature change during diagnostic US though out pregnancy are scarce [19]. There are methodological difficulties to assess the effect of repeated and intense ultrasound in observational studies, since there is no way to measure actual *in situ* exposure in human fetuses, and the actual acoustic output has not been investigated. Also, the reverse causality issue poses a problem in randomized trials where the control group gets ultrasound as clinically needed.

Most prospective studies addressing the safety of the US procedure are available in animal models or tissue cultures [1]. Tarantal and Hendrickx [21] evaluated 30 pregnancies in monkeys, half of which were exposed to ultrasound from gestational day 21 to 152 +/- 2.. Exposures were performed with a commercial real-time sector scanner (ATL, MK 600). The length of exposure was approximately the same as human exposure (10-20 min/exam) although the frequency of the examinations was greater. The scanned fetuses had lower birth-weights and were shorter than the control group. No significant differences were noted between the groups with regard to the rate of abortions, major malformations or stillbirths.

Demonstrable harmful effects of ultrasound in humans have not been shown. A total of 2743 women with single pregnancies were randomized to either a protocol of ultrasound imaging and continuous wave Doppler studies at 18, 24, 28, 34 and 38 weeks gestation (i.e.

beyond the first trimester), the intensive group, or to a protocol of a single ultrasound scan at 18 weeks and further imaging examinations only as clinically indicated (the regular group) [22,23]. Babies in the repeated ultrasound group tended to be shorter and there was a small increase in the proportion of growth-restricted offspring [22,23]. Generally repeated ultrasound, and specifically Doppler studies are performed among fetuses suspected to have intrauterine growth restriction (IUGR), thus the argument of reverse causality is only relevant in the control group where suspected IUGR got more Doppler, thus causing the deleterious effect of Doppler to be underestimated. However, no significant differences indicating deleterious effects of the multiple ultrasound studies were noted after 8 years of follow-up of the children's' development [24].

Moore et al [25] compared the birth weights of 1598 exposed and 944 unexposed single live births. Neonates that were exposed to more than one US scan had lower birth weight. However, a more detailed analysis found the women who were scanned more than once during their pregnancy to be high-risk group in comparison to the patients that were not scanned during pregnancy. Lyons et al [26] evaluated 149 siblings, of whom one was exposed to US scan during pregnancy, and the other was not. Exposure of fetuses to ultrasound did not significantly affect either height or weight in childhood up to 6 years of age.

Acoustic output of the ultrasound machines and length of exposure, however, were not described in any of the above studies [21-26]. Thus, the actual acoustic exposure of the fetuses was not known. Most women today undergo several ultrasound scans during pregnancy. As there is no way to measure actual *in situ* exposure in human fetuses, the only way to attempt and quantify exposure is monitoring the machine output using the thermal and mechanical indices.

First Trimester Ultrasound

Sensitivity of the fetus to external insults changes noticeably during different stages of pregnancy, the greatest being during organogenesis in the first trimester [1,4]. Thus, our group had conducted a single-blinded, observational study to investigate acoustic output during the first trimester [27]. First trimester patients were randomly selected from those scheduled for viability scans. These exams were performed to verify gestational age, viability and number of gestational sac, i.e. when a clear medical indication was present. Doppler studies were not performed during these examinations. The scans were performed on several ultrasound machines, to detect possible variation between manufacturers, such as iU22 (Philips Medical Systems, Bothell, WA), Prosound alfa-10 (Aloka, Wallingford, CT), and Voluson 730 (General Electrics, Milwaukee, WI). Patients were recruited in the department of Ob/Gyn at Rush University Medical Center, and University of Chicago, Chicago Il. Data were collected by an obstetrician present during the entire exam or from video recordings of the entire examinations. The sonographers performing the examinations were unaware of the objectives. Data included duration of the exam, changes in TI and MI and specific time duration spent at each value of MI and TI. A total of 52 first trimester examinations were evaluated. Mean gestational age was 8.9 ± 1.9 weeks. Mean duration of the DUS examinations was 8.1± 1.4 minutes. During the examinations there were 178 MI variations (mean 0.9±0.3), and 167 TI variations (with mean of 0.2±0.1). Accordingly, it seems that first trimester gray scale ultrasound is associated with negligible rise in the thermal index [27].

Doppler Studies

Doppler evaluation of the maternal and fetal circulation by color imaging and pulsed wave spectral Doppler has brought new applications for routine obstetric ultrasound use [28-31]. Investigation of flow in the uterine artery, umbilical vessels, fetal arteries and veins are integral to the modern assessment of the mother and fetus at risk [29]. Abnormal waveforms from Doppler ultrasound may indicate poor fetal prognosis. Doppler ultrasound in high-risk pregnancy, and specifically those complicated by hypertension or IUGR, is associated with a reduction in perinatal deaths and fewer admissions to hospital [28].

Data regarding Doppler effects on living tissues are inconclusive [1,22,24]. The only adverse effect that can be attributed to the use of obstetric Doppler is probably an increase in the occurrence of IUGR [22]. Newnham et al [22] performed a randomized control trial including more than 2,800 parturients and found an increased risk of IUGR when exposed to frequent Doppler examinations. However, since this outcome variable was not the primary goal of the study, a chance occurrence cannot be excluded. Again, after 8 years of follow-up, the authors concluded that growth and measures of developmental outcome in childhood are similar to those in children who had received a single prenatal scan [24]. Special concern should be given to the indication of the Doppler analyses since IUGR is one of the major reasons for these tests. Extensive use of Doppler testing should raise concerns regarding fetal well being especially if performed during early pregnancy since higher levels of acoustic energy than conventional B-mode imaging (gray scale imaging) are generated [1,2,28-30].

The higher energy required for Doppler studies brought Deane and Lees [28] to investigate exposures in pregnant women undergoing Doppler studies. Their study focused on graphical displays of Doppler investigations of fetal and maternal circulation between 24 and 34 weeks of pregnancy. New sample graphical displays were generated showing duration of each mode, thermal and mechanical index levels and overall elapsed time of scan and modes. The authors suggested the use of their new display of acoustic indices in order to monitor exposures and to ensure safe use of ultrasound in obstetrics. However, the display is not in clinical use. With increased utilization of Doppler in first trimester screening, for instance for analysis of the ductus venosus [31] and tricuspid valve regurgitation [32] this becomes particularly relevant since prolonged examination periods are often necessary to obtain adequate measurements.

Recently, our group compared acoustic output between B-mode and Doppler studies [33]. A total of 63 examinations were evaluated (figure 1). The TI was significantly higher in the pulsed wave Doppler (mean 1.5±0.5, range 0.9-2.8) and color flow imaging studies (mean 0.8±0.1, range 0.6-1.2) as compared to B-mode ultrasound (mean 0.3±0.1, range 0.1-0.7; $P<0.01$; Table 1, Figure 2). During the examination 190 B-mode MI variations were recorded (mean 1.1±0.1), which were comparable to the 31 color flow Doppler studies (mean 1.0±0.1; $P=0.09$), but higher than the 190 pulsed-wave Doppler MI variations (mean 0.9±0.2; $P<0.001$). In conclusion, increased acoustic output levels, as expressed by TI levels, are reached while performing obstetrical Doppler studies. In particular, TI levels may reach 1.5 and above. Thus, Doppler procedures should be performed with caution and be as brief as possible during obstetrical ultrasound [33].

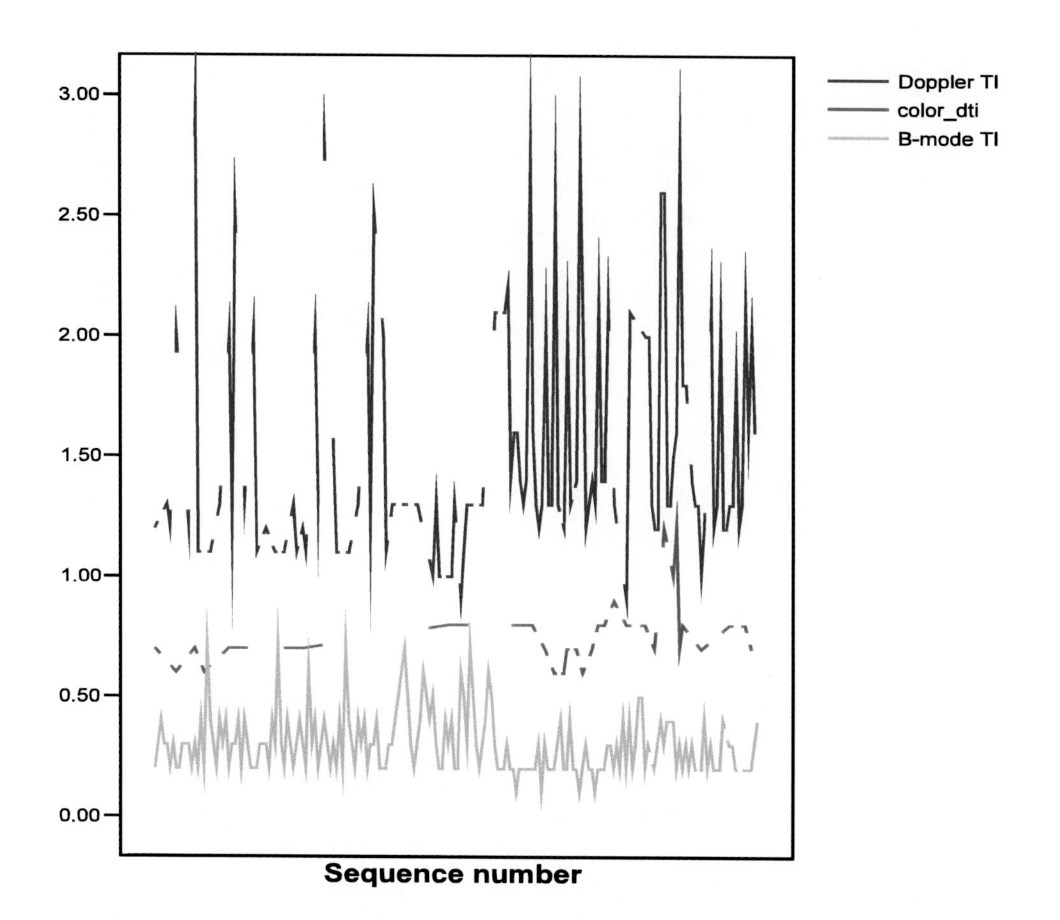

Figure 1. Distribution of all TI variations during Doppler and B-mode ultrasound studies. Adopted from Sheiner et al. An increased thermal index can be achieved when performing Doppler studies in obstetrical ultrasound. J Ultrasound Med 2007;26:71-6.

Table 1. Acoustic output during B-mode and Doppler ultrasound studies

Characteristics	B-Mode (n=190)	Color Doppler (n=31)	Pulsed wave (n=118)	P-value
TI mean SD	0.3±0.1	0.8±0.1	1.5±0.5	<0.01
MI mean SD	1.1±0.1	1.0±0.1	0.9±0.2	<0.01

Adopted from Sheiner E, et al. An increased thermal index can be achieved when performing Doppler studies in obstetrical ultrasound. J Ultrasound Med 2007;26:71-6.

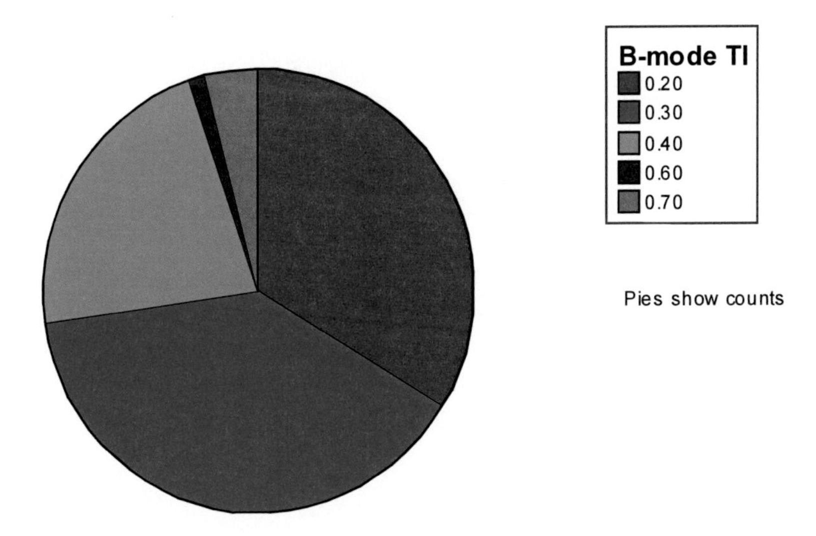

Figure 2 a. Pie distribution of TI during B-mode ultrasound.

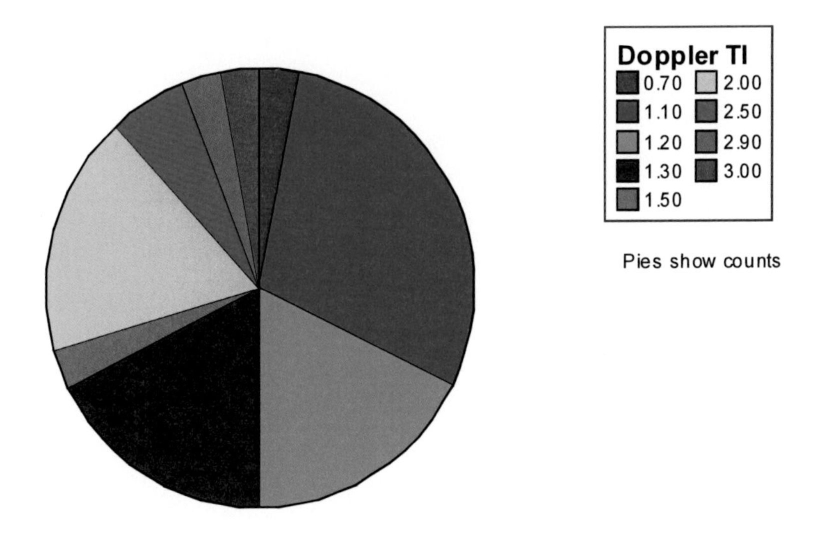

Figure 2 b. Pie distribution of TI during Doppler ultrasound.

3D/4D Ultrasound

Three-dimensional (3D) ultrasound is gaining popularity in prenatal diagnosis and will certainly acquire an increasing role in imaging of the fetus [34,35]. This post imaging reconstruction is a helpful tool, adding to the detection of a wide range of anomalies; mainly those involving the face, skeleton and extremities [36-40]. Ultra fast acquisition enables live 3D, or real time update of the 3D volume, also know as four-dimensional (4D) ultrasound [34,35].

While there are no studies regarding the safety of 3D ultrasound, the short acquisition time and post processing analysis are thought to decrease exposure, although yet no study had examined the acoustic output of this procedure. Nevertheless, it is now widely performed in non-medical facilities, to provide images for the family photo album, for entertainment and certainly not for diagnostic purposes ("keepsake ultrasound").

It is important to remember that although most volume manipulations are performed offline or on the ultrasound machine after the patient's scan is completed, the exam is still performed in addition to the 2D scan, resulting in actual increased scanning time. Furthermore, the need to obtain a "perfect" angle for the 3D reconstruction, may add additional scanning time.

A recent study compared acoustic output of conventional two-dimensional (2D) and 3D/4D ultrasound during pregnancy [41]. Mean TI during the 3D (0.27±0.1) and 4D examinations (0.24±0.1) was comparable to the TI during the B-mode scanning (0.28±0.1; P=0.343; Table 1). The MI indices during the 3D volume acquisitions were significantly lower than the two dimensional B-mode ultrasound studies (0.89±0.2 vs. 1.12±0.1; P=0.002). The 3D volume acquisitions added 2.0±1.8 minutes of actual ultrasound scanning time (i.e. not including data processing and manipulation, nor 3D display, which are all post-processing steps). The 4D added 2.2±1.2 minutes.

The authors concluded that basically, acoustic exposure levels during 3D/4D ultrasound examination, as expressed by TI are comparable to the two-dimensional B-mode ultrasound. However, it is very difficult to evaluate the additional scanning time needed to choose an adequate scanning plane and to acquire a diagnostic 3D volume [41].

Table 2. Acoustic output during 2-D and 3-D ultrasound studies

Characteristics	B-Mode	Three-D	Four-D	P-value
TI mean SD	0.28±0.1	0.27±0.1	0.24±0.1	NS
MI mean SD	1.12±0.1	0.89±0.2	1.11±0.2	0.018

Adopted from Sheiner E, et al. Comparison between acoustic output indices in 2D and 3D/4D ultrasound in obstetrics. Ultrasound Obstet Gynecol 2007; 29:326-8.

Knowledge of Ultrasound End-users in Safety Issues

A concern always arises regarding ultrasound end-users' knowledge in safety issues. There is no requirement for obstetric ultrasound certification in either USA or Europe, in order to perform such examinations [42]. Thus, there is an agreement that all courses on obstetrical and gynecological ultrasound approved by the International Society of Ultrasound in Obstetric and Gynecology (ISUOG) should discuss safety issues [42]. Unfortunately, the perceived safety of DUS is broadly accepted and, accordingly, the topic attracts little attention [43]. It seems that the FDA's upper limit regulatory control is being viewed by ultrasound end-users as safety limits, which is certainly not the case.

In a recent editorial, Marsal [42] distributed a questionnaire to professionals using ultrasound for fetal examinations in Europe. Only 22% could explain what TI is, and only 11% could give a correct explanation for MI. Merely 28% of the responders correctly indicated where, on their own machines, they can find the information about the acoustic indices. While acknowledging the limitation that the questionnaire was distributed only among European ultrasound users, Marsal [42] concluded that the output display standard failed to provide a practical useful platform during obstetric examination.

Ultrasound should be performed only when there is a valid medical indication [44]. Currently, although only 2-3 ultrasound examinations are recommended for low risk pregnancies, pregnant women actually undergo more examinations and can even ask for souvenir photos or videos of the unborn child, requiring prolonged scanning, in shopping malls [19,44].

In order to determine end-users knowledge regarding safety aspects of diagnostic ultrasound (US) during pregnancy in the US, a questionnaire was distributed to ultrasound end-users [45]. One hundred thirty end-users completed the questionnaires (63% response rate). Sixty-three percent were physicians (n=84), the majority of them obstetricians (81.7%). About 18% of participants routinely performed Doppler during the 1st trimester. Fifty percent of end-users thought the number of ultrasound exams in low-risk pregnancy should be limited, to 2-3 (2.6±0.9). Almost 70% disapproved of "keep-sake/entertainment" US. While 32.2% of the participants were familiar with the term TI, only 17.7% actually gave the correct answer to the question on the nature of TI. About 22% were familiar with the term MI, but only 3.8% described it properly. Almost 80% of end-users did not know where to find the acoustic indices. Only 20.8% were aware they are displayed on the US monitor during the exams. End-users with higher knowledge on safety issues thought there should be limitations in the number of ultrasound exams in low-risk pregnancies (OR=3.3; 95% CI 1.1-10.0; p=0.028). Likewise, these end-users were more likely to respond that ultrasound might have adverse effects during pregnancy (OR=3.2; 95% CI 1.1-12.5; p=0.045).

The questionnaire proved that even in the USA, ultrasound end-users are poorly informed regarding safety issues during pregnancy [45]. Further efforts in the realm of education and training are needed in order to improve knowledge of end-users about the acoustic output of the machines and safety issues.

Discussion

Our studies give perspective regarding the acoustic output, as expressed by the thermal and mechanical indices, during routine ultrasound studies, as well as on end-users knowledge regarding safety issues [27,33,41,45]. To date, to the best of our knowledge, no such studies have been performed in pregnant women to evaluate acoustic output during routine obstetrical exams. Importantly, most data indicating lack of adverse effects on human fetuses are based on older studies, using lower intensities. The major findings of our studies are that during routine obstetrical ultrasound MI and TI are generally low [27,33,41]. Nevertheless, higher levels, particularly TI levels above 1.5, can be achieved, although they account for only a very small proportion of exam time [33]. Of major importance, ultrasound end-users are poorly informed regarding safety issues during pregnancy [45].

It is assumed that using modern US machines there is only a negligible rise in temperature, usually less than 1°C [2,46]. Harmful effects have not been shown so far, and although a small unexplained increase in the proportion of growth-restricted offspring was seen following repeated ultrasound scans [22-23], no significant differences indicating deleterious effects of multiple ultrasound studies were noted after 8 years of follow-up [24].

Nevertheless, there is always a concern regarding ultrasound end-users' knowledge of safety issues. While acknowledging the limitation that his questionnaire was distributed only among European ultrasound users, Marsal [42] concluded that the output display standard failed to provide a practical useful platform during obstetric examination.

Unfortunately, our study does not show a better picture. Similarly to their European counterparts, professional ultrasound end-users in the USA show poor knowledge on safety issues, at least among obstetricians, since only 2.4% of the physician respondents defined themselves as radiologists. Most did not answer correctly the TI question, and even less- the MI question. Only 21% knew where to find the TI/MI during the ultrasound exam.

The purpose of the output display standard was to provide the capability for end-users of DUS to operate their machines at higher levels, in order to increase diagnostic capabilities. The ODS did not specify any upper limits. The manufacturers are obliged to provide information on safety indices (i.e. the TI/MI values), but the responsibility for the ultrasound output energy is, ultimately, the end-users'. Ultrasound end-users should be familiar with the output energy, how to control it and accordingly how to use the machine in a safe manner. Nevertheless, if the end-users are not familiar with the acoustic indices or where to find them, one can assume they won't be able to control them.

End-users were divided on the issue of limiting the routine US in low risk pregnancy. Interestingly, these with higher knowledge thought there may be adverse effects to ultrasound and thought there should be limitations on the number of exams performed in low-risk pregnancies. Nevertheless, more than 30% of the ultrasound professionals actually approve "keep-sake" ultrasound, without any clinical indication. These procedures are generally performed by sonographers, but no significant differences in knowledge were noted between physicians and sonographers. Obviously, if professional end-users show such poor knowledge on safety issues, one cannot expect end-users in the shopping malls, performing souvenir ultrasound, to have a better understanding of safety topics. Non-medical fetal ultrasound is of great concern and Wax and co-authors [47] recently pointed out that although most end-users do not favor non-indicated examinations, they do not think that sonographers performing these exams should be reported or punished [47].

US appears to be safe, however the potential exists to produce possible bioeffects with increased power levels associated with localized heating that in a sensitive area, such as the developing fetus, in the first trimester, could lead to possible adverse effects [27]. According to our results, the TIS during routine first trimester ultrasound scanning is basically low, and, if one relies on this criterion alone, it would seem that ultrasound is indeed a safe procedure [27]. Although calculated values of the indices may vary from manufacturer to manufacturer and fluctuate with changes in equipment settings, these indices are subjected to international standards. Indeed, the indices were recently found comparable between different machines [19].

Spectral and color Doppler produce higher acoustic energy levels than B-mode [33]. In particular, TI levels may reach 1.5-2.0 and even above [33]. Exposure to Doppler ultrasound can significantly heat biologic tissue because of the relatively high intensities used and the need to hold the transducer motionless during the exam [30]. Significant temperature increases can occur at or near to bone in the fetus from the second trimester, if the beam is held immobile for more than 30 seconds in some pulsed Doppler applications [48]. The threshold for irreversible damage in the fetal brain is exceeded when a temperature increase of 4° C is maintained for 5 min [49]. Fortunately, this threshold of 5 minutes is generally above Doppler fetal exposure in clinical use. The mean duration of Doppler in the present study was 0.9 minutes, although the longest examination took 4 minutes (range of 0.2-4 minutes). Moreover, in prolonged evaluation, different areas of the vessel are studied and generally not one area is continuously exposed. Still, the end-user should be familiar with the fact that while the use of B-mode grey-scale imaging is not contra-indicated on thermal grounds, some pulsed Doppler equipment has the potential to produce biologically significant temperature increases, specifically at interfaces between bone and soft tissue [50].

Routine examination by Doppler modality during the first trimester is rarely indicated [46]. Thus, it was of concern to find that 17.7% of professional end-users perform Doppler studies routinely during the first trimester. Nevertheless, this should not prevent the use of this mode when clinically indicated (and 42.3% actually do so) provided that the end-user has adequate knowledge of the acoustic output. Unfortunately, this is not the case according to the questionnaire. With increased utilization of Doppler in first trimester screening, for instance for analysis of the ductus venosus [31] and tricuspid valve regurgitation [32], this becomes particularly relevant since prolonged examination periods are often necessary to obtain adequate measurements.

The results presented here are particularly worrisome, given published data on bioeffects in different animal models and the recent report on the effects of ultrasound in neuronal migration in mice embryos [51]. It should, however, be noted that direct correlation with human scanning should not be inferred from that study, secondary to major differences in scanning protocols.

A critical value of the 3D/4D images is the ability to actually see a "real" image of the fetus [34,35]. Mothers who had 3D ultrasound images of the fetus tended to show their images to a greater number of people compared to mothers who had 2D scans alone [52]. Accordingly, it was suggested that 3D might have a greater impact on the maternal-fetal bonding process [52]. The parents, who want keepsake images of the developing fetus, mostly demand it as part of the regular anatomy survey. Sometimes, they even patronize one of the many non-medical companies offering 3D scans to the public.

Acoustic output, as expressed by TI and MI indices during 3D/4D scans is comparable to the 2D ultrasound. Indeed, 3D/4D ultrasound uses computer reconstructions of 2D images obtained through sweeps across the region of interest [34,35]. Thus, it is logical that the level of energy is comparable to the 2D scanning.

Calculated values of the indices may vary from manufacturer to manufacturer and fluctuate with changes in equipment settings. Thus, in order to strengthen the results, we examined acoustic indices variations in 3 different ultrasound machines. There were minimal deviations between the machines, and in general all showed comparable results of 2D and 3D/4D scans.

It is important to remember that although most volume manipulations are performed offline or on the ultrasound machine after the patient's scan is completed, the exam is still performed in addition to the 2D scan. In addition, the scan starts in the 2D mode, in order to select the best view of the region of interest. The time taken in order to select this best plane, which will be sweeped through for the 3D reconstruction was not calculated as part of the 3D exam. Therefore, actual exposure time of the fetus to ultrasound may significantly increase.

The equations for the calculation of TI and MI are complex and these indices are intended to be information available in real-time to the end-users, which is why we chose to analyze their values only, without detailing other parameters of acoustic output or exposure (including time spent at each TI). TIs are reasonable worst-case estimate of the temperature rise resulting from the exposure and can, thus, be used to assess the potential for harm via a thermal mechanism, the higher the TI, the higher this potential. Calculations are also based on steady-state temperature, the ultimate temperature reached after prolonged exposure. This time can be extremely short with a narrow beam and good tissue perfusion (as is the case in late 1st trimester scanning). It is longer (up to 5 minutes) with lesser perfusion (such as the first weeks of gestation when the maternal-fetal circulation may not yet be completely established). When bone is present, this time is very short (approximately 30sec). However, experimental data has shown that this worst-case elevation of temperature may be a gross underestimation, by as much as 150 percent for the TIB and 180 percent for the TIS (and, more rarely, an overestimation) [53]. There are many modifying factors, such as body habitus, perfusion, distance to transducer and presence or absence of bone or long fluid-path [54].

At a time where people can ask for souvenir photos of the unborn child, requiring prolonged scanning, end-users as well as pregnant women should be aware to the acoustic output of the machines. Increased awareness of safety issues should lead to adherence to the as low as reasonably achievable principle in general [55] and using Doppler studies in particular [33].

Acknowledgements

We are indebted to the Fulbright Foundation for its support of Dr Eyal Sheiner.

References

[1] Hershkovitz R, Sheiner E, Mazor M. Ultrasound in obstetrics: a review of safety. *Eur J Obstet Gynecol Reprod Biol* 2002;101:15-8.

[2] Abramowicz JS. Ultrasound in obstetrics and gynecology: is this hot technology too hot? *J Ultrasound Med* 2002;21:1327-33.

[3] Demianczuk NN, Van Den Hof MC, Farquharson D, Lewthwaite B, Gagnon R, Morin L, Salem S, Skoll A; Diagnostic Imaging Committee of the Executive and Council of the Society of Obstetricians and Gynecologists of Canada. The use of first trimester ultrasound. *J Obstet Gynaecol Can.* 2003;25:864-75.

[4] Cavicchi TJ, O'Brien Jr. WD. Heat generated by ultrasound in an absorbing medium. *J Acoust Soc Am.* 1984;70:1244–1245.

[5] Nyborg WL, Steele RB. Temperature elevation in a beam of ultrasound. *Ultrasound Med Biol* 1983; 9:611–620.

[6] Flynn HG. Physics of acoustic cavitation in liquids. In: Mason WP, editor. *Physical acoustics,* vol. 1B. New York: Academic Press, 1964. p. 57.

[7] Child SZ, Hartman C, Schery LA, Cartensen EL. Lung damage from exposure to pulsed ultrasound. *Ultrasound Med Biol* 1990;16:817–825.

[8] Salvesen KA. Ultrasound and left-handedness: a sinister association. *Ultrasound Obstet Gynecol* 2002;19:217-221.

[9] American Institute of Ultrasound in Medicine: Mechanical Bioeffects from diagnostic ultrasound: AIUM Consensus statements. Section 8. *J ultrasound Med* 2000,19: 149-153.

[10] National Council on Radiation Protection and Measurements (NCRP). Exposure criteria for medical diagnostic ultrasound: II. *Criteria based on all known mechanisms. Bethesda,* MD, NCRP Report no. 140, 2002.

[11] O'Brien WD Jr. Ultrasound bioeffect issues related to obstetric sonography and related issues of the output display standard. In: Fleicher AC, Manning FA, Jeanty P, Romero R. *Sonography in Obstetrics and Gynecology, Principles and Practice,* 1996 pp: 17-33.

[12] Abramowicz JS, Kossoff G, Marsal K, ter Haar G. Literature review by the ISUOG Bioeffects and Safety Committee. *Ultrasound Obstet Gynecol* 2002; 19:318-9.

[13] Miller MW, Brayman A, Abramowicz JS. Obstetric ultrasonography, a biophysical consideration of patient safety: the "rules" have changed. *Am J Obstet Gynecol* 1998; 179:241–254.

[14] Nyborg WL. Acoustic streaming. In: Mason WP, ed. *Physical acoustics*, vol. 1B. New York: Academic Press; 1965;265.

[15] Cavicchi TJ, O'Brien WD Jr. Heat generated by ultrasound in an absorbing medium. *J Acoust Soc Am* 1984;70:1244-1245.

[16] Edwards MJ. Apoptosis, the heat shock response, hyperthermia, birth defects, disease and cancer. Where are the common links? *Cell Stress Chaperones* 1998;3:213-220.

[17] O'Brien WD Jr. Ultrasound dosimetry and interaction mechanisms. In: Greene MW, ed. Non ionizing radiation: *Proceedings of the second international non ionizing radiation workshop.* Vancouver BC: Canadian Radiation Protection association; 1992:151.

[18] Barnett SB ed. WFUMB (The World Federation for Ultrasound in Medicine and Biology) symposium on safety of ultrasound in medicine. Recommendation on the safe use of ultrasound. *Ultrasound Med Biol* 1998;24: suppl1: xv-xvi.

[19] Sheiner E, Freeman J, Abramowicz JS. Acoustic output as measured by mechanical and thermal indices during routine obstetrical ultrasound. *J Ultrasound Med* 2005; 24:1665-70.

[20] AIUM/NEMA: American Institute of Ultrasound in Medicine/National Electrical Manufacturers Association Standard for real-time display of thermal and mechanical acoustic output indices on diagnostic ultrasound equipment. *Laurel,* MD: American Institute of Ultrasound in Medicine, 1992.

[21] Tarantal AF, Hendrickx AG. Evaluation of the bioeffect of prenatal ultrasound exposure in the Cynomolgus Macaque (Macaca Fascicularis). II. Growth and behavior during the first year. *Teratology* 1989;39:149-162.

[22] Newnham JP, Evans SF, Michael CA, Stanley FJ, Landau LI. Effects of frequent ultrasound during pregnancy: a randomized controlled trial. *Lancet* 1993;342:887-91.

[23] Evans S, Newnham J, MacDonald W, Hall C. Characterization of the possible effect on birthweight following frequent prenatal ultrasound examinations. *Early Hum Dev* 1996;45:203-14.

[24] Newnham JP, Doherty DA, Kendall GE, Zubrick SR, Landau LL, Stanley FJ. Effects of repeated prenatal ultrasound examinations on childhood outcome up to 8 years of age: follow-up of a randomized controlled trial. *Lancet* 2004;364:2038-44.

[25] Moore RM jr, Diamond EL, Cavalieri RL. The relationship of birth weight and intrauterine diagnostic ultrasound exposure. *Obstet Gynecol* 1988;71:513-517.

[26] Lyons EA, Dyke C, Toms M, Cheang M. In utero exposure to diagnostic ultrasound: a 6 year follow-up. *Radiology* 1988;166:687-690.

[27] Sheiner E, Shoham-Vardi I, Pombar X, Hussy MJ, Strassner HT, Abramowicz JS. First trimester ultrasound: Is the fetus exposed to high levels of acoustic energy? *J Clinical Ultrasound* 2007; 35:245-9.

[28] Deane C, Lees C: Doppler obstetric ultrasound: a graphical display of temporal changes in safety indices. *Ultrasound Obstet Gynecol* 2000;15:418-423.

[29] Neilson JP, Alfirevic Z. Doppler ultrasound for fetal assessment in high-risk pregnancies. *Cochrane Database Syst Rev* 2000;2:CD000073.

[30] Barnett SB, Maulik D; International Perinatal Doppler Society. Guidelines and recommendations for safe use of Doppler ultrasound in perinatal applications. *J Matern Fetal Med* 2001;10:75-84.

[31] Borrell A, Gonce A, Martinez JM, Borobio V, Fortuny A, Coll O, Cuckle H.First-trimester screening for Down syndrome with ductus venosus Doppler studies in addition to nuchal translucency and serum markers. *Prenat Diagn* 2005;25:901-905.

[32] Falcon O, Auer M, Gerovassili A, Spencer K, Nicolaides KH. Screening for trisomy 21 by fetal tricuspid regurgitation, nuchal translucency and maternal serum free beta-hCG and PAPP-A at 11 + 0 to 13 + 6 weeks.*Ultrasound Obstet Gynecol* 2006;27:151-155.

[33] Sheiner E, Shoham-Vardi I, Pombar X, Hussy MJ, Strassner HT, Abramowicz JS. An increased thermal index can be achieved when performing Doppler studies in obstetrical ultrasound. *J Ultrasound Med* 2007;26:71-6.

[34] Benacerraf BR, Benson CB, Abuhamad AZ, Copel JA, Abramowicz JS, Devore GR, Doubilet PM, Lee W, Lev-Toaff AS, Merz E, Nelson TR, O'Neill MJ, Parsons AK, Platt LD, Pretorius DH, Timor-Tritsch IE. Three- and 4-dimensional ultrasound in obstetrics and gynecology: proceedings of the American institute of ultrasound in medicine consensus conference. *J Ultrasound Med.* 2005;24:1587-97.

[35] Timor-Tritsch IE, Platt LD. Three-dimensional ultrasound experience in obstetrics. *Curr Opin Obstet Gynecol.* 2002;14:569-75.

[36] Dyson RL, Pretorius DH, Budorick NE, et al. Three-dimensional ultrasound in the evaluation of fetal anomalies. *Ultrasound Obstet Gynecol* 2000; 16:321–328.

[37] Lee W, McNie B, Chaiworapongsa T, et al. Three-dimensional ultrasonographic presentation of micrognathia. *J Ultrasound Med* 2002; 21:775–781.

[38] Mangione R, Lacombre D, Carles D, Guyon F, Saura R, Horovitz J. Craniofacial dysmorphology and three-dimensional ultrasound: a prospective study on practicability for prenatal diagnosis. *Prenat Diagn* 2003; 23:810–818.

[39] Krakow D, Williams J, Poehl M, Rimoin D, Platt L. Use of three-dimensional ultrasound imaging in the diagnosis of prenatal-onset skeletal dysplasias. *Ultrasound Obstet Gynecol* 2003; 21:467–472.

[40] Merz E, Weber G, Bahlmann F, Miric-Tesanic D. Application of transvaginal and abdominal 3D ultrasound for the detection or exclusion of malformations of the fetal face. *Ultrasound Obstet Gynecol* 1997; 9:1–7.

[41] Sheiner E, Hackmon R, Shoham-Vardi I, Pombar X, Hussy MJ, Strassner HT, Abramowicz JS. Comparison between acoustic output indices in 2D and 3D/4D ultrasound in obstetrics. *Ultrasound Obstet Gynecol* 2007; 29:326-8.

[42] Marsal K, The output display standard: has it missed its target? *Ultrasound Obstet Gynecol* 2005;25:211-4.

[43] Kossoff G. Contentious issues in safety of diagnostic ultrasound. *Ultrasound Obstet Gynecol.* 1997;10:151-5.

[44] Bly S, Van den Hof MC; Diagnostic Imaging Committee, Society of Obstetricians and Gynaecologists of Canada. Obstetric ultrasound biological effects and safety. *J Obstet Gynaecol Can* 2005;27:572-80.

[45] Sheiner E, Shoham-Vardi I, Abramowicz JS. What do clinical users know regarding safety of ultrasound during pregnancy? *J Ultrasound Med* 2007; 26:319-25.

[46] Abramowicz JS, Kossoff G, Marsal K, ter Haar G. International Society of Ultrasound in Obstetrics and Gynecology (ISUOG) Safety and Bioeffects Committee: safety statement. *Ultrasound Obstet Gynecol* 2000; 16:594–596.

[47] Wax JR, Cartin A, Pinette MG, Blackstone J. Nonmedical fetal ultrasound: knowledge and opinions of Maine obstetricians and radiologists. *J Ultrasound Med* 2006;25:331-5.

[48] Barnett SB, Rott HD, ter Haar GR, Ziskin MC, Maeda K. The sensitivity of biological tissue to ultrasound. *Ultrasound Med Biol* 1997;23:805-812.

[49] Barnett SB. Intracranial temperature elevation from diagnostic ultrasound. *Ultrasound Med Biol* 2001;27:883-888.

[50] Barnett SB, Kossoff G, Edwards MJ. Is diagnostic ultrasound safe? Current international consensus on the thermal mechanism. Med J Aust 1994;160:33-37.

[51] Ang ES Jr, Gluncic V, Duque A, Schafer ME, Rakic P. Prenatal exposure to ultrasound waves impacts neuronal migration in mice. *Proc Natl Acad Sci* U S A. 2006;103:12661-2.

[52] Ji E, Pretorius D, Newton R, Uyan K, Hull AD, Hollenbach K, Nelson TR. Effects of ultrasound on maternal-fetal bonding: a comparison of two- and three-dimensional imaging. *Ultrasound Obstet Gynecol* 2005; 25:473–477.

[53] Shaw A, Pay NM, Preston RC: Assessment of the likely thermal index values for pulsed Doppler ultrasonic equipment-Stages II and III: experimental assessment of scanner/transducer combinations, *Report CMAM 12,* 1998, National Physical Laboratory, Teddington, Middlesex, UK.

[54] American Institute of Ultrasound in Medicine: *Medical Ultrasound Safety*, AIUM, Laurel, MD. American Institute of Ultrasound in Medicine; 1994.

[55] Nyborg WL. History of the American Institute of Ultrasound in Medicine's efforts to keep ultrasound safe. *J Ultrasound Med* 2003; 22:1293–1300.

INDEX

A

B

C

D

I

N

O

P

Q

R

S

T

U

V

W

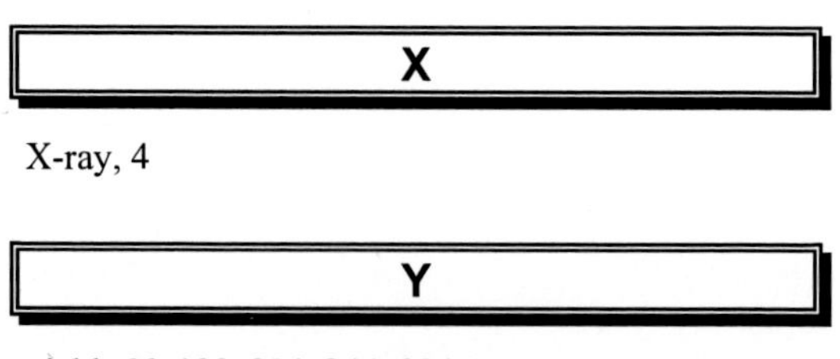

X

Y